Spectroscopy: Theory and Applications

Spectroscopy: Theory and Applications

Editor: Charles Burke

NY RESEARCH PRESS

New York

Published by NY Research Press
118-35 Queens Blvd., Suite 400,
Forest Hills, NY 11375, USA
www.nyresearchpress.com

Spectroscopy: Theory and Applications
Edited by Charles Burke

Cataloging-in-Publication Data

Spectroscopy : theory and applications / edited by Charles Burke.
    p. cm.
Includes bibliographical references and index.
ISBN 978-1-63238-842-1
1. Spectrum analysis. 2. Spectroscope. 3. Chemistry, Analytic--Qualitative. I. Burke, Charles.
QD95 .S64 2022
543.5--dc23

Contents

Permissions

List of Contributors

Index

Preface

The nature of interaction between matter and electromagnetic radiation can be understood with the technique of spectroscopy. It is an exploratory tool used in the fields of astronomy, physics and chemistry to study the composition, electronic structure and physical structure of matter at the atomic and molecular scales, even over astronomical distances. Spectroscopy can be studied with variations in the type of radiative energy used to interact with matter, the nature of the interaction and the type of material used. Depending on these interactions, spectroscopy can be of different types, such as microwave spectroscopy, infrared spectroscopy, electron spectroscopy, absorption spectroscopy, emission spectroscopy, atomic spectroscopy, molecular spectroscopy, etc. It may be used for varied purposes, such as measuring toxic compounds in blood or different compounds in food, exploring electronic structures and conducting elemental analysis, etc. Spectrometers, spectral analyzers and spectrophotometers are some common devices used to achieve spectral measurement. This book unravels the recent studies in spectroscopy. Also included in this book is a detailed explanation of the various techniques and applications of this field. For all those who are interested in this technology, this book can prove to be an essential guide.

This book has been the outcome of endless efforts put in by authors and researchers on various issues and topics within the field. The book is a comprehensive collection of significant researches that are addressed in a variety of chapters. It will surely enhance the knowledge of the field among readers across the globe.

It gives us an immense pleasure to thank our researchers and authors for their efforts to submit their piece of writing before the deadlines. Finally in the end, I would like to thank my family and colleagues who have been a great source of inspiration and support.

Editor

MS-Based Analytical Techniques: Advances in Spray-Based Methods and EI-LC-MS Applications

Federica Bianchi (ID),[1] **Nicolò Riboni** (ID),[1,2] **Veronica Termopoli** (ID),[3] **Lucia Mendez** (ID),[4] **Isabel Medina,**[4] **Leopold Ilag** (ID),[2] **Achille Cappiello,**[3] **and Maria Careri**[1]

[1]Department of Chemistry, Life Sciences, and Environmental Sustainability, University of Parma, Parco Area delle Scienze 17/A, 43124 Parma, Italy
[2]Department of Environmental Science and Analytical Chemistry, Stockholm University, 10691 Stockholm, Sweden
[3]Department of Pure and Applied Sciences, LC-MS Laboratory, Piazza Rinascimento 6, 61029 Urbino, Italy
[4]Instituto de Investigaciones Marinas, Spanish National Research Council (IIM-CSIC), Eduardo Cabello 6, 36208 Vigo, Spain

Correspondence should be addressed to Federica Bianchi; federica.bianchi@unipr.it

Academic Editor: Gauthier Eppe

Mass spectrometry is the most powerful technique for the detection and identification of organic compounds. It can provide molecular weight information and a wealth of structural details that give a unique fingerprint for each analyte. Due to these characteristics, mass spectrometry-based analytical methods are showing an increasing interest in the scientific community, especially in food safety, environmental, and forensic investigation areas where the simultaneous detection of targeted and nontargeted compounds represents a key factor. In addition, safety risks can be identified at the early stage through online and real-time analytical methodologies. In this context, several efforts have been made to achieve analytical instrumentation able to perform real-time analysis in the native environment of samples and to generate highly informative spectra. This review article provides a survey of some instrumental innovations and their applications with particular attention to spray-based MS methods and food analysis issues. The survey will attempt to cover the state of the art from 2012 up to 2017.

1. Introduction

Mass spectrometry (MS) is one of the most powerful techniques for the detection and identification of organic and inorganic compounds. Being able to provide both molecular weight and structural information [1], it is widely used in analytical laboratories for academic research, industrial product development, and regulatory compliance as well as for proteomic or metabolomic studies, DNA characterization, drug discovery, environmental monitoring, food analysis, forensics, and homeland security.

A plethora of analytical MS-based methods based on the use of both stand-alone instruments and mass spectrometers coupled to different separation techniques such as gas and liquid chromatography (GC and LC) or capillary electrophoresis (CE) have been developed and validated in order to analyze complex matrices. Interesting review articles and book chapters dealing with advances in ionization for mass spectrometry have been lately published [2–9].

Recently, the advent of ambient MS technology paved the way for the development of a great variety of applications and innovations characterized by high throughput: the challenge of analyzing samples in their native state without sample treatment encouraged the development of new techniques among which are the spray-based ionization ones including desorption electrospray ionization (DESI) [10], paper spray ionization (PSI) [11], laser ablation electrospray ionization (LAESI) [12], and easy ambient sonic-spray ionization (EASI) [13].

Novel materials and new instrumental configurations are under study to enhance the performance of the different ion sources. Safety risks can be identified at the early stages through nontargeted monitoring technologies. Furthermore, the variety of fragmentation strategies that can be combined in new instrumentation overall enhances work in the omics fields, particularly proteomics and metabolomics.

Although MS-based methods are getting progressively more powerful, reliable, and easily available, the main drawbacks are still related to sample complexity and preparation, mass accuracy, often requiring the use of high-resolution mass spectrometry (HRMS) to guarantee the univocal identification of the targeted compounds, and the need of high-throughput and screening analyses when a great number of samples have to be analyzed.

The aim of the proposed special issue is to cover the aspects regarding emerging features of MS-based techniques focusing on innovative LC-MS studies and ambient MS with particular attention to the spray-based ionization techniques. New materials, prototypes, and instrumental configurations able to increase the performance of the developed methods will be presented and discussed. Finally, an overview of the most recent MS-based methods in food analysis will be given. This survey will attempt to cover the state of the art from 2012 up to 2017.

2. Advances in LC-MS

Electrospray ionization (ESI) is the technique of choice to produce ions suitable for mass analysis. ESI spectra typically are characterized by single protonated or deprotonated molecular ion $(M + H)^+$, $(M - H)^-$, and/or adduct ions. The low fragmentation is a limitation in compound characterization through the use of reliable electronic libraries, making necessary the use of multistage MS (MS/MS, MS^n) or HRMS to compensate the limited structural information. LC-ESI with triple quadrupole (QqQ) MS is the most used technique for qualitative and quantitative determination of targeted nonvolatile compounds in forensic and food applications [14, 15]. In 2016, Remane et al. reviewed the literature on applications of LC-MS/MS in clinical, forensic toxicology, and doping control since 2006 [16]. It must be noted that ESI response is strictly affected by the mobile phase and sample composition as well as by the presence of coeluting interfering compounds, which may interfere with the ionization process. These phenomena are known as "matrix effects" (MEs) and can alter the response of the analytes causing either signal suppression or enhancement [17, 18]. The occurrence of ME introduces some critical analytical shortcomings in quantitative analysis by LC-MS such as reduced sensitivity, nonlinear response, and low precision. In addition, the physicochemical properties of the analytes can play an issue in the instrument response, thus introducing additional limitations. The combination of powerful MS detectors with LC has solved many problems in structural elucidation of unknown hazardous compounds [19]. In this context, HRMS is capable of providing full spectral information by adding high mass resolving power and accuracy to achieve selectivity and capability for accurate mass measurements [20–22]. HRMS is characterized by higher mass resolution, defined as the mass difference between two mass spectral peaks that can be clearly

distinguished [23], and higher mass accuracy (even better than 1 ppm). In addition, high mass resolving power allows discrimination between isobaric interference and ion of interest, leading to an accurate mass measurement even with a complex background. These features increase MS selectivity for the screening of nontargeted compounds in complex matrices, providing a list of possible elemental compositions. Fu et al. have shown how important and efficient is the use of nontargeted screening with LC-HRMS to ensure quality and safety of food [24], whereas Mattarozzi and coworkers exploited the capability of HRMS for the rapid determination of melamine from melamine tableware [25]. On the downside, HRMS-based methods generate complicated data that must be processed for "total ion fragment spectra" to obtain high-quality mass spectral information. Moreover, the mass of protonated or deprotonated molecules is not sufficient to prevent unambiguous compound identification. Hence, the use of spectra-matching approaches that utilize fragmentation ions could be added to achieve additional information on the detected compounds [26, 27]. False positives and false negatives are the major obstacles when screening complex samples. False negatives can occur due to very low concentrations, matrix interferences and suppression, and weak or no ionization. Due to these disadvantages, the practical application of LC-MS and LC-HRMS is still far from the immediacy and simplicity of GC-MS. Taking into account that electron ionization (EI) allows us to obtain characteristic and highly reproducible fragmentation of the analytes, a considerable effort has been devoted by the scientific community to increase compatibility between LC and EI-based MS to develop reliable, easy-to-use, and flawless interfaces. Moreover, implementation of EI fragmentation to LC-amenable compounds could pave the way for many new fields of research.

Recent developments in miniaturized mass spectrometers have enabled these developments to be carried out to portable on-scene detection. In the next paragraphs, some of the most popular and promising techniques are described: among them are LC-MS based on EI interfaces and spray-based ionization techniques.

2.1. LC-MS Based on Electron Ionization-Mass Spectrometry Interfaces. It is known that EI is ideal for the detection of a large number of GC-amenable compounds, but with an appropriate combination of several measures, it can become suitable for many LC-amenable small molecules having MW up to approx. 600 Da. These compounds can be efficiently converted into the gas phase, fast enough to be ionized before any decomposition process.

For analysis of small-to-medium molecules, the coupling between LC and EI-MS represents a valid strategy for overcoming the main disadvantages related to ESI ionization and the use of costly and complicated techniques involving HRMS instrumentations. Furthermore, EI provides a rich fragmentation pattern with a significant amount of structural information allowing a unique automated identification with structures at the isomer level [28]. Hence, the ability of EI for tentative identification of GC-amenable compounds is unparalleled even without HRMS.

On the downside, the coupling between LC and MS based on EI represents a significant issue in the field of analytical chemistry. The reason may be explained by the antagonist conditions of operating. The first one typically works at ambient temperature and uses very high pressure for the efficient separation of the analytes, which are sometimes dissolved in a complex mobile phase. The second one operates at a very high vacuum and high temperature. Therefore, the effort of achieving and maintaining the high vacuum required for mass spectrometry is in contrast with the intrinsic nature of HPLC, predominantly operating at high solvent flow rates. Also, the low tolerance of mass spectrometers for nonvolatile mobile phase components contrasts with an HPLC dependence on nonvolatile buffers to achieve high-resolution separations.

Since the year 2000, a few groups of researchers are working on the development of an efficient EI-based LC-MS interface.

Cappiello and his group played a significant role in the innovation and improvement of LC-EI-MS interfacing and designing a series of systems characterized by steadily increasing performance. Firstly, they presented a prototype of the LC-MS interface called Direct-EI [29–31] based on direct coupling of a low flow rate nano-HPLC with a high-vacuum EI source. The interface governs the direct introduction of a liquid-phase sample into the EI source of the mass spectrometer and the complete conversion of the liquid effluent to the gas phase prior to a conventional electron-assisted ionization. The core of the interface is represented by the nebulizer, which consists of a fused silica/PEEK capillary, to guarantee a sufficient thermal insulation. This interface was used in many different applications, not only in combination with chromatography but also in direct analysis, as a universal detector for the targeted compound. However, both nebulization and vaporization take place inside the ion source, leading to some drawbacks linked to capillary blockings. This concern is mainly due to premature evaporation of the solvent making the analysis very difficult under routine conditions. To meet the challenges of analyzing nontargeted compounds exploiting full potential of EI and the quantification of target compounds at low concentration in complex matrices, Termopoli et al. presented a new, robust, efficient interfacing mechanism coming from the ground up [32]. The new interface is called "liquid-EI" (LEI). The interface is completely independent from the rest of the instrumentation and can be adapted to any gas chromatography-mass spectrometry system, as an add-on for a rapid LC-MS conversion. Secondly, with some little tricks, it can be used with any HPLC system. Nanopumps and capillary pumps allow direct connection, and conventional HPLC needs the use of a two-way splitter to reduce the column flow rate to a level that is compatible with the interface, which is normally between 0.5 and 1 μl/min. In an LEI interface, the vaporization of the LC eluate is carried out at atmospheric pressure inside a suitable, independent microchannel right before entering the ion source, called the vaporization microchannel, representing the core of the interface. It is designed to uncouple and separate the atmospheric pressure found at the end of the HPLC system with the high-vacuum zone of the ion source. This specific place is narrow enough to prevent vacuum from entering into the spray region allowing us to have an atmospheric pressure zone where the vaporization process takes place. A removable silica-deactivated liner ensures a perfect conversion into a gas phase before entering the mass spectrometer. A narrower fused silica capillary, called the inlet, penetrates in the first portion of the liner and releases the LC eluate. An inert gas flow surrounds the gas phase through the vaporization microchannel and helps high boiling compounds to vaporize. Figure 1 shows a complete layout of the LC-MS system equipped with the LEI interface.

The rapid vaporization offered by the lined microchannel reduces the chance of thermal decomposition and capillary blockings, broadening the range of suitable applications, especially those regarding nontargeted analytes. Remarkable results were achieved in different conditions and applications.

Over the past years, Seemann and his group developed the supersonic molecular beams (SMB) LC-MS interface [34]. Their studies started from the knowledge that standard emission energy (70 eV) used in EI is not ideal for many NIST library compounds that have a weak (below 2% relative abundance) or no molecular ion. This issue is a critical point when very large and thermally labile compounds are analyzed. Furthermore, these analytes are usually less volatile and require higher EI ion source temperatures with further intra-ion source degradation, resulting in weaker molecular ion production. To achieve a reliable EI-based sample identification, a more intense production of molecular ion is needed. Thus, the best ionization method should provide the informative library searchable EI fragments combined with enhanced molecular ions, especially for the compounds that are not included in the commercially available EI libraries. Taking into account these considerations, they present a novel concept of the LC-SMB-MS system, based on the use of supersonic molecular beams, as a medium for electron ionization of vibrationally cold sample molecules in a fly-through ion source. It is able to generate library searchable EI spectra and a more intense molecular ion.

The LC-SMB-MS apparatus is schematically shown in Figure 2.

A thorough evaluation of the interface, comprising identification of unknown compounds using obtained library searchable EI mass spectra, enhanced production of molecular ions, demonstration of the absence of matrix effects, simultaneous determination of semipolar and nonpolar compounds with reasonable detection limit, and low-cost instrumentation, was provided by that research group. The group of Seemann demonstrates the feasibility of the SMB interface as a valid tool in the analysis of unknown compounds and as a low-cost LC-EI-MS system.

A third group of researchers, headed by Rigano, presented a new nano-LC-EI-MS for the determination of free fatty acids (FFAs) in mussels [34]. A selective and sensitive nano-LC-EI-MS analytical method to investigate the FFA profile in marine organisms and to monitor marine sentinels

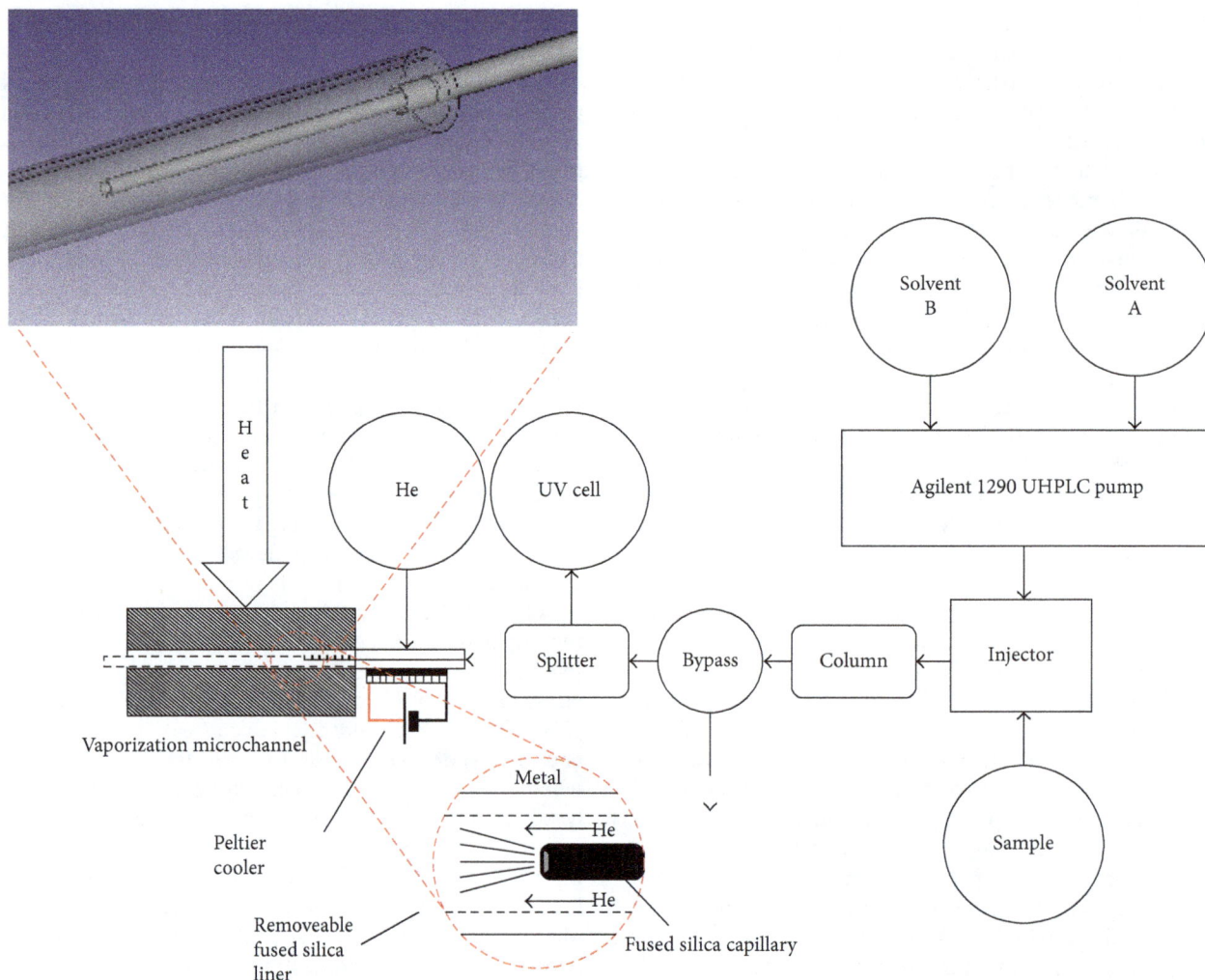

FIGURE 1: Global layout of the fully assembled system; the LEI interface, in gray, is between the UHPLC system and the MS detector. In the red circle, the vaporization zone is highlighted. Reprinted with permission from [33].

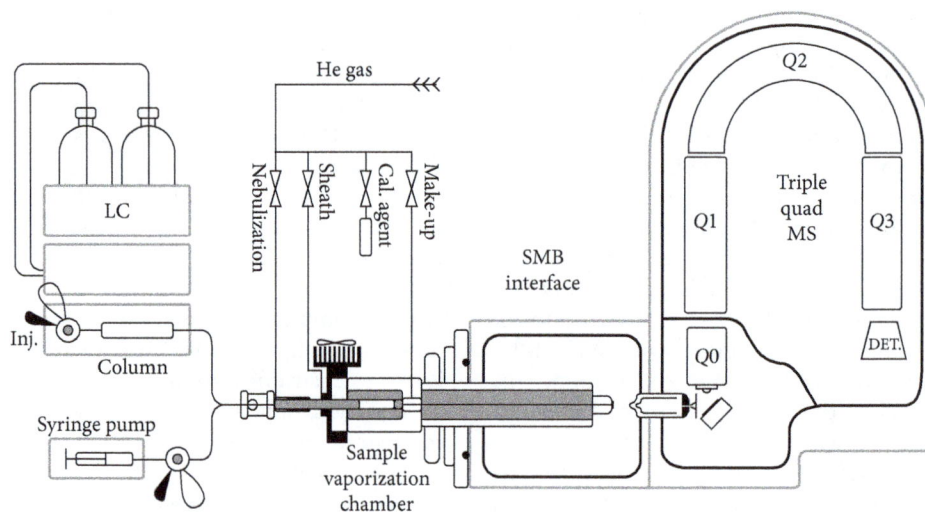

FIGURE 2: EI-LC-MS with the SMB system outline. The liquid is introduced either from the HPLC system after its column or from a syringe pump to the heated vaporization chamber through a pneumatic nebulizer. The helium nebulization gas enters the SMB interface through a nebulization gas line, sheath gas line, and nozzle make-up gas line. Reprinted with permission from [35].

for the assessment of environmental pollution effects was developed [35]. FFAs are minor components of the lipidome, and they are usually analyzed by GC after a derivatization step, such as methylation or trimethylsilylation, is performed to convert FA into less-polar and more-volatile moieties and improve their separation [9, 36]. However, the derivatization step, if not properly selected, can modify the FA profile due to nonhomogeneous derivatization efficiency among different compounds (saturated, unsaturated, and polyunsaturated fatty acids). In addition, oxidation or isomerization products can be generated. Relative to this issue, LC can benefit over GC techniques from direct injection of FFAs in their intact form, without any pretreatment. On the other hand, direct coupling with EI-MS can benefit from the highly informative, repeatable, and reliable MS fragmentation. Drawing conclusions of the several attempts made by each group, they are following a distinctive pathway to obtain a common goal, the development of a more useful and universal LC-EI-MS interface.

Regarding Direct-EI LC-MS, recent studies have been carried out also to increase the inertness of the electron ionization ion source by developing new materials [37, 38].

As already stated, the vaporization surface of an electron ionization MS source is a key parameter for the detection and characterization of targeted and untargeted analytes: it is known that difficulties in the vaporization process arise when compounds characterized by high molecular weight and/or polarity have to be analyzed, thus requiring both the use of inert ion sources to reduce the interactions of the analytes with the stainless steel ion source and the use of high source temperatures to promote analyte vaporization. In this field, Magrini et al. [37] proved that the use of a commercially available ceramic coating is able to improve the detection of high molecular weight and high boiling compounds like polycyclic aromatic hydrocarbons (PAHs) and hormones. More recently, Riboni et al. [38] were able to increase the inertness of the electron ionization ion source by developing different sol-gel coatings based on silica, titania, and zirconia. Again, the developed coatings were tested for the Direct-EI LC-MS determination of PAHs and steroids. The best performances in terms of both signal peak intensity and peak width were obtained by using the silica-based coating, obtaining detection limits in the low ng/ml range with a good precision (RSD <9% for PAHs and <11% for hormones). No problems associated to ion cleaning were observed after prolonged use.

3. Advances in Spray-Based Ionization Techniques

Nowadays, there is also a growing interest in the development of real-time analytical technologies capable of allowing the direct detection of trace analytes in complex samples, especially in their native environment. The development of a new class of techniques, better known as "ambient ionization techniques," has introduced a revolution in the ionization field. These techniques are able to generate ions directly from native environment of the sample at ambient pressure, without any tedious sample preparation steps or laborious time-consuming chromatographic separation.

Technically, spray-based ionization techniques are based on the use of electrospray droplets to extract the analytes from the sample and transfer them to the mass spectrometer. The most common spray-based ionization technique is DESI, in which a high-velocity pneumatically assisted ESI source generates charged microdroplets by the application of a proper potential on the ESI needle. The spray is directed towards the sample where the impact of the primary droplets with the substrate leads to the formation of a micrometer-sized thin solvent film, able to solubilize the analytes at the liquid-solid interface. Secondary droplets containing the analytes expelled by the film solvent are generated, and then, desolvation and ionization in the gas phase occur, as in the traditional ESI analysis. Finally, the ions are collected by the MS inlet.

In addition to DESI, several other techniques like nano-DESI, EASI, and LAESI have been proposed. Probe electrospray ionization (PESI) is another interesting approach based on the use of a solid conductive needle probe that replaces the traditional electrospray capillary for sample introduction. Similarly, PSI is a technique based on the loading of the sample onto a triangular piece of paper from which ions are generated by applying a high voltage in the presence of a proper solvent [11]. Spray-based methods are suitable for the analysis of different compounds, from small analytes, such as explosives [39–43], drugs [44–47], and food contaminants [48–50], to larger molecules such as lipids [51–53], peptides [54, 55], and proteins [56, 57].

3.1. Desorption Electrospray Ionization-Mass Spectrometry. DESI-MS is usually applied for surface desorption/ionization of analytes deposited on a probe material (PTFE, PMMA, glass, etc.) or directly from the sample surface. DESI-MS and DESI imaging have been successfully applied in different fields, such as forensic science [58], food control [59, 60], and clinical applications [61–63].

The derivatization of metabolites deposited in solution onto a glass plate by dropping the derivatizing reagent on the top of the dried analytes was proposed by Lubin et al. [64]. The authors successfully applied this technique to several samples, demonstrating the possibility of performing multiple and subsequent derivatization steps on the same spot.

An interesting approach was developed by Brown et al. [65] for the MS detection of fleeting reaction intermediates in electrochemical reactions utilizing a new *waterwheel* working electrode setup. The proposed technique allowed us to exploit DESI-MS operating at a low voltage. The new apparatus consisted of a round rotating platinum working electrode that was partially immersed in an aqueous electrolyte solution (Figure 3). During the rotation, a thin layer of liquid film was deposited on the electrode surface, as in a waterwheel. A three-electrode system was set by using a platinum wire counterelectrode and an Ag/AgCl reference electrode, immersed in the reservoir of electrolyte solution. The upper surface of the waterwheel was hit by a spray generated by a custom spray probe, thus allowing the

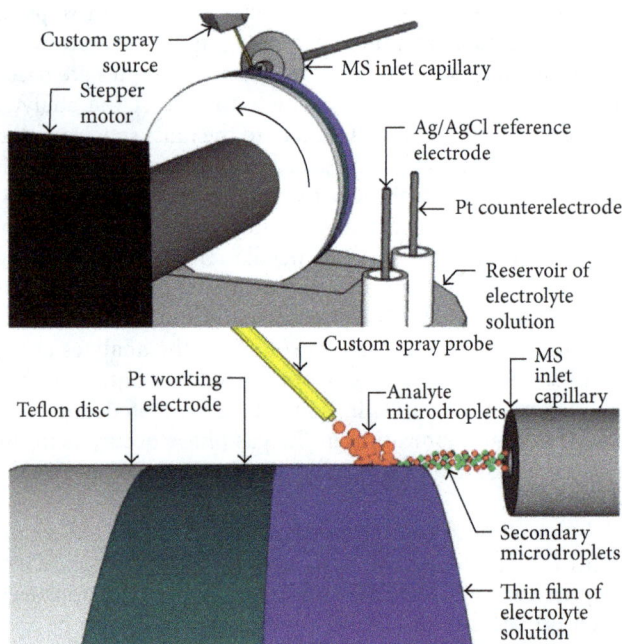

FIGURE 3: Schematic representation of the developed experimental setup. Reprinted with permission from [65].

formation of secondary droplets, analyte ionization, and their collection in the MS inlet. To avoid any electrochemical oxidation or reduction on the electrode surface, no high voltage was applied to the analyte spray, whereas a low potential (few volts) was applied to the platinum rotating electrode.

The authors tested the new apparatus towards the detection of a diimine intermediates during electrochemical oxidation of both uric acid and xanthine.

An MS-electrochemistry coupling was also proposed by Looi et al. [66], who developed a new online electrochemistry-liquid sample desorption electrospray ionization-mass spectrometry (EC-LS DESI-MS) system. In EC-LS DESI-MS, an electrosonic spray ionization source was used to generate a spray directed to the exit of the liquid sample capillary positioned perpendicularly to the spray and the MS inlet. Separately, a thin two-electrode flow-through EC cell was connected to a syringe pump and was used to perform oxidation/reduction processes. The ESSI-generated spray was able to impact the outer surface of the LS capillary, which is continuously coated by sample solution flowing at 200 μl/h, thus allowing the ionization of the analytes. This prototype was developed and tested using N,N-dimethyl-p-phenylenediamine (DMPA). Although oxidation of DMPA was already observed as a result of ionization of DESI-MS in positive mode, by applying a proper voltage to the online electrochemical (EC) cell, it was possible to increase the yields of the oxidation products, thus improving method sensitivity.

Although DESI is usually coupled with high-resolution mass spectrometry, its coupling with LC is possible. A novel splitting method for LC-MS applications, which allows both very fast MS detection of analytes eluted from the LC column and their online collection, was presented by Cai et al.

[67, 68]. In this approach, a PEEK capillary tube with a micro-orifice is used to couple DESI with the UPLC column. By using the proposed instrumental setup, a small amount of LC eluent (few nanoliters) is ionized by DESI with negligible time delay (6~10 ms), whereas the remaining analytes exiting the tube outlet can be collected. In addition, online derivatization using reactive DESI is feasible increasing the charge of proteins and consequently enhancing the ionization yields.

An interesting novel configuration has been recently developed by Ren et al. [69]. The authors developed a method coupling slug-flow microextraction (SFME) and nanoelectrospray ionization for the MS analysis of organic compounds in blood and urine. A disposable glass capillary with a pulled tip for nano-ESI was used to perform the entire extraction and ionization process (Figure 4). Two adjacent liquid plugs were formed by injecting 5 μl of a proper organic solvent and 5 μl of body fluid (urine or blood) into the capillary. Liquid-liquid extraction of the analytes was performed by both moving the capillary and applying a push-and-pull force through air pressure. After the extraction process, a high voltage is applied to the organic solvent plug to generate the nano-ESI for MS analysis.

The proposed method was tested for the extraction and detection of different analytes, namely, methamphetamine, benzoylecgonine, verapamil, amitriptyline, epitestosterone, 6-dehydrocholesterone, 5α-androstane-3β, 17β-diol-16-one, and stigmastadienone. Major analytical features were the reduced consumption of both the organic solvent and sample. The authors demonstrated that a direct derivatization of the extracted analytes in the organic phase was feasible, thus achieving excellent sensitivity with detection limits in the 0.03–0.8 ng/ml range.

Despite its name, nanospray DESI (nano-DESI) is based on a different instrumental configuration compared to the traditional DESI: its setup presents two different silica capillaries, one for solvent delivery and the other devoted to the formation of charged liquid spray in front of the MS inlet. The two capillaries are not in direct contact, thus producing a solvent bridge on the DESI surface. The second nanospray capillary produces a self-aspirating nanospray, which is generated by applying a high voltage between the MS inlet and the primary capillary. In comparison with DESI, nano-DESI is characterized by higher efficiency in liquid transportation and sampling performances.

The capabilities of nano-DESI-MS were tested for the determination of pollutants and organic components in atmospheric fine particles by Cain et al. [70], in environmental aerosol by Tao et al. [71], and in clouds by Boone et al. [72]. In the clinical and pharmaceutical fields, both nano-DESI-MS and nano-DESI-MS imaging proved to be excellent techniques for the analysis of pharmaceuticals, biomolecules, and metabolites [73–77].

A further instrumental innovation has been proposed by Duncan and coauthors [78], who developed a pneumatically assisted nanospray desorption electrospray ionization source. The instrumental setup was based on the introduction of a secondary nebulizer replacing the self-aspirating secondary capillary in order to assist solvent

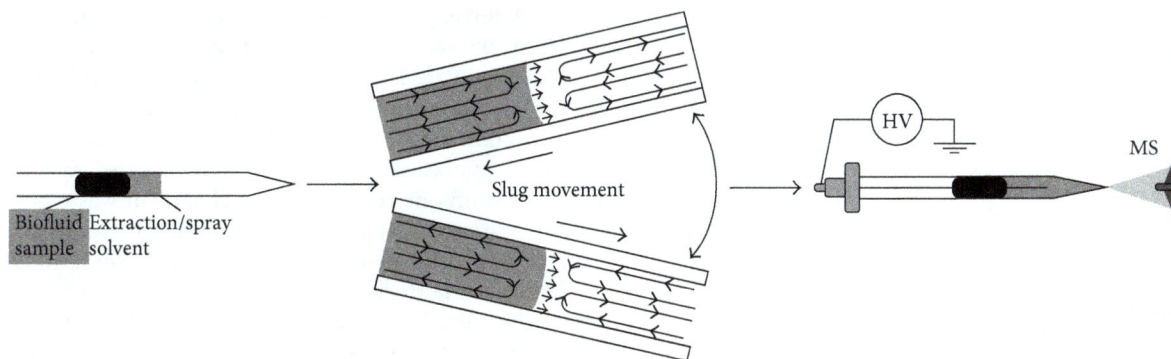

FIGURE 4: Schematic representation of the SFME-nano-ESI sample processing. Reprinted with permission from [69].

flow, to promote desolvation of the analyte, and to increase the distance between the nanospray and the MS inlet (Figure 5).

The developed device was tested for the analysis of rat kidney tissue sections, allowing us to obtain an improvement in sensitivity of about 1–3 orders of magnitude compared to the conventional setup. In addition, ion images characterized by high contrast, suitable for more intricate studies of metabolite distribution in biological samples, were obtained. A more complete desolvation of the analytes and reduced ionization suppression were additional features of the proposed device.

3.2. Extractive Electrospray Ionization-Mass Spectrometry and Laser Ablation Electrospray Ionization-Mass Spectrometry. Extractive electrospray ionization (EESI) has been introduced in 2006 by Chen et al. [79]. It is based on the use of two different sprayers: the ESI sprayer generates a charged solvent spray, whereas the sample sprayer has the function to nebulize the sample solution from an infusion pump. The analytes are ionized in the collision area of the two sprays, and then, they are collected by the MS inlet.

The ionization mechanism of the EESI ion source was studied by Wang et al. [80], and different MS-based methods for the analysis of organic aerosols [81], drugs [82], pesticides [83], amino acids [84], and biomarkers [85] in different matrices were developed in the recent years.

LAESI-MS is another ambient ionization technique developed in 2007 by Nemes and Vertes [12]. Since most cells used for biomedical applications are cultured adherently, the use of LAESI-MS was proposed to analyze adherent cells directly onto the culture surface, thus avoiding chemical modification deriving from their detachment [86]. In order to reduce the LAESI spot size, the authors applied a transmission geometry- (tg-) LAESI and incorporated an objective with a high numerical aperture, thus achieving spot sizes in the 10–20 μm range. This technique (Figure 6) was tested for the analysis of adherent versus suspended mammal cells, highlighting a difference in the metabolite compositions, thus proving that the cell detachment usually performed is able to produce chemical changes. On the contrary, tg-LAESI-MS allowed us to analyze directly the cells in their native state and, due to the smaller spot size, to reduce the sampled cell population by a factor of 20.

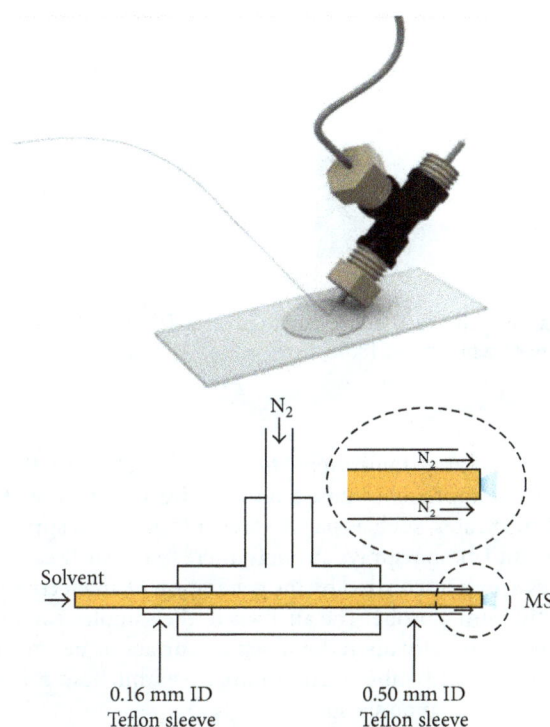

FIGURE 5: Picture and schematic representation of the pneumatically assisted nano-DESI ionization source. Reprinted with permission from [78].

Optical microscopy combined with LAESI-MS has been suggested by Compton et al. [87] in order to acquire both morphological and chemical information from tissue sampling. In the developed instrumental setup, laser ablation occurred inside a chamber placed under an optical microscope: the ablated particulates generated by the laser were transported through a transfer tube by using nitrogen as carrier gas and finally ionized by the ESI spray.

In order to compare the performances of the developed prototype with those of the conventional LAESI-MS, plant tissues were analyzed. In comparison with conventional LAESI, the developed technique was characterized by reduced sensitivity and dynamic range; however, these features were still sufficient for the analysis and characterization of numerous metabolites and lipids in different spatial regions of biological tissues.

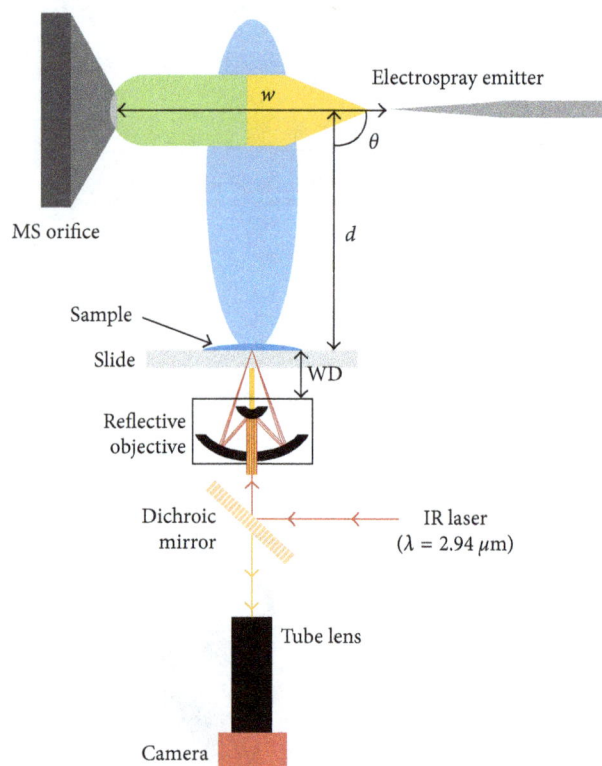

FIGURE 6: Schematic representation of tg-LAESI-MS. Reprinted with permission from [86].

3.3. Paper Spray Ionization-Mass Spectrometry. In the last ten years, approaches based on the direct ionization from solid substrates, such as paper spray, probe-based spray, leaf spray, and tissue spray, strongly increased. All these techniques are characterized by the generation of an electrospray directly from a probe. The analytes in the samples can either be ionized directly upon the substrate surface or be extracted on a probe and subsequently ionized within few minutes, thus boosting analysis speed.

Innovations in PSI-MS have been described by Duarte et al. [88] and Salentijin et al. [89] who developed 3D-printed cartridges in order to obtain a solvent reservoir, thus allowing us to prolong the spray generation from the paper tip. A supporting prototype able to automatically perform PSI-MS analysis of a great number of samples, suitable for high-throughput applications, has been designed by Shen et al. [90]. Finally, the coupling of SFME with PSI-MS for the rapid analysis of macrolide antibiotics at the trace level in biological samples such as whole blood, milk, and other body fluids has been proposed in a study carried out by Deng et al. [91]. The same approach was applied for the detection of perfluorooctanesulfonic acid and perfluorooctanoic acid from *Daphnia magna* body fluids. After the SFME extraction, the organic extract was simply spotted on the PSI paper.

Excellent results were achieved in terms of linearity range (5–500 and 0.5–50 ng/ml range for antibiotics and perfluorocompounds, resp.) and sensitivity (LOQs 0.3–1.3 and 0.03–0.30 ng/ml for antibiotics and perfluorocompounds, resp.). Recovery rates always higher than 85% were obtained.

A novel paper spray cartridge with an integrated solid-phase extraction column has been developed by Zhang and Manicke [92]. The system was designed in order to perform on the same device the extraction, preconcentration, and ionization of the analytes from complex matrices such as blood or plasma. The cartridge was divided into two parts: the bottom one containing the absorbent waste pad and the paper spray substrate and the top one presenting a hole to host the solid-phase extraction (SPE) column. The procedure for performing paper spray analysis is the following: (i) the samples are loaded onto the SPE column (sample volume from $10 \mu l$ up to hundreds of microliters); (ii) the unbounded compounds pass through the SPE column and are absorbed by the waste paper pad; (iii) after sliding the top part of the cartridge to the paper spray substrate, the analytes retained on the SPE column are eluted and analyzed by PSI-MS. The analytical performances in terms of detection and quantitation limits, recovery, and ionization suppression were evaluated for carbamazepine, atenolol, sulfamethazine, diazepam, and alprazolam. The SPE cartridge allowed both the selective enrichment of the targeted analytes from large sample volumes (up to hundreds of μl) and the removal of interfering compounds, thus enhancing signal intensities. Compared to direct PSI, the proposed method allowed us to improve quantitation limits by a factor of 14–70, obtaining limits in the 0.2–7 ng/ml range.

3.4. Wooden-Tip Electrospray Ionization-Mass Spectrometry. Wooden-tip electrospray ionization-mass spectrometry (WT-ESI-MS) is a rapid, in situ, and direct ambient technique based on the use of a wooden tip as a sampling and ionization needle. The tip is dipped into the liquid solution/matrix, and after extraction, it is directly used as an ESI probe by applying a high voltage and spray solvent. The analytes enriched on the tip are desorbed and ionized under ambient and open-air conditions.

This method has been successfully applied by Yang et al. [93] for the analysis of pesticides, toxicants, date rape drugs, and illicit additives in various food samples. The capabilities of untargeted WT-ESI-MS analysis for the identification of the sources of plant materials by using a multivariate statistic approach were exploited by Xin et al. [94], whereas Yang and Deng [95] used an internal standard WT-ESI-MS-based method to obtain the fingerprint mass spectra for rapid quality assessment and control of Shuang-Huang-Lian oral fluid, an herbal preparation registered by Chinese Pharmacopoeia. By using the internal standard and principal component analysis (PCA), it was possible to obtain the fingerprints of samples from different manufacturers. A bamboo pen nib shaped and used for sample loading and an ESI probe for the determination of 4-chloro-amphetamine was developed by Chen et al. [96], resulting in lower detection limits compared to PSI-MS and traditional WT-ESI-MS analyses.

Similarly, a WT-ESI-MS method combined with different nonpolar solvents for the detection of native proteins and protein complexes directly from raw biological samples has been proposed by Hu and Yao [97].

The applicability of field-induced wooden-tip electrospray ionization-mass spectrometry (FI-WT-ESI-MS) for high-throughput analysis of herbal medicines has been proved by Yang et al. [98]. Field-induced ESI was performed by applying a high voltage on the MS inlet, thus allowing the creation of a strong electric field between the capillary emitter and MS inlet to induce ESI from the sample solution. A high-throughput analysis device was developed by the application of sample-contactless high voltage on the MS inlet. In addition, the switch between positive and negative ion detection modes can be readily accomplished, thus providing complete MS information of the analyzed samples. This approach allowed us to boost the analysis speed: 6 s per sample was sufficient to perform the analyses.

The proposed method was applied for the rapid determination of various active ingredients in different raw herbs and herbal medicines. The obtained mass spectra were used as fingerprints for tracing the origins, establishing the authenticity and assessing the quality of herbal medicines.

Very recently, a novel and noninvasive sampling method using a watercolor pen (brush) rinsed with ethanol as a sampling tool to collect analytes from the eyelids of volunteers has been evaluated [99]. The brush was placed between the mass inlet and the ESI plume, thus allowing the desorption and ionization of the analytes. The results achieved proved the suitability of the developed technique for the semiquantitative determination of caffeine and its metabolites in eyelid samples.

3.5. Miscellaneous. Following the method developed by Pan et al. [100], based on the use of a single probe inserted into a single cell for sampling intracellular compounds by real-time MS analysis, Chen et al. [52] described a novel method for single cell analysis and lipid profiling by combining drop-on-demand inkjet cell printing and probe electrospray ionization-mass spectrometry (PESI-MS). Droplets containing single cells were generated from a cell suspension by inkjet sampling, precisely dripped onto the tungsten tip of the ESI needle, and sprayed under a high-voltage electric field. Cellular lipid fingerprints were then obtained by MS detection. The analytes from eight types of cells were detected, and PCA analysis was performed in order to differentiate the samples. The proposed platform was demonstrated to be suitable for the direct MS profiling of single-cell lipid species without derivatization or the labeling procedure.

A method for the direct characterization of metals in solid samples using electrospray laser desorption ionization-mass spectrometry (ELDI-MS) has been developed by Shiea et al. [101]. The main advantages over classic approaches were related both to the absence of sample pretreatment and to the presence of very short analysis time. Mixtures of metal-EDTA complexes were applied on a stainless steel plate and submitted to ELDI-MS analysis. The capabilities of the technique were initially tested by spotting the metal-EDTA complexes on a solid probe and performing laser ablation of the material. The ablated analytes, present as EDTA complexes, were ionized in the electrospray plume.

Further experiments were carried out by deposing the sample on the probe surface while maintaining the EDTA in the ESI spray solvent (functional electrospray). Excellent results in terms of sensitivity were achieved, thus proving method reliability for the rapid analysis of metal substrates without sample preparation.

An electrostatic spray ionization (ESTASI) method for the analysis of samples deposited in or onto an insulating substrate has been proposed by Qiao and coworkers [102]. In this study, the ionization of the analytes is induced by a capacitive contactless coupling between the electrode and the sample: by applying a pulse high voltage to the electrode, an electrostatic charging of the sample occurs, leading to a bipolar spray pulse. When the applied voltage is positive, the bipolar spray pulse consists first of cations, followed by anion production. The instrumental setup can be modified in order to obtain ion emission from samples in a silica capillary, in a disposable pipette tip, and in a polymer microchannel or deposited as droplets onto an insulating poly(methyl methacrylate) plate presenting wells or hydrophilic patches. This technique proved to be suitable for the analysis of peptides and proteins.

4. Materials for Spray-Based Ionization Techniques

The development of new materials is a field of increasing interest in order to enhance the performances of analytical methods. Both high selectivity and increased ionization efficiency are demanded to detect analytes at trace levels in complex matrices and to shorten analysis time. Different studies have been published dealing with the development of novel surfaces for ambient MS. A brief overview on the most recent materials developed for spray-based ionization techniques with particular attention to DESI-MS and PSI-MS applications is described in the next paragraphs.

4.1. Materials for DESI-MS. In 2005, Takats et al. [103] demonstrated that the DESI ion source is strongly influenced by the dielectric constants of both the substrate material and the spray.

The effect of surface chemistry in the DESI ionization mechanism has been deeply investigated by Penna and coworkers [104]: the performances of different glass substrates obtained by the sol-gel technology and functionalized by using different alkylsilanes were tested and compared in terms of ionization efficiency. The substrates were characterized in terms of surface free energy and wettability. Owing to their different polarity, melamine, tetracycline, and lincomycin were used as model compounds. A significant decrease in the ionization efficiency was observed when more hydrophilic surfaces were used, thus demonstrating the pivotal role of both hydrophobicity and wettability to increase the performances of DESI-MS experiments.

More recently, a 3D printed polylactic (PLA) supports in order to detect insulin and gentamicin in chitosan gels have been proposed by Elviri and coauthors [105]. By using 3D printing, hemispherical wells were created, thus allowing us

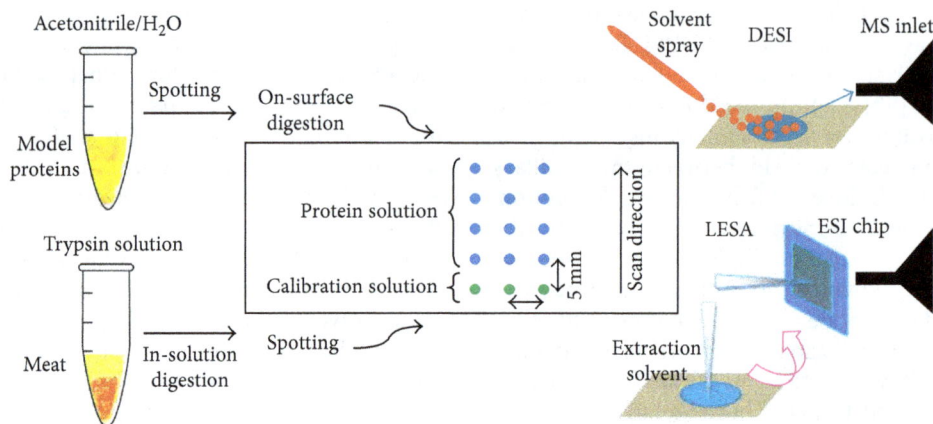

FIGURE 7: Schematic representation of the procedure proposed by Montowska et al. Reprinted with permission from [106].

to obtain DESI-MS responses five times higher than those achieved by using PTFE commercial slides. Improvement in terms of signal stability was also achieved, thus suggesting the capability of 3D printing technology to improve the desorption step in DESI-MS.

Novel substrates have been proposed also for proteomics and peptidomics: the capability of Permanox slides for both DESI-MS and liquid extraction surface analysis-mass spectrometry (LESA-MS) of skeletal muscle proteins obtained from a mixture of standard proteins and raw meat has been discussed in a recent study [106]. The proposed method is schematically reported in Figure 7.

In both cases, good responses were obtained with LESA-MS characterized by higher sensitivity and stability with respect to the DESI-MS approach. Finally, multivariate data analysis allowed the correct discrimination among different meat classes.

Rapid and simple analyses represent a key parameter in proteomics: an interesting study was carried out by Dulay et al. in 2015 [107]. Two hybrid organic-inorganic organosiloxane polymers functionalized by immobilized trypsin (T-OSX) for the on-surface and in situ digestion of four model proteins, that is, melittin, cytochrome c, myoglobin, and bovine serum albumin, were developed and tested under DESI-MS and nano-DESI-MS conditions. The silica polymers were obtained via sol-gel technique using methyltrimethoxysilane and dimethyl-dimethoxysilane as precursors. The OSX polymers were derivatized with trimethoxysilylbutyraldehyde and functionalized with different amounts of trypsin. In both cases, despite the low enzyme-to-substrate ratios, the achieved results proved the suitability of the developed substrates to allow on-surface protein digestion followed by direct DESI-MS analyses obtaining sequence coverages in the 65–100% range.

Taking into account that DESI-MS analyses can be performed also on liquid samples, an improvement of the apparatus commonly used for this kind of analyses has been proposed [108] by replacing the sample transfer silica capillary with a trap column filled with chromatographic stationary phase materials, such as C_4 and C_{18}. The proposed system proved to be suitable for trace analyses of both organic compounds and biomolecules such as proteins/peptides in complex matrices in the presence of high salt content. Another interesting feature was related to the covalent functionalization of the inner wall of the sample transfer capillary with enzymes, thus allowing the fast and online digestion of proteins.

Noticeable applications of DESI-MS are based on its coupling with new sampling devices and extraction and separation techniques to develop methods for high-throughput analyses.

The microfabrication of ultrathin-layer chromatography (UTLC) plates via conformal low-pressure chemical vapor deposition of silicon nitride onto patterned carbon nanotube (CNT) scaffolds, acting as surface templates, has been described by Kanyal et al. [109]. The plates were heated and oxidized to both remove the CNTs and convert Si_3N_4 into silica; finally, the plates were hydroxylated in aqueous ammonium hydroxide. The resulting UTLC phases did not show any distortion of the microfeatures and were characterized by a higher robustness in comparison to high-pressure TLC plates. Good results in terms of chromatographic performances were observed obtaining a faster separation when mixtures of lipophilic, water-soluble, and fluorescent dyes are to be analyzed. A strong reduction in terms of mobile-phase consumption and an enhanced lifetime were observed. Finally, the UPTLC plates were submitted to both DESI-MSI and direct analysis in real-time (DART)-MS analyses, showing a good compatibility with common ambient desorption and ionization techniques.

In the same year, Ewing et al. [110] described the DESI-MS detection of the low vapor pressure nerve agent simulant triethyl phosphate. The analyte was previously adsorbed onto silica gel, forming a very fine particulate that was collected by using a sticky screen sampler. The device was characterized by a stainless steel screen presenting a partially polymerized polydimethylsiloxane (PDMS) coating. The quantitative collection of the particulate sample from a contaminated surface was achieved by interfacing the sticky screen sampler with a bioaerosol collector. Finally, the sticky screen was placed onto a moveable platform mounted in front of the DESI-MS instrument, thus allowing a reproducible sample introduction system able to minimize sampling errors. DESI-MS analyses performed directly on the PDMS coating allowed us to obtain a very low detection limit suitable for trace detection.

Electrospun nylon-6 nanofiber mats for DESI-MS analysis and imprint imaging have been developed by Hemalatha et al. [111]. The nanofibers were developed by needleless electrospinning: nylon-6 was dissolved in formic acid and the solution was electrospun at room temperature. Uniform mats of varying thicknesses composed of ~200 nm diameter fibers were obtained: the properties of these materials can be tuned by varying spinning conditions and surface functionalization. As model compounds, dyes and the extract of periwinkle flower were spotted onto the nylon nanofiber mat, thus obtaining a uniform coating of the fibers. DESI-MS analysis allowed us to obtain spectra without polymer interference and reproducible DESI-MS images. The authors also demonstrated that compounds of interest could be incorporated into nanofibers during their formation by using as model compounds the crude methanol extract of periwinkle flower and tetraphenylphosphonium bromide (TPPB). The major metabolites of periwinkle were identified by DESI-MS, even though the spectrum was different compared to that obtained by spotting the sample. TPPB was detected with no nylon interference. The authors demonstrated the possibility of imprinting patterns made of printing inks, plant parts, and fungal growth on fruits on the nanofiber mats. Metabolites were identified by DESI-MS. The results highlighted that electrospun nanofiber mats could be considered as smart surfaces to capture diverse classes of compounds for rapid detection or to imprint imaging under ambient conditions.

4.2. Materials for PSI-MS. Although traditional paper spray ionization is performed on the filtering and chromatographic paper [11, 112–115], researchers have focused their attention to the development of new functionalized substrates in order to obtain substrates characterized by enhanced selectivity and sensitivity.

Very recently, Lai et al. [116] compared the ionization performances of different paper-like substrates obtained from both natural fibers and synthetic fibers, namely, gampi paper, Tengujou paper (natural), polycarbonate, polylactic acid, and poly-1-lactic acid (PLLA) (synthetic), with those of traditional chromatographic paper for the analysis of designer drugs. The surface characterization of the developed materials showed the presence of different surface morphologies able to affect PSI capabilities: gampi paper and PLLA nanofibers, characterized by a tough and extremely thin structure, were able to promote signal enhancement, thus allowing us to reach lower limits of detection. These findings could be explained by taking into account the reduced thickness of the used papers: by operating under the described conditions, a rapid evaporation of the sample molecules occurs, thus increasing the speed of the ionization process.

The analytical performances of paper with paraffin barrier (PS-PB) for the PSI-MS detection of hydroxymethylfurfural (HMF) and sugars like glucose and xylose in sugarcane liquors have been tested by Colletes et al. [117]. Microfluidic hydrophobic channels were obtained using paraffin barriers on paper substrates, thus delimiting

a region for inducing the sample inlet into the mass spectrometer. The proposed stamping method allowed rapid prototyping of microfluidic paper-based analytical devices, without the need of sophisticated instrumentation. Different types of papers were investigated: an increase of the PSI-MS responses of xylose and glucose as a function of the decrease of porosity of the paper substrate was observed. PS-PB showed the best performance compared to the conventional paper and paper with two rounded corners. PS-PB was applied to detect sugars and their inhibitors in liquors from a second-generation ethanol process, thus obtaining excellent results in terms of linearity (over two orders of magnitude) and limits of detection and quantification.

Another interesting study carried out by Zhang et al. [118] proposed the use of commercially available silica-coated paper in order to increase the PSI-MS responses of therapeutic drugs in dried blood spots. The presence of silica gel particles in the cellulosic framework of the silica-coated paper produced a reduced diffusion of the blood through the substrate, thus leading to a higher percentage of blood sample available on the top side of the substrate. By operating under the optimized conditions, that is, by using dichloromethane/isopropanol (9 : 1 v/v) as a spray solvent mixture, limits of quantitation of about 0.1 ng/mL were achieved, with a sensitivity gain of 5–50-fold in comparison to chromatography papers.

CNT-coated filter paper for low-potential PSI-MS analysis of different organic molecules has been used by Narayanan and coauthors [119]: by applying a voltage of 3 V, full-range mass spectra similar to those obtained by conventional ESI at 3 kV could be recorded. The advantage of the proposed material relies on the use of very mild conditions for the ionization of the analytes, thus allowing the detection of compounds that could be easily oxidized. The performances of the proposed analytical method were assessed for a wide range of volatile and nonvolatile compounds, such as amino acids, antibiotics, and pesticides in different matrices.

Very recently, Wei et al. [120] synthesized graphene oxide (GO) nanosheet-modified N^+-nylon membrane (GOM) for the extraction of malachite green (MG), a highly toxic disinfectant, and its metabolites in liquid samples and fish meat. GO nanosheets are characterized by a very high surface area (~800 m^2/g), suitable for MG adsorption via π-π stacking and electrostatic interactions. GOM was obtained by self-deposition of GO thin films onto N^+-nylon membranes. The material was tested both as a direct spray ionization substrate and as an LDI-MS probe. The latter application resulted in no significant response, whereas the coupling between GOM and direct spray ionization allowed the quantitation of MG and its metabolites at nanomolar levels with extraction recoveries higher than 98%.

An improvement of the performances of PSI-MS in terms of sensitivity has also been achieved in a study dealing with the use of a paper substrate functionalized with urea [121]. Triangles of chromatography paper were treated with 1-[3-(trimethoxysilyl)propyl]urea to create an anion capture layer. The authors demonstrated that the urea-modified paper is an excellent substrate for PSI-MS since it is able to reduce ionization suppression caused by anions and highly polar compounds in the negative-ion MS mode. These findings are of pivotal importance for the analysis of biological samples like urine, blood, and plant extracts.

A selective substrate for PSI-MS, based on the use of molecularly imprinted polymers (MIPs), has been proposed by Pereira et al. [122]. More precisely, a membrane spray ionization method, combining MIP extraction and PSI-MS analysis, was developed and tested for the determination of diuron and 2,4-dichlorophenoxyacetic acid from apple, banana, and grape methanolic extracts. Being used as PSI substrates, MIPs were synthesized directly on a cellulose membrane: the bulk of the MIP was made by ethylene glycol dimethacrylate, using monuron and 2,4,5-trichlorophenoxyacetic acid as templating agents. After extraction, the MIP membranes were washed to remove matrix interferences and tested as PSI-MS substrates using methanol as a spray solvent. The use of these novel materials allowed us to obtain signal intensities of the targeted analytes far higher than those obtained by nonimprinted polymers with detection limits in the μg/l range.

Bills and Manicke [123] developed a disposable paper spray cartridge containing a plasma fractionation membrane to perform on-cartridge plasma fractionation from whole blood samples. Three commercially available blood fractionation membranes, made of different materials, ranging from polymers to natural and synthetic fibers, that is, Vivid polysulfone membrane, NoviPlex plasma fractionation card, and CytoSep, were tested. Even though all the materials were capable of interacting with plasma samples with low levels of cell lysis, difficulties in terms of drug extraction were observed. In particular, Vivid polysulfone membrane and NoviPlex plasma fractionation card exhibited a high binding capability (over 30%) for all the tested drugs, whereas CytoSep showed a lower binding affinity (<17%) only for two out of five drugs. A drawback of the developed device was also related to the poor fractionation efficiency, as measured by the red blood cell content in the fractionated plasma. Quantitative analysis of plasma using PSI-MS provided results closed to those obtained by HPLC-MS without the need of offline extraction or chromatography separation.

A new zero volt-paper spray ionization (ZV-PSI) has been developed recently by Wleklinski et al. [124]: this approach is based on the generation of the electrospray by the action of the pneumatic force of the vacuum at the MS inlet. ZV-PSI analyses were performed over a large variety of samples, including tributylamine, cocaine, terabutylammonium iodide, 3,5-dinitrobenzoic acid, fludioxonil, and sodium tetraphenylborate. In comparison to classic PSI-MS, the achieved results showed a strong decrease of signal intensities for all the investigated analytes. Although the range of analytes useful for ZV-PSI-MS analysis resulted to be very similar to standard PSI-MS, differences in mass spectra were obtained. The observed behavior was related to the ionization mechanism of the proposed approach, which is strongly related to the effects of analyte surface activity. By using a Monte Carlo simulation, the mechanism regarding the formation of ions from initially uncharged droplets was also explained, thus allowing us to predict detection limits very closed to those observed experimentally and to calculate the relative surface activity of both positive and negative ions.

4.3. Materials for Wooden-Tip Electrospray Ionization-Mass Spectrometry. Surface-coated wooden-tip electrospray ionization-mass spectrometry (SCWT-ESI-MS) is a new technique based on the use of a functionalized wooden needle, acting both as extraction/enrichment phase and ESI probe. By using this strategy, the tip is coated by a proper sorbent for highly selective enrichment of targeted compounds from complex matrices, thus making it suitable for analyses at ultratrace levels. Luan suggested the use of a SCWT-ESI-MS technique to detect different analytes in complex matrices [125, 126].

The SCWT-ESI-MS technique has been applied for the detection of perfluorinated compounds (PFCs) in complex environmental and biological samples at ultratrace level [125]. Sharp wooden tips were functionalized via the silanization process by using octadecyltrimethoxysilane and n-octadecyldimethyl[3-(trimethoxysilyl)propyl]ammonium chloride (OTPAC), in order to obtain two different extractive phases, characterized by long alkyl chain. The two phases were tested for the extraction of PFCs spiked water. The OTPAC coating was characterized by the best enrichment capabilities: the extraction is performed not only by the reversed phase, but also by exploiting the ion exchanged adsorption mechanism. Morphological studies of the tip showed a high probe porosity, thus increasing the surface area of the material, and presence of microchannels for transport of the solvent to achieve ambient ionization MS analysis. After method optimization, the probe was tested for the detection of eight different PCFs both in pure water and in complex matrices, that is, lake water, river water, whole blood, and milk. The achieved results proved that the SCWT probe is characterized by outstanding enrichment capabilities, thus being able to enhance method sensitivity by approximately 4000–8000-folds and 100–500-folds in aqueous samples and in whole blood and milk samples, respectively. Method validation resulted in good linearity (two orders of magnitude), excellent quantitation limits (in the 0.21–1.98 ng/l range), and accuracy (recovery rates in the 89–112% range).

Similarly, a SCWT-ESI-MS-based method has been tested for the rapid and sensitive analysis of trace fluoroquinolone and macrolide antibiotics in water [126]. The wooden probe was functionalized via silanization and sulfonation reactions in order to obtain a sulfo-C_8-chain coating able to interact with the analytes with both reversed phase and ion exchange mechanisms. The SCWT-ESI-MS method was then optimized and tested for the extraction of four fluoroquinolone and two macrolide antibiotics in water at trace levels. Method sensitivity allowed us to obtain detection and quantitation limits in the 1.8–4.5 and 5.9–15.1 ng/l range, respectively. Again, linearity was verified over two orders of magnitude: good precision (RSD <14.3%) and recovery rates in the 93.6–112.6% range were other features of the developed method. Finally, the developed method was applied for the analysis of the targeted antibiotics in tap and river water samples.

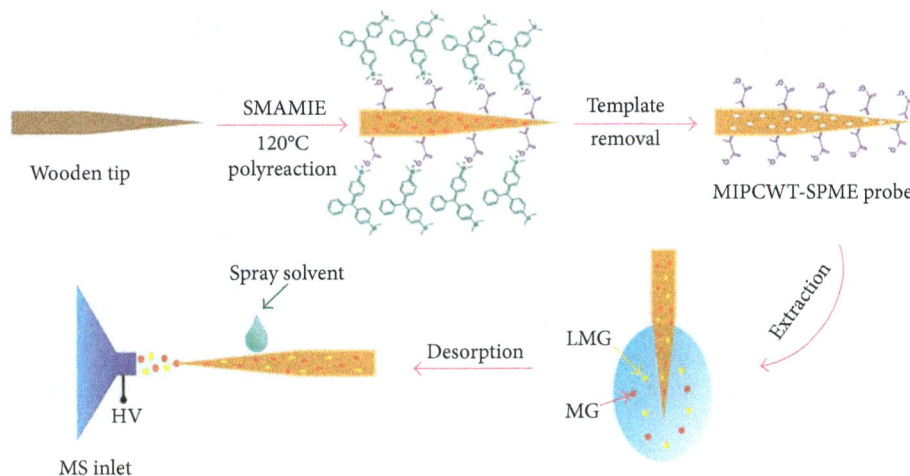

FIGURE 8: Schematic representation of the instrumental setup proposed by Huang. Reprinted with permission from [127].

An interesting approach based on the use of molecularly imprinted polymers as coating for SCWT-ESI-MS has been discussed in a recent study [127]. The coating was synthesized by applying a silicone-modified acrylate molecularly imprinted emulsion onto the tip surface (Figure 8). The developed material was tested for the detection at trace levels of malachite green and its metabolite leucomalachite green in aqueous samples. The MIP-SCWT probe exhibited high enrichment capabilities, allowing us to obtain detection limits at low μg/l levels. In addition, a good linearity was obtained for both the compounds (three orders of magnitude). The method proved to be suitable for high-throughput analysis and was successfully applied for the analysis of tap water, river water, and fish samples.

4.4. Miscellaneous. Similar to PSI, aluminum-foil mass spectrometry (Al-ESI-MS) was recently developed by Hu et al. [128]. This technique is based on the use of a household aluminum foil to obtain the spray ionization of the analytes. The Al foil was cut into triangles, which were folded symmetrically to obtain a mini-reservoir for the sample solution, and was connected to the high voltage supply of the mass spectrometer. The proposed technique was tested for the direct analysis of a wide variety of complex matrices, namely, energetic beverages, urine, skincare and medical creams, and herbal medicines. The inert, hydrophobic and impermeable surface of the Al foil allowed effective on-target extraction of solid samples and on-target sample clean-up, that is, removal of salts, adulterants, and detergents from proteins and peptides. Being Al an excellent heat conductor, the direct monitoring of thermal reactions, such as thermal denaturation of proteins, can be performed in an easy way by Al-ESI-MS.

In a different research study, ESTASI was applied to identify and quantify different compounds from silica gel surfaces, via direct coupling with TLC [129]. The sample spots separated by TLC were analyzed by ESTASI-MSI. The analyses were performed on both drug molecules, using normal phase TLC, and dyes using reversed phase C18 TLC plates, thus guaranteeing the analyses of compounds characterized by very different polarity. After sample separation, the hydrophobic substrate was coated with chlorotrimethylsilane to form hydrophobic surfaces, suitable for ESTASI analysis. Both TLC plates were considered ideal substrates for in situ characterization of samples by using ESTASI-MS, with efficient analyte extraction and separation. In addition, the capability of removing interfering compounds such as salts increased method performance, thus allowing the detection of the investigated analytes at trace levels.

5. MS-Based Approaches for Food Analysis

The demand for safe and high-quality foods has significantly increased in recent years. Food safety and quality have become of greater importance, and the governments of many countries have increased the amount of relevant legislation and demands for food authentication [130]. In consequence, the development of more robust, efficient, cost-effective, and powerful analytical methodologies is continuously needed in order to face these requirements. MS is one of the most suitable techniques because it is featured by excellent specificity, sensitivity, and throughput [131]. MS has been widely used in food safety and quality analysis, and recent advances in MS can provide faster and more accurate methods able to offer better qualitative and quantitative results. Additionally, coupling mass analyzers with separation techniques, such as liquid chromatography (LC-MS) and gas chromatography (GC-MS), have significantly improved food analysis for screening, identification, structural characterization, and quantitation purposes.

One of the most challenges in the application of MS in food analysis, especially in detection of contaminants, is sample preparation because foods are considered very complex matrices in which some natural components can negatively influence the analysis of the targeted compounds. Traditional methods for sample extraction include solid-liquid extraction (SLE), liquid-liquid extraction (LLE), and solid-phase extraction (SPE). More recent is the use of solid-phase microextraction (SPME), pressurized liquid extraction

(PLE), and QuEChERS (quick easy cheap effective rugged safe) [132]. The introduction of high-resolution mass spectrometers, which provide extremely high selectivity and sensitivity, or other emerging MS strategies such as ambient-ionization MS, direct food analysis, and matrix-assisted lased desorption ionization-time of flight-mass spectrometry (MALDI-TOF-MS) profiling and imaging, has strongly reduced sample preprocessing.

According to the PubMed database, since 2012 about 20000 publications dealing with developed MS-based methods for food safety and quality purposes are available. In this section, we do not intend to provide an exhaustive revision of all published studies, but an overview of the most important MS techniques proposed to evaluate food safety and quality.

5.1. MS-Based Approaches for Food-Safety Assessment. The main purpose of food analysis is to ensure food safety, thus requiring the development of accurate and reliable methodologies for the detection of microbial-related spoilage, determination of allergens, detection of environmental contaminants as well as banned external compounds, or the assessment of the occurrence of natural toxins. These methods are strongly influenced by current legislation, which establishes the requirements that an analytical method must meet for an unequivocal identification and quantification of a controlled substance in food samples, as well as the maximum residue limits (MRLs) on certain substances [133].

Being able to allow the quantification of known compounds with great selectivity and sensitivity, tandem MS detection is one of the most frequently utilized analytical approaches to determine contaminants in foods. Triple quadrupole (QqQ) mass spectrometers, running under multiple reaction monitoring (MRM) mode, are the most popular instruments for detecting contaminants in food. This detection procedure allows us to verify the compliance with European legislation on banned and controlled substances in foods [133].

Since 2010, numerous researches have used this methodology to detect pesticides in several fruits and vegetables, such as tomato [134–137], orange [138], mandarin [139], rice and red pepper [139–141], avocado [142, 143], apples and cucumbers [144, 145], mango [146], tea [147], lettuce [148], grains and cereals [149–152], soybean products [153], groundnut oil [154], and wines [155]. The same methodology was also used in the identification and quantification of veterinary drug residues in shellfish [156], meat [157, 158], eggs and milk [158], and contaminants from food contact materials [159].

Triple quadrupoles in MRM mode is also the most-extended approach to detect and quantify toxins and pathogens in food. These pathogens can contaminate foods directly or indirectly, through the productions of toxins. The control of toxin and pathogen levels is extremely important, since the consequences on health due to their contamination of food may be very serious. Following this approach, food products such as nuts [160], maize [161], shellfish [162], tomato [163], beer [164], and multicereal baby food [165] were analyzed.

Almost all these applications combine QqQ-MS with LC or GC separation methods. In some cases, LC- and GC-based techniques were also coupled with other types of MS analyzers such as ion traps (ITs) or quadrupole-linear ion traps (Q-LITs), TOF, or Q-TOF to determine food contaminants [166–168].

Multidimensional procedures allowed us to increase resolving power and separation capabilities that can be beneficial for subsequent MS-based detection, considering that the targeted compounds can reach the detector more separated in time. This is the case of comprehensive two-dimensional gas chromatography (GC × GC) that has been coupled to a TOF-MS analyzer to determine dioxin-related pollutants in complex food samples [169], to screen 68 pesticide residues in oilseed [170], or to detect and to quantify different polychlorobiphenyls (PCBs), polybrominated diphenyl ethers (PBDEs), and PAHs in fish samples [171].

More recently, HRMS analyzers, typically Q-TOF, Orbitrap-MS, and Fourier-transform ion cyclotron resonance- (FT-ICR-) MS, have been used in the field of food safety. These instruments are characterized by high resolution (100,000–1,000,000 FWHM) and high mass precision (1-2 ppm, allowing discrimination between isobaric interferences and ions of interest), thus making possible the screening of unknown compounds with a full MS scan and the construction of databases for targeted compounds. For instance, UPLC-Orbitrap-MS was used to create a database of more than 350 compounds in honey [172]. These databases included different classes of pesticides and veterinary drugs and allowed simultaneous screening of analytes and identification and quantitation of detected compounds in different honey samples. Similar UPLC-HRMS approaches have been lately used to create an accurate-mass database including the fragmentation of more than 600 different food contaminants, such as pesticides, veterinary drug residues, mycotoxins, and perfluoroalkyl substances [173]. Since the particle size of the stationary phase in UPLC is significantly lower than that observed in HPLC, UPLC yields higher speed, better resolution, increased sensitivity, and better peak capacity. Additional examples are related to the development of LC-Orbitrap-MS-based methods for pesticide screening in vegetables and fruits [174], as well as for the analysis of 18 selected mycotoxins in baby food [175].

Ambient MS-based techniques have also been widely applied for food safety purposes: different ionization techniques have been used like LAESI-MS to detect neurotoxin domoic acid in shellfish [176], DESI-MS for the rapid detection of shellfish poisoning toxins in mussels [39], and PSI-MS for the determination of pesticides in fruits and vegetables [114].

5.2. MS-Based Approaches for Food Quality Assessment. Besides food safety, food quality is one of the main concerns of the modern food industry. Food quality encompasses multiple factors, since food authentication and adulteration of food characteristics include food ingredients, such as lipids, proteins, oligosaccharides, vitamins, and carbohydrates, and additives, such as preservatives, antioxidants, and chemicals used for flavor, color, and odor. As a consequence, the evaluation of food quality usually represents a very complex task that needs to consider multiple aspects to achieve the appropriate food quality. Food composition, aroma, flavor, or nutritional properties are among the most important features that need to be evaluated in food quality assessments.

Several MS-based approaches have been used to determine food quality. The most recent publications have mainly used nontargeted MS-based approaches, which very often included the coupling LC-MS and/or GC-MS.

Among LC-MS analytical methods, LC-HRMS techniques have been used for quality evaluation of raw turmeric form different regions [177], for the discrimination of grapes according to plant sterol content [178], for the analysis of the metabolome of the Graciano *Vitis vinifera* wine variety [179], and for the investigation of the quality and authenticity of saffron [180] and strawberries . Moreover, methods based on the UPLC combined with HRMS were developed to assess the authentication and the evaluation of possible adulterations in saffron [181] and fruits juices [182, 183]. The last two methods rely on the feasibility of the application of the UPLC-QToF platform to perform both nontargeted and targeted methods to select potential biomarkers, which should make it possible to develop a targeted method (less sophisticated instrumentation and simpler data analysis) for routine analysis. Similarly, the combination of nontargeted and targeted methods was reported for the qualitative analysis of curcuminoids in turmeric [184]. In this case, nontargeted analysis was performed by using LC-QTOF-MS/MS and the targeted approach by LC-QTRAP-MS/MS.

In the LC-MS-based approaches devoted to food quality assessment, it is noteworthy to highlight the use of hydrophilic interaction chromatography (HILIC), especially in metabolomics approaches. HILIC columns allow profiling highly polar and hydrophilic compounds providing complementary metabolic information to reversed-phase LC. In spite of several caveats associated to HILIC, such as variability in retention times, low peak efficiency, and long re-equilibration times after gradient elution, this methodology has been successfully used to analyze contamination and degradation of infant formulas [185], to separate and detect marine toxins [186], or to identify biomarkers of meat quality [187].

GC-MS-based approaches have also been widely used to evaluate food quality. In these approaches, GC was coupled to a huge diversity of mass analyzers: from simple MS instruments, like quadrupole (Q) [188–192], IT [193], to high-resolution instruments [194–198], as well as hybrid analyzers [199–201]. Studies of the effect of volatile compounds for the classification of saffron based on the concentration of biomarkers [188], classification of olive oils on the basis of their quality parameters [200], the establishment of differences between wine grape cultivars [194], or the detection of milk or meat adulteration [78, 190] are only some of examples relying on the use of GC-MS platforms in food quality analysis.

Besides the much more common LC-MS and GC-MS platforms to assess food quality, comprehensive two-dimensional GC [202] and CE methods [186] coupled to TOF analyzers have also been applied. GC×GC allowed the creation of a panel of biomarkers of rice flavor quality through establishing associations between volatile metabolites and perception of rice aroma [202]. These results are valuable for breeding programs since they can be used to choose pleasant rice aromas. In the latter, the feasibility of using a polymer-coated-capillary for the separation of anionic metabolites in both orange juice and wine has been demonstrated [186]. It offers a complementary coverage of the metabolome of these samples to those provided by other analytical techniques.

In addition to spray-based ionization techniques [203–206], mass spectrometry imaging (MSI) is another useful technique for food safety and quality control through monitoring the spatial distribution of bioactive components and contaminants in food samples. Until recently, MSI was largely performed with MALDI. MALDI-TOF-MSI was successfully applied to investigate the distribution of toxic glycoalkaloids in potato tubers [207], to identify the site of capsaicin in *Capsicum* fruits [208], and to observe both the tissue site of 10 anthocyanin species in blueberries [209] and the posttranslational modified sites in the alpha-melanocyte-stimulatin hormone for carp and goldfish pituitary tissue, as well as their ratio change under different environmental conditions [210]. Although MALDI-MSI can detect compounds in a tissue section without extraction, purification, separation, or labeling, the slow speed of the analysis and the need for matrix deposition in MALDI-MS are critical disadvantages in food imaging applications because they involve analyte diffusion able to affect the original molecular distribution. The development of various ambient ionization techniques revived interest in MSI because of the direct surface sampling in front of a mass spectrometer with submillimeter resolution and no sample preparation. These techniques permit rapid, direct measurement of compounds present on the condensed sample phase and have become potential analytical tools for direct profile-imaging analysis in an atmospheric pressure environment, thus being particularly useful for food-quality control purposes.

Although the application of these techniques in food analysis is not yet fully established, some examples can be found in literature. As an example, ELDI-MS was applied to obtain the molecular profiling and spatial distribution of particular active components in two edible fungi species [211] as well as alpha-solanine and alpha-chaconine in potato [212].

DESI-MSI was used to reveal the spatial distribution of chlorogenic acids and sucrose across the bean endosperm [213], as well as the spatial and temporal distribution of rohitukine and related compounds during various stages of seed development [214]. LAESI allowed macroscopic and microscopic imaging of pesticides, mycotoxins, and plant metabolites in different matrices [215].

6. Conclusions

MS-based techniques represent a highly valuable tool for environmental, bioanalytical, food safety, and food-quality control purposes and their application in these fields strongly increased in the past years. Despite the numerous advantages of MS-based methods, one of the most challenging aspects is still related to the analysis of complex matrices for the detection of nontarget or unknown compounds. The creation of detailed libraries of compounds, including MS-based information such as accurate mass, isotopic patterns, and collision-induced fragmentation, is strongly demanded.

Innovations in ambient MS allowed the development of analytical methods characterized by high-throughput and minimal sample preparation, thus allowing the analysis of samples in their native ambient. However, an important feature to be discussed when ambient MS methods are used is related to the concentration of the detected compounds on the sample surface that might not represent the actual concentration in the whole sample, thus not matching the requirements of current legislation or official methods of analysis.

Nomenclature

CE:	Capillary electrophoresis
CNT:	Carbon nanotube
DART:	Direct analysis in real time
DESI:	Desorption electrospray ionization
DMPA:	*N,N*-dimethyl-*p*-phenylenediamine
EASI:	Easy ambient sonic-spray ionization
EC:	Electrochemical
EESI:	Extractive electrospray ionization
EI:	Electron ionization
ELDI:	Electrospray laser desorption ionization
ESI:	Electrospray ionization
ESTASI:	Electrostatic spray ionization
FA:	Fatty acid
FFA:	Free fatty acid
FI:	Field-induced
FT-ICR:	Fourier-transform ion cyclotron resonance
GC:	Gas chromatography
GC×GC:	Two-dimensional gas chromatography
GO:	Graphene oxide
GOM:	Graphene oxide membrane
HMF:	Hydroxymethylfurfural
HILIC:	Hydrophilic interaction liquid chromatography
HPLC:	High-performance liquid chromatography
HRMS:	High-resolution mass spectrometry
LIT:	Linear ion trap
IT:	Ion trap
LAESI:	Laser ablation electrospray ionization
LC:	Liquid chromatography
LEI:	Liquid-electron ionization
LESA:	Liquid extraction surface analysis
LLE:	Liquid-liquid extraction
LS:	Liquid sample
LOQ:	Limit of quantitation
MALDI:	Matrix-assisted laser desorption/ionization
ME:	Matrix effect
MG:	Malachite green
MIP:	Molecularly imprinted polymer
MRL:	Maximum residue limit
MRM:	Multiple reaction monitoring
MS:	Mass spectrometry
MSI:	Mass spectrometry imaging
MS/MS:	Tandem mass spectrometry

OTPAC:	Octadecyldimethyl[3-(trimethoxysilyl)propyl] ammonium chloride
PAHs:	Polycyclic aromatic hydrocarbons
PB:	Paraffin barrier
PBDE:	Polybrominated diphenyl ether
PCB:	Polychlorobiphenyl
PDMS:	Polydimethylsiloxane
PESI:	Probe electrospray ionization
PFC:	Perfluorinated compound
PMMA:	Poly(methyl methacrylate)
PLA:	Polylactate
PLE:	Pressurized liquid extraction
PLLA:	Poly-l-lactic acid
PSI:	Paper spray ionization
PTFE:	Polytetrafluoroethylene
Q:	Quadrupole
QqQ:	Triple quadrupole
QuEChERS:	Quick easy cheap effective rugged safe
RSD:	Relative standard deviation
SCWT:	Surface-coated wooden tip
SFME:	Slug-flow microextraction
SLE:	Solid-liquid extraction
SMB:	Supersonic molecular beam
SPE:	Solid-phase extraction
SPME:	Solid-phase microextraction
TLC:	Thin-layer chromatography
TOF:	Time of flight
TPPB:	Tetraphenylphosphonium bromide
UPLC:	Ultraperformance liquid chromatography
UTLC:	Ultrathin-layer chromatography
WT:	Wooden tip
ZV-PSI:	Zero volt-paper spray ionization.

Conflicts of Interest

The authors declare that they have no conflicts of interest.

References

[1] T. Kind and O. Fiehn, "Advances in structure elucidation of small molecules using mass spectrometry," *Bioanalytical Reviews*, vol. 2, no. 1–4, pp. 23–60, 2010.

[2] R. Chen, J. Deng, L. Fang et al., "Recent applications of ambient ionization mass spectrometry in environmental analysis," *Trends in Environmental Analytical Chemistry*, vol. 15, pp. 1–11, 2017.

[3] M. Smoluch, P. Mielczarek, and J. Silberring, "Plasma-based ambient ionization mass spectrometry in bioanalytical sciences," *Mass Spectrometry Reviews*, vol. 35, no. 1, pp. 22–34, 2016.

[4] X. Ma and Z. Ouyang, "Ambient ionization and miniature mass spectrometry system for chemical and biological analysis," *TrAC Trends in Analytical Chemistry*, vol. 85, pp. 10–19, 2016.

[5] C. W. Klampfl and M. Himmelsbach, "Direct ionization methods in mass spectrometry: an overview," *Analytica Chimica Acta*, vol. 890, pp. 44–59, 2015.

[6] C. Y. Shi and C. H. Deng, "Recent advances in inorganic materials for LDI-MS analysis of small molecules," *Analyst*, vol. 141, no. 10, pp. 2816–2826, 2016.

[7] B. B. Schneider, E. G. Nazarov, F. Londry, P. Vouros, and T. R. Covey, "Differential mobility spectrometry/mass spectrometry history, theory, design optimization, simulations, and applications," *Mass Spectrometry Reviews*, vol. 35, no. 6, pp. 687–737, 2015.

[8] T. J. Kauppila, J. A. Syage, and T. Benter, "Recent developments in atmospheric pressure photoionization-mass spectrometry," *Mass Spectrometry Reviews*, vol. 36, no. 3, pp. 423–449, 2017.

[9] P. M. Peacock, W. J. Zhang, and S. Trimpin, "Advances in ionization for mass spectrometry," *Analytical Chemistry*, vol. 89, no. 1, pp. 372–388, 2017.

[10] Z. Takàts, J. M. Wiseman, B. Gologan, and R. G. Cooks, "Mass spectrometry sampling under ambient conditions with desorption electrospray ionization," *Science*, vol. 306, no. 5695, pp. 471–473, 2004.

[11] H. Wang, J. Liu, R. G. Cooks, and Z. Ouyang, "Paper spray for direct analysis of complex mixtures using mass spectrometry," *Angewandte Chemie International Edition*, vol. 49, no. 5, pp. 877–880, 2010.

[12] P. Nemes and A. Vertes, "Laser ablation electrospray ionization for atmospheric pressure, in vivo, and imaging mass spectrometry," *Analytical Chemistry*, vol. 79, no. 21, pp. 8098–8106, 2007.

[13] L. Li and K. A. Schug, "On- and off-line coupling of separation techniques to ambient ionization mass spectrometry," *LCGC North America*, vol. 9, no. 4, pp. 8–14, 2011.

[14] A. Herrmann, J. Rosen, D. Jansson, and K. E. Hellenas, "Evaluation of a generic multi-analyte method for detection of >100 representative compounds correlated to emergency events in 19 food types by ultrahigh-pressure liquid chromatography-tandem mass spectrometry," *Journal of Chromatography A*, vol. 1235, pp. 115–124, 2012.

[15] F. T. Peters, "Recent advances of liquid chromatography-(tandem) mass spectrometry in clinical and forensic toxicology," *Clinical Biochemistry*, vol. 44, no. 1, pp. 54–65 2011.

[16] D. Remane, D. K. Wissenbach, and F. T. Peters, "Recent advances of liquid chromatography–(tandem) mass spectrometry in clinical and forensic toxicology—an update," *Clinical Biochemistry*, vol. 49, no. 13-14, pp. 1051–1071, 2016.

[17] F. T. Peters and D. Remane, "Aspects of matrix effects in applications of liquid chromatography-mass spectrometry to forensic and clinical toxicology-a review," *Analytical and Bioanalytical Chemistry*, vol. 403, no. 8, pp. 2155–2172, 2012.

[18] F. Gosetti, E. Mazzucco, D. Zampieri, and M. C. Gennaro, "Signal suppression/enhancement in high-performance liquid chromatography tandem mass spectrometry," *Journal of Chromatography A*, vol. 1217, no. 25, pp. 3929–3937, 2010.

[19] A. K. Malik, C. Blasco, and Y. Pico, "Liquid chromatography-mass spectrometry in food safety," *Journal of Chromatography A*, vol. 1217, no. 25, pp. 4018–4040, 2010.

[20] A. Kaufmann, "The current role of high-resolution mass spectrometry in food analysis," *Analytical and Bioanalytical Chemistry*, vol. 403, no. 5, pp. 1233–1249, 2013.

[21] F. Hernandez, J. V. Sancho, M. Ibanez, E. Abad, T. Portoles, and L. Mattioli, "Current use of high-resolution mass spectrometry in the environmental sciences," *Analytical and Bioanalytical Chemistry*, vol. 403, no. 5, pp. 1251–1264, 2012.

[22] M. M. Gomez-Ramos, C. Ferrer, O. Malato, A. Aguera, and A. R. Fernandez-Alba, "Liquid chromatography-high-resolution mass spectrometry for pesticide residue analysis in fruit and vegetables: screening and quantitative studies," *Journal of Chromatography A*, vol. 1287, pp. 24–37, 2013.

[23] A. G. Marshall and C. L. Hendrickson, "High-resolution mass spectrometers," *Annual Review of Analytical Chemistry*, vol. 1, no. 1, pp. 579–599, 2008.

[24] Y. Fu, C. Zhao, X. Lu, and G. Xu, "Nontargeted screening of chemical contaminants and illegal additives in food based on liquid chromatography-high resolution mass spectrometry," *TrAC Trends in Analytical Chemistry*, vol. 96, pp. 89–98, 2017.

[25] M. Mattarozzi, M. Milioli, C. Cavalieri, F. Bianchi, and M. Careri, "Rapid desorption electrospray ionization-high resolution mass spectrometry method for the analysis of melamine migration from melamine tableware," *Talanta*, vol. 101, pp. 453–459, 2012.

[26] B. O. Crews, A. J. Pesce, R. West, H. Nguyen, and R. L. Fitzgerald, "Evaluation of high-resolution mass spectrometry for urine toxicology screening in a pain management setting," *Journal of Analytical Toxicology*, vol. 36, no. 9, pp. 601–607, 2012.

[27] F. Guale, S. Shahreza, J. P. Waltersheid, H. H. Chen, C. Arndt, and A. T. Kelly, "Validation of LC-TOF-MS screening for drugs, metabolites, and collateral compounds in forensic toxicology specimens," *Journal of Analytical Toxicology*, vol. 37, no. 1, pp. 17–24, 2013.

[28] F. W. McLafferty, *Interpretation of Mass Spectra*, University Science Books, Mill Valley, CA, USA, 1980.

[29] A. Cappiello, G. Famiglini, P. Palma, E. Pierini, V. Termopoli, and H. Trufelli, "Direct-EI in LC-MS: towards a universal detector for small-molecule applications," *Mass Spectrometry Reviews*, vol. 30, no. 6, pp. 1242–1255, 2011.

[30] P. Palma, G. Famiglini, H. Trufelli, E. Pierini, V. Termopoli, and A. Cappiello, "Electron ionization in LC-MS: recent developments and applications of the direct-EI LC-MS interface," *Analytical and Bioanalytical Chemistry*, vol. 399, no. 8, pp. 2683–2693, 2011.

[31] A. Cappiello, G. Famiglini, P. Palma, V. Termopoli, F. Capriotti, and N. Cellar, "Identification potential of direct-EI LC–MS interfacing in small-molecule applications," *Journal of Separation Science*, vol. 5, pp. 13–17, 2013.

[32] V. Termopoli, G. Famiglini, P. Palma, M. Piergiovanni, and A. Cappiello, "Atmospheric pressure vaporization mechanism for coupling a liquid phase with electron ionization mass spectrometry," *Analytical Chemistry*, vol. 89, no. 3, pp. 2049–2056, 2017.

[33] F. Rigano, A. Albergamo, D. Sciarrone, M. Beccaria, G. Purcaro, and L. Mondello, "Nano liquid chromatography directly coupled to electron ionization mass spectrometry for free fatty acid elucidation in mussel," *Analytical Chemistry*, vol. 88, no. 7, pp. 4021–4028, 2016.

[34] B. Seemann, T. Alon, S. Tsizin, A. B. Fialkov, and A. Amirav, "Electron ionization LC-MS with supersonic molecular beams-the new concept, benefits and applications," *Journal of Mass Spectrometry*, vol. 50, no. 11, pp. 1252–1263, 2015.

[35] G. A. Harris, A. S. Galhena, and F. M. Fernandez, "Ambient sampling/ionization mass spectrometry: applications and current trends," *Analytical Chemistry*, vol. 83, no. 12, pp. 4508–4538, 2011.

[36] M. Domin and R. Cody, *Ambient Ionization Mass Spectrometry*, Royal Society of Chemistry, Cambridge, UK, 2015.

[37] L. Magrini, G. Famiglini, P. Palma, V. Termopoli, and A. Cappiello, "Boosting the detection potential of liquid chromatography-electron ionization mass spectrometry using a ceramic coated ion source," *Journal of the American Society for Mass Spectrometry*, vol. 27, no. 1, pp. 153–160, 2016.

[38] N. Riboni, L. Magrini, F. Bianchi, M. Careri, and A. Cappiello, "Sol-gel coated ion sources for liquid chromatography-direct electron ionization mass spectrometry," *Analytica Chimica Acta*, vol. 978, pp. 35–41, 2017.

[39] F. Bianchi, A. Gregori, G. Braun, C. Crescenzi, and M. Careri, "Micro-solid-phase extraction coupled to desorption electrospray ionization-high-resolution mass spectrometry for the analysis of explosives in soil," *Analytical and Bioanalytical Chemistry*, vol. 407, no. 3, pp. 931–938, 2014.

[40] N. L. Sanders, S. Kothari, G. Huang, G. Salazar, and R. G. Cooks, "Detection of explosives as negative ions directly from surfaces using a miniature mass spectrometer," *Analytical Chemistry*, vol. 82, no. 12, pp. 5313–5316, 2010.

[41] I. Cotte-Rodríguez, Z. Takats, N. Talaty, H. Chen, and R. G. Cooks, "Desorption electrospray ionization of explosives on surfaces: sensitivity and selectivity enhancement by reactive desorption electrospray ionization," *Analytical Chemistry*, vol. 77, no. 21, pp. 6755–6764, 2005.

[42] N. Talaty, C. C. Mulligan, D. R. Justes, A. U. Jackson, R. J. Noll, and R. G. Cooks, "Fabric analysis by ambient mass spectrometry for explosives and drugs," *Analyst*, vol. 133, no. 11, pp. 1532–1540, 2008.

[43] I. Cotte-Rodriguez, H. Hernandez-Soto, H. Chen, and R. G. Cooks, "In situ trace detection of peroxide explosives by desorption electrospray ionization and desorption atmospheric pressure chemical ionization," *Analytical Chemistry*, vol. 80, no. 5, pp. 1512–1519, 2008.

[44] J. M. Wiseman, D. R. Ifa, Y. Zhu et al., "Desorption electrospray ionization mass spectrometry: imaging drugs and metabolites in tissues," *Proceedings of the National Academy of Sciences*, vol. 105, no. 47, pp. 18120–18125, 2008.

[45] T. J. Kauppila, N. Talaty, T. Kuuranne, T. Kotiaho, R. Kostiainen, and R. G. Cooks, "Rapid analysis of metabolites and drugs of abuse from urine samples by desorption electrospray ionization-mass spectrometry," *Analyst*, vol. 132, no. 9, pp. 868–875, 2007.

[46] T. J. Kauppila, J. M. Wiseman, R. A. Ketola, T. Kotiaho, R. G. Cooks, and R. Kostiainen, "Desorption electrospray ionization mass spectrometry for the analysis of pharmaceuticals and metabolites," *Rapid Communications in Mass Spectrometry*, vol. 20, no. 3, pp. 387–392, 2006.

[47] Z. Lin, S. Zhang, M. Zhao, C. Yang, D. Chen, and X. Zhang, "Rapid screening of clenbuterol in urine samples by desorption electrospray ionization tandem mass spectrometry," *Rapid Communications in Mass Spectrometry*, vol. 22, no. 12, pp. 1882–1888, 2008.

[48] H. Chen, J. Zheng, X. Zhang, M. Luo, Z. Wang, and X. Qiao, "Surface desorption atmospheric pressure chemical ionization mass spectrometry for direct ambient sample analysis without toxic chemical contamination," *Journal of Mass Spectrometry*, vol. 42, no. 8, pp. 1045–1056, 2007.

[49] R. J. Fussell, D. Chan, and M. Sharman, "An assessment of atmospheric-pressure solids-analysis probes for the detection of chemicals in food," *TrAC Trends in Analytical Chemistry*, vol. 29, no. 11, pp. 1326–1335, 2010.

[50] C. Black, O. P. Chevallier, and C. T. Elliott, "The current and potential applications of ambient mass spectrometry in detecting food fraud," *TrAC Trends in Analytical Chemistry*, vol. 82, pp. 268–278, 2016.

[51] E. Hiyama, A. Ali, S. Amer et al., "Direct lipido-metabolomics of single floating cells for analysis of circulating tumor cells by live single-cell mass spectrometry," *Analytical Sciences*, vol. 31, no. 12, pp. 1215–1517, 2015.

[52] F. Chen, L. Lin, J. Zhang, Z. He, K. Uchiyama, and J. M. Lin, "Single-cell analysis using drop-on-demand inkjet printing and probe electrospray ionization mass spectrometry," *Analytical Chemistry*, vol. 88, no. 8, pp. 4354–4360, 2016.

[53] X. Gong, Y. Zhao, S. Cai et al., "Single cell analysis with probe ESI-mass spectrometry: detection of metabolites at cellular and subcellular levels," *Analytical Chemistry*, vol. 86, no. 8, pp. 3809–3816, 2014.

[54] Z. Takats, J. M. Wiseman, D. R. Ifa, and R. G. Cooks, "Desorption electrospray ionization (DESI) analysis of tryptic digests/peptides," *Cold Spring Harbor Protocols*, vol. 2008, no. 5, p. pdb.prot4993, 2008.

[55] S. P. Pasilis, V. Kertesz, G. J. Van Berkel, M. Schulz, and S. Schorcht, "Using HPTLC/DESI-MS for peptide identification in 1D separations of tryptic protein digests," *Analytical and Bioanalytical Chemistry*, vol. 391, no. 1, pp. 317–324, 2008.

[56] G. Parsiegla, B. Shrestha, F. Carriere, and A. Vertes, "Direct analysis of phycobilisomal antenna proteins and metabolites in small cyanobacterial populations by laser ablation electrospray ionization mass spectrometry," *Analytical Chemistry*, vol. 84, no. 1, pp. 34–38, 2012.

[57] Z. P. Yao, "Characterization of proteins by ambient mass spectrometry," *Mass Spectrometry Reviews*, vol. 31, no. 4, pp. 437–447, 2012.

[58] M. Morelato, A. Beavis, P. Kirkbride, and C. Roux, "Forensic applications of desorption electrospray ionisation mass spectrometry (DESI-MS)," *Forensic Science International*, vol. 226, no. 1–3, pp. 10–21, 2013.

[59] C. Ibáñez, V. García-Cañas, A. Valdés, and C. Simó, "Novel MS-based approaches and applications in food metabolomics," *TrAC Trends in Analytical Chemistry*, vol. 52, pp. 100–111, 2013.

[60] M. Castro-Puyana and M. Herrero, "Metabolomics approaches based on mass spectrometry for food safety, quality and traceability," *TrAC Trends in Analytical Chemistry*, vol. 52, pp. 74–87, 2013.

[61] P. Nemes and A. Vertes, "Ambient mass spectrometry for in vivo local analysis and in situ molecular tissue imaging," *TrAC Trends in Analytical Chemistry*, vol. 34, pp. 22–34, 2012.

[62] E. R. St John, M. Rossi, P. Pruski, A. Darzi, and Z. Takats, "Intraoperative tissue identification by mass spectrometric technologies," *TrAC Trends in Analytical Chemistry*, vol. 85, pp. 2–9, 2016.

[63] Y. Yang, Y. Huang, J. Wu, N. Liu, J. Deng, and T. Luan, "Single-cell analysis by ambient mass spectrometry," *TrAC Trends in Analytical Chemistry*, vol. 90, pp. 14–26, 2017.

[64] A. A. Lubin, D. Cabooter, P. Augustijns, and F. Cuyckens, "One drop chemical derivatization – DESI-MS analysis for metabolite structure identification," *Journal of Mass Spectrometry*, vol. 50, no. 7, pp. 871–878, 2015.

[65] T. A. Brown, H. Chen, and R. N. Zare, "Identification of fleeting electrochemical reaction intermediates using desorption electrospray ionization mass spectrometry," *Journal of the American Chemical Society*, vol. 137, no. 23, pp. 7274–7277, 2015.

[66] W. D. Looi, B. Brown, L. Chamand, and A. Brajter-Toth, "Merits of online electrochemistry liquid sample desorption electrospray ionization mass spectrometry (EC/LS DESI MS)," *Analytical and Bioanalytical Chemistry*, vol. 408, no. 9, pp. 2227–2238, 2016.

[67] Y. Cai, D. Adams, and H. Chen, "A new splitting method for both analytical and preparative LC/MS," *Journal of the*

American Society for Mass Spectrometry, vol. 25, no. 2, pp. 286–292, 2014.

[68] Y. Cai, Y. Liu, R. Helmy, and H. Chen, "Coupling of ultrafast LC with mass spectrometry by DESI," *Journal of The American Society for Mass Spectrometry*, vol. 25, no. 10, pp. 1820–1823, 2014.

[69] Y. Ren, M. N. McLuckey, J. Liu, and Z. Ouyang, "Direct mass spectrometry analysis of biofluid samples using slug-flow microextraction nano-electrospray ionization," *Angewandte Chemie International Edition*, vol. 53, no. 51, pp. 14124–14127, 2014.

[70] J. Cain, A. Laskin, M. R. Kholghy, M. J. Thomson, and H. Wang, "Molecular characterization of organic content of soot along the centerline of a coflow diffusion flame," *Physical Chemistry Chemical Physics*, vol. 16, no. 47, pp. 25862–25875, 2014.

[71] S. Tao, X. Lu, N. Levac et al., "Molecular characterization of organosulfates in organic aerosols from Shanghai and Los Angeles urban areas by nanospray-desorption electrospray ionization high-resolution mass spectrometry," *Environmental Science & Technology*, vol. 48, no. 18, pp. 10993–11001, 2014.

[72] E. J. Boone, A. Laskin, J. Laskin et al., "Aqueous processing of atmospheric organic particles in cloud water collected via aircraft sampling," *Environmental Science & Technology*, vol. 49, no. 14, pp. 8523–8530, 2015.

[73] C. Cardoso-Palacios and I. Lanekoff, "Direct analysis of pharmaceutical drugs using nano-DESI MS," *Journal of Analytical Methods in Chemistry*, vol. 2016, Article ID 3591908, 6 pages, 2016.

[74] J. Watrous, P. Roach, B. Heath, T. Alexandrov, J. Laskin, and P. C. Dorrestein, "Metabolic profiling directly from the Petri dish using nanospray desorption electrospray ionization imaging mass spectrometry," *Analytical Chemistry*, vol. 85, no. 21, pp. 10385–10391, 2013.

[75] H. M. Bergman and I. Lanekoff, "Profiling and quantifying endogenous molecules in single cells using nano-DESI MS," *Analyst*, vol. 142, no. 19, pp. 3639–3647, 2017.

[76] H. M. Bergman, E. Lundin, M. Andersson, and I. Lanekoff, "Quantitative mass spectrometry imaging of small-molecule neurotransmitters in rat brain tissue sections using nano-spray desorption electrospray ionization," *Analyst*, vol. 141, no. 12, pp. 3686–3695, 2016.

[77] C. C. Hsu, P. T. Chou, and R. N. Zare, "Imaging of proteins in tissue samples using nanospray desorption electrospray ionization mass spectrometry," *Analytical Chemistry*, vol. 87, no. 22, pp. 11171–11175, 2015.

[78] D. Duncan, H. M. Bergman, and I. Lanekoff, "A pneumatically assisted nanospray desorption electrospray ionization source for increased solvent versatility and enhanced metabolite detection from tissue," *Analyst*, vol. 142, no. 18, pp. 3424–3431, 2017.

[79] H. W. Chen, A. Venter, and R. G. Cooks, "Extractive electrospray ionization for direct analysis of undiluted urine, milk and other complex mixtures without sample preparation," *Chemical Communications*, vol. 42, no. 19, pp. 2042–2044, 2006.

[80] R. Wang, A. J. Gröhn, L. Zhu et al., "On the mechanism of extractive electrospray ionization (EESI) in the dual-spray configuration," *Analytical and Bioanalytical Chemistry*, vol. 402, no. 8, pp. 2633–2643, 2012.

[81] P. J. Gallimore and M. Kalberer, "Characterizing an extractive electrospray ionization (EESI) source for the online mass spectrometry analysis of organic aerosols,"

Environmental Science & Technology, vol. 47, no. 13, pp. 7324–7331, 2013.

[82] G. K. Koyanagi, V. Blagojevic, and D. K. Bohme, "Applications of extractive electrospray ionization (EESI) in analytical chemistry," *International Journal of Mass Spectrometry*, vol. 379, pp. 146–150, 2015.

[83] M. Deng, T. Yu, H. Luo, T. Zhu, X. Huang, and L. Luo, "Direct detection of multiple pesticides in honey by neutral desorption-extractive electrospray ionization mass spectrometry," *International Journal of Mass Spectrometry*, vol. 422, pp. 111–118, 2017.

[84] N. Xu, Z. Q. Zhu, S. P. Yang et al., "Direct detection of amino acids using extractive electrospray ionization tandem mass spectrometry," *Chinese Journal of Analytical Chemistry*, vol. 41, no. 4, p. 523, 2013.

[85] X. Li, X. Fang, Z. Yu et al., "Direct quantification of creatinine in human urine by using isotope dilution extractive electrospray ionization tandem mass spectrometry," *Analytica Chimica Acta*, vol. 748, pp. 53–57, 2012.

[86] R. S. Jacobson, R. L. Thurston, B. Shrestha, and A. Vertes, "In situ analysis of small populations of adherent mammalian cells using laser ablation electrospray ionization mass spectrometry in transmission geometry," *Analytica Chimica Acta*, vol. 87, no. 24, pp. 12130–12136, 2015.

[87] L. R. Compton, B. Reschke, J. Friend, M. Powell, and A. Vertes, "Remote laser ablation electrospray ionization mass spectrometry for non-proximate analysis of biological tissues," *Rapid Communications in Mass Spectrometry*, vol. 29, no. 1, pp. 67–73, 2015.

[88] L. C. Duarte, T. C. de Carvalho, E. O. Lobo-Júnior, P. V. Abdelnur, B. G. Vaza, and W. K. T. Coltro, "3D printing of microfluidic devices for paper-assisted direct spray ionization mass spectrometry," *Analytical Methods*, vol. 8, no. 3, pp. 496–503, 2016.

[89] G. I. J. Salentijn, H. P. Permentier, and E. Verpoorte, "3D-printed paper spray ionization cartridge with fast wetting and continuous solvent supply features," *Analytical Chemistry*, vol. 86, no. 23, pp. 11657–11665, 2014.

[90] L. Shen, J. Zhang, Q. Yang, N. E. Manicke, and Z. Ouyang, "High throughput paper spray mass spectrometry analysis," *Clinica Chimica Acta*, vol. 420, pp. 28–33, 2013.

[91] J. Deng, W. Wang, Y. Yang et al., "Slug-flow microextraction coupled with paper spray mass spectrometry for rapid analysis of complex samples," *Analytica Chimica Acta*, vol. 940, pp. 143–149, 2016.

[92] C. Zhang and N. E. Manicke, "Development of a paper spray mass spectrometry cartridge with integrated solid phase extraction for bioanalysis," *Analytical Chemistry*, vol. 87, no. 12, pp. 6212–6219, 2015.

[93] B. Yang, F. Wang, W. Deng et al., "Wooden-tip electrospray ionization mass spectrometry for trace analysis of toxic and hazardous compounds in food samples," *Analytical Methods*, vol. 7, no. 14, pp. 5886–5890, 2015.

[94] G. Z. Xin, B. Hu, Z. Q. Shi et al., "Rapid identification of plant materials by wooden-tip electrospray ionization mass spectrometry and a strategy to differentiate the bulbs of *Fritillaria*," *Analytica Chimica Acta*, vol. 820, pp. 84–91, 2014.

[95] Y. Yang and J. Deng, "Internal standard mass spectrum fingerprint: a novel strategy for rapid assessing the quality of Shuang-Huang-Lian oral liquid using wooden-tip electrospray ionization mass spectrometry," *Analytica Chimica Acta*, vol. 837, pp. 83–92, 2014.

[96] H. K. Chen, C. H. Lin, J. T. Liu, and C. H. Lin, "Electrospray ionization using a bamboo pen nib," *International Journal of Mass Spectrometry*, vol. 356, pp. 37–40, 2013.

[97] B. Hu and Z. P. Yao, "Detection of native proteins using solid-substrate electrospray ionization mass spectrometry with nonpolar solvents," *Analytica Chimica Acta*, vol. 1004, pp. 51–57, 2017.

[98] Y. Yang, J. Deng, and Z. P. Yao, "Field-induced wooden-tip electrospray ionization mass spectrometry for high-throughput analysis of herbal medicines," *Analytica Chimica Acta*, vol. 887, pp. 127–137, 2015.

[99] Y. W. Liou, K. Y. Chang, and C. H. Lin, "Sampling and profiling caffeine and its metabolites from an eyelid using a watercolor pen based on electrospray ionization/mass spectrometry," *International Journal of Mass Spectrometry*, vol. 422, pp. 51–55, 2017.

[100] N. Pan, W. Rao, N. R. Kothapalli, R. Liu, A. W. G. Burgett, and Z. Yang, "The single-probe: a miniaturized multifunctional device for single cell mass spectrometry analysis," *Analytical Chemistry*, vol. 86, no. 19, pp. 9376–9380, 2014.

[101] C. Shiea, Y. L. Huang, S. C. Cheng, Y. L. Chen, and J. Shiea, "Determination of elemental composition of metals using ambient organic mass spectrometry," *Analytica Chimica Acta*, vol. 968, pp. 50–57, 2017.

[102] L. Qiao, R. Sartor, N. Gasilova et al., "Electrostatic-spray ionization mass spectrometry," *Analytical Chemistry*, vol. 84, no. 17, pp. 7422–7430, 2012.

[103] Z. Takats, J. M. Wiseman, and R. G. Cooks, "Ambient mass spectrometry using desorption electrospray ionization (DESI): instrumentation, mechanisms and applications in forensics, chemistry, and biology," *Journal of Mass Spectrometry*, vol. 40, no. 10, pp. 1261–1275, 2005.

[104] A. Penna, M. Careri, N. D. Spencer, and A. Rossi, "Effects of tailored surface chemistry on desorption electrospray ionization mass spectrometry: a surface-analytical study by XPS and AFM," *Journal of the American Society for Mass Spectrometry*, vol. 26, no. 8, pp. 1311–1319, 2015.

[105] L. Elviri, R. Foresti, A. Bianchera, M. Silvestri, and R. Bettini, "3D-printed polylactic acid supports for enhanced ionization efficiency in desorption electrospray mass spectrometry analysis of liquid and gel samples," *Talanta*, vol. 155, pp. 321–328, 2016.

[106] M. Montowska, W. Rao, M. R. Alexander, G. A. Tucker, and D. A. Barrett, "Tryptic digestion coupled with ambient desorption electrospray ionization and liquid extraction surface analysis mass spectrometry enabling identification of skeletal muscle proteins in mixtures and distinguishing between beef, pork, horse, chicken, and turkey meat," *Analytical Chemistry*, vol. 86, no. 9, pp. 4479–4487, 2014.

[107] M. T. Dulay, L. S. Eberlin, and R. N. Zare, "Protein analysis by ambient ionization mass spectrometry using trypsin-immobilized organosiloxane polymer surfaces," *Analytical Chemistry*, vol. 87, no. 24, pp. 12324–12330, 2015.

[108] S. Cheng, J. Wang, Y. Cai, J. A. Loo, and H. Chen, "Enhancing performance of liquid sample desorption electrospray ionization mass spectrometry using trap and capillary columns," *International Journal of Mass Spectrometry*, vol. 392, pp. 73–79, 2015.

[109] S. S. Kanyal, T. T. Häbe, C. V. Cushman et al., "Microfabrication, separations, and detection by mass spectrometry on ultrathin-layer chromatography plates prepared via the low-pressure chemical vapor deposition of silicon nitride onto carbon nanotube templates," *Journal of Chromatography A*, vol. 1404, pp. 115–123, 2015.

[110] K. J. Ewing, D. Gibson, J. Sanghera, and F. Miklos, "Desorption electrospray ionization–mass spectrometric analysis of low vapor pressure chemical particulates collected from a surface," *Analytica Chimica Acta*, vol. 853, pp. 368–374, 2015.

[111] R. G. Hemalatha, M. A. Ganayee, and T. Pradeep, "Electrospun nanofiber mats as "smart surfaces" for desorption electrospray ionization mass spectrometry (DESI MS)-based analysis and imprint imaging," *Analytical Chemistry*, vol. 88, no. 11, pp. 5710–5717, 2016.

[112] C. H. Lin, W. C. Liao, H. K. Chen, and T. Y. Kuo, "Paper spray-MS for bioanalysis," *Bioanalysis*, vol. 6, no. 2, pp. 1–10, 2014.

[113] Q. Yang, H. Wang, J. D. Maas et al., "Paper spray ionization devices for direct, biomedical analysis using mass spectrometry," *International Journal of Mass Spectrometry*, vol. 312, pp. 201–207, 2012.

[114] H. Evard, A. Kruve, R. Lõhmus, and I. Leito, "Paper spray ionization mass spectrometry: study of a method for fast-screening analysis of pesticides in fruits and vegetables," *Journal of Food Composition and Analysis*, vol. 41, pp. 221–225, 2015.

[115] Z. P. Zhang, X. N. Liu, and Y. J. Zheng, "Ambient ionization-paper spray ionization and its application," *Chinese Journal of Analytical Chemistry*, vol. 42, no. 1, pp. 145–152, 2014.

[116] P. H. Lai, P. C. Chen, Y. W. Liao, J. T. Liu, C. C. Chen, and C. H. Lin, "Comparison of gampi paper and nanofibers to chromatography paper used in paper spray-mass spectrometry," *International Journal of Mass Spectrometry*, vol. 375, pp. 14–17, 2015.

[117] T. C. Colletes, P. T. Garcia, R. B. Campanha et al., "A new insert sample approach to paper spray mass spectrometry: a paper substrate with paraffin barriers," *Analyst*, vol. 141, no. 5, pp. 1707–1713, 2016.

[118] Z. Zhang, W. Xu, N. E. Manicke, R. G. Cooks, and Z. Ouyang, "Silica coated paper substrate for paper-spray analysis of therapeutic drugs in dried blood spots," *Analytical Chemistry*, vol. 84, no. 2, pp. 931–938, 2012.

[119] R. Narayanan, D. Sarkar, R. G. Cooks, and T. Pradeep, "Molecular ionization from carbon nanotube paper," *Angewandte Chemie International Edition*, vol. 53, no. 23, pp. 5936–5940, 2014.

[120] S. C. Wei, S. Fan, C. W. Lien et al., "Graphene oxide membrane as an efficient extraction and ionization substrate for spray-mass spectrometric analysis of malachite green and its metabolite in fish samples," *Analytica Chimica Acta*, vol. 1003, pp. 42–48, 2018.

[121] J. Liu, Y. He, S. Chen, M. Ma, S. Yao, and B. Chen, "New urea-modified paper substrate for enhanced analytical performance of negative ion mode paper spray mass spectrometry," *Talanta*, vol. 166, pp. 306–314, 2017.

[122] I. Pereira, M. F. Rodrigues, A. R. Chaves, and B. G. Vaz, "Molecularly imprinted polymer (MIP) membrane assisted direct spray ionization mass spectrometry for agrochemicals screening in foodstuffs," *Talanta*, vol. 178, pp. 507–514, 2018.

[123] B. J. Bills and N. E. Manicke, "Development of a prototype blood fractionation cartridge for plasma analysis by paper spray mass spectrometry," *Clinical Mass Spectrometry*, vol. 2, pp. 18–24, 2016.

[124] M. Wleklinski, Y. Li, S. Bag et al., "Zero volt paper spray ionization and its mechanism," *Analytical Chemistry*, vol. 87, no. 13, pp. 6786–6793, 2015.

[125] J. Deng, Y. Yang, L. Fang, L. Lin, H. Zhou, and T. Luan, "Coupling solid-phase microextraction with ambient mass spectrometry using surface coated wooden-tip probe for

rapid analysis of ultra trace perfluorinated compounds in complex samples," *Analytical Chemistry*, vol. 86, no. 22, pp. 11159–11166, 2014.

[126] J. Deng, T. Yu, Y. Yao et al., "Surface-coated wooden-tip electrospray ionization mass spectrometry for determination of trace fluoroquinolone and macrolide antibiotics in water," *Analytica Chimica Acta*, vol. 954, pp. 52–59, 2017.

[127] Y. Huang, Y. Ma, H. Hu et al., "Rapid and sensitive detection of trace malachite green and its metabolite in aquatic products using molecularly imprinted polymer-coated wooden-tip electrospray ionization mass spectrometry," *RSC Advances*, vol. 7, no. 82, pp. 52091–52100, 2017.

[128] B. Hu, P. K. So, and Z. P. Yao, "Electrospray ionization with aluminum foil: a versatile mass spectrometric technique," *Analytica Chimica Acta*, vol. 817, pp. 1–8, 2014.

[129] X. Zhong, L. Qiao, B. Liu, and H. H. Girault, "Ambient in situ analysis and imaging of both hydrophilic and hydrophobic thin layer chromatography plates by electrostatic spray ionization mass spectrometry," *RSC Advances*, vol. 5, no. 92, pp. 75395–75402, 2015.

[130] V. Garcia-Canas, M. Herrero, E. Ibanez, and A. Cifuentes, "Food analysis: present, future, and foodomics," *Analytical Chemistry*, vol. 84, no. 23, pp. 10150–10159, 2012.

[131] M. Herrero, C. Simo, V. Garcia-Canas, E. Ibanez, and A. Cifuentes, "Foodomics: MS-based strategies in modern food science and nutrition," *Mass Spectrometry Reviews*, vol. 31, no. 1, pp. 49–69, 2012.

[132] F. Cacciola, P. Donato, M. Beccaria, P. Dugo, and L. Mondello, "Advances in LC-MS for food analysis," *LC GC Europe*, vol. 25, no. 5, pp. 15–24, 2012.

[133] *European Union Commission Decision 2002/657/EC*, 2002.

[134] G. C. R. M. Andrade, S. H. Monteiro, J. G. Francisco, L. A. Figueiredo, R. G. Botelho, and V. L. Tornisielo, "Liquid chromatography–electrospray ionization tandem mass spectrometry and dynamic multiple reaction monitoring method for determining multiple pesticide residues in tomato," *Food Chemistry*, vol. 175, pp. 57–65, 2015.

[135] O. Golge and B. Kabak, "Evaluation of QuEChERS sample preparation and liquid chromatography–triple-quadrupole mass spectrometry method for the determination of 109 pesticide residues in tomatoes," *Food Chemistry*, vol. 176, pp. 319–332, 2015.

[136] F. Diniz Madureira, F. A. da Silva Oliveira, W. R. de Souza, A. P. Pontelo, M. L. Gonçalves de Oliveira, and G. Silva, "A multi-residue method for the determination of 90 pesticides in matrices with a high water content by LC-MS/MS without clean-up," *Food Additives & Contaminants: Part A*, vol. 29, no. 4, pp. 665–678, 2012.

[137] A. Garrido-Frenich, M. M. Martín Fernández, L. Díaz Moreno, J. L. Martínez-Vidal, and N. López-Gutiérrez, "Multiresidue pesticide analysis of tuber and root commodities by QuEChERS extraction and UPLC coupled to tandem MS," *Journal of AOAC International*, vol. 95, no. 5, pp. 1319–1330, 2012.

[138] O. Golge and B. Kabak, "Determination of 115 pesticide residues in oranges by high-performance liquid chromatography–triple-quadrupole mass spectrometry in combination with QuEChERS method," *Journal of Food Composition and Analysis*, vol. 41, pp. 86–97, 2015.

[139] J. Cho, J. Lee, C. U. Lim, and J. Ahn, "Quantification of pesticides in food crops using QuEChERS approaches and GCeMS/MS," *Food Additives & Contaminants: Part A*, vol. 33, no. 12, pp. 1803–1816, 2016.

[140] S. S. Shida, S. Nemoto, and R. Matsuda, "Simultaneous determination of acidic pesticides in vegetables and fruits by liquid chromatography–tandem mass spectrometry," *Journal of Environmental Science and Health, Part B*, vol. 50, no. 3, pp. 151–162, 2015.

[141] C. Rasche, B. Fournes, U. Dirks, and K. Speer, "Multi-residue pesticide analysis (gas chromatography–tandem mass spectrometry detection)–improvement of the quick, easy, cheap, effective, rugged, and safe method for dried fruits and fat-rich cereals–benefit and limit of a standardized apple purée calibration (screening)," *Journal of Chromatography A*, vol. 1403, pp. 21–31, 2015.

[142] L. Han, J. Matarrita, Y. Sapozhnikova, and S. J. Lehotay, "Evaluation of a recent product to remove lipids and other matrix co-extractives in the analysis of pesticide residues and environmental contaminants in foods," *Journal of Chromatography A*, vol. 1449, pp. 17–29, 2016.

[143] B. D. Morris and R. B. Schriner, "Development of an automated column solid-phase extraction cleanup of QuEChERS extracts, using a zirconia-based sorbent, for pesticide residue analyses by LC-MS/MS," *Journal of Agricultural and Food Chemistry*, vol. 63, no. 21, pp. 5107–5119, 2015.

[144] G. Ramadan, M. Al Jabir, N. Alabdulmalik, and A. Mohammed, "Validation of a method for the determination of 120 pesticide residues in apples and cucumbers by LC-MS/MS," *Drug Testing and Analysis*, vol. 8, no. 5-6, pp. 498–510, 2016.

[145] M. A. Zhao, Y. N. Feng, Y. Z. Zhu, and J. H. Kim, "Multi-residue method for determination of 238 pesticides in Chinese cabbage and cucumber by liquid chromatography-tandem mass spectrometry: comparison of different purification procedures," *Journal of Agricultural and Food Chemistry*, vol. 62, no. 47, pp. 11449–11456, 2014.

[146] N. Fleury-Filho, C. A. Nascimento, E. O. Faria, A. R. Crunivel, and J. M. Oliveira, "Within laboratory validation of a multi-residue method for the analysis of 98 pesticides in mango by LC tandem MS," *Food Additives & Contaminants: Part A*, vol. 29, no. 4, pp. 641–656, 2012.

[147] T. Cajka, C. Sandy, V. Bachavolva et al., "Streamlining sample preparation and GC tandem MS analysis of multiple pesticide residues in tea," *Analytica Chimica Acta*, vol. 743, pp. 51–60, 2012.

[148] V. Havolt, S. Goscinny, and M. Deridder, "A simple multi-residue method for the determination of pesticides in fruits and vegetables using a methanolic extraction and ultra-high-performance liquid chromatography-tandem mass spectrometry: optimization and extension of scope," *Journal of Chromatography A*, vol. 1384, pp. 53–66, 2015.

[149] S. Walorczyk and D. Drozdzynski, "Improvement and extension to new analytes of a multi-residue method for the determination of pesticides in cereals and dry animal feed using gas chromatography-tandem quadrupole mass spectrometry revisited," *Journal of Chromatography A*, vol. 1251, pp. 219–231, 2012.

[150] O. Lacina, M. Zachariasova, J. Urbavolva, M. Vaclavikova, T. Cajka, and J. Hajslova, "Critical assessment of extraction methods for the simultaneous determination of pesticide residues and mycotoxins in fruits, cereals, spices and oil seeds employing UPLC tandem MS," *Journal of Chromatography A*, vol. 1262, pp. 8–18, 2012.

[151] J. Wang, W. Chow, and W. Cheung, "Application of a tandem mass spectrometer and core-shell particle column for the determination of 151 pesticides in grains," *Journal of Agricultural and Food Chemistry*, vol. 59, no. 16, pp. 8589–8608, 2011.

[152] Z. He, L. Wang, Y. Peng, M. Luo, W. Wang, and X. Liu, "Multiresidue analysis of over 200 pesticides in cereals using a QuEChERS and gas chromatography-tandem mass spectrometry-based method," *Food Chemistry*, vol. 169, pp. 372–380, 2015.

[153] A. Paleníkova, G. Martínez-Domínguez, F. J. Arrebola, R. Romero-Gonzalez, S. Hrouzkova, and A. Garrido Frenich, "Multifamily determination of pesticide residues in soya-based nutraceutical products by GC/MS-MS," *Food Chemistry*, vol. 173, pp. 796–807, 2015.

[154] S. Chawla, H. K. Patel, K. M. Vaghela et al., "Development and validation of multi residue analytical method in cotton and groundnut oil for 87 pesticides using low temperature and dispersive cleanup on gas chromatography and liquid chromatography-tandem mass spectrometry," *Analytical and Bioanalytical Chemistry*, vol. 408, no. 3, pp. 983–997, 2016.

[155] D. L. Christodoulou, P. Kanari, P. Hadjiloizou, and P. Constantivolu, "Pesticide residues analysis in wine by liquid chromatography-tandem mass spectrometry and using ethyl acetate extraction method: validation and pilot survey in real samples," *Journal of Wine Research*, vol. 26, no. 2, pp. 81–98, 2015.

[156] G. R. Chang, H. S. Chen, and F. Y. Lin, "Analysis of banned veterinary drugs and herbicide residues in shellfish by liquid chromatography-tandem mass spectrometry (LC/MS/MS) and gas chromatography-tandem mass spectrometry (GC/MS/MS)," *Marine Pollution Bulletin*, vol. 113, no. 1-2, pp. 579–584, 2016.

[157] M. E. Dasenaki, C. S. Michali, and N. S. Thomaidis, "Analysis of 76 veterinary pharmaceuticals from 13 classes including aminoglycosides in bovine muscle by hydrophilic interaction liquid chromatography–tandem mass spectrometry," *Journal of Chromatography A*, vol. 1452, pp. 67–80, 2016.

[158] D. Chen, J. Yu, Y. Tao et al., "Qualitative screening of veterinary anti-microbial agents in tissues, milk, and eggs of food-producing animals using liquid chromatography coupled with tandem mass spectrometry," *Journal of Chromatography B*, vol. 1017-1018, pp. 82–88, 2016.

[159] M. Aznar, A. Rodriguez-Lafuente, P. Alfaro, and C. Nerin, "UPLC-Q-TOF-MS analysis of non-volatile migrants from new active packaging materials," *Analytical and Bioanalytical Chemistry*, vol. 404, no. 6-7, pp. 1945–1957, 2012.

[160] B. Skribic, J. Zivancev, and M. Godula, "Multimycotoxin analysis of crude extracts of nuts with ultra-high performance liquid chromatography/tandem mass spectrometry," *Journal of Food Composition and Analysis*, vol. 34, no. 2, pp. 171–177, 2014.

[161] M. Ludovici, C. Ialongo, M. Reverberi, M. Beccaccioli, M. Scarpari, and V. Scala, "Quantitative profiling of oxylipins through comprehensive LC-MS/MS analysis of *Fusarium verticillioides* and maize kernels," *Food Additives & Contaminants: Part A*, vol. 31, no. 12, pp. 2026–2033, 2014.

[162] M. García-Altares, A. Casavolva, V. Bane, J. Diogene, A. Furey, and P. de la Iglesia, "Confirmation of pinnatoxins and spirolides in shellfish and passive samplers from Catalonia (Spain) by liquid chromatography coupled with triple quadrupole and high-resolution hybrid tandem mass spectrometry," *Marine Drugs*, vol. 12, no. 6, pp. 3706–3732, 2014.

[163] Y. Rodriguez-Carrasco, J. Manes, H. Berrada, and C. Juan, "Development and validation of a LC-ESI-MS/MS method for the determination of alternaria toxins alternariol, alternariol methyl-ether and tentoxin in tomato and tomato based products," *Toxins*, vol. 8, no. 11, p. 328, 2016.

[164] Y. Rodriguez-Carrasco, M. Fattore, S. Albrizio, H. Berrada, and J. Manes, "Occurrence of *Fusarium mycotoxins* and their dietary intake through beer consumption by the European population," *Food Chemistry*, vol. 178, pp. 149–155, 2015.

[165] C. Juan, J. Mañes, A. Raiola, and A. Ritieni, "Evaluation of beauvericin and enniatins in Italian cereal products and multicereal food by liquid chromatography coupled to triple quadrupole mass spectrometry," *Food Chemistry*, vol. 140, no. 4, pp. 755–762, 2013.

[166] K. Zhang, J. W. Wong, P. Yang et al., "Protocol for an electrospray ionization tandem mass spectral product ion library: development and application for identification of 240 pesticides in foods," *Analytical Chemistry*, vol. 84, no. 13, pp. 5677–5684, 2012.

[167] M. I. Cervera, T. Portoles, E. Pitarch, J. Beltran, and F. Hernandez, "Application of gas chromatography time-of-flight mass spectrometry for target and non-target analysis of pesticide residues in fruits and vegetables," *Journal of Chromatography A*, vol. 1244, pp. 168–177, 2012.

[168] F. Lambertini, V. Di Lallo, D. Catellani, M. Mattarozzi, M. Careri, and M. Suman, "Reliable liquid chromatography-mass spectrometry method for investigation of primary aromatic amines migration from food packaging and during industrial curing of multilayer plastic laminates," *Journal of Mass Spectrometry*, vol. 49, no. 9, pp. 870–877, 2014.

[169] C. Planche, J. Ratel, F. Mercier, P. Blinet, L. Debrauwer, and E. Engel, "Assessment of comprehensive two-dimensional gas chromatography-time-of-flight mass spectrometry based methods for investigating 206 dioxin-like micropollutants in animal-derived food matrices," *Journal of Chromatography A*, vol. 1392, pp. 74–81, 2015.

[170] X. Wang, P. Li, W. Zhang et al., "Screening for pesticide residues in oil seeds using solid-phase dispersion extraction and comprehensive two-dimensional gas chromatography time-of-flight mass spectrometry," *Journal of Separation Science*, vol. 35, no. 13, pp. 1634–1643, 2012.

[171] K. Kalachova, J. Pulkrabova, T. Cajka, L. Drabova, and J. Hajslova, "Implementation of comprehensive two-dimensional GC-time-of-flight-MS for the simultaneous determination of halogenated contaminants and polycyclic aromatic hydrocarbons in fish," *Analytical and Bioanalytical Chemistry*, vol. 403, no. 10, pp. 2813–2824, 2012.

[172] M. L. Gomez-Perez, P. Plaza-Bolavols, R. Romero-Gonzalez, J. L. Martinez-Vidal, and A. Garrido-Frenich, "Comprehensive qualitative and quantitative determination of pesticides and veterinary drugs in honey using liquid chromatography-Orbitrap high resolution mass spectrometry," *Journal of Chromatography A*, vol. 1248, pp. 130–138, 2012.

[173] P. Perez-Ortega, F. J. Lara-Ortega, J. F. García-Reyes, B. Gilbert-Lopez, M. Trojavolwicz, and A. Molina-Díaz, "A feasibility study of UHPLC-HRMS accurate-mass screening methods for multiclass testing of organic contaminants in food," *Talanta*, vol. 160, pp. 704–712, 2016.

[174] H. G. J. Mol, P. Zomer, and M. de Koning, "Qualitative aspects and validation of a screening method for pesticides in vegetables and fruits based on liquid chromatography coupled to full scan high resolution (Orbitrap) mass spectrometry," *Analytical and Bioanalytical Chemistry*, vol. 403, no. 10, pp. 2891–2908, 2012.

[175] J. Rubert, K. J. James, J. Manes, and C. Soler, "Applicability of hybrid linear ion trap-high resolution mass spectrometry and quadrupole-linear ion trap-mass spectrometry for

mycotoxin analysis in baby food," *Journal of Chromatography A*, vol. 1223, pp. 84–92, 2012.

[176] D. G. Beach, C. M. Walsh, and P. McCarron, "High-throughput quantitative analysis of domoic acid directly from mussel tissue using laser ablation electrospray ionization—tandem mass spectrometry," *Toxicon*, vol. 92, pp. 75–80, 2014.

[177] M. Guijarro-Díez, L. Volzal, M. L. Marina, and A. L. Crego, "Metabolomic fingerprinting of saffron by LC/MS: novel authenticity markers," *Analytical and Bioanalytical Chemistry*, vol. 407, no. 23, pp. 7197–7213, 2015.

[178] L. Millán, M. C. Sampedro, A. Sanchez et al., "Liquid chromatography–quadrupole time of flight tandem mass spectrometry–based targeted metabolomic study for varietal discrimination of grapes according to plant sterols content," *Journal of Chromatography A*, vol. 1454, pp. 67–77, 2016.

[179] M. Arbulu, M. C. Sampedro, A. Gomez-Caballero, M. A. Goicolea, and R. J. Barrio, "Untargeted metabolomic analysis using liquid chromatography quadrupole time-of-flight mass spectrometry for non-volatile profiling of wines," *Analytica Chimica Acta*, vol. 858, pp. 32–41, 2015.

[180] J. Rubert, O. Lacina, M. Zachariasova, and J. Hajslova, "Saffron authentication based on liquid chromatography high resolution tandem mass spectrometry and multivariate data analysis," *Food Chemistry*, vol. 204, pp. 201–209, 2016.

[181] A. Kårlund, U. Moor, G. McDougall, M. Lehtonen, R. O. Karjalainen, and K. Hanhineva, "Metabolic profiling discriminates between strawberry (*Fragaria* × *ananassa* Duch.) cultivars grown in Finland or Estonia," *Food Research International*, vol. 89, pp. 647–653, 2016.

[182] Z. Jandric, D. Roberts, M. N. Rathor, A. Abraham, M. Islam, and A. Cannavan, "Assessment of fruit juice authenticity using UPLC-QToF MS: a metabolomics approach," *Food Chemistry*, vol. 148, pp. 7–17, 2014.

[183] Z. Jandric, M. Islam, D. K. Singh, and A. Cannavan, "Authentication of Indian citrus fruit/fruit juices by untargeted and targeted metabolomics," *Food Control*, vol. 72, pp. 181–188, 2017.

[184] S. Jin, C. Song, S. Jia et al., "An integrated strategy for establishment of curcuminoid profile in turmeric using two LC–MS/MS platforms," *Journal of Pharmaceutical and Biomedical Analysis*, vol. 132, pp. 93–102, 2017.

[185] K. Inoue, C. Tanada, T. Sakamoto et al., "Metabolomics approach of infant formula for the evaluation of contamination and degradation using hydrophilic interaction liquid chromatography coupled with mass spectrometry," *Food Chemistry*, vol. 181, pp. 318–324, 2015.

[186] M. Mattarozzi, M. Milioli, F. Bianchi et al., "Optimization of a rapid QuEChERS sample treatment method for HILIC-MS2 analysis of paralytic shellfish poisoning (PSP) toxins in mussels," *Food Control*, vol. 60, pp. 138–145, 2016.

[187] A. K. Subbaraj, Y. H. Brad-Kim, K. Fraser, and M. M. Farouk, "A hydrophilic interaction liquid chromatography–mass spectrometry (HILIC-MS) based metabolomics study on colour stability of ovine meat," *Meat Science*, vol. 117, pp. 163–172, 2016.

[188] G. Aliakbarzadeh, H. Sereshti, and H. Parastar, "Pattern recognition analysis of chromatographic fingerprints of *Crocus sativus* L. secondary metabolites towards source identification and quality control," *Analytical and Bioanalytical Chemistry*, vol. 408, no. 12, pp. 3295–3307, 2016.

[189] F. R. Pinu, S. de Carvalho-Silva, A. P. Trovatti Uetanabaro, and S. G. Villas-Boas, "Vinegar metabolomics: an explorative study

of commercial balsamic vinegars using gas chromatography-mass spectrometry," *Metabolites*, vol. 6, no. 3, p. 22, 2016.

[190] P. Scano, A. Murgia, F. M. Pirisi, and P. Caboni, "A gas chromatography-mass spectrometry-based metabolomic approach for the characterization of goat milk compared with cow milk," *Journal of Dairy Science*, vol. 97, no. 10, pp. 6057–6066, 2014.

[191] L. L. Monti, C. A. Bustamante, S. Osorio et al., "Metabolic profiling of a range of peach fruit varieties reveals high metabolic diversity and commonalities and differences during ripening," *Food Chemistry*, vol. 190, pp. 879–888, 2016.

[192] M. N. A. Khalil, M. I. Fekry, and M. A. Farag, "Metabolome based volatiles profiling in 13 date palm fruit varieties from Egypt via SPME GC-MS and chemometrics," *Food Chemistry*, vol. 217, pp. 171–181, 2017.

[193] I. Akhatou, R. Gonz alez-Domínguez, and A. Fern andez-Recamales, "Investigation of the effect of genotype and agronomic conditions on metabolomic profiles of selected strawberry cultivars with different sensitivity to environmental stress," *Plant Physiology and Biochemistry*, vol. 101, pp. 14–22, 2016.

[194] A. Cuadros-Ivolstroza, S. Ruíz-Lara, E. Gonz alez, A. Eckardt, L. Willmitzer, and H. Pena-Cortes, "GC-MS metabolic profiling of Cabernet Sauvignon and Merlot cultivars during grapevine berry development and network analysis reveals a stage- and cultivar-dependent connectivity of primary metabolites," *Metabolomics*, vol. 12, no. 2, p. 39, 2016.

[195] B. Khakimov, R. J. Mongi, K. M. Sørensen, B. K. Ndabikunze, B. E. Chove, and S. B. Engelsen, "A comprehensive and comparative GC-MS metabolomics study of non-volatiles in Tanzanian grown mango, pineapple, jackfruit, baobab and tamarind fruits," *Food Chemistry*, vol. 213, pp. 691–699, 2016.

[196] J. Welzenbach, C. Neuhoff, C. Looft, K. Schellander, E. Tholen, and C. Große-Brinkhaus, "Different statistical approaches to investigate porcine muscle metabolome profiles to highlight new biomarkers for pork quality assessment," *PLoS One*, vol. 11, no. 2 article e0149758, 2016.

[197] G. Min-Lee, D. Ho-Suh, E. Sung-Jung, and C. Hwan-Lee, "Metabolomics provides quality characterization of commercial gochujang (fermented pepper paste)," *Molecules*, vol. 21, no. 7, p. 921, 2016.

[198] D. E. Lee, G. R. Shin, S. Lee et al., "Metabolomics reveal that amino acids are the main contributors to antioxidant activity in wheat and rice gochujangs (Korean fermented red pepper paste)," *Food Research International*, vol. 87, pp. 10–17, 2016.

[199] E. J. Gu, D. W. Kim, G. J. Jang et al., "Mass-based metabolomic analysis of soybean sprouts during germination," *Food Chemistry*, vol. 217, pp. 311–319, 2017.

[200] C. Sales, M. I. Cervera, R. Gil, T. Portoles, E. Pitarch, and J. Beltran, "Quality classification of Spanish olive oils by untargeted gas chromatography coupled to hybrid quadrupole-time of flight mass spectrometry with atmospheric pressure chemical ionization and metabolomics-based statistical approach," *Food Chemistry*, vol. 216, pp. 365–373, 2017.

[201] D. K. Trivedi, K. A. Hollywood, N. J. W. Rattray et al., "Meat, the metabolites: an integrated metabolite profiling and lipidomics approach for the detection of the adulteration of beef with pork," *Analyst*, vol. 141, no. 7, pp. 2155–2164, 2016.

[202] V. D. Daygon, S. Prakash, M. Calingacion et al., "Understanding the jasmine phenotype of rice through metabolite profiling and sensory evaluation," *Metabolomics*, vol. 12, no. 4, p. 63, 2016.

[203] L. Di Donna, D. Taverna, S. Indelicato, A. Napoli, G. Sindona, and F. Mazzotti, "Rapid assay of resveratrol in red wine by paper spray tandem mass spectrometry and isotope dilution," *Food Chemistry*, vol. 229, pp. 354–357, 2017.

[204] H. V. Pereira, V. S. Amador, M. M. Sena, R. Augusti, and E. Piccin, "Paper spray mass spectrometry and PLS-DA improved by variable selection for the forensic discrimination of beers," *Analytica Chimica Acta*, vol. 940, pp. 104–112, 2016.

[205] J. A. Reis Teodoro, H. V. Pereira, M. M. Sena, E. Piccin, J. J. Zacca, and R. Augusti, "Paper spray mass spectrometry and chemometric tools for a fast and reliable identification of counterfeit blended Scottish whiskies," *Food Chemistry*, vol. 237, pp. 1058–1064, 2017.

[206] A. K. Meher and Y. C. Chen, "Analysis of volatile compounds by open-air ionization mass spectrometry," *Analytica Chimica Acta*, vol. 966, pp. 41–46, 2017.

[207] M. Ha, J. H. Kwak, Y. Kim, and O. P. Zee, "Direct analysis for the distribution of toxic glycoalkaloids in potato tuber tissue using matrix-assisted laser desorption/ionization mass spectrometric imaging," *Food Chemistry*, vol. 133, no. 4, pp. 1155–1162, 2012.

[208] S. Taira, S. Shimma, I. Osaka et al., "Mass spectrometry imaging of the capsaicin localization in the *Capsicum* fruits," *International Journal of Biotechnology for Wellness Industries*, vol. 1, pp. 61–66, 2012.

[209] Y. Yoshimura, H. Evolmoto, T. Moriyama, Y. Kawamura, M. Setou, and N. Zaima, "Visualization of anthocyanin species in rabbiteye blueberry *Vaccinium ashei* by matrix-assisted laser desorption/ionization imaging mass spectrometry," *Analytical and Bioanalytical Chemistry*, vol. 403, no. 7, pp. 885–1895, 2012.

[210] A. Yasuda, Y. Tatsu, and Y. Shigeri, "Characterization of triacetyl-α-melanocyte-stimulating hormone in carp and goldfish," *General and Comparative Endocrinology*, vol. 175, no. 2, pp. 270–276, 2012.

[211] M.-Z. Huang, S.-C. Cheng, S.-S. Jhang et al., "Ambient molecular imaging of dry fungus surface by electrospray laser desorption ionization mass spectrometry," *International Journal of Mass Spectrometry*, vol. 325–327, pp. 172–182, 2012.

[212] S. S. Jhang, M.-Z. Huang, and J. Shiea, "Ambient molecular imaging of toxins within a sprouted potato slice by ELDI/MS," in *Proceedings of the 60th ASMS Conference on Mass Spectrometry and Allied Topics*, Vancouver, BC, Canada, May 2012.

[213] R. Garrett, C. M. Rezende, and D. R. Ifa, "Revealing the spatial distribution of chlorogenic acids and sucrose across coffee bean endosperm by desorption electrospray ionization-mass spectrometry imaging," *LWT-Food Science and Technology*, vol. 65, pp. 711–717, 2016.

[214] P. M. Kumara, A. Srimany, G. Ravikanth, R. U. Shaanker, and T. Pradeep, "Ambient ionization mass spectrometry imaging of rohitukine, a chromone anti-cancer alkaloid, during seed development in *Dysoxylum binectariferum* Hook.f (Meliaceae)," *Phytochemistry*, vol. 116, pp. 104–110, 2015.

[215] M. W. Nielen and T. A. van Beek, "Macroscopic and microscopic spatially-resolved analysis of food contaminants and constituents using laser-ablation electrospray ionization mass spectrometry imaging," *Analytical and Bioanalytical Chemistry*, vol. 406, no. 27, pp. 6805–6815, 2014.

Study of Methanol Extracts from Different Parts of *Peganum harmala* L. using ¹H-NMR Plant Metabolomics

Yinping Li,[1,2] **Qing He** (iD),[3] **Shushan Du** (iD),[4] **Shanshan Guo,**[4] **Zhufeng Geng** (iD),[4,5] **and Zhiwei Deng** (iD)[5]

[1]*College of Chemistry, Beijing Normal University, Beijing 100875, China*
[2]*College of Chemistry and Chemical Engineering, Xinjiang Normal University, Urumqi, Xinjiang 830000, China*
[3]*School of Chemical Engineering and Technology, Tianjin University, Tianjin 300350, China*
[4]*Beijing Key Laboratory of Traditional Chinese Medicine Protection and Utilization, Faculty of Geographical Science, Beijing Normal University, Beijing, China*
[5]*Analytic and Testing Center, Beijing Normal University, Beijing 100875, China*

Correspondence should be addressed to Zhufeng Geng; gengzhufeng@bnu.edu.cn and Zhiwei Deng; dengzw@bnu.edu.cn

Academic Editor: Eduardo Dellacassa

A nuclear magnetic resonance- (NMR-) based metabolomics method was used to identify differential metabolites of methanol extracts obtained from six parts of *Peganum harmala* L. (*P. harmala*), namely, the root, stem, leaf, flower, testa, and seed. Two multivariate statistical analysis methods, principal component analysis (PCA) and partial least squares-discriminant analysis (PLS-DA), were combined to clearly distinguish among the *P. harmala* samples from the six different parts. Eleven differential components were screened by the PLS-DA loading plot, and the relative contents were calculated by univariate analysis of variance. Chemometric results showed significant differences in the metabolites of the different parts of *P. harmala*. The seeds contained large amounts of harmaline, harmine, and vasicine compared to other organs. The acetic acid, proline, lysine, and sucrose contents of the roots were significantly higher than those of the other parts. In the testa, the vasicine, asparagine, choline, and 4-hydroxyisoleucine contents were clearly dominant. The obtained data revealed the distribution characteristics of the metabolomes of the different *P. harmala* parts and provided fundamental knowledge for the rational development of its medicinal parts.

1. Introduction

P. harmala is the only salt-tolerant perennial herb in the *Peganum* genus of the family Zygophyllaceae [1]. It has been used as a popular traditional ethnodrug for a long time due to its phytotherapeutic value [2]. Traditionally, the seeds have been used to relieve pain, to promote blood circulation, and to treat rheumatism and illnesses such as cough and asthma [3, 4]. The whole plant has been used for pain relief and as a rheumatism treatment [5]. *P. harmala* extract is a rich source of bioactive substances [6], including large amounts of primary and secondary metabolites [7]. The whole plant and its seeds contain several kinds of alkaloids [8]. The alkaloid content of the seeds

accounts for 3.92–7% of the dry weight [9]. The main alkaloid components are derivatives of quinazoline and β-carboline [10], which have exhibited anticancer and antibacterial activities in pharmacological studies [4, 10]. Because these metabolites located in different parts of the plant can vary greatly in type and quantity, the resulting pharmacological activities and antibacterial effects are significantly different [11]. Previous studies mainly focused on the extraction and activities of alkaloid compounds in *P. harmala* [12]. Only few alkaloids (β-carbolines and quinazoline derivatives) isolated from the different organs in *P. harmala* have been investigated [4, 13, 14], and no comprehensive studies of the metabolome of each of its parts have been performed.

Plant metabolomics is a method for investigating the dynamic changes in small-molecule metabolites or components in plants, and it has played an increasingly important role in explaining plant growth and reproduction [15, 16]. However, it is challenging to use metabolomics to comprehensively annotate metabolites and analyse the physiological and ecological roles of metabolomes [17]. As a rule, the structure of a compound is determined by spectral methods after being isolated and purified by various chromatographic techniques [13]. This process is time-consuming and laborious. At present, nuclear magnetic resonance (NMR) metabolomics technology offers a fast and sensitive method to detect distinctive signals and resolve the structures of compounds in a mixture by comparing with available data [18]. This technique has been widely applied in pharmacology, pharmacodynamics, and pharmacokinetics studies because of its advantages, including simple preparation, its unbiased nature, and its ability to qualitatively and quantitatively detect multiple metabolic components simultaneously [19].

In the current study, an NMR-based plant metabolomics method was used to analyse fingerprint spectra of methanol extracts obtained from different parts of *P. harmala*. A multivariate unsupervised analysis method, namely, principal component analysis (PCA), was employed to distinguish between the metabolomes of six different *P. harmala* parts. To enlarge the difference found in the PCA model and detect influential variables, partial least squares-discriminant analysis (PLS-DA), a supervised pattern recognition method, was used to recognize the characteristic differential metabolites among the groups classified according to different organs of plants. The relative contents of the major metabolites were analysed to help in explaining their ecological significance. This investigation of the characteristic components of the metabolomes of different *P. harmala* organs provides a molecular-level understanding of the distribution pattern of the metabolomes in this plant, which is expected to facilitate the rational development of medicinal plant resources.

2. Materials and Methods

NaH_2PO_4 and Na_2HPO_4 were used to prepare the buffer (pH = 6). Methanol and NaN_3 were used to inhibit the activity of the decomposing enzymes in the samples (analytical pure, Beijing Chemical Plant). A 1 mmol/L trimethylsilylpropanoic acid (TSP) solution was prepared using D_2O (99.9% deuterated, Cambridge Isotope Laboratories, USA). The deuterated 3-(trimethylsilyl)propionate sodium (TSP, 99% purity, J&K Scientific Co., Ltd.) solution was used as the internal standard. Milli-Q deionized water was used in the experiments. The *P. harmala* plant samples were collected in the Changji prefecture of Xinjiang, China. The flowers, stems, roots, testas, seeds, and leaves of the samples were collected, dried in air, ground, dried to a constant weight, and stored in a desiccator until use.

Sample preparation: first, 100 mL of methanol was added to a precisely weighed 2.0000 g crushed and dried plant sample three times for ultrasonic extraction. After centrifugation for 10 minutes, the supernatants were combined and subjected to rotary evaporation. The sample was freeze-dried to obtain the extract. Then, 0.2 mL of the deuterated TSP solution (1 mmol), 0.3 mL of a phosphate buffer (pH = 6.0), and 0.2 mL of NaN_3 (10 mmol) solution were added to 10 mg of the extract, which was precisely weighed and mixed well. After sonication and centrifugation, 0.6 mL of the supernatant was pipetted into a 5 mm tube for NMR analysis. Five parallel preparations were performed for each sample.

NMR experiments: ^1H-NMR measurements were performed with a 500 MHz NMR spectrometer at 298K. ^1H and ^{13}C nuclear resonance frequencies were 500 and 125 MHz, respectively. For the one-dimensional hydrogen spectra, the noesygppr1d pulse sequence with suppressed water peaks was used with the following parameters: a water peak suppression power of 41 dB (Bruker nomenclature), pulse delay time of 2 s, scan number of 128, spectrum width of 12 ppm, pulse time of 9.8 μs, sampling time of 2.72 s, relaxation time of 2.0 s, sampled data point number of 32,768, and free-induction decay (FID) resolution of 0.18 Hz. The FID was processed with an exponential window function with a widening factor of 0.3 Hz. The baseline adjustment and phase correction were all performed manually. The methanol-extracted, water-soluble metabolites were determined by NMR using the presaturation technique to suppress the water peaks; TSP was used as the internal standard, and deuterated water was used to lock the field. The ^1H-NMR spectra were heavily superimposed, and it was difficult to identify many of the signals. Two-dimensional nuclear magnetic resonance experiments, including correlated spectroscopy (COSY), heteronuclear single quantum coherence (HSQC) spectroscopy, and heteronuclear multiple bond correlation (HMBC) spectroscopy, enabled the facile identification of the metabolites from the collected signals of the methanol extracts of the different *P. harmala* plant parts. For the COSY spectra, the spectral width was 5000 Hz, the number of sampling points was 400 (F1) × 4096 (F2), and the number of time increments was 24. For the HSQC spectra, the ^1H and ^{13}C spectral widths were 5000 Hz and 22638 Hz, respectively; the number of sampling points was 256 (F1) × 2048 (F2), and the number of time increments was 60. For the HMBC spectra, the ^1H and ^{13}C spectral widths were 5000 Hz and 30184 Hz, respectively; the number of sampling points was 256 (F1) × 4096 (F2), and the number of time increments was 112.

NMR data analysis: all the ^1H-NMR spectra were processed using the TopSpin 3.2 software (Bruker Biospin). The baseline and phase were calibrated manually. After the chemical shifts were calibrated using TSP (δ_H = 0.00), the spectra were imported into the MestReNova software (version 8.0.1, Mestrelab Research, Santiago de Compostela, Spain) for data processing. After calibrating the phase and adjusting the baseline, the spectrum was integrated in the range of δ_H 0.5–10 ppm with an integration interval of 0.02 ppm. However, the spectrum was not integrated in the range of δ_H 4.71–5.05 ppm (residual water peak). The sum of the integral areas of the spectrum was normalized to generate an Excel data file, which was imported into MATLAB

(Umetrics, Umea, Sweden). PCA and PLS-DA were performed after mean centre processing, and the differential metabolites were screened using a variable importance factor (VIP) of >1 in the loading model. The effectiveness of the PLS-DA model was validated by a permutation test. The relative contents of some metabolites were analysed by analysis of variance (ANOVA) and t-tests. Duncan's new multiple range test was used to correct the p value [20].

3. Results and Discussion

3.1. ^{1}H-NMR Fingerprint Peak Assignment for the Methanol Extracts of the Different Parts of P. harmala. The fingerprint spectra of the polar extracts of *P. harmala* revealed the presence of 23 metabolites, 19 primary and 4 secondary (Figure 1, Table 1). The 19 primary metabolites were three carbohydrate compounds, five organic acids, eight amino acids, and the last three were compounds of other types. The four secondary metabolites were vasicine, vasicinone, harmine, and harmaline. These twenty-three compounds were identified in the methanol extracts of *P. harmala* by comparing the chemical shifts and coupling splitting values in the nuclear magnetic spectra and the relevant information from the two-dimensional NMR spectra to related literature data [21] and the standard spectra of amino acids, organic acids, and sugar compounds in the Human Metabolome Database [22] (HMDB) (http://www.hmdb.ca/). The ^{1}H-NMR spectra of methanol extracts from different organs of *P. harmala* and the characteristic signals of the identified metabolites are shown in Figure 1. The main amino acids identified in the organic acid and amino acid region (δ_H 0.5–3.0 ppm) were isoleucine, valine, threonine, alanine, proline, lysine, 4-hydroxyisoleucine, and asparagine. The main organic acids identified in this region included acetic acid, succinic acid, and malic acid. The signals of the sugar compounds in the range of δ_H 3.0–6.0 ppm overlapped considerably. However, diagnostic anomeric proton signal of sucrose and glucose could be easily identified. Meanwhile, the characteristic signals of nitrogen-containing metabolites, such as choline, phosphorylcholine, and betaine, were clearly observed in this region, as shown in Table 1.

In the aromatic region, vasicine, vasicinone, harmaline, and harmine were identified in the methanol extracts of *P. harmala*. Moreover, a comparison with the literature [23–25] and analysis of the two-dimensional COSY, HMBC, and HSQC spectra further confirmed the existence of these compounds, and the peak assignments are shown in Table 1. In addition to alkaloids described above, signal related to formic acid was also identified in the aromatic region of the spectra.

3.2. Multivariate Analysis of the ^{1}H-NMR Data. The ^{1}H-NMR fingerprint spectra of the methanol extracts of the six parts of *P. harmala* are shown in Figure 1. A visual observation of the spectra revealed that the presence of many types of metabolites in stems, leaves, roots, and flowers, especially amino acids, with relatively high contents, whereas the metabolite levels in the testas and seeds were

low. To further determine the potential differential metabolites, a multivariate statistical analysis (PCA) of the nuclear magnetic data of the polar metabolites of the six different *P. harmala* parts was performed, and the results are shown in Figure 2.

PCA is the most commonly performed unsupervised pattern recognition method in metabolomics research. The latent variable information can be determined from massive data sets to reduce the data dimensionality [26]. The sample classification information can be obtained from the score plot. In the current work, PCA was undertaken on the obtained data, and then satisfied results were generated. As shown in the PCA score plot in Figure 2, the first three principal components (PC1, PC2, and PC3) accounted for 92.07% of the original variable information (PC1: 62.77%, PC2: 19.81%, and PC3: 9.49%). The metabolites of the six *P. harmala* parts were significantly different, and the separation trend was obvious. Along the first ordination axis, the seeds and roots were clearly distinguished from the other plant parts and exhibited a positive correlation. In the PC2 direction of the score plot, the stems and roots were clearly distinguished from the extracts of the other plant parts, exhibiting a negative correlation.

To identify the differential metabolites of the six parts, the root sample was used as the control and the samples of the other parts were compared to it and sorted to identify the metabolites that contribute the most to the group classification. PLS-DA was applied to the ^{1}H-NMR data of the stem, testa, flower, leaf, and seed to obtain score and loading plots (Figure 3). The score plot of A1, B1, C1, D1, and E1 in Figure 3 shows that the root was completely separated from the other parts. The parameters R2X, R2Y, and Q2 of the five models are shown in Figure 3 and were all greater than 0.9, indicating that the models had strong predictive power [27, 28]. Permutation tests were performed to PLS-DA models to verify that these parameters generated by the established models were not overfitted. Therefore, the models were validated, showing that the results were reliable. The loading plot can be used to determine the variables contributing to the classification, and depending on the levels of their contributions, the variables that can be considered as potential biomarkers can be discovered. Based on the A2, B2, C2, D2, and E2 groups shown in the PLS-DA loading plot, the significant differential metabolites included acetic acid (6), asparagine (11), lysine (5), proline (7), choline (12), phosphocholine (13), 4-hydroxyisoleucine (8), sucrose (15), vasicine (17), harmine (20), and harmaline (21).

3.3. Screening and Univariate Analysis of the Potential Characteristic Metabolites. The data with VIP>1 in the PLS-DA loading plot of the five methanol extracts of *P. harmala* were analysed to obtain the most significant metabolites for their classification, which included acetic acid, asparagine, choline, harmine, harmaline, 4-hydroxyisoleucine, phosphocholine, proline, lysine, sucrose, and vasicine. Specifically, to obtain the relative contents of the metabolites, which were subjected to univariate analysis of variance, their characteristic peaks were compared with the peak area of the

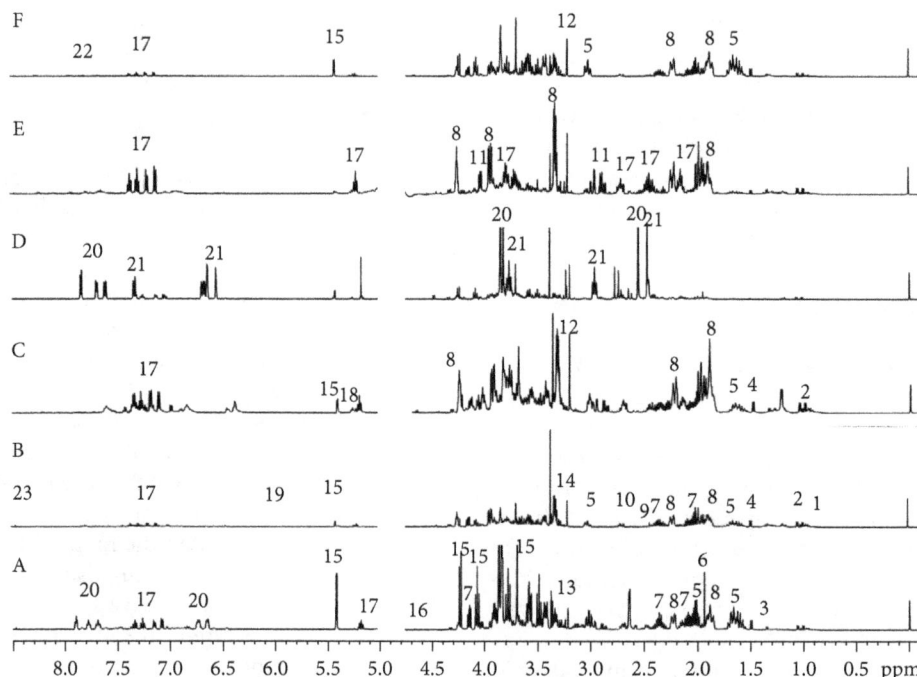

FIGURE 1: Typical 500 MHz ^{1}H-NMR spectra of the metabolites of the different parts of *P. harmala*: (A) root, (B) leaf, (C) flower, (D) seed, (E) testa, and (F) stem. Key metabolites: 1, leucine; 2, valine; 3, threonine; 4, alanine; 5, lysine; 6, acetic acid; 7, proline; 8, 4-hydrox-yisoleucine; 9, succinic acid; 10, malic acid; 11, asparagine; 12, choline; 13, phosphorylcholine; 14, betaine; 15, sucrose; 16, β-glucose; 17, vasicine; 18, α-glucose; 19, maleic acid; 20, harmine; 21, harmaline; 22, vasicinone; 23, formic acid.

internal standard. The calculated relative contents of the metabolites in the methanol extracts of the different *P. harmala* parts are shown in Table 2. All the data were analysed by ANOVA and *t*-tests. Differences were considered to be statistically significant at $p < 0.05$. The data in Table 2 show that the acetic acid, proline, lysine, sucrose, and vasicine contents of the *P. harmala* roots were significantly higher than those of the other parts. In addition to vasicine, the asparagine, choline, and 4-hydroxyisoleucine contents in the testas were clearly dominant. The seeds contained large amounts of harmaline, harmine, and phosphocholine. However, the differences in the metabolomes of the *P. harmala* stems, leaves, and flowers were not significant, which is consistent with the results of the multivariate analysis.

3.4. Biological Significance of the Characteristic Metabolites. The results showed that the sucrose, proline, acetic acid, betaine, and lysine contents of the *P. harmala* roots were much higher than those of the other parts. Of these metabolites, sucrose, proline, betaine, and acetic acid are osmotic regulators in plants, controlling osmotic adjustment [29]. The roots of a plant are the vegetative organ responsible for the physiological functions of nutrient absorption, production, and transport [30]. Sucrose is also used as an energy carrier in plants and is the main source of carbon and energy for plant growth and development. *P. harmala* roots are well suited to performing the main nutrient functions for plant growth and development and to contributing to plant metabolism, and sucrose provides energy for the growth,

development, and reproduction of the plant [31]. The proline and betaine contents can increase under abiotic stress conditions, such as exposure to drought, high salinity, high temperatures, and heavy metals [32]. Besides to serve as valuable sources of nitrogen and carbon, proline and betaine prevent plant damage caused by osmotic stress and act as free radical scavengers [33–35]. It is generally believed that proline is mostly synthesized in the roots and most of the product is transported to the above-ground parts. The acidic substances in the roots can enhance the effectiveness of the nutrients in the rhizosphere soil, promote the acidification, chelation, ion exchange, and reduction in the insoluble nutrients in the plant, and participate in circulation and energy flow of the nutrient elements through the plant [36].

The experimental results showed that the alkaloid contents of the *P. harmala* seeds, including the harmine, harmaline, and phosphocholine contents, were most abundant compared to other parts [37]. Alkaloids are secondary metabolites that are generated when plants resist invasion from the environment and defend against external factors. They do not directly participate in plant growth, development, and reproduction, but they can improve the adaptability of the plant to adverse environments. The seed of *P. harmala* is the reproductive organ and is rich in alkaloids. To germinate and grow in an arid environment, the plant must strongly compete for water, nutrients, and space. The alkaloids in *P. harmala* seeds must be washed out by rainwater before germination. When these compounds enter the soil environment, they might inhibit the germination of other plant seeds and the growth of their seedlings through allelopathy [38]. The results demonstrate a higher

TABLE 1: ^{1}H-NMR assignments of major metabolites in *P. harmala* extracts.

No.	Metabolite	δ_H (multiplicity, J)	Sample
1	Isoleucine	1.02 (d, 7.06), 0.95 (t, 7.15)	B[b]
2	Valine	1.05 (d, 7.0), 0.99 (d, 7.0)	B[b]
3	Threonine	4.25 (m), 1.33 (d, 6.55)	A[b]
4	Alanine	3.57 (m), 1.48 (d, 7.3)	B[b]
5	Lysine	3.6 (m), 1.65, 1.89 (m), 2.25 (m), 3.02, 3.4 (m)	F[b]
6	Acetic acid	1.92 (s)	A[b]
7	Proline	3.30–3.35 (m), 2.31–2.37 (m), 1.95–2.00 (m), 4.12 (dd, 8.63, 6.56)	A[b]
8	4-Hydroxyisoleucine	4.25 (m), 2.167 (m), 3.835 (m), 2.22, 1.98 (m), 1.88 (m)	E[a]
9	Succinic acid	2.43 (s), 2.43 (s)	F[b]
10	Malic acid	2.71 (dd, 2.9, 15.62), 4.31 (dd, 3.1, 10.2)	F[b]
11	Asparagine	4.03 (dd, 4.4, 7.25), 2.94 (m), 2.84 (m)	E[b]
12	Choline	4.07, 3.21 (s)	C[b]
13	Phosphorylcholine	3.23 (s)	A[b]
14	Betaine	3.27 (s)	B[b]
15	Sucrose	3.69 (s), 4.22 (d, 8.65), 4.06 (t, 8.89), 3.91 (m), 3.87 (d, 3.15), 5.42 (d, 3.8), 3.59 (m), 3.78 (t, 9.48), 3.49 (t, 9.28), 3.91(m)	A[b]
16	β-Glucose	4.65 (d, 8.0)	A[b]
17	Vasicine	3.77, 3.69 (m),2.15, 2.71 (m), 5.22 (t, 8.80), 7.11 (m), 7.36 (m), 7.28 (m), 7.19 (d, 7.5)	E[a]
18	α-Glucose	5.24 (d, 3.83)	C[b]
19	Maleic acid	6.01 (s)	B[b]
20	Harmine	6.64 (d, 1.9), 6.69 (dd, 8.75, 2.25), 7.84 (d, 5.64), 2.55 (s), 7.69 (d, 5.42), 7.61 (d, 8.59), 3.84 (s)	D[a]
21	Harmaline	6.46 (dd, 8.85, 1.94), 6.37 (d, 1.66), 7.0 (d, 8.64), 2.47 (s), 2.96(t, 8.83), 3.22(t, 8.83), 3.81(s)	D[a]
22	Vasicinone	4.11, 3.91 (m), 1.99, 2.45 (m), 4.99 (t, 6), 7.71 (m), 7.93 (m), 7.64 (m), 8.25 (d, 6)	F[a]
23	Formic acid	8.46 (s)	A[b]

Note. "a" was determined by the ^{1}H-NMR, COSY, HSQC, and HMBC spectra. "b" was determined by the ^{1}H-NMR and HSQC spectra. s, singlet; d, doublet; t, triplet; m, multiplet; A, root; B, leaf; C, flower; D, seed; E, testa; F, stem.

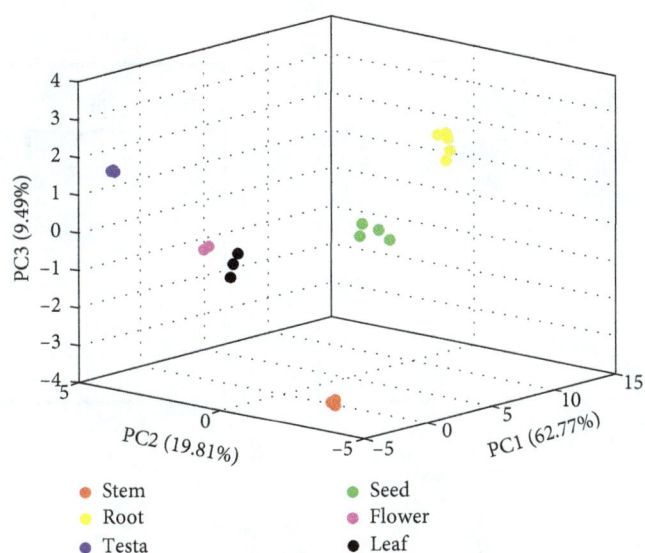

FIGURE 2: PCA score plot of the NMR spectra of the methanol extracts of the six different *P. harmala* parts ("red," "yellow," "blue," "green," "magenta," and "black" dots represent the stem, root, testa, seed, flower, and leaf of the plant, respectively).

(a)

(b)

(c)

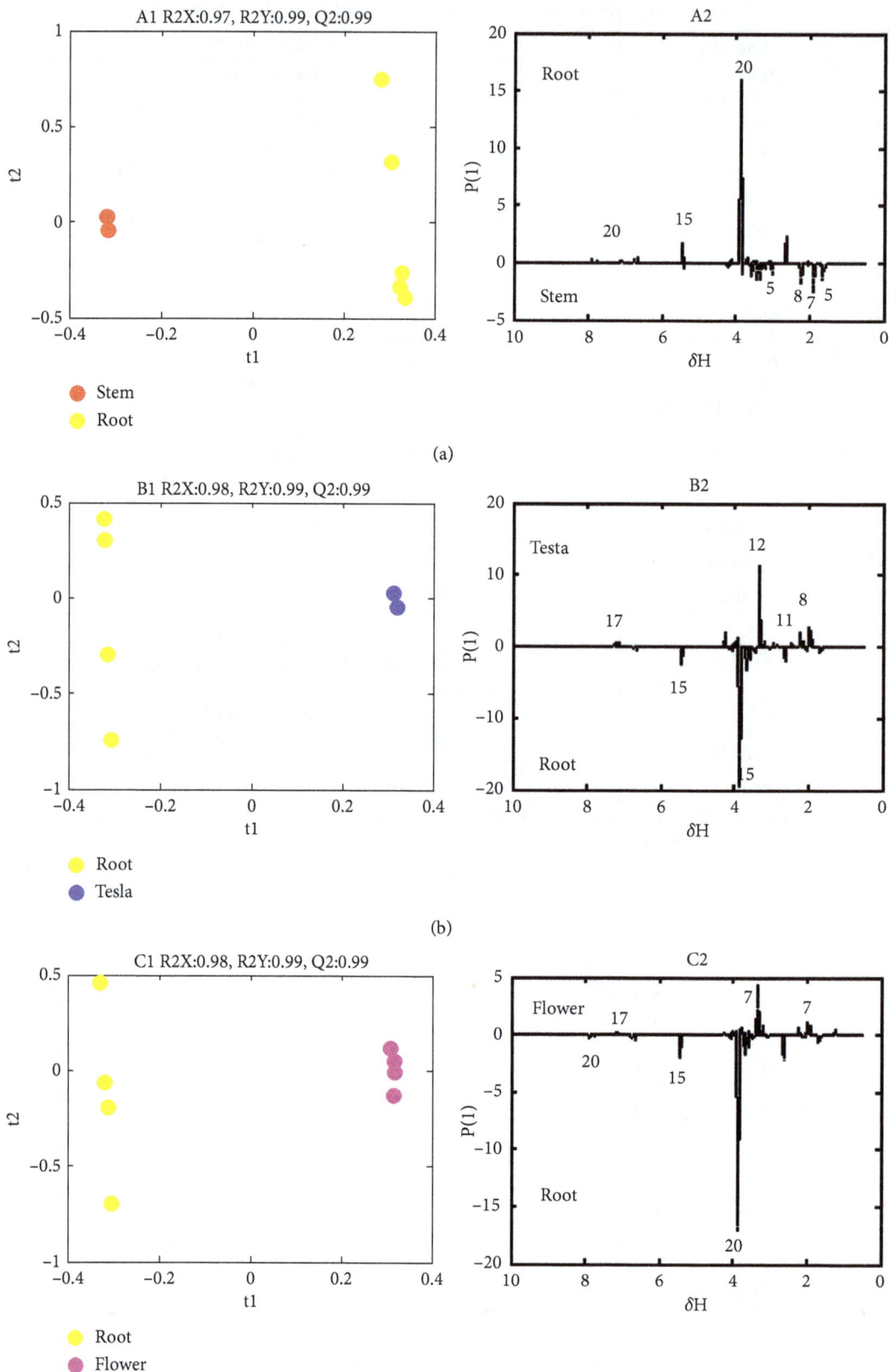

FIGURE 3: Continued.

(d)

(e)

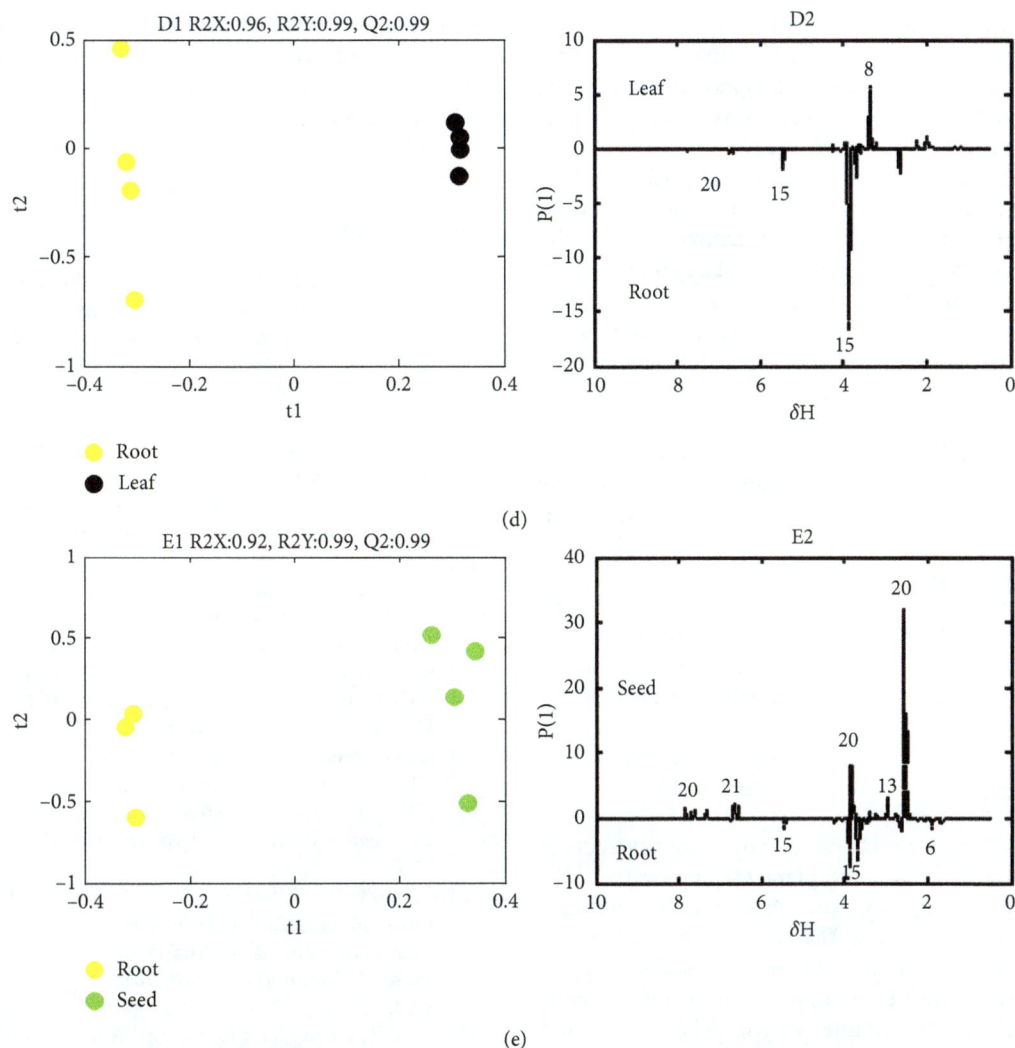

FIGURE 3: PLS-DA score plot (A1-E1) and linear loading plot (A2-E2) of the methanol extracts of the different *P. harmala* parts ("red," "yellow," "blue," "green," "magenta," and "black" dots represent the methanol extracts of the stem, root, testa, seed, flower, and leaf of the plant, respectively; t1 and t2 are the scoring points for the first and second principal components, respectively; δ_H is the ^{1}H-NMR chemical shift; P[1] is the loading point of the first principal component. The numbers represent different metabolites, as shown in Table 2).

TABLE 2: Relative quantification of principal metabolites in different parts of *P harmala* (mg/g).

Compounds	Stem	Root	Testa	Seed	Flower	Leaf
Proline	37.88 ± 18.95bc	86.53 ± 10.27a	26.70 ± 2.14c	—	55.10 ± 7.88b	55.93 ± 2.39b
Lysine	153.63 ± 5.41b	180.02 ± 12.46a	—	—	95.87 ± 12.23c	68.85 ± 4.08d
Asparagine	17.77 ± 1.75c	17.96 ± 3.32c	123.84 ± 12.11a	29.37 ± 29.68c	76.03 ± 9.36b	32.03 ± 4.48c
4-Hydroxyisoleucine	204.85 ± 9.81b	189.29 ± 16.19b	364.89 ± 50.07a	—	195.70 ± 15.17b	160.35 ± 8.92b
Acetic acid	1.85 ± 0.17c	7.77 ± 0.98a	4.39 ± 0.85b	—	4.18 ± 0.54b	2.68 ± 0.09c
Sucrose	113.024 ± 8.45b	339.19 ± 35.69a	13.29 ± 4.92d	63.28 ± 2.0c	60.08 ± 5.91c	40.63 ± 1.63c
Vasicine	17.19 ± 9.8b	64.31 ± 8.44ab	124.13 ± 71.58a	—	68.36 ± 6.69ab	26.69 ± 2.56b
Harmine	—	19.27 ± 1.68b	—	40.77 ± 5.98a	—	—
Harmaline	—	—	—	58.46 ± 9.61	—	—
Choline	4.24 ± 0.17c	2.67 ± 0.18d	8.59 ± 0.96a	2.31 ± 1.45d	6.12 ± 0.51b	3.31 ± 0.19cd
Phosphorylcholine	0.55 ± 0.13d	2.92 ± 0.20b	3.56 ± 0.52b	5.50 ± 1.33a	0.97 ± 0.14cd	1.99 ± 0.13bc

Note. The numbers are the contents of the compounds in the methanol extracts of the different *P. harmala* parts (mean value ± standard deviation). "—" indicates that the compound was not detected. Duncan's new multiple range test was used to calculate the p value (when $p < 0.05$, different letters within same row represent subsets with significant differences; the order of the letters indicates the range of sample concentrations from large to small).

concentration of phosphocholine in *P. harmala* seeds than other organs. The phosphate in the plant fluid exists mainly in the form of phosphocholine, a phosphorus carrier that accounts for 5–20% of the total phosphorus in plants, and is a key component of the plant cell membrane [39].

The data show an increase in the level of vasicine, asparagine, choline, and 4-hydroxyisoleucine contents in testa than the other groups. The testa is a protective layer on the outside of the plant seed that protects the seed embryo from mechanical damage and prevents pests, disease invasion, and water loss [40]. Its main physiological role is to control the germination of the seed by enhancing its dormancy, limit unfavourable biochemical activities during seed storage, and protect the seed and root from animal foraging and microbiological and viral invasion [41]. Harmaline and choline in the *P. harmala* testa might play a role in these defensive functions. 4-Hydroxyisoleucine is a unique nonprotein amino acid in plants that exhibits physiological and chemical activities that are responsible for glucose and lipid metabolism [42]. The amino acid asparagine is a soluble form of nitrogen that is predominantly absorbed directly by the plant roots. These compounds in the testa of *P. harmala* are closely related to its self-protection capabilities [42].

4. Conclusions

In this study, NMR was used to analyse methanol extracts of six parts of *P. harmala* and the plant metabolomics technique aimed at investigating the characteristic chemical composition of these extracts. The results showed that the differential metabolites in the different *P. harmala* parts included sucrose, harmine, proline, lysine, betaine, acetic acid, harmaline, vasicine, choline, 4-hydroxyisoleucine, and asparagine. From the quantitative point of view, the relative contents of these 11 metabolites in the methanol extracts of the six *P. harmala* parts studied were determined. The results revealed the significant differences in the types of metabolites in the *P. harmala* roots, seeds, and testas. The NMR-based plant metabolomics method established in this research can provide a detailed metabolomic profile of biological samples from different plant parts and allow the holistic analysis of metabolome. Resulting in improved understanding of *P. harmala* metabolism, the results can provide important data to aid in plant ecophysiological studies and facilitate the development of *P. harmala* medicinal resources.

Conflicts of Interest

The authors declare that they have no conflicts of interest.

Acknowledgments

We wish to thank Professor Jin Li, School of Life Sciences of Xinjiang Normal University, for sample identification. This work was supported by the National Natural Science Foundation of China (Nos. 21165081 and 21465023) and the National Key Research and Development Program (No. 2016YFC0500805).

References

[1] M. Z. Ahmed and M. A. Khan, "Tolerance and recovery responses of playa halophytes to light, salinity and temperature stresses during seed germination," *Flora-Morphology, Distribution, Functional Ecology of Plants*, vol. 205, no. 11, pp. 764–771, 2010.

[2] M. A. Aziz, A. H. Khan, M. Adnan, and I. Izatullah, "Traditional uses of medicinal plants reported by the indigenous communities and local herbal practitioners of Bajaur Agency, federally administrated tribal areas, Pakistan," *Journal of Ethnopharmacology*, vol. 198, pp. 268–281, 2017.

[3] M. C. Niroumand, M. H. Farzaei, and G. Amin, "Medicinal properties of *Peganum harmala* L. in traditional Iranian medicine and modern phytotherapy: a review," *Journal of Traditional Chinese Medicine*, vol. 35, no. 1, pp. 104–109, 2015.

[4] S. Li, K. Wang, C. Gong et al., "Cytotoxic quinazoline alkaloids from the seeds of *Peganum harmala*," *Bioorganic and Medicinal Chemistry Letters*, vol. 28, no. 2, pp. 103–106, 2018.

[5] L. Bournine, S. Bensalem, S. Fatmi et al., "Evaluation of the cytotoxic and cytostatic activities of alkaloid extracts from different parts of *Peganum harmala* L. (Zygophyllaceae)," *European Journal of Integrative Medicine*, vol. 9, pp. 91–96, 2017.

[6] D. Khlifi, R. M. Sghaier, S. Amouri et al., "Composition and anti-oxidant, anti-cancer and anti-inflammatory activities of Artemisia herba-alba, Ruta chalpensis L. and *Peganum harmala* L.," *Food and Chemical Toxicology*, vol. 55, pp. 202–208, 2013.

[7] S. Li, X. Cheng, and C. Wang, "A review on traditional uses, phytochemistry, pharmacology, pharmacokinetics and toxicology of the genus Peganum," *Journal of Ethnopharmacology*, vol. 203, pp. 127–162, 2017.

[8] M. Moloudizargari, P. Mikaili, S. Aghajanshakeri, M. Asghari, and J. Shayegh, "Pharmacological and therapeutic effects of *Peganum harmala* and its main alkaloids," *Pharmacognosy Reviews*, vol. 7, no. 14, p. 199, 2013.

[9] I. Apostolico, L. Aliberti, L. Caputo et al., "Chemical composition, antibacterial and phytotoxic activities of *Peganum harmala* seed essential oils from five different localities in Northern Africa," *Molecules*, vol. 21, no. 9, p. 1235, 2016.

[10] A. Daoud, J. Song, F. Xiao, and J. Shang, "B-9-3, a novel β-carboline derivative exhibits anti-cancer activity via induction of apoptosis and inhibition of cell migration in vitro," *European Journal of Pharmacology*, vol. 724, pp. 219–230, 2014.

[11] J. Asgarpanah and F. Ramezanloo, "Chemistry, pharmacology and medicinal properties of *Peganum harmala* L," *African Journal of Pharmacy and Pharmacology*, vol. 6, no. 22, pp. 1573–1580, 2012.

[12] F. Lamchouri, M. Zemzami, A. Jossang et al., "Cytotoxicity of alkaloids isolated from *Peganum harmala* seeds," *Pakistan Journal of Pharmaceutical Sciences*, vol. 26, no. 4, pp. 699–706, 2013.

[13] X. Ma, D. Liu, H. Tang et al., "Purification and characterization of a novel antifungal protein with antiproliferation and anti-HIV-1 reverse transcriptase activities from *Peganum*

harmala seeds," *Acta Biochimica et Biophysica Sinica*, vol. 45, no. 2, pp. 87–94, 2013.

[14] M. H. J. A. Mahmoudian, "Toxicity of *Peganum harmala*: review and a case report," *Iranian Journal of Pharmacology and Therapeutics*, vol. 1, pp. 1–4, 2002.

[15] H. K. Kim, Y. H. Choi, and R. Verpoorte, "NMR-based metabolomic analysis of plants," *Nature Protocols*, vol. 5, no. 3, pp. 536–549, 2010.

[16] R. D. Hall, "Plant metabolomics: from holistic hope, to hype, to hot topic," *New Phytologist*, vol. 169, no. 3, pp. 453–468, 2006.

[17] O. A. H. Jones, M. L. Maguire, J. L. Griffin et al., "Metabolomics and its use in ecology," *Austral Ecology*, vol. 38, no. 6, pp. 713–720, 2013.

[18] T. W. M. Fan and A. N. Lane, "Applications of NMR spectroscopy to systems biochemistry," *Progress in Nuclear Magnetic Resonance Spectroscopy*, vol. 92-93, pp. 18–53, 2016.

[19] J. L. Markley, R. Bruschweiler, A. S. Edison et al., "The future of NMR-based metabolomics," *Current Opinion in Biotechnology*, vol. 43, pp. 34–40, 2017.

[20] D. B. Duncan, "Multiple range and multiple F tests," *Biometrics*, vol. 11, no. 1, pp. 1–42, 1955.

[21] Y. Li, Q. He, Z. Geng et al., "NMR-based metabolomic profiling of *Peganum harmala* L. reveals dynamic variations between different growth stages," *Royal Society Open Science*, vol. 5, no. 7, article 171722, 2018.

[22] D. S. Wishart, D. Tzur, C. Knox et al., "HMDB: the human metabolome Database," *Nucleic Acids Research*, vol. 35, pp. 521–526, 2007.

[23] B. S. Jaw, Y. Bai, M. S. Puar, K. K. Dubose, and S. William Pelletier, "¹H- and ¹³C-NMR assignments for some pyrroloe [2,1b] quinazoline alkaloids of adhatoda vaszca," *Journal of Natural Products*, vol. 57, no. 7, pp. 953–962, 1994.

[24] X. Zheng, Z. Zhang, G. Chou et al., "Acetylcholinesterase inhibitive activity-guided isolation of two new alkaloids from seeds of *Peganum nigellastrum* bunge by an in vitro TLC-bioautographic assay," *Archives of Pharmacal Research*, vol. 32, no. 9, pp. 1245-1251, 2009.

[25] S. B. Mhaske and N. P. Argade, "Concise and efficient synthesis of bioactive natural products pegamine, deoxy-vasicinone, and (−)-Vasicinone," *Journal of Organic Chemistry*, vol. 66, no. 26, pp. 9038–9040, 2001.

[26] M. Maulidiani, A. Mediani, F. Abas et al., "1 H NMR and antioxidant profiles of polar and non-polar extracts of persimmon (Diospyros kaki L.) – metabolomics study based on cultivars and origins," *Talanta*, vol. 184, pp. 277–286, 2018.

[27] A. Basoglu, I. Sen, G. Meoni, L. Tenori, and A. Naseri, "NMR-based plasma metabolomics at set intervals in newborn dairy calves with severe sepsis," *Mediators of Inflammation*, vol. 2018, Article ID 8016510, 12 pages, 2018.

[28] S. Kim, B. Shin, D. K. Lim et al., "Expeditious discrimination of four species of the Panax genus using direct infusion-MS/MS combined with multivariate statistical analysis," *Journal of Chromatography B*, vol. 1002, pp. 329–336, 2015.

[29] M. R. W. Parvaiz Ahmad, *Physiological Mechanisms and Adaptation Strategies in Plants Under Changing Environment*, Springer, New York, NY, USA, 2014.

[30] C. M. Iversen, "Using root form to improve our understanding of root function," *New Phytologist*, vol. 203, no. 3, pp. 707–709, 2014.

[31] L. Hu, L. Chen, L. Liu et al., "Metabolic acclimation of source and sink tissues to salinity stress in bermudagrass (*Cynodon dactylon*)," *Physiologia Plantarum*, vol. 155, no. 2, pp. 166–179, 2015.

[32] Z. Li, X. Duan, X. Min, and Z. Zhou, "Methylglyoxal as a novel signal molecule induces the salt tolerance of wheat by regulating the glyoxalase system, the antioxidant system, and osmolytes," *Protoplasma*, vol. 254, no. 5, pp. 1995–2006, 2017.

[33] M. BRIENS and F. LARHER, "Osmoregulation in halophytic higher-plants-a comparative-study of soluble carbohydrates, polyols, betaines and free proline," *Plant, Cell and Environment*, vol. 5, no. 4, pp. 287–292, 1982.

[34] Z. Li, X. Duan, X. Min, and Z. Zhou, "Methylglyoxal as a novel signal molecule induces the salt tolerance of wheat by regulating the glyoxalase system, the antioxidant system, and osmolytes," *Protoplasma*, vol. 254, no. 5, pp. 1995–2006, 2017.

[35] M. Ashraf and M. R. Foolad, "Roles of glycine betaine and proline in improving plant abiotic stress resistance," *Environmental and Experimental Botany*, vol. 59, no. 2, pp. 206–216, 2007.

[36] W. Jia and W. J. Davies, "Modification of leaf apoplastic pH in relation to stomatal sensitivity to root-sourced abscisic acid signals," *Plant Physiology*, vol. 143, no. 1, pp. 68–77, 2006.

[37] T. Herraiz, H. Guillén, V. J. Arán, and A. Salgado, "Identification, occurrence and activity of quinazoline alkaloids in *Peganum harmala*," *Food and Chemical Toxicology*, vol. 103, pp. 261–269, 2017.

[38] H. Shao, X. Huang, Y. Zhang, and C. Zhang, "Main alkaloids of *Peganum harmala* L. And their different effects on dicot and monocot crops," *Molecules*, vol. 18, no. 3, pp. 2623–2634, 2013.

[39] H. Hayashi, T. Chen, and N. Murata, "Transformation with a gene for choline oxidase enhances the cold tolerance of Arabidopsis during germination and early growth," *Plant, Cell and Environment*, vol. 21, pp. 232–239, 1998.

[40] T. Steinbrecher and G. Leubner-Metzger, "The biomechanics of seed germination," *Journal of Experimental Botany*, vol. 68,, no. 4, pp. 765–783, 2016.

[41] Z. Zhou, W. Bao, and N. Wu, "Dormancy and germination in rosa multibracteata hemsl. & E. H. Wilson," *Scientia Horticulturae*, vol. 119, no. 4, pp. 434–441, 2009.

[42] P. Zandi, S. K. Basu, L. B. Khatibani et al., "Fenugreek (*Trigonella foenum-graecum* L.) seed: a review of physiological and biochemical properties and their genetic improvement," *Acta Physiologiae Plantarum*, vol. 37, no. 1, 2015.

α-Amylase and α-Glucosidase Inhibitory Activities of Chemical Constituents from *Wedelia chinensis* (Osbeck.) Merr. Leaves

Nguyen Phuong Thao [ID],[1] Pham Thanh Binh,[1] Nguyen Thi Luyen,[1,2] Ta Manh Hung,[3] Nguyen Hai Dang,[1,2] and Nguyen Tien Dat [ID][2,4]

[1]*Advanced Center for Bio-Organic Chemistry, Institute of Marine Biochemistry (IMBC), Vietnam Academy of Science and Technology (VAST), 18 Hoang Quoc Viet, Caugiay, Hanoi, Vietnam*
[2]*Graduate University of Science and Technology, VAST, 18 Hoang Quoc Viet, Caugiay, Hanoi, Vietnam*
[3]*National Institute of Drug Quality Control (NIDQC), 48 Hai Ba Trung, Hoankiem, Hanoi, Vietnam*
[4]*Center for Research and Technology Transfer, VAST, 18 Hoang Quoc Viet, Caugiay, Hanoi, Vietnam*

Correspondence should be addressed to Nguyen Phuong Thao; thaonp@imbc.vast.vn

Academic Editor: Federica Pellati

As part of an ongoing search for new natural products from medicinal plants to treat type 2 diabetes, two new compounds, a megastigmane sesquiterpenoid sulfonic acid (**1**) and a new cyclohexylethanoid derivative (**2**), and seven related known compounds (**3–9**) were isolated from the leaves of *Wedelia chinensis* (Osbeck.) Merr. The structures of the compounds were conducted via interpretation of their spectroscopic data (1D and 2D NMR, IR, and MS), and the absolute configurations of compound **1** were determined by the modified Mosher's method. The MeOH extract of *W. chinensis* was found to inhibit α-amylase and α-glucosidase inhibitory activities as well as by the compounds isolated from this extract. Furthermore, compound **7** showed the strongest effect with IC_{50} values of 112.8 ± 15.1 g/mL (against α-amylase) and 785.9 ± 12.7 g/mL (against α-glucosidase). Compounds **1**, **8**, and **9** showed moderate α-amylase and α-glucosidase inhibitory effects. Other compounds showed weak or did not show any effect on both enzymes. The results suggested that the antidiabetic properties from the leaves of *W. chinensis* are not simply a result of each isolated compound but are due to other components such as the accessibility of polyphenolic groups to α-amylase and α-glucosidase activities.

1. Introduction

In recent years, the number of diabetic patients is rapidly rising in most parts of the world, especially in developing Southeast Asian countries [1, 2]. The control of blood glucose concentrations near the normal range is mainly based on the use of oral hypoglycaemic/antihyperglycaemic agents and insulin. However, all of these treatments have limited efficacy and are associated with undesirable side effects [3, 4], leading to increasing interest in the use of medicinal plants for the alternative management of type 2 diabetes mellitus. An effective suggestion for type 2 diabetes management is the inhibition of α-glucosidase and α-amylase [5].

Natural health-care products derived from medicinal plants or herbs have been developed as alternative or complementary treatments for many common disorders [6]. Several medicinal plants have been the useful sources of novel biologically active compounds. Many pharmaceutical agents have been discovered by screening natural products from plants, many of which have been developed as new leads for pharmaceuticals [7]. Predominantly herbal drugs have been widely used globally for diabetic treatment over thousands of years due to their traditional acceptability and lesser side effects. Therefore, screening of α-amylase (1,4-α-D-glucan glucanohydrolases; EC. 3.2.1.1) and α-glucosidase (α-D-glucoside glucohydrolase, EC 3.2.1.20) inhibitors in medicinal plants has received much attention [6, 7].

Among the traditionally used important medicinal plants, the genus *Wedelia* (Asteraceae) contains approximately 107 species in the world, and among them, about 6 species are in

Vietnam. They are all herbal plants and distributed in tropical and subtropical regions of Asia and Pacific Islands [8]. *Wedelia chinensis* (Osbeck.) Merr. (Asteraceae) is a deciduous shrub, widely distributed in several Asian countries such as China, Japan, and mainland Vietnam. The leaves, stems, and fruits of this species have been traditionally used in fold medicine for the treatment of chin cough, diarrhoea, diphtheria, faucitis, hemorrhoids, and injuries due to falls, jaundice, and pertussis and are often consumed as tea in the form of infusion [9]. Phytochemically, up to date, a number of secondary metabolites have been identified, including various types of compounds belonging to chemical classes of flavonoids, diterpenoids, and triterpenoids. Besides, several other compounds including common saponins and phytosterol derivatives were also reported in the species, but they appear to have a more limited distribution. Recently, several studies have demonstrated the exploration of pharmacological potential, such as analgesic [10], androgen-suppressing [11], anticancer [12, 13], antibacterial, anticonvulsant, antifungal [14], antioxidant [13], anti-inflammatory [10, 15], anti-osteoprotic [16], antiulcerogenic, antistress [17], hepatoprotective [18], and sedative activities [19]. However, the investigation about the chemical constituents of *W. chinensis* is not sufficient compared to the other plants in the genus *Wedelia*.

As part of an ongoing search for new biologically active natural products from Vietnamese medicinal plants, we found that a MeOH extract of *W. chinensis* leaves showed significant *in vitro* α-amylase and α-glucosidase inhibitory activities. Previously, there have been no reports on either the extracts or isolated components from this species against α-amylase and α-glucosidase activities. Reported herein are the isolation and structural elucidation of a new megastigmane sesquiterpenoid sulfonic acid (**1**), a new cyclohexylethanoid derivative (**2**), and 7 known compounds (**3–9**) as well as the evaluation of α-amylase and α-glucosidase activities.

2. Materials and Methods

2.1. General Experimental Procedures. Optical rotations were measured using a JASCO P-2000 polarimeter (JASCO, Oklahoma, OK, US). IR spectra were obtained on a Bruker TENSOR 37 FT-IR spectrometer (Bruker, Billerica, MA, USA). The ^{1}H and ^{13}C, HMQC, HMBC, NOESY/ROESY, and COSY NMR spectra were recorded on a 500 MHz Bruker DRX spectrometer (Bruker, Tupper Hall, CA, USA), and the chemical shift (δ) was expressed in ppm with reference to the TMS signals. The HRESIMS were obtained using an Agilent 6550 iFunnel Q-TOF LC/MS system (Emeryville, CA, USA). Medium-pressure liquid chromatography (MPLC) was carried out on a Biotage-Isolera One system. Column chromatography (CC) was conducted using 65–250 or 230–400 mesh silica gel (Sorbent Technologies, Atlanta, GA, USA), porous polymer gel (Diaion® HP-20, 20–60 mesh, Mitsubishi Chemical, Tokyo, Japan), Sephadex™ LH-20 (Supelco, Bellefonte, PA, USA), octadecyl silica (ODS, 50 μm, Cosmosil 140 C_{18}-OPN, Nacalai Tesque), and YMC RP-C_{18} resins (30–50 μm, Fuji

Silysia Chemical). Analytical thin-layer chromatography (TLC) systems were performed on precoated silica gel 60 F_{254} plates (1.05554.0001, Merck) and RP-18 F_{254S} plates (1.15685.0001, Merck), and the isolated compounds were visualized by spraying with 10% H_2SO_4 in water and then heating for 1.5–2 minutes. All procedures were carried out with solvents purchased from commercial sources, which were used without further purification.

2.2. Plant Material. The leaves of *W. chinensis* (Osbeck.) Merr. were collected from Ba Dinh, Ha Noi, Vietnam, in April 2017 and taxonomically identified by Professor Tran Huy Thai (Institute of Ecology and Biological Resources). A voucher specimen (NCCB-2016.55.01) was deposited at the Herbarium of Institute of Marine Biochemistry and Institute of Ecology and Biological Resources, Vietnam Academy of Science and Technology.

2.3. Compounds. From the methanolic extract of *W. chinensis*, 9 compounds (**1–9**) were isolated and structurally elucidated. Stock solutions of tested compounds in DMSO were prepared kept at −20°C and diluted to the final concentration in fresh media before each experiment. To not affect the cell growth, the final DMSO concentration did not exceed 0.5% in all experiments.

2.4. Extraction and Isolation. The dried leaves of *W. chinensis* (4.4 kg) were cut into pieces and extracted with 95% aqueous MeOH (3 × 6.5 L) under ultrasonic agitation at 90 Hz. The methanol solution was removed of solvent under a vacuum and was filtered through a Büchner funnel to produce a dried brown extract (160 g, A). Since the MeOH extract significantly reduced α-amylase and α-glucosidase inhibitory activities, it was suspended in distilled water and partitioned between EtOAc (1 L × 3) to obtain EtOAc (16.9 g, B) and aqueous soluble fractions (W).

The EtOAc fraction was separated on silica gel MPLC (column: Biotage® SNAP Cartridge, KP-SIL, 340 g) using the mobile phase of *n*-hexane-acetone (gradient 30 : 70, 50 : 50, 70 : 30, 0 : 100, 15 mL/min, 90 min) to give six fractions (B-1 to B-6). This MPLC procedure was repeated 5 times using the same conditions before further isolation. Fraction B-3 was chromatographed by Sephadex® LH-20 CC (Φ25 mm, L 1250 mm) eluted with acetone-H_2O (gradient 95 : 5, 70 : 30, 50 : 50, v/v) to give three subfractions (B-3.1 to B-3.3) and further purified by YMC RP-C_{18} CC (Φ15 mm, L 700 mm) using acetone-H_2O (1 : 2) as the eluent to pomonic acid (**8**, crystalline powder, 11.2 mg) and pomolic acid (**9**, crystalline powder, 15.4 mg). Next, fraction B–6 was chromatographed over a silica gel CC (Φ12 mm, L 600 mm) eluted with *n*-hexane-EtOAc (2 : 1) to obtain jaceosidin (**7**, pale yellow crystals, 56.8 mg).

The H_2O fraction was separated using a Diaion HP-20 column (Φ100 mm, L 500 mm) and was eluted with a gradient solvent mixture of MeOH-H_2O (gradient 25 : 75, 50 : 50, 65 : 35, 75 : 25, to pure MeOH, stepwise) to yield five fractions (W-1 to W-5), based on TLC analysis. The

TABLE 1: ^{1}H and ^{13}C NMR spectroscopic data of **1** and **2** in CD$_3$OD.

Position	1		2	
	$\delta_C{}^a$	$\delta_H{}^b$ mult. (*J* in Hz)	$\delta_C{}^a$	$\delta_H{}^b$ mult. (*J* in Hz)
1	35.5, C	—	79.4, C	—
2	37.9, CH$_2$	1.47 dd (3.5, 12.5) 1.84 t (12.5)	83.3, CH	4.00 dd (6.0, 9.5)
3	75.5, CH	4.60 ddd (3.0, 3.5, 12.5)	37.5, CH$_2$	1.39 dd (6.0, 13.5) 1.80 m
4	71.6, CH	4.27 dd (1.0, 3.0)	75.4, C	—
5	69.4, C	—	29.2, CH$_2$	1.53 ddd (4.5, 9.5, 13.5) 1.66 m
6	71.3, C	—	30.5, CH$_2$	1.93 ddd (4.0, 11.5, 13.5) 2.09 ddd (4.0, 7.5, 13.5)
7	125.6, CH	5.92 dd (1.0, 16.5)	36.0, CH$_2$	1.76 ddd (4.0, 8.0, 12.5) 2.22 ddd (8.0, 9.5, 12.5)
8	139.3, CH	5.69 dd (6.0, 16.5)	66.4, CH$_2$	3.95 ddd (3.0, 8.0, 9.5) 4.05 dd (8.0, 8.5)
9	68.5, CH	4.31 dd (6.0, 12.5)	74.3, CH	3.51 q (6.5)
10	23.7, CH$_3$	1.24 d (6.0)	17.0, CH$_3$	1.15 d (6.5)
11	24.8, CH$_3$	1.03 s	—	—
12	29.5, CH$_3$	1.13 s	—	—
13	17.1, CH$_3$	1.28 s	—	—

^{a}125 MHz; ^{b}500 MHz. Assignments were made using the HMQC, HMBC, COSY, and NOESY spectra.

fractionation W-4 was separated via silica gel CC (Φ30 mm, L 750 mm) and eluted repeatedly with CH$_2$Cl$_2$-MeOH (0 → 100%) to yield five subfractions (W-4.1 to W-4.5). Subfraction W-4.1 was subjected to a silica gel CC (Φ20 mm, L 800 mm with a solvent mixture of CH$_2$Cl$_2$-MeOH, 9 : 1) and passed over a Sephadex LH-20 column (Φ15 mm, L 950 mm) and then through an open YMC RP-C$_{18}$ silica gel column (Φ15 mm, L 800 mm, 65 → 100%, H$_2$O-MeOH) to afford wednenol (**2**, colorless oil, 12.5 mg), cleroindicin E (**3**, colorless oil, 15.2 mg), and rengyol (**4**, white solid, 46.6 mg). Finally, when the same steps were repeated as above, wednenic (**1**, white powder, 10.6 mg), cornoside (**5**, brown oil, 28.7 mg), and benzyl β-D-glucopyranoside-2-sulfate (**6**, amorphous white powder, 86.1 mg) were also obtained by purifying subfraction W-4.5 on YMC RP-C$_{18}$ silica gel (Φ20 mm, L 700 mm) and followed by passing over a Sephadex LH-20 column (Φ15 mm, L 900 mm) using mixtures of MeOH-H$_2$O (1 : 5).

2.5. Physical and Spectroscopic Data of Compounds.
Wednenic (**1**): white powder; $[\alpha]_D^{24}$ −26.5 (*c* 0.25, MeOH); IR ν_{max} (KBr) 3395, 2968, 2930, 1648, 1580, 1511, 1372, 1226, 1163, 1075, and 1040 cm^{-1}; HRESIMS (positive-ion mode) m/z 345.0987 [M + H]$^+$ (cald. for C$_{13}$H$_{23}$O$_7$S, 345.0984) and 367.0801 [M + Na]$^+$ (cald. for C$_{13}$H$_{22}$NaO$_7$S, 367.0803) and (negative-ion mode) m/z 225.1482 [M − SO$_4$Na]$^-$ (cald. for C$_{13}$H$_{21}$O$_3$, 225.1490); for ^{1}H NMR (CD$_3$OD, 500 MHz) and ^{13}C NMR (CD$_3$OD, 125 MHz) spectroscopic data, see Table 1.

Wednenol (**2**): colorless oil; $[\alpha]_D^{24}$ −47.2 (*c* 0.22, MeOH); IR ν_{max} (KBr) 3341, 2950, 1455, 1270, 1168, and 920 cm^{-1}; HRESIMS (positive-ion mode) m/z 185.1170 [M + H]$^+$ (cald. for C$_{10}$H$_{17}$O$_3$, 185.1177); for ^{1}H NMR (CD$_3$OD, 500 MHz) and ^{13}C NMR (CD$_3$OD, 125 MHz) spectroscopic data, see Table 1.

Benzyl β-D-glucopyranoside-2-sulfate (**6**): amorphous white powder; $[\alpha]_D^{24}$ 60.8 (*c* 0,25, MeOH); IR ν_{max} (KBr): 3595, 3100, 2952, 2850, 1647, 1575, 1511, 1362, 1228, 1153, 1055, and 1010 cm^{-1}; ESIMS (negative-ion mode) m/z 349,1 [M − H]$^-$; ^{1}H NMR (500 MHz, CD$_3$OD): δ_H 3.31 (1 H, m, H-5′), 3.44 (1 H, t, *J* = 9.0 Hz, H-3′), 3.66 (1 H, t, *J* = 9.0 Hz, H-4′), 3.71 (1 H, d, *J* = 1.5, 12.0 Hz, H-6′b), 3.91 (1 H, d, *J* = 1.5, 12.0 Hz, H-6′a), 4.17 (1 H, t, *J* = 8.0 Hz, H-2′), 4.53 (1 H, d, *J* = 7.5 Hz, H-1′), 4.74 (1 H, d, *J* = 12.0 Hz, H-7b), 4.93 (1 H, d, *J* = 12.0 Hz, H-7a), 7.25 (1 H, d, *J* = 7.5 Hz, H-4), 7.33 (2 H, d, *J* = 7.5 Hz, H-3 and H-5), and 7.47 (2 H, br t, *J* = 7.5 Hz, H-2 and H-6); ^{13}C NMR (125 MHz, CD$_3$OD): δ_C 138.9 (C-1), 128.8 (C-2 and C-6), 129.1 (C-3 and C-5), 128.4 (C-4), and 71.5 (C-7); *Glc*: 101.0 (C-1′), 81.4 (C-2′), 77.4 (C-3′), 71.5 (C-4′), 77.6 (C-5′), and 62.6 (C-6′).

*2.6. Preparation of (S)- and (R)-MTPA Ester Derivatives of **1**.*
Compound **1** (3.0 mg) was dissolved in 2.5 mL of anhydrous CH$_2$Cl$_2$. Dimethylaminopyridine (35 mg), triethylamine, and (R)-MTPA chloride (30 μL) were then added in sequence. The reaction mixture was stirred for 3 h at room temperature and then quenched by the addition of 1 mL of aqueous MeOH. The solvents were removed under vacuum, and the residue was passed through a small silica gel column using CH$_2$Cl$_2$-MeOH (100 : 1) as the eluent to provide the (S)-MTPA ester of **1** (**1a**, 1.2 mg). The (R)-MTPA derivative (**1b**, 1.5 mg) was prepared with (S)-MTPA chloride and purified in the same manner.

(S)-MTPA ester derivative of **1** (**1a**): 1 H·NMR (CD3OD, 500 MHz): δ_H 7.613–7.421 (10 H, overlap, aromatic protons), 3.630 (3 H, s, OCH3), 1.501 (1H, dd, *J* = 3.5, 12.5 Hz, H-2a), 1.859 (1 H, t, *J* = 12.5 Hz, H-2b), 4.693 (1 H, dt, *J* = 3.5, 12.5 Hz, H-3), 5.767 (1 H, d, *J* = 3.5 Hz, H-4), 6.418 (1 H, dd, *J* = 1.0, 16.5 Hz, H-7), 5.901 (1 H, dd, *J* = 6.0, 16.5 Hz, H-8), 5.983 (1 H, dd, *J* = 6.0, 12.5 Hz, H-9), 1.431 (3 H, d, *J* = 6.0 Hz,

H-10), 1.202 (3 H, s, H-11), 1.050 (3 H, s, H-12), and 1.315 (3 H, s, H-13).

(R)-MTPA ester derivative of **1** (**1b**): 1 H·NMR (CD3OD, 500 MHz): δ_H 7.613–7.401 (10 H, overlap, aromatic protons), 3.58 (3 H, s, OCH3), 1.516 (1 H, dd, $J = 3.5$, 12.5 Hz, H-2a), 1.878 (1 H, t, $J = 12.5$ Hz, H-2b), 4.676 (1 H, dt, $J = 3.5$, 12.5 Hz, H-3), 5.767 (1 H, d, $J = 3.5$ Hz, H-4), 6.395 (1 H, dd, $J = 1.0$, 16.5 Hz, H-7), 5.846 (1 H, dd, $J = 6.0$, 16.5 Hz, H-8), 5.982 (1 H, dd, $J = 6.0$, 12.5 Hz, H-9), 1.463 (3 H, d, $J = 6.0$ Hz, H-10), 1.174 (3 H, s, H-11), 1.014 (3 H, s, H-12), and 1.328 (3 H, s, H-13).

2.7. Assay for α-Amylase Inhibition.

The porcine pancreas α-amylase (A3176, Sigma-Aldrich) enzyme inhibitory activity was carried out according to the standard method with minor modifications [20–22]. The substrate was prepared by boiling 100 mg potato starch in 5 mL phosphate buffer (pH 7.0) for 5 min and then cooling to room temperature. The sample (2 mL dissolved in DMSO) and substrate (50 mL) were mixed in 30 mL of 0.1 M phosphate buffer (pH 7.0). After 5 min preincubation, 5 mg/mL α-amylase solution (20 mL) was added, and the solution was incubated at 37°C for 15 min. The reaction was stopped by adding 50 mL of 1 M·HCl, and then, 50 mL of iodine solution was added. The absorbances were measured at 650 nm by a microplate reader. Acarbose was used as a positive control. The inhibitory activity was calculated by the following equation: α-amylase inhibitory activity (%) = $(1 - A/A0) \times 100$, where A is the absorbance of the sample and A0 is the absorbance of the blank, respectively. The IC_{50} value was calculated by GraphPad Prism.

2.8. Assay for α-Glucosidase Inhibition.

The yeast α-glucosidase (G0660, Sigma-Aldrich) inhibition assay was performed using the substrate p-nitrophenyl-α-D-glucopyranoside according to the previously described method [21–23]. Briefly, samples and acarbose were prepared by dissolving at 2 mg/mL (with extracts) or 0.8 mM (with pure compound) in DMSO, and 0.5 U/mL α-glucosidase (40 mL) was mixed in 120 mL of 0.1 M phosphate buffer (pH 6.8). After 5 min preincubation, 5 mM p-nitrophenyl-α-D-glucopyranoside solution (40 mL) was added, and the solution was incubated at 37°C for 30 min. Pipette the following reagents into a 96-well plate. Each concentration of samples was carried out in triplicate. The absorbance of released 4-nitrophenol was measured at 405 nm by using a microplate reader (xMark, Bio-Rad, USA). Acarbose was used as the positive control. The inhibitory activity was calculated by the following equation: α-glucosidase inhibitory activity (%) = $(1 - A/A0) \times 100$, where A is the absorbance of the sample and A0 is the absorbance of the blank, respectively. The IC_{50} value was calculated by GraphPad Prism.

2.9. Data Expression and Statistical Analysis.

Data were expressed as mean value ± standard deviation (SD) of blood glucose. Data were evaluated using two-way ANOVA followed by Dunnett's multiple comparison test, and groups were considered significantly different if $P < 0.05$. All data are presented as mean ± SD.

3. Results and Discussion

3.1. Identification of Compounds 1–9.

A MeOH extract from the leaves of *W. chinensis* was suspended in H_2O and fractionated successively with EtOAc-soluble fraction, and then, each fraction was evaluated for α-amylase and α-glucosidase activities. The EtOAc-soluble fraction and water layer were chosen for subsequent studies, which resulted in the isolation of two new compounds (**1–2**), together with seven known compounds (**3–9**; Figure 1). Moreover, compounds **1–6** were reported for the first time from this species and from the genus *Wedelia*.

Wednenic (**1**) was obtained as a white powder with a negative optical rotation ($[\alpha]_D^{24} -26.5$, c 0.25, MeOH), and the molecular formula, $C_{13}H_{22}O_7S$, was determined by HRESIMS, with a protonated molecular ion peak at m/z 345.0987 $[M + H]^+$ and a sodium adduct molecular ion peak at m/z 367.0801 $[M + Na]^+$. The fragment ion peak at m/z 225.1482 $[M - SO_4Na]^-$ in the (−)HRESIMS spectrum showed the presence of a sulfate group in **1**. A detailed assessment of the NMR data indicated that **1** is a megastigmane, a compound class known as components of plant species.

The ^1H NMR spectrum (Table 1) of **1** exhibited signals for four methyl groups [δ_H 1.24 (d, $J = 6.0$ Hz, H-10), 1.03 (s, H-11), 1.13 (s, H-12), and 1.28 (s, H-13)], a pair of methylene protons [δ_H 1.47 (dd, $J = 3.5$, 12.5 Hz, H-2a) and 1.84 (t, $J = 12.5$ Hz, H-2b)], three oxygenated methines [δ_H 4.60 (ddd, $J = 3.0$, 3.5, 12.5 Hz, H-3), 4.27 (dd, $J = 1.0$, 3.0 Hz, H-4), and 4.31 (dd, $J = 6.0$, 12.5 Hz, H-9)], and two olefinic protons [δ_H 5.92 (dd, $J = 1.0$, 16.5 Hz, H-7) and 5.69 (dd, $J = 6.0$, 16.5 Hz, H-8)]. The larger coupling constant ($J = 16.5$ Hz) between H-7 and H-8 indicates that the geometry of the Δ^7 double bond is E. The ^{13}C NMR and DEPT spectra (Table 1) showed 13 carbon signals of four methyls [δ_C 23.7 (C-10), 24.8 (C-11), 29.5 (C-12), and 17.1 (C-13)], a methylene [(δ_C 37.9 (C-2)], an oxygenated methine bearing a sulfate group [δ_C 75.5 (C-3)], two oxygenated methines [δ_C 71.6 (C-4) and 68.5 (C-9)], and three nonprotonated carbons [δ_C 35.5 (C-1), 69.4 (C-5), and 71.3 (C-6)], together with a pair of *trans*-olefinic methine carbons [δ_C 125.6 (C-7) and 139.3 (C-8)]. Based on these data, a megastigmane sesquiterpenoid sulfonic acid has been determined for **1**.

Interpretation of the COSY and HSQC spectra of **1** revealed the presence of two partial structures, "C-2/C-3/C4" and "C-7/C-8/C-9/C-10" (Figure 2). These two partial structures were connected through a nonprotonated carbon (C-6) on the basis of HMBC correlations of H-4, H-7, H-8, H-12, and H-13 to C-6. The HMBC correlations from H_3-11 and H_3-12 to C-1, C-2, and C-6 indicated that C-2, C-11, C-12, and C-6 were all connected with C-1. The chemical shifts and coupling constants of **1** were in good agreement with those of (3S,4S,5R,6S,9S,7E)-megastigman-7ene-5,6-epoxy-3,4,9-triol 9-O-β-D-glucopyranoside recorded in the same deuterated solvent [24] but quite different from the data of the C-4. The substitution at C-3 and C-4 was

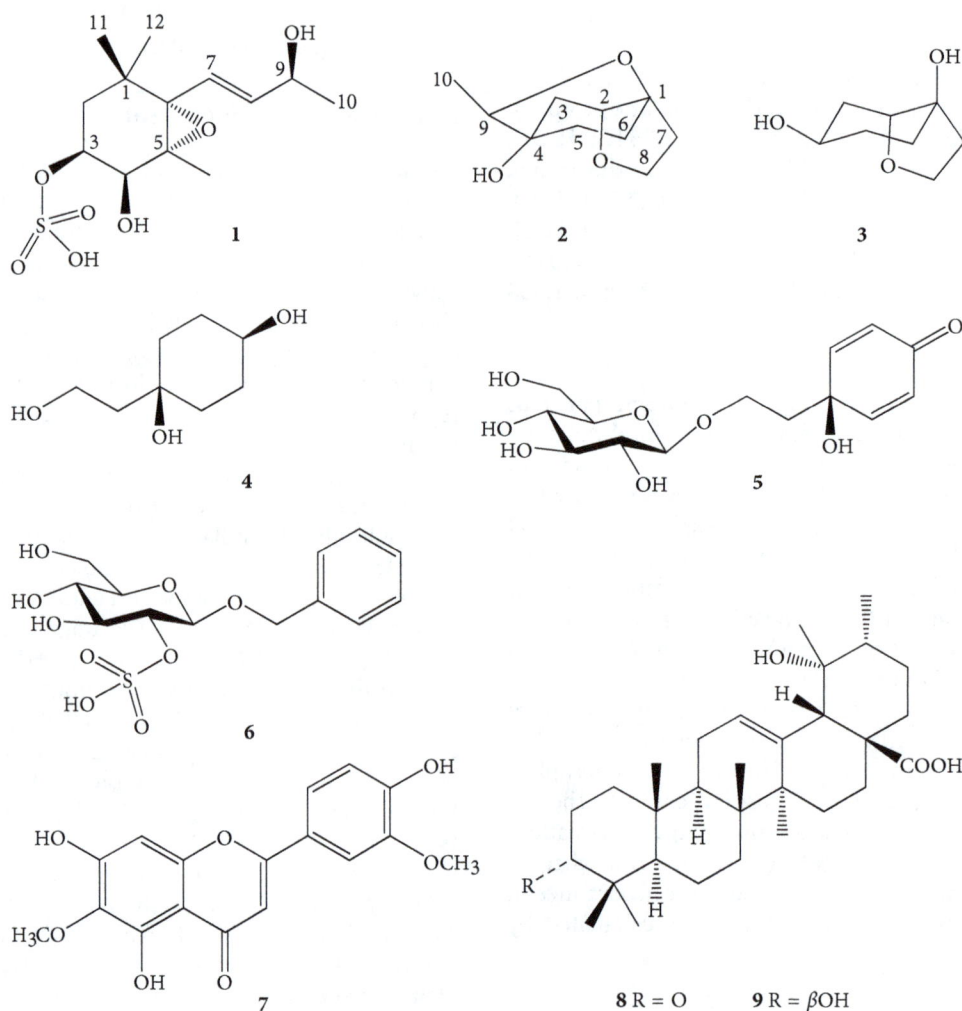

FIGURE 1: Structures of compounds 1–9 isolated from *W. chinensis.*

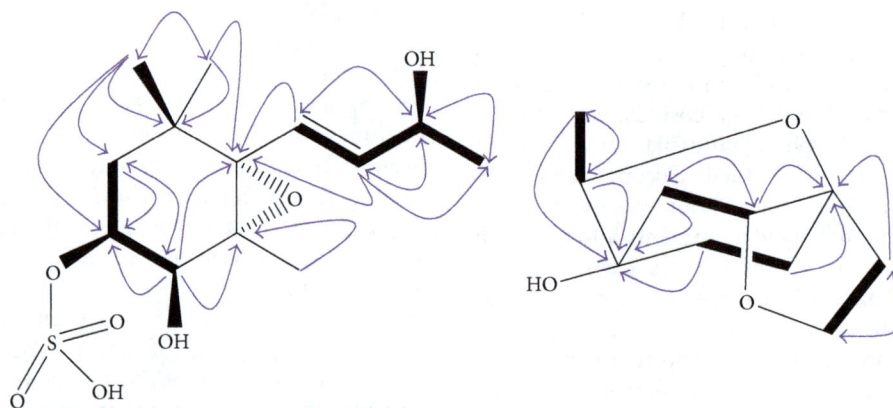

FIGURE 2: Key HMBC (⟶) and COSY (▬) correlations of 1 and 2.

tentatively identified as a sulfonate group and a hydroxyl group based on the HMBC correlations, which supported the total structure of 1 (Figure 2).

The relative configuration of 1 was deduced from analysis of coupling constants and the NOESY spectrum (except for C-9), which were both consistent with a chair conformation for the cyclohexane ring. The α-orientation of H-3 was deduced from the cross-peak of H-3 to H-2a/H$_3$-12 and H-7 and H-2b to H-11 and H-8 in the NOESY spectrum, and the larger coupling constant $J = 12.5$ Hz of H-3 (δ_H 4.60) with one of the H-2a protons (δ_H 1.84) indicated that H-3 was axial. Moreover, the NOESY correlation between H-4

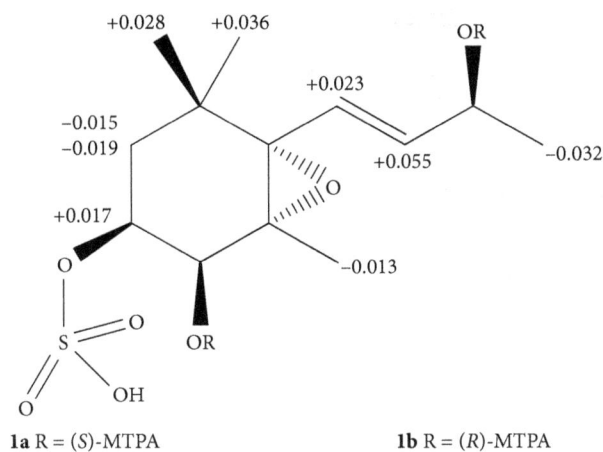

1a R = (S)-MTPA   1b R = (R)-MTPA

FIGURE 3: $\Delta\delta_H$ $_{(S-R)}$ values (in ppm) for MTPA esters of 1.

and H-3 indicated that H-4 and H-3 were axial bonds, which established its α-orientation. Additionally, the coupling constants at H-3 [δ_H 4.60 (1 H, ddd, J = 3.0, 3.5, 12.5 Hz)] and H-4 [δ_H 4.27 (1 H, dd, J = 1.0, 3.0 Hz)] of 1 were in good agreement with those of (3S,4S,5R,6S,9S,7E)-megastigman-7ene-5,6-epoxy-3,4,9-triol 9-O-β-D-glucopyranoside [δ_H 3.79 (1 H, ddd, J = 3.0, 3.0, 12.0 Hz, H-3) and 3.88 (1 H, dd, J = 1.0, 3.0 Hz, H-4)] recorded in the same deuterated solvent. These findings confirmed that two compounds have the same configurations at C-3 and C-4 [24]. The chemical shifts of C-5 (δ_C 69.4) and C-6 (δ_C 71.3) corresponded well to the similar signals observed in the ^{13}C NMR spectra (in CD$_3$OD) with (3S,4S,5R,6S,9S,7E)-megastigman-7ene-5,6-epoxy-3,4,9-triol 9-O-β-D-glucopyranoside [δ_C 69.7 (C-5) and 71.6 (C-6)] [24] and the major difference from the ^{13}C NMR spectrum (in CD$_3$OD) with 5S,6R configurations of (3S,4S,5S,6R,7E,9S)-5,6-epoxy-3,4,9-trihydroxy-7-megastigmen-3-O-β-D-glucopyranoside (komaroveside C) [δ_C 68.2 (C-5) and 70.4 (C-6)] [25]. These data clearly indicated the presence of 5R,6S configurations in 1. To determine the absolute configuration of C-9 in 1, (S)- and (R)-MTPA esters (1a and 1b) were prepared. Significant $\Delta\delta$ values ($\Delta\delta = \delta_{S\text{-MTPA-ester}} - \delta_{R\text{-MTPA-ester}}$) were observed for the proton signal adjacent to C-9, as shown in Figure 3. According to the rule of the modified Mosher's method [26, 27], the absolute configuration at C-9 in 1 was assigned S-form. On the basis of the abovementioned data, the structure of 1 was elucidated to be (3S,4R,5R,6S,9S,7E)-megastigman-5,6-epoxy-7-ene,4,9-diol,3-sulfonic acid.

A molecular formula of C$_{10}$H$_{16}$O$_3$ was established for wednenol (2) based on the presence of 10 signals in its ^{13}C NMR spectrum and the HRESIMS protonated molecular ion peak at m/z 185.1170 [M + H]$^+$ (cald. 185.1178). In the ^{1}H NMR spectrum, signals were observed for a secondary methyl group [δ_H 1.15 (d, J = 6.5 Hz, H-10)], two oxygen-bearing methine groups [δ_H 4.00 (dd, J = 6.0, 9.5 Hz, H-2) and 3.51 (q, J = 6.5 Hz, H-9)], oxygenated methylene groups [δ_H 3.95 (ddd, J = 3.0, 8.0, 9.5 Hz, H-8a) and 4.05 (dd, J = 8.0, 8.5 Hz, H-8b)], and four methylene groups (Table 1). Its ^{13}C NMR spectrum exhibited ten carbon signals, including a methyl [δ_C 17.0 (C-10)], two methines [δ_C 83.3 (C-2) and

74.3 (C-9)], an oxygenated methylene [δ_C 66.4 (C-8)], and four methylenes [δ_C 37.5 (C-3), 29.2 (C-5), 30.5 (C-6), and 36.0 (C-7)], suggesting that 2 was a cyclohexylethanoid derivative [28].

Analysis of the ^{1}H and ^{13}C NMR data of 2 (Table 1) suggested that this compound shares several structural similarities with cleroindicin E (3) [28], except for major differences in the resonances associated with the additional methyl and oxygenated methine groups. The ^{1}H-^{1}H COSY spectrum of 2 was observed to exhibit proton correlations between H-5 and H-6, between H-7 and H-8, between H-2 and H-3, and between H-9 and H-10 (Figure 2). In the HMBC spectrum, cross-peaks between H-10 (δ_H 1.15) and C-4 (δ_C 75.4), and H-10 (δ_H 1.15) and C-9 (δ_C 74.3), between H-9 (δ_H 3.51) and C-3 (δ_C 37.5), C-4 (δ_C 75.4), and C-5 (δ_C 29.2) clearly indicated the positions of the methyl and oxygenated methine groups. Moreover, the C-1 resonance was shifted downfield, from δ_C 75.8 in 3 to δ_C 79.4 in 2. The downfield shift of C-1 may be explained by the presence of a C-1–C-9 ether linkage in 2. These findings suggested that one cyclohexylethanoid unit in 2 should have a 1,2,4-trioxygenated-cyclohexylethanoid structure. Detailed analysis of the other COSY and HSQC spectra unambiguously identified the planar structure of 2 (Figure 2). In the NOESY spectrum of 2, the doublet of doublets belonging to H-2 [δ_H 4.00 (1H dd, J = 6.0, 9.5 Hz)] was observed, which correlated with H-3a and H-7a, and the coupling constant J = 6.0 Hz of H-2 (δ_H 4.00) with one of the H-3a protons (δ_H 1.39) indicated that H-2 was equatorial orientation. Additionally, the NOESY correlation of H-3a and H-9 and their coupling constant (J = 6.5 Hz) suggested that H-9 was also an equatorial bond. Thus, the structure of wednenol was elucidated as 2.

On comparison of their physical and spectroscopic data with published values, the known compounds were identified as cleroindicin E (3) [29], rengyol (4) [30], cornoside (5) [28], benzyl β-D-glucopyranoside-2-sulfate (6) [31], jaceosidin (7) [32], pomonic acid (8) [33], and pomolic acid (9) [34].

3.2. α-Amylase and α-Glucosidase Inhibitory Activities of Compounds 1–9. The inhibitory effects of the isolated compounds against porcine pancreas α-amylase and yeast α-glucosidase were evaluated in comparison with the antidiabetic acarbose. α-Glucosidase is the key catalyzing enzyme involved in the process of carbohydrate digestion and glucose release. Inhibition of α-glucosidase is one very effective way of delaying glucose absorption and lowering the postprandial blood glucose level, which can potentially suppress the progression of DM.

The isolated flavonoid (7) showed the most active α-amylase and α-glucosidase inhibitory activities with IC$_{50}$ values of 112.8 ± 15.1 and 785.9 ± 12.7 μg/mL, respectively (Table 2 and Figure 4). This is in agreement with a recent report of the α-amylase and α-glucosidase inhibitory activities in other flavonoids [35–37]. Compounds 1, 8, and 9 showed moderate inhibitory effects against α-amylase and α-glucosidase, when compared with those of a standard

TABLE 2: Inhibitory effects of selected compounds against α-amylase and α-glucosidase activities (IC$_{50}$ ± SD, μg/mL).

Compounds[a]	α-Amylase	α-Glucosidase
1	436.8 ± 28.6	915.6 ± 36.5
7	112.8 ± 15.1	785.9 ± 12.7
8	420.7 ± 25.2	—
9	395.6 ± 18.3	821.4 ± 55.2
Acarbose[b]	124.0 ± 21.3	642.6 ± 46.4

[a]Compounds were tested in a set of experiments three times. For different versus control group, $P < 0.05$. [b]Acarbose was used as a positive control. (−): no inhibition (less than 10% inhibition).

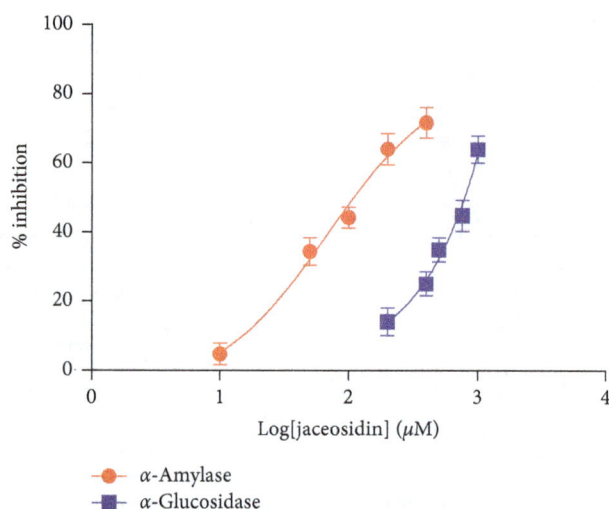

FIGURE 4: IC$_{50S}$ of 7 against α-amylase and α-glucosidase. Activity data points (absorbance) were plotted as mean ± SD ($n = 3$).

reference drug, acarbose, with IC$_{50}$ values of 124.0 ± 21.3 μg/mL (against α-amylase) and 642.6 ± 46.4 μg/mL (against α-glucosidase). In contrast, other compounds (2–6) showed weak or did not show any effect on both enzymes. These results indicated that the megastigmane, flavonoids, and triterpenoids exhibited high inhibitory activity but were not sufficient to clarify the structure-activity relationships between derivatives. Further research is required to clarify their potential selective α-amylase and α-glucosidase activities.

4. Conclusion

In conclusion, the present work reported for the first time the α-amylase and α-glucosidase inhibitory effects of W. chinensis, in support of their ethnomedicinal use for diabetes. This report partly defines the reason on why these medicinal plants possess antidiabetic properties and could provide a scientific warrant for their application as health supplementary herbal products for diabetes treatment and prevention.

Conflicts of Interest

The authors declare that there are no conflicts of interest regarding the publication of this paper.

Acknowledgments

This research was funded by the Vietnam National Foundation for Science and Technology Development (NAFOSTED) under Grant no. 104.01-2016.55.

References

[1] http://www.who.int/diabetes/facts/world_figures/en/index5.html.

[2] Addressing Asia's fast growing diabetes epidemic," Bulletin of the World Health Organization, vol. 95, no. 8, pp. 550-551, 2017.

[3] Y. I. Kwon, E. Apostolidis, and K. Shetty, "Inhibitory potential of wine and tea against α-amylase and α-glucosidase for management of hyperglycemia linked to type 2 diabetes," Journal of Food Biochemistry, vol. 32, no. 1, pp. 15–31, 2008.

[4] R. K. Campbell, J. R. White Jr., and B. A. Saulie, "Metformin: a new oral biguanide," Clinical Therapeutics, vol. 18, no. 3, pp. 360–371, 1996.

[5] A. J. Krentz and C. Bailey, "Oral antidiabetic agents: current role in type 2 diabetes mellitus," Drugs, vol. 65, no. 3, pp. 385–411, 2005.

[6] M. Ekor, "The growing use of herbal medicines: issues relating to adverse reactions and challenges in monitoring safety," Frontiers in Pharmacology, vol. 4, no. 177, pp. 1–10, 2013.

[7] A. G. Atanasov, B. Waltenberger, E.-M. Pferschy-Wenzig et al., "Discovery and resupply of pharmacologically active plant-derived natural products: a review," Biotechnology Advances, vol. 33, no. 8, pp. 1582–1614, 2015.

[8] "V. The Plant List," http://www.theplantlist.org/.

[9] The Wealth of India, A Dictionary Indian Raw Materials and Industrial Products, The Wealth of India, Raw Materials, Council of Scientific and Industrial Research, New Delhi, India, 2005.

[10] H. Wagner, B. Geyer, Y. Kiso, H. Hikino, and G. S. Rao, "Coumestans as the main active principles of the liver drugs Eclipta alba and Wedelia calendulacea," Planta Mededica, vol. 34, no. 5, pp. 370–374, 1986.

[11] F. M. Lin, L. R. Cheu, E. H. Lin, F. C. Ke, and M. J. Tsai, "Compounds from W. chinensis synergistically suppress androgen activity and growth in prostate cancer," Carcinogenesis, vol. 28, no. 12, pp. 2521–2529, 2007.

[12] C. H. Tsai, S. F. Tzeng, S. C. Hsieh et al., "Development of a standardized and effect-optimized herbal extract of Wedelia chinensis for prostate cancer," Phytomedicine, vol. 22, no. 3, pp. 406–414, 2015.

[13] A. Manjamalai and B. Grace, "Antioxidant activity of essential oils from Wedelia chinensis (Osbeck) in vitro and in vivo lung cancer bearing C57BL/6 mice," Asian Pacific Journal of Cancer Prevention, vol. 13, no. 7, pp. 3065–3071, 2012.

[14] I. Nomani, A. Mazumder, and G. S. Chakraborthy, "Wedelia chinensis (Asteraceae)-an overview of a potent medicinal herb," International Journal of PharmTech Research, vol. 5, no. 3, pp. 957–964, 2013.

[15] W.-C. Lin, C.-C Wen, Y.-H. Chen et al., "Integrative approach to analyze biodiversity and anti-inflammatory bioactivity of *Wedelia* medicinal plants," *PLoS One*, vol. 10, no. 6, article e0129067, 2015.

[16] A. Shirwaikar, R. G. Prabhu, and S. Malini, "Activity of *Wedelia calendulacea* Less. in post-menopausal osteoporosis," *Phytomedicine*, vol. 13, no. 1-2, pp. 43–48, 2006.

[17] N. Verma and R. L. Khosa, "Effect of *Costus speciosus* and *Wedelia chinensis* on brain neurotransmitters and enzyme monoamine oxidase following cold immobilization stress," *Journal of Pharmaceutical Sciences*, vol. 1, no. 2, pp. 22–25, 2009.

[18] P. Murugaian, V. Ramamurthy, and N. Karmegam, "Hepatoprotective activity of *Wedelia calendulacea* against acute hepatotoxicity in rats," *Research Journal of Agriculture and Biological Sciences*, vol. 4, no. 6, pp. 685–687, 2008.

[19] T. Prakash, N. R. Rao, and A. H. M. V. Swamy, "Neuropharmacological studies on *Wedelia calendulacea* Less stem extract," *Phytomedicine*, vol. 15, no. 11, pp. 959–970, 2008.

[20] R. Kusano, S. Ogawa, Y. Matsuo, T. Tanaka, Y. Yazaki, and I. Kouno, "α-Amylase and lipase inhibitory activity and structural characterization of acacia bark proanthocyanidins," *Journal of Natural Products*, vol. 74, no. 2, pp. 119–128, 2011.

[21] E. A. H. Mohamed, M. J. A. Siddiqui, L. F. Ang et al., "Potent α-glucosidase and α-amylase inhibitory activities of standardized 50% ethanolic extracts and sinensetin from *Orthosiphon stamineus* Benth as anti-diabetic mechanism," *BMC Complementary and Alternative Medicine*, vol. 12, no. 176, 2012.

[22] T. T. H. Hanh, N. M. Chau, L. H. Tram et al., "Inhibitors of α-glucosidase and α-amylase from *Cyperus rotundus*," *Pharmaceutical Biology*, vol. 52, no. 1, pp. 74–77, 2014.

[23] M.S. Ali, M. Jahangir, S. S. Hussan, and M. I. Choudhary, "Inhibition of alpha-glucosidase by oleanolic acid and its synthetic derivatives," *Phytochemistry*, vol. 60, no. 3, pp. 295–299, 2002.

[24] Q. Yu, K. Katsunami, H. Otsuka, and Y. Takeda, "Staphylionosides A-K: megastigmane glucosides from the leaves of *Staphylea bumalda* DC," *Chemical and Pharmaceutical Bulletin*, vol. 53, no. 7, pp. 800–807, 2005.

[25] I. K. Lee, K. H. Kim, S. Y. Lee, S. U. Choi, and K. R. Lee, "Three new megastigmane glucopyranosides from the *Cardamine komarovii*," *Chemical and Pharmaceutical Bulletin*, vol. 59, no. 6, pp. 773–777, 2011.

[26] I. Ohtani, T. Kusumi, Y. Kashman, and H. Kakisawa, "Highfield FT NMR application of Mosher's method. The absolute configurations of marine terpenoids," *Journal of the American Chemical Society*, vol. 113, no. 11, pp. 4092–4096, 1991.

[27] J. M. Seco, E. Quiñoá, and R. Riguera, "The assignment of absolute configuration by NMR," *Chemical Reviews*, vol. 104, no. 1, pp. 17–118, 2004.

[28] T. Hase, Y. Kawamoto, K. Ohtani, R. Kasai, K. Yamasaki, and C. Picheansoonthon, "Cyclohexylethanoids and related glucosides from *Millingtonia hortensis*," *Phytochemistry*, vol. 39, no. 1, pp. 235–241, 1995.

[29] M. Honzumi, T. Kamikubo, and K. Ogasawara, "A stereocontrolled route to cyclohexylethanoid natural products," *Synlett*, vol. 1998, pp. 1001–1003, 1998.

[30] H. Abdullahi, E. Nyandat, C. Galeffi, I. Messana, M. Nicoletti, and G. B. M. Bettolo, "Cyclohexanols of *Halleria lucida*," *Phytochemistry*, vol. 25, no. 12, pp. 2821–2823, 1986.

[31] K. Ohtani, R. Kasai, K. Yamasaki et al., "Lignan glycosides from stems of *Salvadora persica*," *Phytochemistry*, vol. 31, no. 7, pp. 2469–2471, 1992.

[32] A. Tapia, J. Rodriguez, C. Theoduloz, S. Lopez, G. E. Feresin, and G. Schmeda-Hirschmann, "Free radical scavengers and antioxidants from *Baccharis grisebachii*," *Journal of Ethnopharmacology*, vol. 95, no. 2-3, pp. 155–161, 2004.

[33] D. L. Cheng and X. P. Cao, "Pomolic acid derivatives from the root of *Sanguisorba officinalis*," *Phytochemistry*, vol. 31, no. 4, pp. 1317–1320, 1992.

[34] C. Hata, M. Kakuno, K. Yoshikawa, and S. Arihara, "Triterpenoid saponin of aquifoliaceous plants.V. Ilexosides XV-XIX from the barks of *Ilex crenata* Thunb," *Chemical and Pharmaceutical Bulletin*, vol. 40, no. 8, pp. 1990–1992, 1992.

[35] D. F. Pereira, L. H. Cazarolli, C. Lavado et al., "Effects of flavonoids on α-glucosidase activity: potential targets for glucose homeostasis," *Nutrition*, vol. 27, no. 11-12, pp. 1161–1167, 2011.

[36] A. H. Zeid, M. AliFarag, M. A. AzizHamed, Z. A. AzizKandil, R. HassanEl-Akad, and H. MohamedEl-Rafie, "Flavonoid chemical composition and antidiabetic potential of *Brachychiton acerifolius* leaves extract," *Asian Pacific Journal of Tropical Biomedicine*, vol. 7, no. 5, pp. 389–396, 2017.

[37] P. V. A. Babu, D. Liu, and E. R. Gilbert, "Recent advances in understanding the anti-diabetic actions of dietary flavonoids," *Journal of Nutritional Biochemistry*, vol. 24, no. 11, pp. 1777–1789, 2013.

4

New Advances in Toxicological Forensic Analysis using Mass Spectrometry Techniques

Noroska Gabriela Salazar Mogollón (iD),[1,2] **Cristian Daniel Quiroz-Moreno,**[1]
Paloma Santana Prata,[2] **Jose Rafael de Almeida,**[1] **Amanda Sofía Cevallos,**[1]
Roldán Torres-Guiérrez,[1] **and Fabio Augusto**[2]

[1]*Ikiam-Universidad Regional Amazónica, Km 7 Via Muyuna, Tena, Napo, Ecuador*
[2]*Institute of Chemistry, State University of Campinas, Cidade Universitária Zeferino Vaz, 13083-970 Campinas, SP, Brazil*

Correspondence should be addressed to Noroska Gabriela Salazar Mogollón; gaby867@gmail.com

Academic Editor: Veronica Termopoli

This article reviews mass spectrometry methods in forensic toxicology for the identification and quantification of drugs of abuse in biological fluids, tissues, and synthetic samples, focusing on the methodologies most commonly used; it also discusses new methodologies in screening and target forensic analyses, as well as the evolution of instrumentation in mass spectrometry.

1. Introduction

The development of mass spectrometry methods has offered new possibilities for forensic toxicology analyses, where the identification and quantification of drugs of abuse are the most concerning issues in the forensic science [1]. The prevalence of drug addiction and abuse in the population worldwide is significantly high, resulting in one of the main causes of high criminal activities [2]. The excessive use of psychotropic substances, natural drugs, hallucinogens, and most recently "new psychoactive substances," which are designed from skeletons of some natural drugs previously known, are the main focus of the development of new analytical methodologies, where mass spectrometry has had a key role [3, 4]. When a toxicological analysis needs to identify and quantify metabolites from unknown drugs, a screening can be performed by coupling different chromatography techniques, such as liquid and gas chromatography to mass spectrometry. In cases where an increment in the signal/noise ratio (S/N) is necessary, and the structure of the compound is known (target analysis) [5, 6] an additional selectivity can be provided using tandem mass spectrometry (MS/MS) in ion products or selected reaction monitoring (SRM). This latter one is the most widely used because of its increase in the specificity, selectivity, and detectability; however, the analyses become too time-consuming when a previous chromatographic separation and sample preparation are required [6, 7].

Moreover, ionization mass spectrometry techniques such as direct analysis in real time (DART), desorption electrospray ionization (DESI), low-temperature plasma (LTP), desorption atmospheric-pressure photoionization (DAPPI), paper spray (PS), touch spray mass spectrometry (TS-MS), more recently in toxicological analysis laser diode thermal desorption (LDTD), and atmospheric solids analysis probe (ASAP) have gained popularity as they can be used with less or even without sample preparation [8–10]. Nevertheless, depending on the matrix sample, compounds with identical patterns of fragmentation cannot be identified; this is the reason why more development in mass spectrometry needs to be conducted in order to provide relevant information that can help solve a crime [11]. In this sense, this review presents the main current applications of mass spectrometry for the control of drugs of abuse and the discovery of synthetic drugs in biological and synthetic matrices; besides, methodological limitations as well as innovative methodologies to enhance forensic toxicology analysis are discussed, examining the current literature in the past eight years.

2. Chromatography and Mass Spectrometry

2.1. Conventional MS Methods. The coupling of chromatography techniques with mass spectrometry has been widely used in drugs of abuse analysis, especially when the screening of the sample is needed, having separation techniques such as gas chromatography-mass spectrometry (GC-MS), liquid chromatography-mass spectrometry (LC-MS), liquid chromatography-mass spectrometry in tandem (LC-MS/MS) and, more recently, two-dimensional gas chromatography-MS (GC × GC-MS) as the most commonly used.

Normally, the analyses of nonobjective analytes, after the chromatographic separation, have the same steps to follow. First, a scan is performed by the mass spectrometer in order to identify or recognize some compounds of interest; then, it is necessary to perform a selected ion monitoring (SIM) [12, 13] in order to increase the sensitivity and selectivity of the analysis, in which only fragments of a specific group of molecules are monitored, resulting in an increased S/N. Consequently, this technique is the most widely used in quantitative analysis of compounds.

The focus in this section is on the most recent and innovative analyses that have been performed using mass spectrometry coupled to chromatographic techniques, including all the new methodologies developed in toxicological analyses. Table 1 provides a summary with their advantages and disadvantages, and Figure 1 shows these methodologies as well.

2.1.1. GC-MS. The advances in techniques using MS coupled to gas chromatography have not been very significant due to the type of analyte that can be analyzed using this chromatographic technique (low molecular weight, volatiles). Even though high molecular weight compounds can be derivatized and analyzed by GC, the sample treatment is not appealing for the forensic toxicological analysis of drugs of abuse where the quickness of analysis is fundamental. For this reason, most of the advances using GC-MS focus on the resolution and separation capacity during the analysis. However, toxicological analysis methods in various matrices are well established and widely used in the analysis of drugs, in order to confirm forensic toxicology from samples of blood, urine, saliva, and hair, among others, during specific screening analysis, demonstrating high selectivity, detectability, and robustness.

In this sense, in order to improve the detection and identification of compounds using GC-MS, negative and positive modes of analysis in MS have been integrated, taking advantage of the stability of the fragments after a positive or negative ionization. For instance, Wu et al. used GC-MS with electron impact ionization and negative chemical ionization (GC/NCI-MS) and the traditional GC-MS with electron ionization mass spectrometry (GC/EI-MS) to analyze opiates, amphetamines, and ketamines in human hair. These analyses were capable of providing more sensitivity at low concentration of pictogram (pg) using only 25 mg from the sample and improving the detection of compounds during the analysis owing to the electronegative moieties. The strategy also avoided wrong results and misinterpretations obtaining lower limits of detection, in comparison with the use of only traditional GC/EI-MS in mode SIM; therefore, NCI can serve as a complementary technique in order to improve the sensitivity during the analysis [14].

The use of a miniaturized analytical method is the aim during the development of new analytical methods, and the analysis of drugs of abuse by mass spectrometry is not the exception. Most recently, GC-MS methodologies have used cold EI to analysis of heroin and cocaine [15, 16]. Here, the GC has an interface known as a supersonic molecular beam (SMB) where the ionization vibration cold sample is in an axial fly-through ion source configuration (Figure 1(a)), providing mass spectra with enhanced molecular ions that are compatible with reference libraries, and the range of compounds are amenable to GC-MS compounds. Additionally, this configuration allowed the increment of the flow rate in GC-MS without declines in the sensitivity in the analysis in the EI source, since the fly-through ion source sensitivity is fully independent on the column flow rate; therefore, column flow rate increase is automatically offset by a corresponding reduction in the helium make-up gas flow rate, the supersonic nozzle backing pressure, and the SMB flow rate are stabilized. The authors considered this aspect during the determination of heroin and cocaine in paper money and composite drug powders using column flow programming as a tool to further reduce the time of analysis. With this method, the time of analysis decreased, allowing the use of a column flow from 1 mL/min to 32 mL/min and the use of relatively small column dimension (5 m 0.25 mm) [55].

However, if the analysis aims to identify target compounds, and if specific fragments of a molecule are known, it is possible to increase the S/N with the use of mass spectrometry in tandem (MS/MS). GC-MS/MS is commonly used in SRM and *product ion scan* modes with collision-induced dissociation (CID). On the one hand, an ion precursor is generated into the collision cell during SRM mode, and then one ion product is monitored—this monitoring is also called transitions—this mode is widely used in quantitative analysis because of its selectivity. On the other hand, *product ion scan* consists of scanning product ions once the molecules are fragmented in the collision cell, generating, as a result, high reliability results due to the specificity of the monitored transitions. This method is generally used for transition optimization and the creation of libraries in MS/MS. Thus, these analyses can obtain an unequivocal identification of the eluted analyte. For example, this method identified methamphetamines in blood and urine with a simple and quick LLE and derivatization, as well as managed to differentiate between them [56].

Versace et al. used GC-MS to perform a screening of unknown compounds without an excessive sample preparation in urine samples and GC-MS/MS with the purpose of increasing the specificity using SRM transitions, identifying

TABLE 1: Main modifications and modes of analysis applied in mass spectrometry in forensic toxicological analysis.

| Type of MS analysis | Ionization techniques in mass spectrometry coupled to separations techniques | | References |
	Advantages	Disadvantages	
Negative chemical ionization	(i) It provides more sensitivity at low concentration (pg) based on the stability of electronegative moieties. (ii) Avoids wrong interpretations of correct results reducing time consumption.	(i) Better results are provided when the technique is combined with EI-MS in order to obtain more structural information. (ii) This method requires an additional reagent for the ionization; methane is commonly used.	[14]
Cold electrospray ionization	(i) It can be considered as a miniaturized analytical method because of the interface that it uses and the supersonic molecular beams through analysis with short columns and high column flow rates. (ii) Can provide enhanced molecular ions to much larger and more polar compounds with GC, using the same library to EI-MS (NIST). (iii) The flow rate can be increased up to 100 mL/min, and its fly-through ion source sensitivity is fully independent from the column flow rate. (iv) In this method, the nozzle flow rate is constant; as a result, the cold EI fly-through ion source is unaffected by the column flow rate, unlike any other ion source. (v) The use of GC-MS with cold EI has no limitations for the column used.	(i) Additional instrumentation is required.	[15, 16]
Surface-activated chemical ionization	(i) The ionization of solutes occurs upon the polarization of neutral, solvent molecules, which makes it a highly sensitive method. (ii) The electrostatically charged surface increases the ESI ionization efficiency. (iii) When it is used with ESI, the efficiency of proton-transfer ionization reactions is enhanced by the polarization of neutral solvent molecules or by charged solute molecules induced by the proximity of the charged surface. (iv) The solvent and the analyte ions are better focused towards the analyzer. (v) The increase in signal intensity provides an increase in sensitivity, because there is a reduction in the chemical noise observed in the mass analyzer.	(i) SACI is used to maximize the sensitivity in the analysis of highly polar compounds, but data about less polar compounds have not been revealed until now.	[17]

Type of MS analysis	Advantages	Disadvantages	References
	Ionization techniques in mass spectrometry coupled to separations techniques		
	New methods of analysis used in mass spectrometry		
Dynamic multiple reaction monitoring mode of analysis	(i) This method monitors the analytes only around the expected retention time, decreasing the number of concurrent MRM transitions, allowing both the cycle and the dwell time, which can be optimized in order to obtain higher sensitivity, accuracy, and reproducibility. (ii) dMRM allows the monitoring of more MRM transitions in a single run without compromising data quality. (iii) The dwell time is intelligently optimized by association with the delta retention time. Additionally, information about delta retention time and retention time are key to maximize the dwell time and increasing sensitivity. (iv) This method gives the possibility of applying simultaneous quantifications of multiple components.	(i) It is necessary to maintain the analyte analysis in the same polar mode since a switch of polarity within a single run would reduce the sensitivity and accuracy of quantification with the applied MS instrumentation. (ii) The retention time must be informed, optimized, and defined with reference standards using established chromatographic conditions if it is possible. If the retention time drifts, this might result in an incomplete peak definition and quantitation. (iii) It is necessary to optimize the MS conditions for the all transitions.	[17–20]
	Ambient ionization techniques in mass spectrometry		
Desorption electrospray ionization-mass spectrometry	(i) Direct analysis with high-velocity nebulizing gas. (ii) The selectivity and sensitivity of this technique can be increased by a pretreatment sample.	(i) During analysis of drugs in biological matrix with a high amount of salt, the suppression ionization effect is elevated. (ii) The ion source geometry affects the dynamic of the splashing mechanism resulting in changes in droplet size, charge, and analyte dissolution extent. (iii) A high velocity of nebulization can mechanically ablate delicate samples/powders.	[8, 21–23]
Desorption atmospheric-pressure photoionization	(i) Matrix with high salt content do not provide an elevated suppression of ionization.	(i) High suppression ionization can be found depending of the biological matrix. (ii) Sample preparation is commonly needed in order to avoid suppression of ionization.	[8, 21–25]
Direct analysis in real time	(i) It is commonly used in the analysis of drugs of low molecular weight; therefore, its sensitivity depends on analyte volatility. (ii) The geometrical configuration of the ion source is simple and robust for its operation. (iii) Pretreatment of sample can increase the selectivity of the analysis in complex biological samples.	(i) Compounds of high molecular weight may need derivatization. (ii) Its sensitivity depends of the temperature of the ionization region; therefore, the higher the temperature is the higher the risk of damage is. (iii) Its reproducibility depends on the position of the sample inside the ion source, which represents a big problem in the quantification of the analysis.	[4, 8, 11, 26–35]

TABLE 1: Continued.

| | Ionization techniques in mass spectrometry coupled to separations techniques | | |
Type of MS analysis	Advantages	Disadvantages	References
Low-temperature plasma	(i) It is possible to perform direct analysis without sample preparation. (ii) The instrumentation is simple, and its configuration provides low consumption of discharge gas and the possibility of using air as the discharge gas. (iii) High sensitivity and sensitivity can be obtained without pretreatment of the samples.	(i) This technique is exclusively used with small organic molecules with low to moderate polarity.	[8, 36]
Matrix-assisted laser desorption electrospray ionization	(i) It can be coupled to mass spectrometric imaging (MSI) in order to obtain the distribution spectra of the target. (ii) A mode of analysis called "dynamic pixel" can be used to obtain an imaging method that is faster to do a screening of the compounds. (iii) The analysis does not need sample preparation. This method is based on a direct analysis over the sample. (iv) The sensitivity of the analysis can be improved using a specific matrix. For example, umbelliferone matrix obtained better results in the analysis of methamphetamine in hair than the common matrices CHCA or DHB. (v) The technique has been tested along with MAMS, and it is possible to cause reproducibility of the signal with this technology.	(i) Quantitative analysis has not been carried out until this present date.	[8, 37–41]
Metal-assisted secondary ion mass spectrometry	(i) It can be coupled to mass spectrometric imaging (MSI) in order to obtain the distribution spectra of the target. (ii) The limits of detection are lower than those obtained with MALDESI and also compared with the ones with LC-MS/MS. (iii) It is not necessary to perform preparation of the sample.	(i) Quantitative analysis has not been carried out.	[8, 42, 43]
Paper spray	(i) This technique can analyze a wide range of molecules, from small to large biomolecules. (ii) The use of a pretreatment of the sample can enhance the sensitivity of the analysis.	(i) It has a high matrix effect on most of the drugs. (ii) The paper can extract impurities from the surface and cause the suppression of ionization.	[8, 44–51]
High-performance ion mobility spectrometry	(i) Methods of introduction of samples such as a chromatographic separation can be used to minimize the suppression of ionization.	(i) Direct analysis can result in suppression of ionization.	[50, 51]

TABLE 1: Continued.

Type of MS analysis	Advantages	Disadvantages	References
	Ionization techniques in mass spectrometry coupled to separations techniques		
Differential mobility spectrometry-mass spectrometry separation	(i) Separation conditions of the target analysis can be selectively transmitted into a mass spectrometer. (ii) It can be considered as an ionization technique coupled to a separation method that has a small interface which gives results in few seconds.	(i) This technique is rarely being implemented in commercial devices, and it is not known yet whether it can be used to establish profiles of drug mixtures in complex biological samples.	[52]
Touch spray	(i) The substrate (medical swabs) used can serve as a sample collection tool; thus, ionization helps in the analysis of solid or liquid samples without pretreatment. (ii) The TS-MS can allow noninvasive and direct sampling of neat oral fluids.	(i) The drying step of this substrate represents the most time-consuming part of the analytical protocol.	[9]
Laser diode thermal desorption	(i) The method is completely automatic.	(i) It is not possible to perform a simple interchange between negative and positive modes of ionization. (ii) Effects of interferences in complex biological samples must be explored, and more sample preparation is necessary before the liquid samples are transferred towards the capillary surface.	[53]
Atmospheric solids probe analysis	(i) It is possible to perform analysis of solids and liquids easily. (ii) This design allows the possibility of positive/negative switch during the analysis.	(i) Effects of interferences in complex biological samples must be explored with more detail. (ii) This technique provides better sensitivity during the analysis of small-molecule drugs, decreasing the analysis of the high-molecule compounds.	[53, 54]

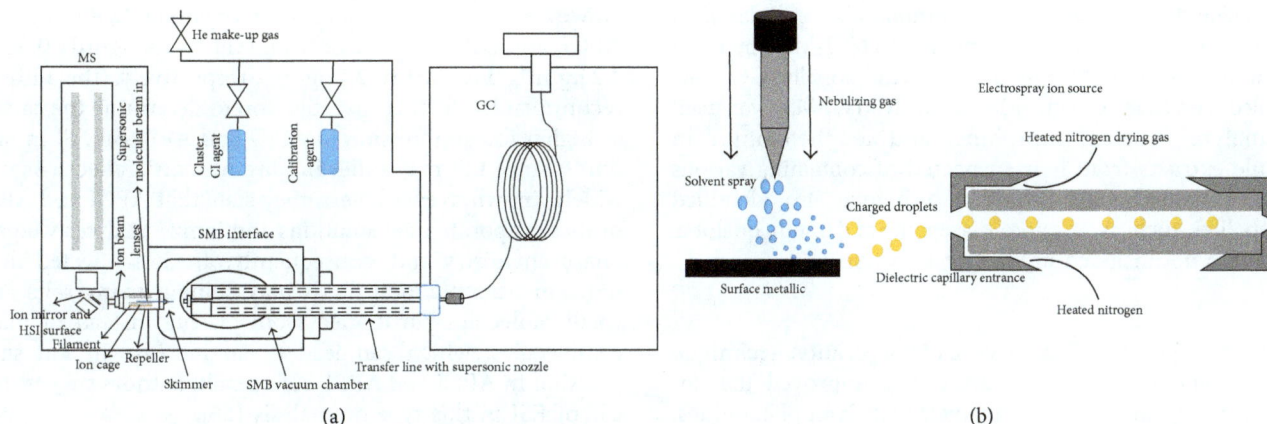

FIGURE 1: Main advances in mass spectrometry coupled to chromatographic technique in toxicological analysis are (a) gas chromatography interface-supersonic molecular beam with ionization vibration cold sample to mass spectrometry and (b) electrospray ionization and surface-activated chemical ionization to mass analyzed coupled with liquid chromatography.

54 drugs (i.e., 11-nor-9-carboxy-Δ^9-tetrahydrocannabinol, cocaine, hydrocodone, and flurazepam) [57], while Emidio et al. developed a new methodology to determine cannabinoids in hair using 10 mg of sample and headspace solid-phase microextraction (HS-SPME) and GC ion trap/tandem mass spectrometry. Here, CID was used to adjust the breakage of the cannabinoid fragment (ion precursor) and improve the detectability of the technique, demonstrating an excellent linearity range between 0.1 and 8.0 ng/mg with a limit of quantification (LOQ) of

0.007–0.031 ng/mg and 0.012–0.062 ng/mg, which are smaller than the cutoff value established by the Society for Toxicological and Forensic Chemistry (GTFCh) [58].

On the other hand, ethyl glucuronides (a biomarker of ethanol), commonly used in the detection of chronic and excessive alcohol consumption, were identified using MS/MS operating in NCI-MS and SRM mode, obtaining differences between teetotalers and moderate drinkers, according to the current cutoff (i.e., 7 pg/mg hair). In this case, the use of negative mode provided an enhanced sensitivity in low concentration samples which were combined with the specificity of the fragments in the SRM analysis.

Therefore, a better analytical selectivity and S/N were achieved along with long-term markers for the detection of chronic and excessive alcohol consumption [6].

In the same manner, GC-MS/MS has been used to differentiate among important isomers such as methoxyethylamphetamines and monomethoxydimethylamphetamines, as synthetic drugs without derivatization. Using CID and SRM, the specificity of the fragments obtained provided intensity differences in product ions among the isomers, enabling mass spectrometric differentiation of the isomers [59]. One the contrary, GC × GC as a previous treatment can be used in order to increase sensitivity, detection, separation, and resolution. In this sense, GC × GC-MS was used in the determination of cannabinoid-like drugs in 1 mL of postmortem blood, which present a challenge due to the matrix interferences in endogenous lipophilic compounds, proteins, drug degradation/formation, and production of artefacts. With this technique, a limit of detection of 0.25 ng/mL for 11-hydroxy-Δ^9-tetrahydrocannabinol was obtained [60].

The same methodology was applied in oral fluid samples, where it is common to have a small volume of samples, and the concentration of some drugs is usually lower, which can complicate the analysis [61]. The compounds were identified using GC × GC-MS with cold trapping and NCI-MS, obtaining a limit of detection of 0.5 ng/mL [61]. Additionally, GC × GC coupled with time-of-flight-MS (GC × GC-TOFMS) was used to analyze codeine, morphine, and amphetamines in sample extracts from hair suspected of containing various drug compounds. The analytical technique also identified metabolites such as cocaine, diazepam, and methaqualone, which are not included in the target analysis [62].

2.1.2. LC.

The use LC as a versatile separation technique (volatile and nonvolatile analytes) has improved the detection and quantification of analytes such as amphetamines, benzodiazepines, hallucinogens, cannabinoids, opiates, cocaine, designer drugs, pharmaceutical products, or illicit drugs in several matrices.

Ultrahigh-performance liquid chromatography has been used along with tandem mass spectrometry (UHPLC-MS/MS) operating in SRM mode in order to establish an individual ion transition ratio to each analyte. Thus, each analyte is infused into the electrospray capillary, and the declustering potential was adjusted to maximize the intensity of the protonated molecular species $[M + H]^+$. The

signals were optimized using a source block temperature of 500°C and an on-spray in the determination of Δ^9-tetrahydrocannabinol, cannabidiol, and cannabinol in 50 mg of 179 hair samples. This method allowed the identification of one new synthetic cannabinoid, obtaining an LOQ of around 0.07 pg/mg and 18 pg/mg in the analysis [63].

High-resolution (HR) MS has been used successfully along with LC in the drug of abuse analysis. The resolving power and the high mass accuracy obtained with HRMS were advantageous in the analysis of complex matrices and data acquisition in a targeted and nontargeted manner in order to decrease the number of interferences caused by biological matrix in the drugs analysis. In this case, the authors used UHPLC-HR-TOFMS to analyze cannabinoids and cathinones in 1 mL of urine. During the analysis, a broad-band collision-induced dissociation (bbCID) was used with the purpose of providing a confirmation-level screening featuring both high sensitivity and wide scope. The precursor ions were fragmented in the collision cell without preselection, and the analysis allowed the identification of 75 compounds with cannabinoids spectra database with cutoff concentration values of 0.2–60 ng/mL and cathinones 0.7–15 ng/mL, respectively [64, 65].

The methods of ionization are fundamental to ensure the correct ionization of the sample and, therefore, the identification of the compounds, especially in complex matrices. Wang et al. performed an analysis of cocaine and their metabolites under three types of ionization such as electrospray ionization (ESI), atmospheric-pressure chemical ionization (APCI), and atmospheric-pressure photoionization (APPI) in order to evaluate the chemical suppression during the analysis of 17 illicit drugs in 100 μL of oral fluids using UHPLC-MS/MS in mode SRM. The authors found that ESI presents the smallest ion suppression for all cocaine metabolites analyzed facing the APPI and APCI mode. However, the method developed obtained LOQs in ESI, APCI, and APPI in a range from 0.11 to 1.9 ng/mL, 0.02 to 2.2 ng/mL, and 0.02 to 2.1 ng/mL, respectively. The authors recommended further investigation to determine the causes of higher ion suppression in APCI and APPI on ESI in oral fluids, since ESI may suffer important matrix effects, as it is widely known. Nevertheless, they state that APCI and APPI probes evaporate inlet solutions and ionize analytes via gas-phase chemistry and, consequently, are less affected than ESI. For example, oral fluids may contain many salts and small molecules partitioned from plasma instead of macromolecules, which can lead to an increase in ion suppression in APCI and APPI. As a result, authors suggest the use of ESI in this type of analysis [66].

Liquid chromatography has also taken advantage the benefits of negative mode in mass spectrometry. In this case, LC coupled to (HR)-MS was used along with Orbitrap technology in the analysis of metabolites of drugs such as cocaine, ephedrine, and morphine in urine. The analyses were performed in full-scan mode with positive/negative switching, and subsequently making use of a selective screening through data dependent acquisition (DDA) mode, resulting in a fast analysis. Additionally, the risk of false-negative results caused by ion suppression or isomer overlapping could be reduced by including metabolites and artefacts, as well as recording in the positive and negative modes [67].

Recently, new methods of analyses in MS/MS have been coupled with LC techniques. Dynamic multiple reaction monitoring (dMRM) has been used in toxicological analysis, and it is recognized by the use of a timetable based on the retention time for each analyte. Such technique monitors the analytes only around the expected retention time, and decreases the number of concurrent SRM transitions, also known as multiple reaction monitoring (MRM), allowing both the cycle time and the dwell time to be optimized to the highest sensitivity, accuracy, and reproducibility [18]. For example, a quantitative LC-MS/MS method has been developed for the simultaneous determination of 17 antipsychotic drugs in human postmortem brain tissue; these drugs are of forensic interest because they have been associated with sudden death cases.

In this method, the analysis was performed operating on dMRM mode, using ESI+. Calibration curves prepared in the spiked brain tissue were linear in the range 20–8000 ng/g ($R^2 > 0.993$) for all drugs, except olanzapine [19]. Besides, LC-MS/MS in dMRM mode was used by Shah et al. in order to identify around 200 drugs/metabolites, such as meth-amphetamine, amphetamines, ephedrine, and cocaine in hair samples [20]. This method proved an interesting alternative for fast analysis of drugs. All these analyses were performed in one chromatographic run (i.e., 8 min), showing a high sensitivity and accuracy [20].

With the purpose of identifying cannabinoids, such as Δ^9-tetrahydrocannabinol, Conti et al. coupled LC along with two ionization systems, electrospray ionization, and surface-activated chemical ionization (ESI-SACI-MS) to several types of mass analyzer (ion trap, triple quadrupole, and Orbitrap) to improve the detection of 11-nor-9-carboxy-tetrahydrocannabinol in biological samples (urine and hair) operating in SRM mode. This coupling consists in a metallic surface that keeps a fixed voltage and that is inserted into a commercial ESI source (Figure 1(b)). This electrostatically charged surface is able to improve the ESI ionization efficiency, and it increases the ion focusing efficiency towards the mass spectrometric analyzer. The authors show that the sensitivity provided was better with SACI-ESI than with the classical ESI approach alone [17, 68].

Furthermore, an ultrafast and sensitive microflow liquid chromatography-MS/(MFLC-MS/MS) was used to quantify hallucinogens such as LSD and their metabolites in 500 µL of plasma in order to miniaturize and accelerate the analysis of drugs of abuse using LC techniques; this coupling is known by its decrease in run numbers, a higher ionization yield, and reduced ion suppression/enhancement effects. Here, the MS ion trap operated in *product ion scan* and SRM mode in order to perform the quantification along with a dynamic fill-time trap; this method allowed sensitive detection and fast analysis, obtaining LOQs corresponding to 0.01 ng/mL for all analytes [69].

3. Current Analytical Approaches to Target Analyses

3.1. Mass Spectrometry. Mass spectrometry is the preferred technique when the aim is to perform a quick analysis directly on the sample. The main difference between chromatography-MS methods is the sample introduction. MS instrumentation is assembled by ion sources, mass analyzer, and detector [13]. However, the challenge in forensic analysis is the possibility of decreasing the time and cost of analysis per sample. A primary analytical focus in toxicology is determining the presence or absence of drug metabolites in biological samples. In this sense, the use of ambient ionization technique mass spectrometry has allowed the analysis of the entire sample without an excessive preparation of sample. These techniques make possible the concept of open-air surface analysis directly under ambient conditions, being particularly useful for surface analysis of solids, avoiding many, if not all, sample preparation steps typically required [8].

HRMS is widely employed in the coupling with ambient mass spectrometry currently because of its capability of measuring accurate masses and differentiating among compounds with identical nominal masses, providing a comprehensive full-scan MS and MS/MS for the search for any analyte without sample pretreatment. This provides accurate m/z values that can be used to generate chemical formulas with high mass accuracy (<5 ppm mass error) [70]. Therefore, HRMS can be theoretically applied in different configurations with interchangeable ionization sources and sophisticated data acquisition capabilities, making HRMS one of the preferred techniques for the analysis of new drugs [71].

Ambient ionization mass spectrometry can be divided depending on the desorption technique, which will be discussed in the next section about the most used procedures in the analysis of drugs of abuse. Table 1 and Figure 2 present the advantages and disadvantages of these techniques or procedures.

3.1.1. Ambient Ionization Technique Mass Spectrometry: Desorption by Solid-Liquid Extraction. In these techniques, the desorption occurs by solid-liquid extraction followed by ESI-like ion-production mechanisms; the ionization can be performed by DESI (Figure 2(a)) or DAPPI (Figure 2(b)) [8].

DESI-MS is used in forensics due to its ability for in situ analysis. However, direct analyses that involve DESI and DAPPI are less common in toxicological analysis due to the interferences which can be caused by the suppression of ionization product of the matrix effects in the samples; therefore, adding an additional step in the preparation of the sample is often required. For instance, matrix-suppression effects were studied within direct analysis of benzodiazepines and opioids from 1 mg/mL of urine with DESI-MS and DAPPI-MS [21]. The authors found that the urine matrix affects the ionization mechanism of the opioids in DAPPI-MS and favors the proton transfer over charge exchange reaction. However, the sensitivity of the drugs in the solvent matrix was at the same level in DESI-MS and DAPPI-MS with limits of detection of 0.05–6 µg/mL, along with a decrease in sensitivity for the urine matrix that was higher with DESI (typically 20–160-fold) than with DAPPI (typically 2–15-fold), indicating better matrix tolerance in DAPPI over

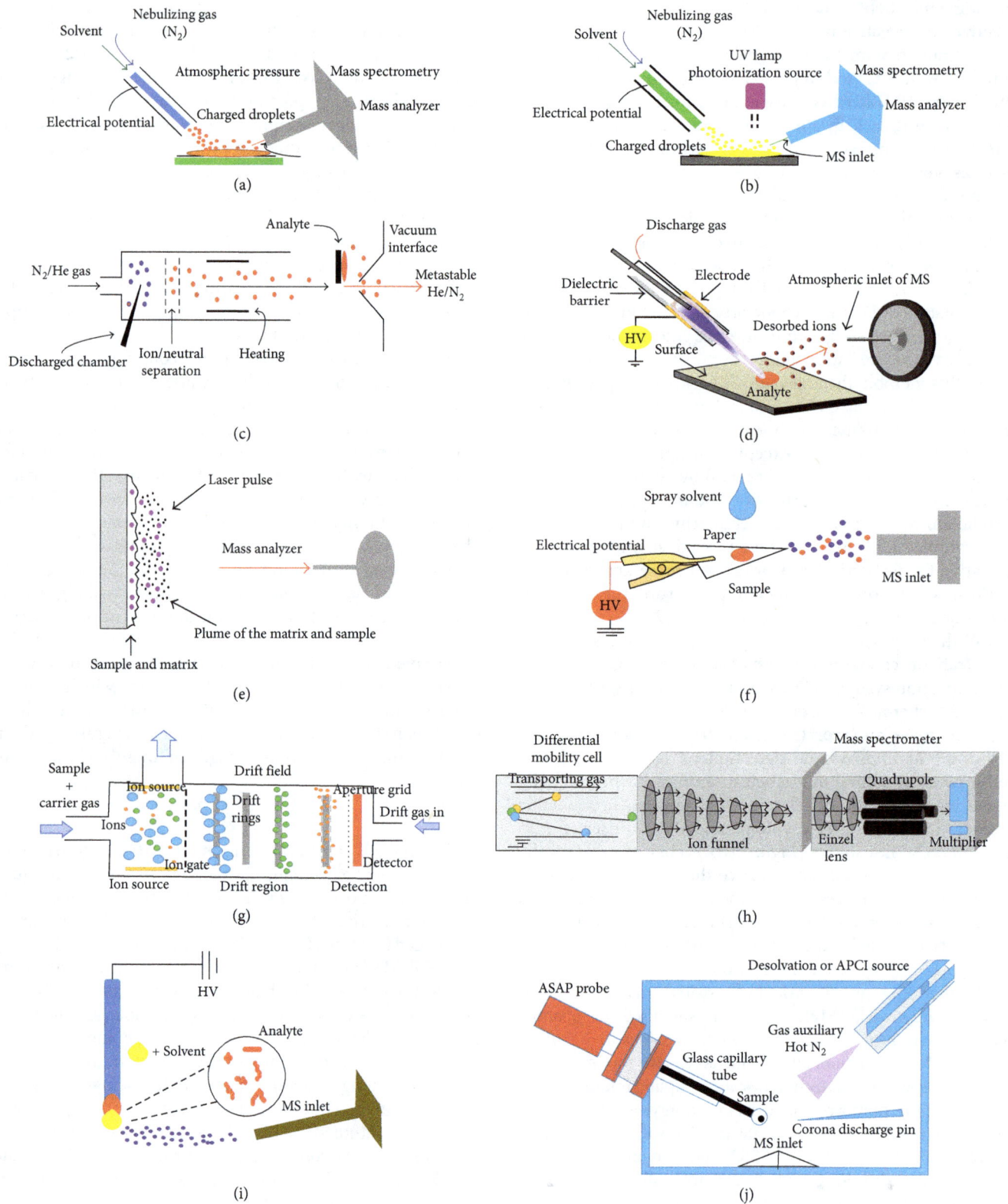

(a)

(b)

(c)

(d)

(e)

(f)

(g)

(h)

(i)

(j)

FIGURE 2: Continued.

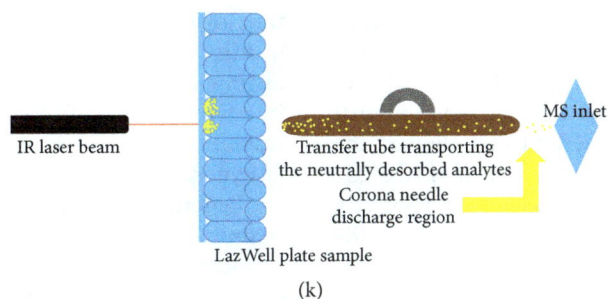

(k)

FIGURE 2: Main ambient ionization techniques used in toxicological forensic analysis. (a) DESI-MS, (b) DAPPI-MS, (c) DART-MS, (d) LTP-MS, (e) MALDI-MS, (f) PS-MS, (g) HPIMS, (h) IMS-MS, (i) TS, (j) ASAP-MS, and (k) LDTD-MS.

DESI. This illustrates that urine contains high concentrations of salts in DAPPI, and the salts in the urine samples are not efficiently evaporated from the sampling surface which do not significantly interfere with the ionization [21].

DESI-MS also allowed the analysis of common drugs in urine samples such as pethidine, diphenhydramine, nortriptyline, and methadone using pretreatment sample by liquid-phase microextraction (LPME). These selective extraction capabilities of three-phase LPME provided a significant reduction in the matrix effects observed in direct aqueous LPME extracts [22]. However, some drugs such as Δ^9-tetrahydrocannabinol and cannabidiol that have identical fragmentation spectra presented significant interference, resulting in the impossibility of an unequivocal identification of each other [22]. Additionally, DESI-MS/MS has been used coupled with solid-phase extraction (SPE) in order to analyze clenbuterol in urine specimens to detect doping; the authors mentioned that the suppression effects were minimized by SPE using DESI-MS/MS [23].

Moreover, DAPPI-MS has been used coupled with quadurople-ion trap MS and MS/MS mode to analyze directly herbal products such as *Catha edulis*, *Phycybe* mushoms, opium, designer drugs in tablets, confiscated drug samples of several forms as tablets, blotter paper, plant resin, and powder forms that contain meta-chlorophenylpiperazine, 3-fluoromethamphetamine, methylenedioxypyrovalerone, amphetamines, phenazepam, buprenorphine, and methylone. DAPPI-MS proved a specific analysis without sample preparation [24], showing that it is sufficient in most criminal cases where the main purpose is to do a qualitative screening [25].

3.1.2. Thermal or Chemical Sputtering Neutral Desorption. These techniques involve metastable and reactive ions, in which the species react with the analyte directly or indirectly through proton- and charge-transfer reactions [8]. In the analysis of drugs of abuse, DART (Figure 2(c)) and LTP (Figure 2(d)) can be used; the former is the most preferred in toxicology forensic analysis and uses a negatively biased point-to-plane atmospheric-pressure glow discharge at lower currents, physically separated from the ionization region by one or several electrodes. The metastable species are formed within the discharge supporting gas that typically is He or N_2, generating protonated water clusters. The main advantage of DART is the analysis of samples in solid, gas,

and liquid states, handling polar and nonpolar analytes with masses below 1 kDa [26].

Compounds are identified by combining information about elemental compositions from exact masses and isotopic abundances with fragment-ion mass spectra obtained by collisional activation. In toxicological forensic analysis, DART has been widely used in the detection of small drugs, but its quantitation remains a big problem due to its minor reproducibility, which depends on the position of the sample inside the ion source, making the number of drugs that can be quantified very limited [27]. However, when it is necessary to obtain more details in the identification matrix based on the natural products, high temperatures of ionizing gas can be used during the analysis, benefitting the resolution of more complex spectra [28].

DART achieved the detection of γ-hydroxy butyrate without any sample preparation or other illicit synthetic cannabinoid products coupling mass spectrometry DART-MS with CID analysis. The use of fragments obtained by CID provided a sensitive and specific detection, increasing the limit of detection to identify individual components and showing the ions related to each synthetic cannabinoid, since the $[M + H]^+$ precursor ions were still present in the mass spectra [29]. Thus, an unambiguous differentiation of each species could be accomplished. $[M + H]^+$ precursor ions could also be used as a complement in the analysis of drugs through screening in order to identify new and unknown drugs.

DART-TOFMS detected alprazolam, which is one of the ingredients of the "Houston Cocktail," containing hydrocodone/acetaminophen, and achieved an analysis with high mass accuracy [30]. Habala et al. identified six synthetic cannabinoids in methanolic extracts from solid herbal material using a DART source coupled with a hybrid ion trap—LTQ ORBITRAP—mass analyzer, discovering that the leaves have a greater concentration than the stems of the plant material [11].

In the identification of new psychoactive substances (NPS), DART-MS has had an important role. Gwak and Almirall [4] performed a screening of 35 NPS in urine using DART coupled with hybrid TOFMS and ion mobility spectrometry (IMS), identifying synthetic cathinones with a single phenethylamine as the most common compounds. The analytes detected had an error within ±5 ppm, but isomeric compounds could not be differentiated. Similarly,

DART-TOFMS was used in order to detect synthetic cannabinoid in botanical matrices like *Coriandrum sativum*, *Ocimum basilicum*, and *Mentha spicata* [28]. In this research, intensive sample preparation was not required, just methanol dissolution, which is a method that allowed the identification of the synthetic cannabinoids such as AM-251 and JWH-015. Although botanical samples exhibit relatively complex mass spectral profiles, this did not prohibit the identification of the target compounds. Additionally, the DART-TOFMS analyses were conducted with different ionizing gas (helium) temperatures in order to determine the optimum desorption temperature, and it was observed that higher temperatures had the additional benefit of yielding more complex spectra that could permit a more detailed identification of the plant matrix based on the natural products [28].

Grange and Sovocool [31] developed a methodology for the extraction and clean-up of drugs in smoke deposited on household surfaces so as to determine the exposure of the patients to drugs of abuse using DART-TOFMS. A field sample carrier and an auto sampler were used to minimize the time per analysis. The sampling was performed just with cotton swab wipes with isopropanol, finding a quantification of each drug of around $0.025 \mu g/100 \, cm^2$. However, Δ^9-tetrahydrocannabinol and nicotine had m/z 315 and m/z 163 interferences, respectively. The authors found that this interference could be a sugar unit from the cellulose of the cotton-swabs. In spite of this interference, the method is highly recommended for the analysis of residues in clandestine drug laboratories [31].

Moreover, phenethylamine, a synthetic drug known by its effects similar to LSD and its sublingual consumption via blotter paper, was analyzed directly in the sample by DART-TOFMS. This was studied in blotter paper street samples, and the results can be used in preliminary identifications, since this technique is extremely fast and advantageous for the quick screening of unknown street samples in crime laboratories [32]. Poklis et al. used DART coupled with HRMS in the analysis of legal purchases on the Internet under the name "Raving Dragon Novelty Bath Salts and Raving Dragon Voodoo Dust" and found out that they contain methylone and pentedrone, respectively, which can be identified as unsupervised drug market [33].

More recently, DART has been developed using SPME-fiber format for coupling nanogold surfaces with mass spectrometry in order to perform an effective drug capture in toxicological matrices like methamphetamine, diazepan, and alprozolam in human plasma. The authors used LC-MS/MS and DART-MS/MS, in this case, coupling antibodies to nanogold-coated wires. An antibody with cross reactivity to multiple drugs was used for simultaneous extraction of a mixture of drugs. The immunoaffinity nanogold is known by its possibility of eliminating chemical noise. The limits of detection achieved with DART-MS/MS were comparable to those observed with LC-MS/MS [34].

Different mass analyzers have been used to evaluate sensitivity and selectivity in the detection of Δ^9-tetrahydrocannabinol (THC) from intact hair samples using DART [35]. The mass analyzers evaluated were an Orbitrap, a quadrupole-Orbitrap, a triple quadrupole, and a quadrupole time-of-flight (QTOF). The authors found that only the quadrupole-Orbitrap in high-resolution mode achievement distinguished THC in hair samples from endogenous isobaric interferences [35]. Those are important data since when the resolution in the mass analyzers is low, the risk of obtaining false/positive is high.

Different from DART, LTP was developed for direct sampling ionization in chemical analysis using mass spectrometry. The plasma here is generated by dielectric barrier discharge (DBD) and a discharge of gas at low flow rate ($<500 \, mL/min$), and a high-voltage to sustain the plasma in an ambient environment [8]. This technique has proved a powerful tool in direct analyses, exclusively with small organic molecules with low to moderate polarity. For this reason, it is not commonly used in the analysis of illicit high molecular weight drugs as it limits the analysis of unknown drugs. However, LTP proved effective in the analysis of stomach fluid content of a diseased dog suspected to have died from ingestion of insecticide. Direct sampling ionization was applied in MS analysis and protonated Terbufos, and Terbufos sulfoxide were observed [72]. These two compounds are common in Terbufos-based insecticides, which were suspected to be the cause of the death of the dog.

Furthermore, the analysis of drugs of abuse in urine and 25 mg of hair extract samples were systematically investigated, where several drugs such as amphetamine, benzoylecgonine, caffeine, cannabidiol, cocaine, codeine, diazepam, ephedrine hydrochloride, heroin, ketamine, methadone, methamphetamine, morphine, and Δ^9-tetrahydrocannabinol were identified obtaining a limit of detection of around 10 ng/mL without any sample preparation [36].

3.1.3. Laser Desorption/Ablation. In these techniques, the analytes are desorbed or ablated from a surface by an IR or UV laser with or without a matrix (Figure 2(e)). The sample is subsequently merged with an electrospray droplet cloud or a plasma stream, depending on the ionization source used [8]. When the source excites an exogenous matrix that cocrystallizes and has energy absorbent capabilities, it can coat the sample surface to be analyzed. Then, a laser adds excess energy to the matrix-sample complex, where the matrix absorbs laser energy to pass it to the sample and, finally, to produce ions from analytes; this technique is called matrix-assisted laser desorption electrospray ionization (MALDESI) [73]. In contrast to MALDESI, metal-assisted secondary ion mass spectrometry (MetA-SIMS) procedure adds small amounts of metals onto sample surface to enhance mass spectra analysis [42]. These two methods may provide an image coupled with ionization mass spectrometric imaging (MSI), which is a powerful technique to obtain spatial information (distribution) of compound mass spectra.

Porta et al. used MALDESI coupled to MSI in order to monitor the distribution of cocaine and its metabolites in 12 mL of extracts of intact single hair samples from chronic users. The acquisitions were performed applying rastering mode in the SRM mode on a MALDI triple quadrupole linear-fitted ion trap. The time of analysis of an intact single hair sample of 6 cm was of 6 min approximately. Cocaine and its metabolites were identified and quantified, and the results were obtained with a limit of detection of 5 ng/mL, becoming an excellent methodology to detect cocaine consumption [37]. In the same manner, matrix-assisted laser desorption/ionization-mass spectrometry imaging (MALDI-IMS) was used to rapidly screen longitudinally sectioned drug user hair samples for cocaine and its metabolites; using continuous raster imaging, the optimization of the spatial resolution and raster speed were performed on cocaine-contaminated intact hair samples. Besides, the MALDI-MS/MS images showed the distribution of the most abundant cocaine, using *product ion scan* as a mode of analyzing. With this method, it is possible to obtain mass spectra with the main fragment of the molecule target.

An SRM experiment was also performed using the "dynamic pixel" imaging method to screen for cocaine and a range of its metabolites, in order to differentiate between contaminated hairs and drug users. Therefore, these methods are important when the imaging information on drug distribution is necessary, for example, in human hair without extensive sample preparation, or when labelling techniques are required. However, it should be noted that it only provides qualitative data about administered drugs, through a pixelated representation [38]. MALDESI has also been coupled with HRMS during the identification of 74 drug samples which were detected using the ionic liquid matrix N,N diisopropylethylammonium α-cyanohydroxycinnamate. This method allowed the identification of new designer drugs, which come from the use of safrole as a precursor for the synthesis. Nevertheless, the result obtained presents weaker resolution and lower sensitivities, leading to lower peak intensities. The authors affirm that this limitation is a consequence of the matrix, since this can be related to the formation of adducts, but the matrix may be enhanced by adding specific cations and anions. Further investigations to improve ionization through matrix additives are still necessary in the field. Another limitation of this methodology is the impossibility to distinguish drug position isomers, such as methamphetamine and 4-methyl-amphetamine, as well as structural elucidation of unknown compounds. The authors recommend the combination between this methodology and bioinformatics software tools which provide untargeted compound searches, even if respective HRMS spectra are not included in a library just based on the precursor ion fingerprinting [39].

On the contrary, MALDI-MSI and MALDI-Fourier transform ion cyclotron resonance (FTICR-MS) have also been used for mapping and direct detection of methamphetamine in longitudinal sections of the single hair sample in positive mode, in which umbelliferone was used as a matrix. This matrix has the advantage of being hydrophobic and capable of assisting in the ionization of methamphetamine in hair. The authors observed that the detection and sensitivity provided by this matrix is higher than α-cyano-4-hydroxycinnamic acid (CHCA) or 2,5-dihydroxybenzoic acid (DHB). In addition, the distribution semi-quantitative of methamphetamine can be performed.

This method enhances the detection and sensitivity of target drugs embedded in a hair matrix, achieving a detection level down to nanogram per milligram; for this reason, the authors compared the results with the obtained by LC-MS/MS, but in this case with less sample amount required [40].

More recently, Kernalléguen et al. [41] have made possible the semiquantification of cocaine and its metabolites (benzoylecgonine, cocaethylene, and ecgonine methyl ester) in hair, using microarrays for MS and MALDI-MS/MS. So far, it is well known that the inhomogeneous MALDI matrix crystallization and laser shot-to-shoot variability make the quantitation more difficult; therefore, the authors used a high-throughput MALDI method, along with an innovative high-density microarray for mass spectrometry (MAMS) technology. This technology consists of a sample preparation slide containing lanes of hydrophilic spots, and an automated slider which drags a sample droplet over several small spots, with the purpose of achieving homogeneous crystallization of the matrix-analyte mixture and, therefore, to a reproducible signal. In this manner, it was possible to establish a calendar of consumption in only 1 mg of hair with a great correlation, becoming an excellent methodology when urgent results are required [41].

However, metal MetA-SIMS was used to determine the differentiation between systemic exposure and external contamination that remains in the hair because of exposure to drugs after following the protocols of decontamination (hair wash) [43]. The authors reached a comparison of the results among MetA-SIMS, MALDESI-MS, and LC-MS/MS, showing that there is still cocaine detected after the washes of decontamination, using MetA-SIMS. MALDESI was in turn inefficient for forensic hair analysis since no cocaine was detected after decontaminating the samples. LC-MS/MS detected 5 ng/10 mg in the sample after the washing. Finally, the authors concluded that the washing protocols are not reliable, because external cocaine can migrate into the hair, and recommended a simple analysis of images which makes the evaluation of the differences among hair samples contaminated externally and the interpretation of the correct results easier [43].

3.1.4. Other Methods of Ionization. Paper spray (PS) technique was introduced in 2009 and has been used in the development of a wide range of quantitative and qualitative applications. Here, the sample is deposited in the paper with a sharp point, and ions are produced by voltage applied, while the substrate is held by a metal clip in the paper and placed in the front of the inlet of a mass spectrometer. Then, the front mass spectrometer performs the detection after the sample elution, which can be carried out in the same manner of paper chromatography, but with a direct sample injection to the mass spectrometer (Figure 2(f)). In this technique, a wide range of chemicals can be ionized by paper spray, from small molecules to large biomolecules [8].

Paper spray ionization coupled to high-resolution tandem mass spectrometry (PSI-HR-MS/MS) have also been used in order to validate a screening of drugs in urine such as codeine-6-glucuronide, diclofenac, among others, and to validate a comprehensive urine screening. Nevertheless, the procedure showed high matrix effects for most drugs, but also acceptable limits of identification that have the potential of reducing workload. However, the authors recommend its implementation as a promising alternative to conventional procedures, but they warned that there is a risk of false positive/negative results caused by mixed spectra during the detection of low concentrations. Therefore, some problems should be solved before implementing it in routine analysis [44].

Thin-layer chromatography (TLC) has been also used as an introductory sampling method combined with PS-MS to analyze cocaine and its adulterants in $10\,\mu$L of sample. This analysis obtained promising results in which the limit of detection was reduced five thousand times ($1.0\,\mu$g/mL), showing an $R^2 > 0.999$ that is another indicator of the reliability of this technique, and the possibility to be implemented in routine analyses [45]. In the same manner, simultaneous analyses of methamphetamines, cocaine, morphine, and Δ^9-tetrahydrocannabinol were performed in a single blood spot by PS-MS in only 2 minutes, with minimal sample preparation through the extraction of the compounds by solvents [46].

PS-MS has also been used in positive ionization mode to obtain chemical profiles of illicit drugs such as blotter papers containing extracts and leaves of natural cannabinoids and synthetic cannabinoids; here, 1 mg of blotter paper was used as the PS ionization source. For this reason, the authors recommend to be careful with the low sensitivity of this technique that was observed to possibly occur due to an ionic suppression process, caused by the matrix effect (extracted impurities from the surface of the blotter paper). The results provided a limit of detection of around 0.17 ppb [47].

PS-MS/MS has been used in targeted drug screening using an Orbitrap QMS, one in positive mode and the other in negative mode. In the positive ion mode, over 130 drugs and drug metabolites in postmortem samples were semi-quantitatively determined, proving an adequate method in postmortem analysis. In the analysis in negative mode, an ion-screening method was also developed for a small panel of barbiturates and structural analogs. This method showed good qualitative agreement with LC–MS-MS; the true positive rate of paper spray MS/MS was 92%, and the true negative rate was over 98%. This result shows that this technique possesses the necessary potential for acidic drug detection and screening without sample preparation; however, the authors did not present a list of possible interferences during the analysis [48].

Most recently, PS-MS has been used modifying the paper through molecularly imprinted polymers (MIP) to create a specific site for cocaine analysis in 1 mL of the oral fluid. In this case, the PS was set by holding the membrane connected directly to the outlet probe of the ESI with a 0.5 mm wire using an alligator-type clip and applying a voltage of 4 V, obtaining an LOQ of 100 ng/mL, and becoming a promising method to analyze cocaine [49].

High-performance ion mobility spectrometry (HPIMS) has been used along with electrospray ionization to detect codeine and morphine in urine samples without extra sample pretreatment (Figure 2(g)). However, issues of charge suppression in the presence of drug mixtures interfering with matrix components were observed, so the authors recommended considering some previous steps before sample preparation. For instance, the authors introduced a sample into a drift tube via pulse Bradbury–Neilson ion gate and operated it in positive mode, and the ions passed to desolvation to be separated [50]. This method achieved a resolving power double than the currently accepted method without an excessive necessity of sample preparation [51].

Ion mobility-based separation methods can be combined with mass spectrometry (IMS-MS) in order to minimize chemical suppression caused by interference and the use of chromatography separations to targeted applications (Figure 2(h)). The interface has only a few centimeters in length and operates in seconds; besides, it can be adapted to any MS system using atmospheric-pressure ionization-targeted applications. In this analysis, a miniature differential ion mobility filter is used and placed in front of the entrance of the mass spectrometer, and a solution of 10 ng/mL of the sample was introduced using infusion introduction of ions created by electrospray ionization source coupled with ion trap MS/MS. This method allowed the characterization of samples in 30 seconds, reducing case backlogs in the targeted analysis of analytes of interest, showing the range of quantification of around 0.01–10 ng/uL of cocaine [52].

Recently, a new method has been developed coupling microfluidics with a miniature mass spectrometer in order to quantify cocaine in urine samples. This method is able to deliver droplets of solvents to dried urine samples, separating droplets of $80\,\mu$L of extracts, then performing splits from the hydrophilic dried urine zones and driving them to the destination electrode for analysis. The LOQ for cocaine was 40 ng/mL [74].

Another recent method of direct analysis is touch spray (TS). In this technique, the sample is transferred to a substrate with subsequent ionization; in this manner, the substrate can serve both as the means for the sample collection, ionization, and as straightforward handling analysis of either solid or liquid samples without pretreatment (Figure 2(i)). Using TS-MS coupled with MS/MS, drugs of abuse like Δ^9-tetrahydrocannabinol and buprenorphine were identified in spiked oral fluid using medical swabs directly, providing limits of detection of around 50 ng/mL, which are sought by international forensic and toxicological societies. This adaptation of medical swabs for TS-MS analysis allows noninvasive and direct sampling of neat oral fluids; however, the authors affirm that the drying step represents the most time-consuming part of the analytical protocol, but the potential of the technique is high in terms of specificity, selectivity, and sensitivity [9].

More recently, laser diode thermal desorption (LDTD) and atmospheric solids analysis probe (ASAP) have been coupled with HRMS using APCI ionization in order to generate high-quality data from multiple samples with none or minimal sample preparation, with the purpose of identifying synthetic cannabinoids/cathinones through full-MS and MS/MS experiments. In ASAP, a melting-point capillary tube is used to introduce the sample into a stream of heated nitrogen gas, which results in the sample being desorbed from the capillary [53], and the desorbed sample is then ionized by a corona discharge needle. During ASAP-MS analysis, it was possible to examine solid and liquid samples transferred to the capillary surface (Figure 2(j)); whereas in the LDTD-MS analysis, the samples were extracted by a solvent. This method uses a specially designed 96-well plate with stainless alloy steel inserts, where the sample is thermally desorbed from the stainless steel by an infrared laser which forms neutral gas-phase molecules [54] (Figure 2(k)). These gas-phase molecules are carried into the mass spectrometer inlet by compressed air. Before they enter the mass spectrometer inlet, a corona discharge needle ionizes the neutral molecules.

This LDTD-APCI-MS method results in a completely automated analysis with low sampling times. The authors recommended the use of both methods of ambient ionization, which allow rapid experiments from a single sample introduction. However, when performing the optimization, they verified that the simplicity of ASAP design allows it to be easily switched between API techniques and possible positive/negative switching for a single sample introduction, which provides many possibilities of optimization during the analysis. More studies in this field are required, especially in possible interference of suppression of ionization [75].

4. Conclusions

Mass spectrometry is the most important technique used in toxicological forensic analysis. MS coupled with chromatography are the preferred techniques to identify new drugs or metabolites through screening analysis, providing excellent results in limit of detection, precision, accuracy, and sensitivity, although it may be a time-consuming process. Direct techniques with MS (with less sample preparation) are more likely to be used in target analysis or in routine qualitative analysis. However, sample complexity complicates the identification among compounds with similar fragmentation patterns, along with the problems caused by ionization chemical suppression. As a result, recent developments in MS are concerned with the necessity of creating new software in order to help improve simplicity and robustness in the identification of drugs. There is a growing necessity to develop more innovative methodologies to reduce time consumption in the analyses, enhance sensitivity, and finally move forward towards greener chemistry.

Conflicts of Interest

The authors declare that there are no conflicts of interest regarding the publication of this paper.

References

[1] C. Moore, L. Marinetti, C. Coulter, and K. Crompton, "Analysis of pain management drugs, specifically fentanyl, in hair: application to forensic specimens," *Forensic Science International*, vol. 176, no. 1, pp. 47–50, 2008.

[2] O. Beck, "Exhaled breath for drugs of abuse testing—evaluation in criminal justice settings," *Science and Justice*, vol. 54, no. 1, pp. 57–60, 2014.

[3] H. H. Lee, J. F. Lee, S. Y. Lin, and B. H. Chen, "Simultaneous identification of abused drugs, benzodiazepines, and new psychoactive substances in urine by liquid chromatography tandem mass spectrometry," *Kaohsiung Journal of Medical Sciences*, vol. 32, no. 3, pp. 118–127, 2016.

[4] S. Gwak and J. R. Almirall, "Rapid screening of 35 new psychoactive substances by ion mobility spectrometry (IMS) and direct analysis in real time (DART) coupled to quadrupole time-of-flight mass spectrometry (QTOF-MS)," *Drug Testing and Analysis*, vol. 7, no. 10, pp. 884–893, 2015.

[5] C. Poole, *Gas Chromatography*, Elsevier, New York, NY, USA, 1st edition, 2012.

[6] D. Cappelle, H. Neels, M. Yegles et al., "Gas chromatographic determination of ethyl glucuronide in hair: Comparison between tandem mass spectrometry and single quadrupole mass spectrometry," *Forensic Science International*, vol. 249, pp. 20–24, 2015.

[7] M. Chèze, A. Lenoan, M. Deveaux, and G. Pépin, "Determination of ibogaine and noribogaine in biological fluids and hair by LC-MS/MS after Tabernanthe iboga abuse. Iboga alkaloids distribution in a drowning death case," *Forensic Science International*, vol. 176, no. 1, pp. 58–66, 2008.

[8] M. Domin and R. Cody, *Ambient Ionization Mass Spectrometry*, Royal Society of Chemistry, Cambridge, UK, 2014.

[9] V. Pirro, A. K. Jarmusch, M. Vincenti, and R. G. Cooks, "Direct drug analysis from oral fluid using medical swab touch spray mass spectrometry," *Analytica Chimica Acta*, vol. 861, pp. 47–54, 2015.

[10] H. Wang, J. Liu, R. G. Cooks, and Z. Ouyang, "Paper spray for direct analysis of complex mixtures using mass spectrometry," *Angewandte Chemie*, vol. 122, no. 5, pp. 889–892, 2010.

[11] L. Habala, J. Valentová, I. Pechová, M. Fuknová, and F. Devínsky, "DART–LTQ ORBITRAP as an expedient tool for the identification of synthetic cannabinoids," *Legal Medicine*, vol. 20, pp. 27–31, 2016.

[12] M. Carson and S. Kerrigan, "Quantification of suvorexant in urine using gas chromatography/mass spectrometry," *Journal of Chromatography B*, vol. 1040, pp. 289–294, 2017.

[13] J. Greaves and J. Roboz, *Mass Spectrometry for the Novice*, CRC Press, Boca Raton, FL, USA, 2008.

[14] Y.-H. Wu, K. Lin, S.-C. Chen, and Y.-Z. Chang, "Integration of GC/EI-MS and GC/NCI-MS for simultaneous quantitative determination of opiates, amphetamines, MDMA, ketamine, and metabolites in human hair," *Journal of Chromatography B*, vol. 870, no. 2, pp. 192–202, 2008.

[15] A. Amirav, A. Gordin, M. Poliak, and A. B. Fialkov, "Gas chromatography-mass spectrometry with supersonic molecular beams," *Journal of Mass Spectrometry*, vol. 43, no. 2, pp. 141–163, 2008.

[16] T. Alon and A. Amirav, "How enhanced molecular ions in Cold EI improve compound identification by the NIST library," *Rapid Communications in Mass Spectrometry*, vol. 29, no. 23, pp. 2287–2292, 2015.

[17] M. Conti, V. Tazzari, M. Bertona, M. Brambilla, and P. Brambilla, "Surface-activated chemical ionization combined with electrospray ionization and mass spectrometry for the analysis of cannabinoids in biological samples. Part I: analysis of 11-nor-9-carboxytetrahydro-cannabinol," *Rapid Communications in Mass Spectrometry*, vol. 25, no. 11, pp. 1552–1558, 2011.

[18] J. Liang, W.-y. Wu, G.-x. Sun et al., "A dynamic multiple reaction monitoring method for the multiple components quantification of complex traditional Chinese medicine preparations: Niuhuang Shangqing pill as an example," *Journal of Chromatography A*, vol. 1294, pp. 58–69, 2013.

[19] M. C. Sampedro, N. Unceta, A. Gómez-Caballero et al., "Screening and quantification of antipsychotic drugs in human brain tissue by liquid chromatography-tandem mass spectrometry: application to postmortem diagnostics of forensic interest," *Forensic Science International*, vol. 219, no. 1–3, pp. 172–178, 2012.

[20] I. Shah, A. Petroczi, M. Uvacsek, M. Ránky, and D. P. Naughton, "Hair-based rapid analyses for multiple drugs in forensics and doping: application of dynamic multiple reaction monitoring with LC-MS/MS," *Chemistry Central Journal*, vol. 8, no. 1, p. 73, 2014.

[21] N. M. Suni, P. Lindfors, O. Laine et al., "Matrix effect in the analysis of drugs of abuse from urine with desorption atmospheric pressure photoionization-mass spectrometry (DAPPI-MS) and desorption electrospray ionization-mass spectrometry (DESI-MS)," *Analytica Chimica Acta*, vol. 699, no. 1, pp. 73–80, 2011.

[22] J. Thunig, L. Flø, S. Pedersen-Bjergaard, S. H. Hansen, and C. Janfelt, "Liquid-phase microextraction and desorption electrospray ionization mass spectrometry for identification and quantification of basic drugs in human urine," *Rapid Communications in Mass Spectrometry*, vol. 26, no. 2, pp. 133–140, 2012.

[23] Z. Lin, S. Zhang, M. Zhao, C. Yang, D. Chen, and X. Zhang, "Rapid screening of clenbuterol in urine samples by desorption electrospray ionization tandem mass spectrometry," *Rapid Communications in Mass Spectrometry*, vol. 22, no. 12, pp. 1882–1888, 2008.

[24] T. J. Kauppila, A. Flink, M. Haapala et al., "Desorption atmospheric pressure photoionization–mass spectrometry in routine analysis of confiscated drugs," *Forensic Science International*, vol. 210, no. 1–3, pp. 206–212, 2011.

[25] T. J. Kauppila, V. Arvola, M. Haapala et al., "Direct analysis of illicit drugs by desorption atmospheric pressure photoionization," *Rapid Communications in Mass Spectrometry*, vol. 22, no. 7, pp. 979–985, 2008.

[26] R. B. Cody and J. A. Larame, "Versatile new ion source for the analysis of materials in open air under ambient conditions," *Analytical Chemistry*, vol. 77, no. 8, pp. 2297–2302, 2005.

[27] E. S. Chernetsova and G. E. Morlock, "Determination of drugs and drug-like compounds in different samples with direct analysis in real time mass spectrometry," *Mass Spectrometry Reviews*, vol. 35, no. 5, pp. 875–883, 2011.

[28] R. A. Musah, M. A. Domin, M. A. Walling, and J. R. E. Shepard, "Rapid identification of synthetic cannabinoids in herbal samples via direct analysis in real time mass spectrometry," *Rapid Communications in Mass Spectrometry*, vol. 26, no. 9, pp. 1109–1114, 2012.

[29] R. A. Musah, M. A. Domin, R. B. Cody, A. D. Lesiak, A. John Dane, and J. R. E. Shepard, "Direct analysis in real time mass spectrometry with collision-induced dissociation for structural analysis of synthetic cannabinoids," *Rapid Communications in Mass Spectrometry*, vol. 26, no. 19, pp. 2335–2342, 2012.

[30] W. C. Samms, Y. J. Jiang, M. D. Dixon, S. S. Houck, and A. Mozayani, "Analysis of alprazolam by DART-TOF mass spectrometry in counterfeit and routine drug identification cases," *Journal of Forensic Sciences*, vol. 56, no. 4, pp. 993–998, 2011.

[31] A. H. Grange and G. W. Sovocool, "Detection of illicit drugs on surfaces using direct analysis in real time (DART) time-of-flight mass spectrometry," *Rapid Communications in Mass Spectrometry*, vol. 25, no. 9, pp. 1271–1281, 2011.

[32] M. K. McGonigal, J. A. Wilhide, P. B. Smith, N. M. Elliott, and F. L. Dorman, "Analysis of synthetic phenethylamine street drugs using direct sample analysis coupled to accurate mass time of flight mass spectrometry," *Forensic Science International*, vol. 275, pp. 83–89, 2017.

[33] J. L. Poklis, C. E. Wolf, O. I. ElJordi, K. Liu, S. Zhang, and A. Poklis, "Analysis of the first- and second-generation raving dragon novelty bath salts containing methylone and pentedrone," *Journal of Forensic Sciences*, vol. 60, pp. S234–S240, 2015.

[34] K. M. Evans-Nguyen, T. L. Hargraves, and A. N. Quinto, "Immunoaffinity nanogold coupled with direct analysis in real time (DART) mass spectrometry for analytical toxicology," *Analytical Methods*, vol. 9, no. 34, pp. 4954–4957, 2017.

[35] W. F. Duvivier, T. A. van Beek, and M. W. F. Nielen, "Critical comparison of mass analyzers for forensic hair analysis by ambient ionization mass spectrometry," *Rapid Communications in Mass Spectrometry*, vol. 30, no. 21, pp. 2331–2340, 2016.

[36] A. U. Jackson, J. F. Garcia-Reyes, J. D. Harper et al., "Analysis of drugs of abuse in biofluids by low temperature plasma (LTP) ionization mass spectrometry," *Analyst*, vol. 135, no. 5, p. 927, 2010.

[37] T. Porta, C. Grivet, T. Kraemer, E. Varesio, and G. Hopfgartner, "Single hair cocaine consumption monitoring by mass spectrometric imaging," *Analytical Chemistry*, vol. 83, no. 11, pp. 4266–4272, 2011.

[38] B. Flinders, E. Beasley, R. M. Verlaan et al., "Optimization of sample preparation and instrumental parameters for the rapid analysis of drugs of abuse in hair samples by MALDI-MS/MS imaging," *Journal of The American Society for Mass Spectrometry*, vol. 28, no. 11, pp. 2462–2468, 2017.

[39] K. M. Ostermann, A. Luf, N. M. Lutsch et al., "MALDI orbitrap mass spectrometry for fast and simplified analysis of novel street and designer drugs," *Clinica Chimica Acta*, vol. 433, pp. 254–258, 2014.

[40] H. Wang and Y. Wang, "Matrix-assisted laser desorption/ionization mass spectrometric imaging for the rapid segmental analysis of methamphetamine in a single hair using umbelliferone as a matrix," *Analytica Chimica Acta*, vol. 975, pp. 42–51, 2017.

[41] A. Kernalléguen, R. Steinhoff, S. Bachler et al., "High-throughput monitoring of cocaine and its metabolites in hair using microarrays for mass spectrometry and matrix-assisted laser desorption/ionization-tandem mass spectrometry," *Analytical Chemistry*, vol. 90, no. 3, pp. 2302–2309, 2018.

[42] E. Cuypers et al., "Article a closer look into the consequences of decontamination procedures in forensic hair analysis using MetA-SIMS analysis," *Analytical Chemistry*, 2016.

[43] E. Cuypers, B. Flinders, C. M. Boone et al., "Consequences of decontamination procedures in forensic hair analysis using metal-assisted secondary ion mass spectrometry analysis," *Analytical Chemistry*, vol. 88, no. 6, pp. 3091–3097, 2016.

[44] J. A. Michely, M. R. Meyer, and H. H. Maurer, "Paper spray ionization coupled to high resolution tandem mass spectrometry for comprehensive urine drug testing in comparison to liquid chromatography-coupled techniques after urine precipitation or dried urine spot workup," *Analytical Chemistry*, vol. 89, no. 21, pp. 11779–11786, 2017.

[45] T. C. De Carvalho, F. Tosato, L. M. Souza et al., "Thin layer chromatography coupled to paper spray ionization mass spectrometry for cocaine and its adulterants analysis," *Forensic Science International*, vol. 262, pp. 56–65, 2016.

[46] R. D. Espy, S. F. Teunissen, N. E. Manicke et al., "Paper spray and extraction spray mass spectrometry for the direct and simultaneous quantification of eight drugs of abuse in whole blood," *Analytical Chemistry*, vol. 86, no. 15, pp. 7712–7718, 2014.

[47] E. Domingos, T. C. de Carvalho, I. Pereira et al., "Paper spray ionization mass spectrometry applied to forensic chemistry–drugs of abuse, inks and questioned documents," *Analytical Methods*, vol. 9, no. 30, pp. 4400–4409, 2017.

[48] J. McKenna, R. Jett, K. Shanks, and N. E. Manicke, "Toxicological drug screening using paper spray high-resolution tandem mass spectrometry (HR-MS/MS)," *Journal of Analytical Toxicology*, vol. 42, no. 5, pp. 300–310, 2018.

[49] L. S. Tavares, T. C. Carvalho, W. Romão, B. G. Vaz, and A. R. Chaves, "Paper spray tandem mass spectrometry based on molecularly imprinted polymer substrate for cocaine analysis in oral fluid," *Journal of The American Society for Mass Spectrometry*, vol. 29, no. 3, pp. 566–572, 2017.

[50] A. J. Midey, A. Patel, C. Moraff, C. A. Krueger, and C. Wu, "Improved detection of drugs of abuse using high-performance ion mobility spectrometry with electrospray ionization (ESI-HPIMS) for urine matrices," *Talanta*, vol. 116, pp. 77–83, 2013.

[51] T. Gabowitcz, D. Ridjosic, and S. Nacson, *Ion Mobility Spectrometer Having Improved Sample Receiving Device, US 2008/0101995 A1*, 2008.

[52] A. B. Hall, S. L. Coy, E. G. Nazarov, and P. Vouros, "Rapid separation and characterization of cocaine and cocaine cutting agents by differential mobility spectrometry-mass spectrometry," *Journal of Forensic Sciences*, vol. 57, no. 3, pp. 750–756, 2012.

[53] E. Jagerdeo, J. A. Clark, J. N. Leibowitz, and L. J. Reda, "Rapid analysis of forensic samples using an atmospheric solid analysis probe interfaced to a linear ion trap mass spectrometer," *Rapid Communications in Mass Spectrometry*, vol. 29, no. 2, pp. 205–212, 2015.

[54] J. Wu, C. S. Hughes, P. Picard et al., "High-throughput cytochrome P450 inhibition assays using laser diode thermal desorption-atmospheric pressure chemical ionization-tandem mass spectrometry," *Analytical Chemistry*, vol. 79, no. 12, pp. 4657–4665, 2007.

[55] A. Amirav, "Fast heroin and cocaine analysis by GC–MS with cold EI: the important role of flow programming," *Chromatographia*, vol. 80, no. 2, pp. 295–300, 2017.

[56] M. K. Woźniak, M. Wiergowski, J. Aszyk, P. Kubica, J. Namieśnik, and M. Biziuk, "Application of gas chromatography–tandem mass spectrometry for the determination of amphetamine-type stimulants in blood and urine," *Journal of Pharmaceutical and Biomedical Analysis*, vol. 148, pp. 58–64, 2018.

[57] F. Versace, F. Sporkert, P. Mangin, and C. Staub, "Rapid sample pre-treatment prior to GC–MS and GC–MS/MS urinary toxicological screening," *Talanta*, vol. 101, pp. 299–306, 2012.

[58] E. S. Emídio, V. de Menezes Prata, and H. S. Dórea, "Validation of an analytical method for analysis of cannabinoids in hair by headspace solid-phase microextraction and gas chromatography–ion trap tandem mass spectrometry," *Analytica Chimica Acta*, vol. 670, no. 1-2, pp. 63–71, 2010.

[59] K. Zaitsu, H. Miyagawa, Y. Sakamoto et al., "Mass spectrometric differentiation of the isomers of mono-methoxyethylamphetamines and mono-methoxydimethylamphetamines by GC–EI–MS–MS," *Forensic Toxicology*, vol. 31, no. 2, pp. 292–300, 2013.

[60] R. Andrews and S. Paterson, "A validated method for the analysis of cannabinoids in post-mortem blood using liquid–liquid extraction and two-dimensional gas chromatography-mass spectrometry," *Forensic Science International*, vol. 222, no. 1-3, pp. 111–117, 2012.

[61] G. Milman, A. J. Barnes, R. H. Lowe, and M. A. Huestis, "Simultaneous quantification of cannabinoids and metabolites in oral fluid by two-dimensional gas chromatography mass spectrometry," *Journal of Chromatography A*, vol. 1217, no. 9, pp. 1513–1521, 2010.

[62] B. Guthery, T. Bassindale, A. Bassindale, C. T. Pillinger, and G. H. Morgan, "Qualitative drug analysis of hair extracts by comprehensive two-dimensional gas chromatography/time-of-flight mass spectrometry," *Journal of Chromatography A*, vol. 1217, no. 26, pp. 4402–4410, 2010.

[63] A. Salomone, E. Gerace, F. D'Urso, D. Di Corcia, and M. Vincenti, "Simultaneous analysis of several synthetic cannabinoids, THC, CBD and CBN, in hair by ultra-high performance liquid chromatography tandem mass spectrometry. Method validation and application to real samples," *Journal of Mass Spectrometry*, vol. 47, no. 5, pp. 604–610, 2012.

[64] M. Sundström, A. Pelander, V. Angerer, M. Hutter, S. Kneisel, and I. Ojanperä, "A high-sensitivity ultra-high performance liquid chromatography/high- resolution time-of-flight mass spectrometry (UHPLC-HR-TOFMS) method for screening synthetic cannabinoids and other drugs of abuse in urine," *Analytical and Bioanalytical Chemistry*, vol. 405, no. 26, pp. 8463–8474, 2013.

[65] E. Partridge, S. Trobbiani, P. Stockham, T. Scott, and C. Kostakis, "A validated method for the screening of 320 forensically significant compounds in blood by LC/QTOF, with simultaneous quantification of selected compounds," *Journal of Analytical Toxicology*, vol. 42, no. 4, pp. 220–231, 2018.

[66] I.-T. Wang, Y.-T. Feng, and C.-Y. Chen, "Determination of 17 illicit drugs in oral fluid using isotope dilution ultra-high performance liquid chromatography/tandem mass spectrometry with three atmospheric pressure ionizations," *Journal of Chromatography B*, vol. 878, no. 30, pp. 3095–3105, 2010.

[67] A. G. Helfer, J. A. Michely, A. A. Weber, M. R. Meyer, and H. H. Maurer, "Orbitrap technology for comprehensive metabolite-based liquid chromatographic–high resolution-tandem mass spectrometric urine drug screening–exemplified for cardiovascular drugs," *Analytica Chimica Acta*, vol. 891, pp. 221–233, 2015.

[68] T. R. Fiorentin, F. B. D'Avila, E. Comiran et al., "Simultaneous determination of cocaine/crack and its metabolites in oral fluid, urine and plasma by liquid chromatography-mass spectrometry and its application in drug users," *Journal of Pharmacological and Toxicological Methods*, vol. 86, pp. 60–66, 2017.

[69] A. E. Steuer, M. Poetzsch, L. Stock et al., "Development and validation of an ultra-fast and sensitive microflow liquid chromatography-tandem mass spectrometry (MFLC-MS/MS) method for quantification of LSD and its metabolites in plasma and application to a controlled LSD administration study in huma," *Drug Testing and Analysis*, vol. 9, no. 5, pp. 788–797, 2017.

[70] F. Xian, C. L. Hendrickson, and A. G. Marshall, "High resolution mass spectrometry," *Analytical Chemistry*, vol. 84, no. 2, pp. 708–719, 2012.

[71] D. Pasin, A. Cawley, S. Bidny, and S. Fu, "Current applications of high-resolution mass spectrometry for the analysis of new

psychoactive substances: a critical review," *Analytical and Bioanalytical Chemistry*, vol. 409, no. 25, pp. 5821–5836, 2017.

[72] J. D. Harper, N. A. Charipar, C. C. Mulligan, X. Zhang, R. G. Cooks, and Z. Ouyang, "Low-temperature plasma probe for ambient desorption ionization," *Analytical Chemistry*, vol. 80, no. 23, pp. 9097–9104, 2008.

[73] J. M. Wiseman, B. Gologan, and R. G. Cooks, "Mass spectrometry sampling under ambient conditions with desorption electrospray ionization," *Science*, vol. 306, no. 5695, pp. 471–474, 2004.

[74] A. E. Kirby, N. M. Lafrenière, B. Seale, P. I. Hendricks, R. G. Cooks, and A. R. Wheeler, "Analysis on the go: quantitation of drugs of abuse in dried urine with digital microfluidics and miniature mass spectrometry," *Analytical Chemistry*, vol. 86, no. 12, pp. 6121–6129, 2014.

[75] E. Jagerdeo and A. Wriston, "Rapid analysis of forensic-related samples using two ambient ionization techniques coupled to high-resolution mass spectrometers," *Rapid Communications in Mass Spectrometry*, vol. 31, no. 9, pp. 782–790, 2017.

5

Multiresidue Method for Quantification of Sulfonamides and Trimethoprim in Tilapia Fillet by Liquid Chromatography Coupled to Quadrupole Time-of-Flight Mass Spectrometry using QuEChERS for Sample Preparation

Kátia S. D. Nunes,[1] Márcia R. Assalin,[2] José H. Vallim,[2] Claudio M. Jonsson,[2] Sonia C. N. Queiroz,[2] and Felix G. R. Reyes ⓘ[1]

[1]Department of Food Science, School of Food Engineering, University of Campinas, Rua Monteiro Lobato 80, 13083-862 Campinas, SP, Brazil
[2]Embrapa Meio Ambiente, P.O. Box 69, 13820-000 Jaguariúna, SP, Brazil

Correspondence should be addressed to Felix G. R. Reyes; reyesfgr@gmail.com

Academic Editor: Gauthier Eppe

A multiresidue method for detecting and quantifying sulfonamides (sulfapyridine, sulfamerazine, sulfathiazole, sulfamethazine, sulfadimethoxine, sulfamethoxazole, and sulfamethoxypyridazine) and trimethoprim in tilapia fillet (*Oreochromis niloticus*) using liquid chromatography coupled to mass spectrometry was developed and validated. The sample preparation was optimized using the QuEChERS approach. The chromatographic separation was performed using a C18 column and 0.1% formic acid in water and acetonitrile as the mobile phase in the isocratic elution mode. Method validation was performed based on the Commission Decision 2002/657/EC and Brazilian guideline. The validation parameters evaluated were linearity ($r \geq 0.99$); limits of detection (LOD) and quantification (LOQ), 1 ng·g^{-1} and 5 ng·g^{-1}, respectively; intraday and interdays precision (CV lower than 19.4%). The decision limit (CCα 102.6–120.0 ng·g^{-1} and 70 ng·g^{-1} for sulfonamides and trimethoprim, respectively) and detection capability (CCβ 111.7–140.1 ng·g^{-1} and 89.9 ng·g^{-1} for sulfonamides and trimethoprim, respectively) were determined. Analyses of tilapia fillet samples from fish exposed to sulfamethazine through feed (incurred samples) were conducted in order to evaluate the method. This new method was demonstrated to be fast, sensitive, and suitable for monitoring sulfonamides and trimethoprim in tilapia fillet in health surveillance programs, as well as to be used in pharmacokinetics and residue depletion studies.

1. Introduction

Brazil is one of the five largest veterinary markets in the world, and aquaculture, in particular fish farming, is the fastest growing sector of animal food production in the country [1, 2]. In fish farming, antimicrobials, including sulfonamides, are used for the treatment of bacterial diseases. Sulfonamides (Figure 1) belong to an important group of synthetic antimicrobial agents that have been used in human and veterinary medicine for over 60 years. Recently, these drugs have been extensively employed in animals intended to produce food for human consumption since it is practically impossible to keep the production environment free of pathogenic organisms. Sulfonamides have become a useful tool for achieving high levels of productivity, thereby contributing to further growth, feed efficiency, and reduced mortality and morbidity [3]. However, sulfonamide residues are a major concern because of their potential risk to human health by development of bacterial resistance and adverse effects, such as allergic reactions, in hypersensitive people [4].

Trimethoprim (Figure 1) is a diaminopyrimidine antimicrobial agent, which is active against a wide range of Gram-positive and Gram-negative microorganisms including *Escherichia coli* and some *Klebsiella*, *Proteus*,

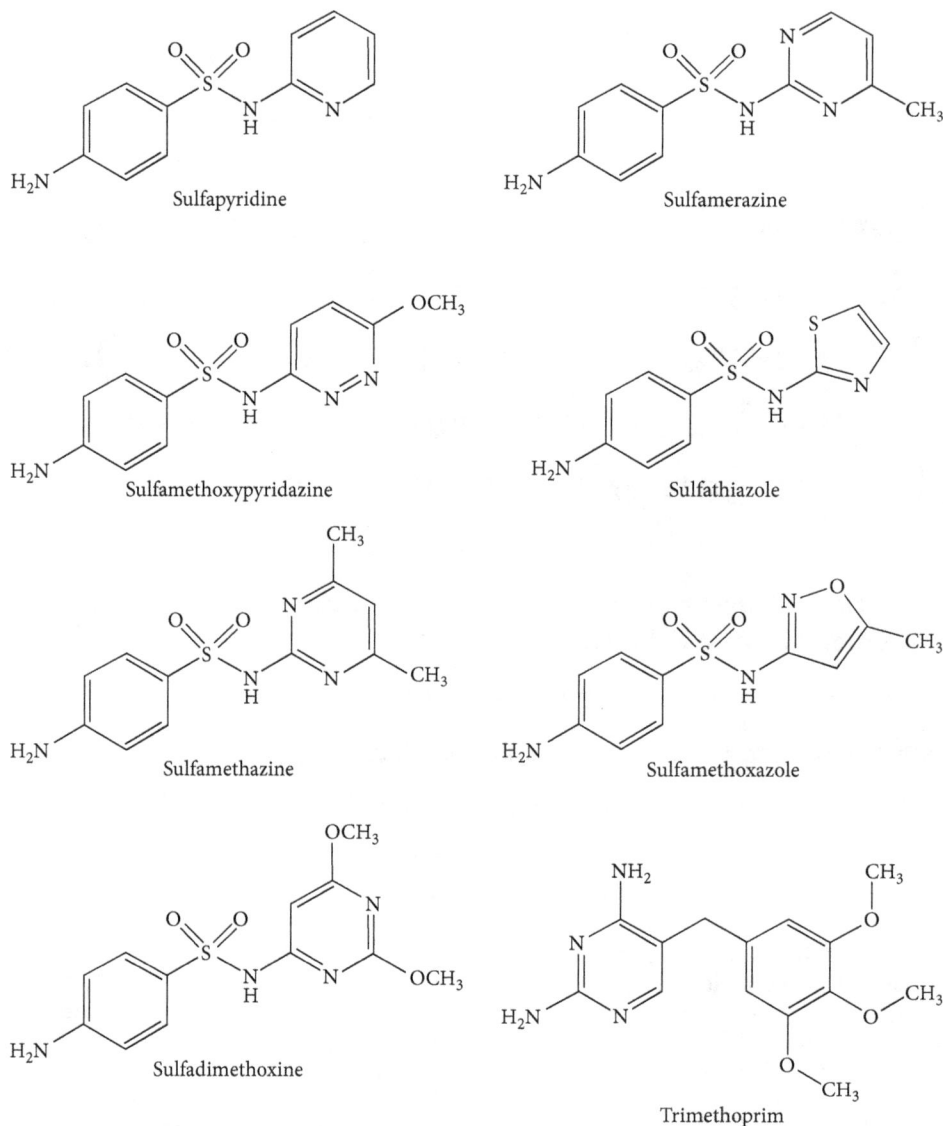

FIGURE 1: Chemical structures of the sulfonamides and trimethoprim.

and *Staphylococcus* species. In veterinary medicine, it is often used in combination with a sulfonamide to increase the antimicrobial activity of the sulfonamides but is excreted faster. Consequently, if no residues of sulfonamide are detectable, no residues of trimethoprim would be expected. Trimethoprim is of low acute mammalian toxicity, and there is no evidence for the potentiation of acute toxicity when it is administered in combination with a sulfonamide [5].

At its 40th session, the Codex Alimentarius Commission reported a maximum residue limit (MRL) value for sulfadimidine (sulfamethazine) of $100 \mu g \cdot kg^{-1}$ in muscle, for species not specified [6]. According to the European Commission Regulation (EU) No. 37/2010 [7], for the muscle of fin fish, the MRL value for individual sulfonamides, or the combined total residues of all substances belonging to the sulfonamide group, is $100 \mu g \cdot kg^{-1}$. In relation to trimethoprim, the MRL value is $50 \mu g \cdot kg^{-1}$. The MRL value relates to the muscle and skin in natural proportions. In Brazil, the use of sulfonamides in farm-raised

fish is not permitted (it does not appear in the legislative framework) and, therefore, its use is considered out of label (prohibited substance). However, for monitoring purposes (and taking actions), the Brazilian National Plan for Control of Residues and Contaminants (PNCRC/Fish) establishes a reference limit of $100 \mu g \cdot kg^{-1}$ for the residue of the individual sulfonamides (sulfachlorpyridazine, sulfadoxine, sulfamerazine, sulfadiazine, sulfamethoxazole, sulfathiazole, sulfamethazine, sulfaquinoxaline, and sulfadimethoxine) or the sum of them. Trimethoprim is not considered under the PNCRC/Fish sampling plan [8].

Studies on the determination of antimicrobial residues in foods of animal origin began in Belgium, the Netherlands, and Luxembourg in the late 1960s and early 1970s. In most European countries, research on residues and their application in inspection of slaughtered animals started later [9]. In relation to the sample preparation step, strategies such as salting out liquid-liquid extraction [10], solid-liquid extraction [11], and microscale matrix solid-phase dispersion [12] have been employed to perform the extraction and cleanup of sulfonamides from fish and other biological matrices. More recently, Ziarrusta et al. [13] used focused ultrasound solid-liquid extraction (FUSLE) for extraction of fluoroquinolones from fish tissues. The FUSLE method improves the extraction yield of target analytes (organic compounds), quantitatively, from biota samples. Regarding the systems of separation and detection, the high performance liquid chromatography-tandem mass spectrometry (HPLC-MS/MS) is an analytical technique that has been used in the determination of veterinary drug residues. In this regard, a few sulfonamide multiresidue methods in food matrices have been described in the literature by this technique [14, 15]. For instance, Abdallah et al. [16] determined sulfonamide residues in sheep, pork, beef, chicken, and dromedary, Nebot et al. [17] in bovine milk, Tsai et al. [18] in different fish species, and Jansomboon et al. [19] in *Pangasius* catfish. Alternatively, a time-of-flight (TOF) mass spectrometer provides high sensitivity and accurate mass measurements (0.005 Da), enabling the detection of low concentrations (ng·g^{-1}) of residues and contaminants in highly complex food matrices [15, 20]. Nevertheless, to our knowledge, there is no reported multiresidue method for the combined quantification of sulfonamides and trimethoprim in tilapia fillet using liquid chromatography coupled to quadrupole time-of-flight mass spectrometry (LC-QTOF/MS).

The aim of this study was to develop and validate a rapid, simple (without the need of solid-phase extraction (SPE) cartridges or similar materials), and reliable multiresidue method for the identification and quantification of sulfonamides and trimethoprim in tilapia fillets (*Oreochromis niloticus*) by LC-QTOF/MS, to be suitable for application in monitoring programmes as well as in pharmacokinetic and residue depletion studies. The sample preparation involved the QuEChERS (Quick, Easy, Cheap, Effective, Rugged, and Safe) approach as described by Lehotay et al. [21]. The validation was conducted in-house based on the Commission Decision 2002/657/EC [22] and Brazilian guideline [23]. To evaluate the precision of the method, analysis of tilapia fillet samples from fish exposed to sulfamethazine through feed (incurred samples) was also conducted.

2. Materials and Methods

2.1. Chemicals and Reagents. The sulfonamide analytical standards (sulfathiazole (STZ), sulfamethoxazole (SMX), sulfamerazine (SMR), sulfamethoxypyridazine (SMPD), sulfadimethoxine (SDMX), sulfapyridine (SP), sulfamethazine (SMZ)), and trimethoprim (TMP) were purchased from Sigma-Aldrich Company Ltd. (St. Louis, MO, USA). All analytical standards had a purity greater than 99.0%. Primary secondary amine (PSA) was obtained from United Chemical Technologies, Inc. (UCT Inc., Bristol, PA, USA), and formic acid (98%) was purchased from Sigma-Aldrich Company Ltd. (St. Louis, MO, USA). Anhydrous magnesium sulfate was supplied by J.T. Baker (Center Valley, PA, USA) and sodium acetate trihydrate from Spectrum Chemical Mfg., Corp. (New Brunswick, NJ, USA). Methanol (MeOH) and acetonitrile (ACN) were obtained from

Honeywell Burdick & Jackson (Muskegon, MI, USA) and J.T. Baker (Center Valley, PA, US), respectively. All solvents were of HPLC grade, and all reagents were of analytical grade. Ultra-pure deionized water was obtained from a Milli-Q Plus water purification system (Millipore, Bedford, MA, USA). Filtration of the aqueous mobile phase was performed using polyvinylidene fluoride (PVDF) membranes, and polytetrafluoroethylene (PTFE) membranes were used for organic mobile-phase filtration, both with 0.22 μm pore size obtained from Millipore (Bedford, MA, USA).

2.2. Instrumentation. The identification and quantitation of sulfonamides and trimethoprim was carried out using an UPLC-Q-TOF system comprising an Acquity UPLC system coupled to a hybrid quadrupole orthogonal time-of-flight (Q-TOF) mass spectrometer (SYNAPT HDMS Q-TOF mass spectrometer) with electrospray source ionization (ESI) in positive mode. The software of acquisition control and data treatment was the MassLynx version 4.1 (Waters Corp., Milford, MA, USA). For sample preparation, the following equipment were used: semianalytical balance (Tecnal; Boulder, CO, USA); analytical balance (Scientech, SA 210; Boulder, CO, USA); tubes stirring vortex type (IKA model MS1 Minishaker, 2700 rpm; Wilmington, DE, USA); refrigerated centrifuge (Thermo Scientific model Heraeus Multifuge 3 L-R; Madison, WI, USA); ultrasonic bath (Elma model Transsonic 660/H; Singen, Baden-Württemberg, Germany); Waring Commercial Blender, model 33BL79 (New Hartford, CT, USA); and Ultra-Turrax IKA, model TP 10N (Wilmington, DE, USA).

2.3. Solution Preparation. Standard stock solutions of SP, STZ, SMZ, SDMX, SMX, SMPD, SMR, and TMP were prepared in acetonitrile at 1000 μg·mL^{-1}, stored in 10 mL bottles, and kept at -20°C. These solutions were used for a maximum period of 1 month. The intermediate standard solutions were prepared daily by dilution of stock solutions in an appropriate buffer solution.

2.4. Blank and Incurred Fish Samples. The blank samples of tilapia (*Oreochromis niloticus*) with no detectable analyte concentration used for the development and validation of the analytical method were provided by a local producer (Rio Doce fish farm, São João da Boa Vista, SP) with a guarantee that the fish were not exposed to the compounds that were the analytical focus of this work. Nonetheless, to ensure the viability of the blank samples, they were analysed, and the chromatograms did not show the presence of any interference at the retention time corresponding to the studied analytes. For validation of the analytical method, blank samples and incurred samples (truly contaminated samples) were used, that is, samples of fish exposed to SMZ through feed, obtained from an experiment conducted at Embrapa Environment, Jaguariuna, SP, Brazil, where tilapia were given SMZ at a dose level of 422 mg·kg^{-1} body weight, for 11 consecutive days. The incurred samples used in this study were from fish slaughtered by thermal shock and immersion in an ice bath, 12 h after stopping medication. All samples were stored in a freezer (−20°C) until analysis [24]. The experiment with fish to obtain the incurred samples was approved by the Ethics Committee on Animal Experiments of Embrapa Environment (Protocol No. 001/2013) [25].

2.5. Sample Preparation by QuEChERS. Tilapia fillet samples were ground using a domestic food processor. Triturated samples (2.5 g) were weighed in a 50 mL polypropylene tube, and ACN (5 mL) was added and then homogenized using a Turrax for 30 seconds. The homogenized sample was then added of 5 mL ACN, the tubes were shaken vigorously by vortexing for 1 min and placed in an ultrasonic bath for 5 min. Next, 2.0 g of anhydrous magnesium sulfate and 0.75 g of sodium acetate were added to the homogenized samples and vortexed for 1 min and centrifuged at 17,500 × g for 10 min, at 5°C. For sample cleanup, an aliquot of 5.0 mL of supernatant was volumetrically pipetted to another tube containing 150 mg of PSA and 0.5 g of anhydrous magnesium sulfate. The tube was subsequently vortexed for 30 seconds and centrifuged at 17,500 × g again for 5 min, at 5°C. A 2.0 mL aliquot of the supernatant was pipetted and transferred to another tube, and the solvent was completely evaporated under nitrogen stream, in an ice bath, to avoid losses of the analytes. Next, the residue was suspended in 0.5 mL of the mobile phase (ACN : 0.1% aqueous formic acid, 95 : 5 v/v). To facilitate the dissolution of analytes, the tubes were placed in ultrasonic bath for 5 min. Finally, the resulting extracts were filtered through a cellulose filter unit (0.22 μm pore size) directly into the vial and injected in the LC-QTOF/MS system. A schematic representation of the sample preparation procedure is shown in Figure 2.

2.6. UPLC-QTOF/MS Conditions. The chromatographic separation was performed on a reversed-phase analytical column Poroshell EC-120 C18 (50 mm × 2.1 mm, 2.7 μm), supplied by Agilent Technologies (Santa Clara, CA, USA) preceded by a similar precolumn (30 mm × 2.1 mm, 2.7 μm). The chromatographic separation was performed at 25°C. The mobile phase was composed of (A) H_2O : acetonitrile : formic acid (95 : 5 : 0.1%, v/v/v) and (B) H_2O : acetonitrile : formic acid (5 : 95 : 0.1%, v/v/v), and the isocratic elution mode was used with 70% (A) and 30% (B). The flow rate was 0.2 mL·min^{-1} with a run time of 4 min and injection volume of 5 μL.

The following ionization conditions were established for the ESI-QTOF/MS system: positive ionization mode, capillary voltage: 2.5 kV, detector voltage: 1.850 kV, sample cone voltage: 20.0 V, extraction cone voltage: 2.0 V, source temperature: 100°C, desolvation gas temperature: 300°C, nitrogen gas flow in the cone: 50 L·h^{-1}, and desolvation flow: 400 L·h^{-1}. The molecules of interest were quantified by monitoring the signal related to the protonated molecular ion m/z $(M + H^+)$. The sulfonamide and trimethoprim identity was confirmed by obtaining the accurate mass of the protonated molecular ion, as well as by the consideration of fragment ions in order to obtain the identification points (IPs) according to Commission Decision 2002/657/EC [22] (Table 1).

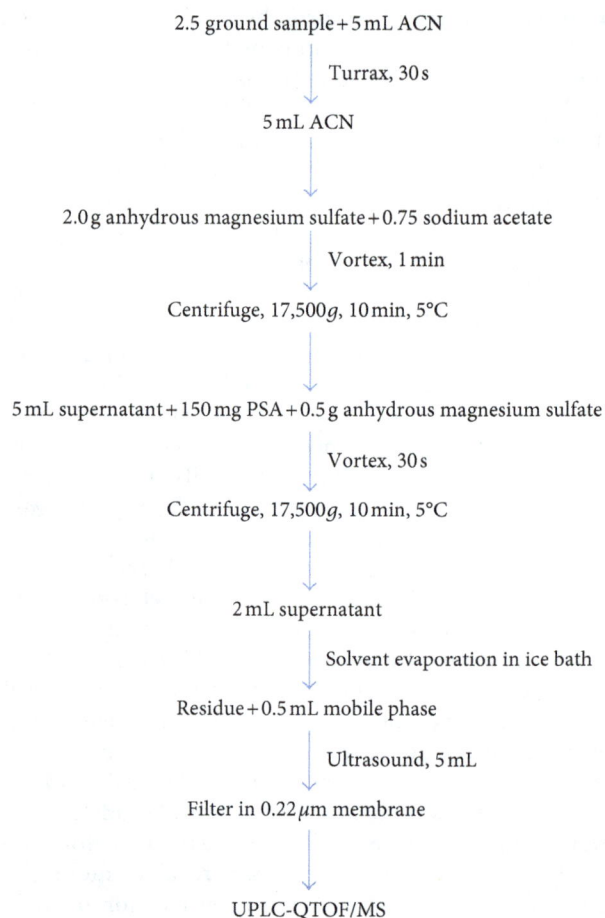

FIGURE 2: Schematic representation of the sample preparation procedure.

2.7. Validation Parameters. The purpose of this step was to establish the performance parameters and the minimum requirements of acceptance that must be satisfied such that the analytical method presented in this study is considered validated. The recommendations of the European Community [22] and the Guide to Analytical Methods Validation of the Brazilian Ministry of Agriculture, Livestock, and Supply [23] were used as reference to perform the method validation.

After optimization of the preparation procedure (extraction and cleanup), the validation of the analytical method was performed. The following validation parameters were evaluated: selectivity; linearity, sensitivity, and matrix effect; precision (intra- and interday); accuracy; and decision limit (CCα) and detection capability (CCβ). The limit of detection (LOD) and limit of quantification (LOQ) were also assessed to evaluate the potential use of the analytical method in pharmacokinetic and residue depletion studies where lower LOD and LOQ are required. Selectivity of the method was evaluated by comparing the chromatograms obtained from blank samples ($n = 10$) and the samples spiked with sulfonamides and trimethoprim standard solutions ($n = 10$). The chromatograms were evaluated for the presence of the analytical signal at the same retention time observed for the mass-to-charge ratio (m/z) of the analytes of interest.

TABLE 1: Elemental composition, retention time, the m/z experimental (precursors and fragment) ions, and mass error determined in standard solution for the studied analytes.

Compound	Molecular formula	Retention time (min)	Monoisotopic mass (Da)	m/z experimental $[M + H]^+$ (Da)	Mass error (ppm)	m/z experimental fragment ion (Da)
Trimethoprim	$C_{14}H_{18}N_4O_3$	0.83	290.1379	291.1460	1.0	123.0592
Sulfapyridine	$C_{11}H_{11}N_3O_2S$	1.08	249.0572	250.0650	0.0	156.0128
Sulfamerazine	$C_{11}H_{12}N_4O_2S$	1.18	264.0681	265.0760	0.4	108.0483
Sulfathiazole	$C_9H_9N_3O_2S_2$	0.99	255.0136	256.0210	1.6	156.0127
Sulfamethazine	$C_{12}H_{14}N_4O_2S$	1.25	278.0837	279.0920	1.4	108.0475
Sulfadimethoxine	$C_{12}H_{14}N_4O_4S$	2.19	310.0736	311.0810	1.3	156.0771
Sulfamethoxazole	$C_{10}H_{11}N_3O_3S$	1.81	253.0521	254.0600	0.4	156.0124
Sulfamethoxypyridazine	$C_{11}H_{12}N_4O_3S$	1.44	280.0630	281.0710	0.7	156.0125

Linearity was established from analytical curves obtained by duplicate analysis of blank samples spiked with trimethoprim and sulfonamides in the following concentrations: 5.0, 12.5, 25.0, 50.0, 75.0, 100.0, 125.0, and 250.0 ng·g^{-1}. The results were analysed by the method of least squares , and the linearity was expressed through the coefficient of determination (R^2) which was adopted as $R^2 \geq 0.99$, as recommended by the Brazilian validation guide [23]. The matrix effect was evaluated by comparing three different concentrations (12.5, 50.0, and 100.0 ng·g^{-1}) of sulfonamides and trimethoprim, prepared in solvent and fortified extracts. The evaluation was done by comparing the area of the analytical signal in solvent with the area of analyte in the fortified extracts. Accuracy was evaluated by recovery tests of the spiked blank matrix at three concentration levels (10.0, 20.0, and 40.0 ng·g^{-1}) with five replicates of each spiked level, during 3 days. The results were expressed as mean values ($n = 15$) of percentage of recoveries. The coefficient of variation (CV%) is also reported.

The precision of the method was determined in two steps: intraday precision (repeatability) and interdays precision (intermediate precision). Repeatability was expressed as the CV% of the results obtained with five replicates at three different concentrations (10.0, 20.0, and 40.0 ng·g^{-1}) analysed on the same day by the same analyst. The intermediate precision was expressed by CV% of the results of three different concentrations with five replicates of each concentration on three different days by the same analyst.

The calculation of the decision limit (CCα) and the detection capability (CCβ) was based on the Commission Decision 2002/657/EC [22]. The decision limit is defined as the lowest concentration level at which the method can discriminate with a statistical certainty of $1-\alpha$ if the analyte is present. For substances with an MRL, the value of α is considered to be 5%. The calculation was performed by analysing 20 blank samples fortified with the analyte at the MRL level. The concentration of the MRL plus 1.64 times the standard deviation corresponds to the CCα ($\alpha = 5\%$). The detection capability (CCβ) is the lowest amount of the substance that can be detected, identified, and/or quantified in a sample with an acceptable error probability (β). For substances with an MRL, the determination of CCβ can be accomplished by the analysis of 20 blank samples fortified with the analyte in the decision limit (CCα). The value of

CCα plus 1, 64 times the standard deviation, corresponds to the CCβ ($\beta = 5\%$).

For each sulfonamide and trimethoprim, the LOD and LOQ were established by analysing the fortified matrix with standard solution of the analytes. LOD was determined based on signal-to-noise approach. Thus, LOD was expressed as the lowest concentration with a signal equal to three times the signal-to-noise ratio. The LOQ was taken as the first level of the analytical curve, which was measured with acceptable precision (CV $\leq$ 20%) [26].

3. Results and Discussion

The representative sulfonamide veterinary drugs were chosen based on a study of their use in fish farming around the world, those monitored by the Brazilian National Plan for Control of Residues and Contaminants (PNCRC/Fish) of the Brazilian Ministry of Agriculture, Livestock, and Food Supply and those used for other animal species that could potentially be illegally employed in fish farming. Thus, sulfamethazine, sulfathiazole, sulfadimethoxine, sulfamerazine, sulfamethoxazole (monitored by the PNCRC/Fish [8]), sulfapyridine, sulfamethoxypyridazine, and trimethoprim (regulated for veterinary use [7], although not regulated for use in fish farming in Brazil) were selected. The maximum residue limit (MRL) adopted for all the sulfonamides (individual or the combined total residues) was 100 μg·kg^{-1}, and 50 μg·kg^{-1} for trimethoprim [7].

3.1. Sample Preparation Based on QuEChERS. Dispersive solid-phase extraction (d-SPE) technique and QuEChERS have been previously used for the determination of veterinary drug residues in animal fluids and tissues [16, 27, 28], but not for the concomitant determination of sulfonamides and trimethoprim in fish fillet. It is well known that the step of sample preparation (extraction of analytes and cleanup of the extract) is crucial. This approach can influence the magnitude of the matrix effect, depending on the amount of endogenous substances from which it is coextracted. Acetonitrile has been widely used in the extraction of analytes from complex matrices as it extracts analytes with few interfering compounds (e.g., low amount of lipophilic coextractives from the sample) and further promotes the precipitation of proteins. This is necessary because the lower the quantity of interfering content present in the extract, the less matrix effect is observed, which leads to a better quality analysis [29].

Kruve et al. [30] reported the minimizing matrix effect in LC-ESI-MS analysis by using extrapolative dilution. It was demonstrated by several tests using QuEChERS sample preparation procedure that the use of a greater volume of acetonitrile for analyte extraction of complex matrices tends to reduce the matrix effect, possibly eliminating the matrix effect if a suitable dilution is achieved. It should be mentioned that although LC-ESI-QTOF/MS technique is very selective, possible interference caused by matrix substances can lead to suppression or an increase in the ionization of the analytes of interest [31]. Thus, this study explores the extraction of sulfonamides and trimethoprim by using QuEChERS procedure making use of acetonitrile as the extracting solvent and extrapolative dilution.

Preliminary studies have shown that for the quantification of sulfonamides and trimethoprim in tilapia fillet using the proportion of acetonitrile : sample 4 : 1 (v/w) showed the best results with fewer coextracts, thus decreasing the presence of interfering compounds. It is noteworthy that although the amount of sample used in this study was four times lower than that used by Lehotay et al. [21], it was possible to achieve an LOQ of $5 \, ng \cdot g^{-1}$ for all analytes, consequently to the LC-ESI-QTOF/MS system used. Literature data show that the LOQ for SDZ was $36 \, ng \cdot g^{-1}$, using 5 g of the sample [32]. Stubbings and Bigwood [33] showed an LOQ for SP, STZ, SMZ, SDMX, SMX, and SMR of $50 \, ng \cdot g^{-1}$, also using 5 g of the sample.

The addition of salts to promote the salting out effect has been shown to enhance the optimization of the analyte recovery percentages in multiresidue methods since it increases the solubility of these molecules in the organic phase [34, 35]. In the QuEChERS approach reported by Lehotay et al. [21], 6 g of anhydrous magnesium sulfate and 2.5 g of sodium acetate trihydrate were used. In the present method for extracting sulfonamides and trimethoprim from tilapia fillet, 2 g of anhydrous magnesium sulfate and 0.75 g of sodium acetate trihydrate were employed. At the cleanup step, PSA and/or C18 were used. Since no significant variation was observed between them in relation to recovery values, we opted for the use of PSA only. This finding may be observed because the fat content in tilapia fillet is low. There are studies in matrices that have considerably higher fat content in which the concomitant use of PSA and C18 is required for a better cleanup of the sample extract [33].

3.2. Identity Confirmation of Analytes.
On the basis of Commission Decision 2002/657/EC [22], the identity confirmation of a substance is performed by a system of identification points (IPs). The mass accuracy of a high-resolution mass spectrometer acquires 2 IPs for the precursor ion and 2.5 for each transition product. The resolution of mass spectrometer used in this study (SYNAPT HDMS Q-TOF) is more than 10,000, which fall within the criteria established by the guide as a high-resolution MS. Under the conditions selected, the protonated molecule and one

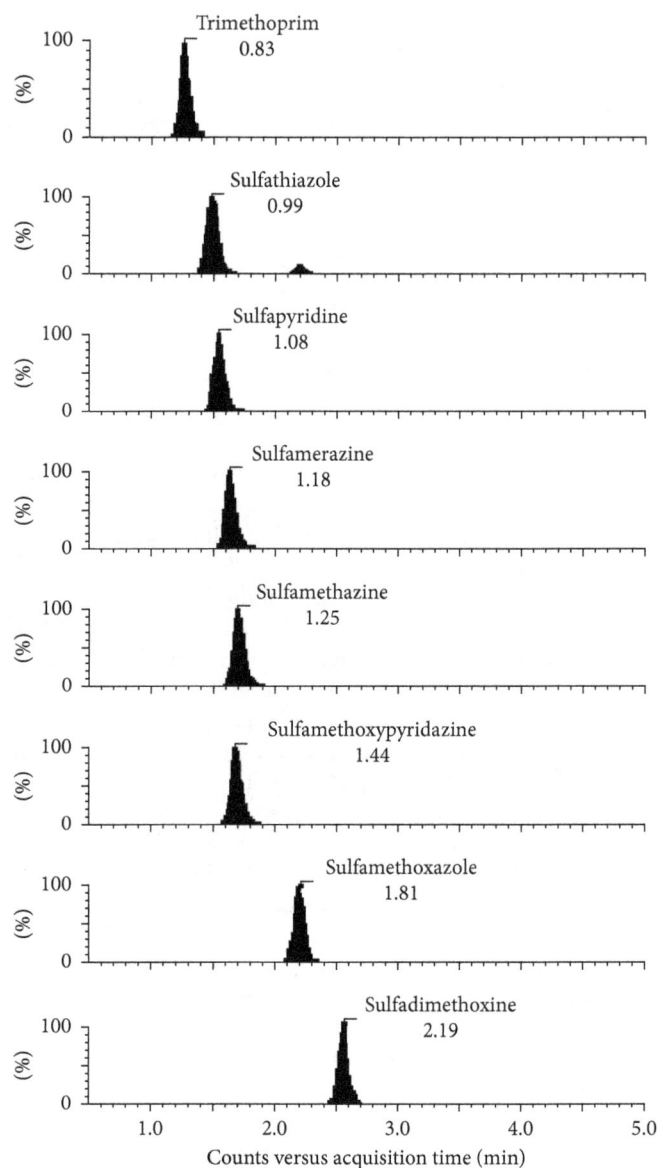

FIGURE 3: Extracted chromatograms of spiked samples with sulfonamides and trimethoprim at concentration of $50.0 \, ng \cdot g^{-1}$.

fragmented ion for each analyte could be monitored, thus reaching the requirements to confirm their identity in accordance with Commission Decision 2002/657/EC [22]. For the quantitative purpose, only the sulfonamides and trimethoprim molecular ions were monitored.

3.3. Analytical Method Validation.
The method selectivity was evaluated by analysing ten samples free of analytes (blank samples) and comparing them to the chromatograms obtained from samples spiked with the sulfonamides and trimethoprim. Peaks for interfering compounds with the same retention times as the analytes of interest with the same m/z were not observed. Therefore, the method performed is satisfactorily selective.

Figure 3 shows the chromatograms of each analyte studied.

To study the linearity, sensitivity, and matrix effect, the analytical results at the following concentrations were compared: 5.0, 12.5, 25.0, 50.0, 75.0, 100.0, 125.0, and 250.0 ng·mL^{-1}. Measurements were carried out for the analytes dissolved in the solvent, in the fortified extract, and in the fortified blank matrix (matrix-matched). The matrix effect, expressed as a percentage, was calculated from the division between the areas obtained for the analyte in solvent and in the fortified extract, at the same concentration level [36]. The highest matrix effect value observed was 18,98%, which is below the maximum acceptable value by the validation guides (20%) [22]. Thus, the matrix effect was considered irrelevant for this method. However, when comparing the analytical curves in extract with the curves in the fortified blank matrix, it was noted that the slope (angular coefficient) of the curve for the matrix-matched sample was much lower, indicating the loss of analytes (sulfonamides and trimethoprim) during sample preparation step (extraction and cleanup). Thus, for the present method a matrix-matched analytical curve must be employed.

Accuracy was evaluated from recovery tests (%), as recommended by the Commission Decision 2002/657/EC when no certified reference material (CRM) is available [22]. The experiment was carried out through the recovery test of the spiked samples at 3 levels (10.0, 20.0, and 40.0 ng·g^{-1}), evaluating each level using 5 independent replicates on 3 consecutive days. Analytes SP, SMR, and TMP had satisfactory recovery values (between 79.5 and 103.6%), SMZ and SMPD showed intermediate recoveries (between 64.6 and 80.0), and STZ, SDMX and SMX exhibit lower recovery values (between 38.4 and 52.9) (Table 2). Low recovery values for sulfonamides have been reported. Won et al. [37] reached a recovery of 58.8% for SDMX after extraction of this molecule from marine products, such as common eel, blue crab, shrimp, and flatfish, among others. Sulfonamides' low recoveries have also been reported in other matrices. Summa et al. [38] report recoveries for SMX and SDM, extracted from eggs, around 60% and 55%, respectively. A review dealing about the presence of sulfonamides in edible tissues reports recoveries of various sulfonamides ranging from 40 to 67% for honey, 45–85% for pork veal, and 57–63% for salmon muscle [39]. Although recovery values found for some sulfonamides were below the percentage established in the validation guide [22], the method has been shown to be precise (CV% found is within the value specified in the validation guide), and the required LOQ was achieved, which leads us to consider that the method is suitable for the intended purpose. Nevertheless, this corroborates the need to use matrix-matched analytical curves for the quantification of the analytes in samples of unknown origin.

The precision of the method was determined through intraday precision (repeatability) and interdays (intermediate precision) at three spiked levels and was expressed as coefficient of variation (CV%). The intraday and interdays precision were evaluated in the concentration levels at 10.0, 20.0, and 40.0 ng·g^{-1}, with 5 replicates at each level. Working in this concentration range, we can ensure the precision and accuracy since in the most dispersive points, the CV is ≤20%. The repeatability (analysed on the same day and same equipment) and the interdays precision (intermediate precision) are shown in Table 3.

For compounds with concentration levels lower than 100 ng·g^{-1}, the Commission Decision 2002/657/EC [22] and Brazilian validation guideline [23] recommend a maximum acceptable CV ≤ 20%. As shown in Table 3, the validation parameters (intraday and interdays precision) meet the specifications recommended by both guides since they recommend a CV ≤ 20%.

The decision limit (CCα) is a parameter that takes into account the precision of the method for establishing a critical reference level, from which we can conclude that a sample is classified as nonconforming with a probability of error of 5%. An additional critical parameter, detection capability (CCβ), is calculated for use with nonconforming samples in order to confirm their concentration, and their identities are confirmed with an error probability of 5% (β = 5%).

The decision limit (CCα) and detection capability (CCβ) values for each of the analyte studied are shown in Table 4. For sulfonamides, values varied from 102.6 to 120.0 µg·kg^{-1} and from 111.7 to 140.1 µg·kg^{-1} for CCα and CCβ, respectively. For trimethoprim, those values were 70.0 and 89.9 µg·kg^{-1}, respectively. Thus, considering the MRL values of 100 µg·kg^{-1} and 50 µg·kg^{-1}, respectively, for sulfonamides and trimethoprim, established by the regulatory framework of the European Union in fin fish [7], we can conclude that the method reported here is suitable for application in surveillance programmes of residues of sulfonamides and trimethoprim in fin fish muscle samples.

The evaluation of LOD and LOQ for the determination of sulfonamides and trimethoprim residues in tilapia fillet was performed using the matrix-matched analytical curve fortified with the analytes. The LOD and LOQ of the method were 1.0 ng·g^{-1} and 5.0 ng·g^{-1} for all sulfonamides and trimethoprim, respectively. The LOQ was validated by analyse of 10 replicates that showed a CV ≤ 20% for all of the analytes. This indicates that due to the low value of LOQ obtained, the method can be used by restrictive regulatory agencies of countries such as Japan [40], which, for the multiresidue method intended for quantification of veterinary drug residues in animal and fishery products, adopt for individual sulfonamides and trimethoprim a LOQ value of 10 ng·g^{-1} and 20 ng·g^{-1}, respectively.

3.4. Analysis of Incurred Samples. To assess the method developed, analysis was performed on genuinely contaminated (incurred) fish samples obtained from an experiment in laboratory where the fishes were exposed to SMZ through the feed. This study was related to the effects of dietary exposure to SMZ on the haematological parameters and hepatic oxidative stress biomarkers in Nile tilapia [25]. The residue of SMZ in the muscle of 10 independent samples analysed in the same day was 1,062.9 ± 53.2 ng·g^{-1} (mean value ± standard deviation), and the precision (CV%) was 5.0%. Due to the high concentration levels, the extract of the samples was diluted prior to injection to adjust the concentration to fit the range of the analytical curve. This corroborates the precision of the method and provides confidence that it is appropriate

TABLE 2: Validation parameters of sulfonamides and trimethoprim.

Validation parameters	Sulfonamides and trimethoprim							
	SP	STZ	SMZ	SDMX	SMX	SMPD	SMR	TMP
Working range (ng·g^{-1})	5–250	5–250	5–250	5–250	5–250	5–250	5–250	5–250
Linearity (R^2)	0.9958	0.9914	0.9922	0.9992	0.9994	0.9935	0.9964	0.9984
Sensibility	1174.07	633.811	2324.3	2670.56	1626.52	2057.08	1345.72	1588.92
Matrix effect (%)								
12.5 ng·g^{-1}	−18.98	−2.07	−11.19	−3.58	−5.22	9.57	−3.68	9.65
50 ng·g^{-1}	1.60	14.71	0.37	−3.28	1.13	−4.15	1.85	−0.12
100 ng·g^{-1}	0.13	−7.02	0.46	1.61	−1.77	3.74	0.79	−0.03
Accuracy (% recovery (CV%))								
10 ng·g^{-1}	83.9 (14.4)	52.9 (19.2)	69.1 (19.4)	49.6 (17.8)	45.4 (11.9)	80.0 (9.4)	79.5 (15.4)	92.0 (13.2)
20 ng·g^{-1}	91.5 (19.3)	38.4 (17.2)	72.4 (11.8)	47.4 (3.4)	41.8 (4.2)	66.8 (9.0)	85.4 (16.7)	88.2 (17.7)
40 ng·g^{-1}	103.6 (19.0)	41.3 (18.6)	68.0 (19.1)	51.1 (4.5)	43.5 (10.0)	64.6 (5.5)	81.2 (15.0)	91.5 (16.2)
LOD (ng·g^{-1})	1.0	1.0	1.0	1.0	1.0	1.0	1.0	1.0
LOQ (ng·g^{-1})	5.0	5.0	5.0	5.0	5.0	5.0	5.0	5.0

SP, sulfapyridine; STZ, sulfathiazol; SMZ, sulfamethazine; SDMX, sulfadimethoxine; SMX, sulfamethoxazole; SMPD, sulfamethoxypyridazine; SMR, sulfamerazine; TMP, trimethoprim; LOD, limit of detection; LOQ, limit of quantitation; CV, coefficient of variation.

TABLE 3: Intraday and interdays precision of sulfonamides and trimethoprim.

Validation parameters	Sulfonamides and trimethoprim							
	SP	STZ	SMZ	SDMX	SMX	SMPD	SMR	TMP
Intraday precision (CV%)								
10 ng·g^{-1}	12.8	11.7	11.6	7.7	6.4	7.1	11.3	8.3
20 ng·g^{-1}	11.2	8.2	8.3	7.7	5.9	13.9	13.6	7.8
40 ng·g^{-1}	12.9	14.9	15.0	12.3	7.9	6.1	14.5	10.6
Interdays precision (CV%)								
10 ng·g^{-1}	14.4	19.2	19.4	17.8	11.9	9.4	15.4	13.2
20 ng·g^{-1}	19.3	17.2	11.8	3.4	4.2	9.0	16.7	17.7
40 ng·g^{-1}	19.0	18.6	19.1	4.5	10.0	5.5	15.0	16.2

SP, sulfapyridine; STZ, sulfathiazol; SMZ, sulfamethazine; SDMX, sulfadimethoxine; SMX, sulfamethoxazole; SMPD, sulfamethoxypyridazine; SMR, sulfamerazine; TMP, trimethoprim; CV, coefficient of variation.

TABLE 4: CCα and CCβ values for sulfonamides and trimethoprim in tilapia fillet.

Validation parameters	Sulfonamides[a] and trimethoprim[b]							
	SP	STZ	SMZ	SDMX	SMX	SMPD	SMR	TMP
Limit of decision (CCα), ng·g^{-1}	119.8	110.9	114.0	102.6	102.9	105.9	120.0	70.0
Detection capability (CCβ), ng·g^{-1}	139.7	121.7	122.0	117.1	111.7	118.2	140.1	89.9

SP, sulfapyridine; STZ, sulfathiazole; SMZ, sulfamethazine; SDMX, sulfadimethoxine; SMX, sulfamethoxazole; SMPD, sulfamethoxypyridazine; SMR, sulfamerazine; TMP, trimethoprim. [a]The MRL value adopted for the calculation of CCα and CCβ for all sulfonamides was 100 ng·g^{-1} [6]. [b]The MRL value adopted for the calculation of CCα and CCβ for TMP was 50 ng·g^{-1} [6].

for the intended purpose and can be used by regulatory agencies in health surveillance programs, as well as in pharmacokinetics and residue depletion studies.

4. Conclusions

A multiresidue method for determination of sulfonamides STZ, SMX, SMR, SMPD, SDMX, SP, and SMZ and trimethoprim (TMP) in tilapia fillet was developed and

validated. The analytes selected were those most frequently used worldwide in fish farming and those with the greatest potential for illegal use. QuEChERS approach with extrapolative dilution was shown to be a simple and inexpensive sample preparation process that can be easily used in routine analysis. Quantitation by liquid chromatography-quadrupole time-of-flight mass spectrometry (LC-ESI-QTOF/MS) showed to be a selective and low detectability method. Thus, the method is suitable for application in monitoring programmes of residues of sulfonamides and trimethoprim in tilapia fillet, even by countries such as Japan that adopt low LOQ values for analytical methods to be used in food for determination of residues of substances such as veterinary drugs. Also, it was shown to be appropriate to be used in pharmacokinetic and residue depletion studies.

Conflicts of Interest

The authors declare that there are no conflicts of interest regarding the publication of this paper.

Acknowledgments

The authors gratefully acknowledge the financial support received from São Paulo Research Foundation-Agilent Technologies (FAPESP-Agilent, 2013/50452-5), the Brazilian Coordination for the Improvement of Higher Education Personnel (PROEX/CAPES, 3300301702P1), and the Brazilian National Council of Technological and Scientific Development (CNPq). The authors also thank Dr. Patricia Aparecida de Campos Braga for her technical assistance in reviewing the manuscript.

References

[1] FAO, "Yearbook 2014 fishery and aquaculture statistics," 2016, http://www.fao.org/fishery/publications/yearbooks/en.

[2] MPA, "Ministério da pesca e aquicultura. Boletim estatístico da pesca e aquicultura Brasília," 2011, http://www.icmbio.gov.br/cepsul/images/stories/biblioteca/download/estatistica/est_2011_bol__bra.pdf.

[3] C. K. V. Nonaka, A. M. G. Oliveira, C. R. Paiva et al., "Occurrence of antimicrobial residues in Brazilian food animals in 2008 and 2009," Food Additives and Contaminants, vol. 29, no. 4, pp. 526–534, 2012.

[4] I. R. Pecorelli, L. Bibi, R. Fiorino, and R. Galarini, "Validation of a confirmatory method for determination of sulphonamides in muscle according to the European Union regulation 2002/657/EC," Journal of Chromatography A, vol. 1032, no. 1-2, pp. 23–29, 2004.

[5] EMEA, "The European Agency for the evaluation of medicinal products. Committee for veterinary medicinal products. Trimethoprim summary report," 2009, http://www.ema.europa.eu/docs/en_GB/document_library/Maximum_Residue_Limits_-_Report/2009/11/WC500015678.pdf.

[6] Codex Alimentarius, "Maximum residue limits (MRLs) and risk management recommendations (RMRs) for residues of veterinary drugs in foods. CAC/MRL 2-2017. Updated at the 40th session of the Codex Alimentarius Commission," 2017, http://www.fao.org/fao-who-codexalimentarius/codex-texts/maximum-residue-limits/en/.

[7] EU, "Council regulation (EU) no. 37/2010 of 22 December 2009, on pharmacologically active substances and their classification regarding maximum residue limits in foodstuffs of animal origin," Official Journal of the European Union, L 15/1, 2010, https://ec.europa.eu/health/sites/health/files/files/eudralex/vol-5/reg_2010_37/reg_2010_37_en.pdf.

[8] MAPA, "Ministério da agricultura, pecuária e abastecimento. National plan for residues and contaminants control - PNCRC. Instrução normativa no. 09, de 21 de fevereiro de," 2017, http://www.agricultura.gov.br/assuntos/inspecao/produtos-animal/plano-de-nacional-de-controle-de-residuos-e-contaminantes/documentos-da-pncrc/pncrc-2017.pdf.

[9] H. F. de Bradander, H. Noppe, K. Verheyden et al., "Residue analysis: future trends from a historical perspective," Journal of Chromatography A, vol. 1216, no. 46, pp. 7964–7976, 2009.

[10] J. Liu, M. Jiang, G. Li, L. Xu, and M. Xie, "Miniaturized salting-out liquid-liquid extraction of sulfonamides from different matrices," Analytica Chimica Acta, vol. 679, no. 1-2, pp. 74–80, 2010.

[11] S. Borràs, R. Companyó, and J. Guiteras, "Analysis of sulfonamides in animal feeds by liquid chromatography with fluorescence detection," Journal of Agriculture and Food Chemistry, vol. 59, no. 10, pp. 5240–5247, 2011.

[12] Q. Shen, R. Jin, J. Xue, Y. Lu, and Z. Dai, "Analysis of trace levels of sulfonamides in fish tissue using micro-scale pipette tip-matrix solid-phase dispersion and fast liquid chromatography tandem mass spectrometry," Food Chemistry, vol. 194, pp. 508–515, 2016.

[13] H. Ziarrusta, N. Val, H. Dominguez et al., "Determination of fluoroquinolones in fish tissues, biological fluids, and environmental waters by liquid chromatography tandem mass spectrometry," Analytical and Bioanalytical Chemistry, vol. 409, no. 27, pp. 6359–6370, 2017.

[14] M. Piatkowska, P. Jedziniak, and J. Zmudzki, "Multiresidue method for the simultaneous determination of veterinary medicinal products, feed additives and illegal dyes in eggs using liquid chromatography-tandem mass spectrometry," Food Chemistry, vol. 197, pp. 571–580, 2016.

[15] O. Lacina, J. Urbanova, J. Poustka, and J. Hajslova, "Identification/quantification of multiple pesticide residues in food plants by ultra-high-performance liquid chromatography-time-of-flight mass spectrometry," Journal of Chromatography A, vol. 1217, no. 5, pp. 648–659, 2010.

[16] H. Abdallah, C. Arnaudguilhem, R. Lobinskib, and F. Jaber, "A multi-residue analysis of sulphonamides in edible animal tissues using QuEChERS extraction and HPLC-MS/MS," Analytical Methods, vol. 7, no. 4, pp. 1549–1557, 2015.

[17] C. Nebot, P. Regal, J. M. Miranda, C. Fente, and A. Cepeda, "Rapid method for quantification of nine sulfonamides in bovine milk using HPLC/MS/MS and without using SPE," Food Chemistry, vol. 141, no. 3, pp. 2294–2299, 2013.

[18] C. Tsai, C. Lin, and W. Wang, "Multi-residue determination of sulfonamide and quinolone residues in fish tissues by high performance liquid chromatography-tandem mass spectrometry (LC-MS/MS)," Journal of Food and Drug Analysis, vol. 20, pp. 674–680, 2012.

[19] W. Jansomboon, S. K. Boontanon, N. Boontanon, C. Polprasert, and C. T. Da, "Monitoring and determination of sulfonamide antibiotics (sulfamethoxydiazine, sulfamethazine, sulfamethoxazole and sulfadiazine) in imported Pangasius catfish products in Thailand using liquid chromatography coupled with tandem mass spectrometry," Food Chemistry, vol. 212, pp. 635–640, 2016.

[20] S. B. Turnipseed, S. B. Clark, J. M. Storey, and J. R. Carr, "Analysis of veterinary drug residues in frog legs and other aquaculture species using liquid chromatography quadrupole time-of-flight Mass Spectrometry," Journal of Agricultural and Food Chemistry, vol. 60, no. 18, pp. 4430–4439, 2012.

[21] S. J. Lehotay, K. Mastovska, A. R. Lightfield, and R. A. Gates, "Multi-Analyst, multi-matrix performance of the QuEChERS approach for pesticide residues in foods and feeds using HPLC/MS/MS analysis with different calibration techniques," Journal of AOAC International, vol. 93, pp. 355–367, 2010.

[22] Commission Decision 2002/657/EC, "Implementing Council Directive 96/23/EC concerning the performance of analytical methods and the interpretation of results," *Official Journal of the European Communities*, L221/8, 2002.

[23] MAPA, "Ministério da agricultura, pecuária e abastecimento. Guia de validação e controle de qualidade analítica: fármacos em produtos para alimentação e medicamentos veterinários," 2011, https://bibliotecaquimicaufmg2010.files.wordpress.com/2012/02/guia-de-validac3a7c3a3o-e-controle-de-qualidade-analc3adtica.pdf.

[24] M. E. Dasenaki and N. S. Thomaidis, "Multi-residue determination of seventeen sulfonamides and five tetracyclines in fish tissue using a multistage LC–ESI–MS/MS approach based on advanced mass spectrometric Techniques," *Analytica Chimica Acta*, vol. 672, no. 1-2, pp. 93–102, 2010.

[25] F. G. Sampaio, M. L. Carra, C. M. Jonsson et al., "Effects of dietary exposure to Sulfamethazine on the hematological parameters and hepatic oxidative stress biomarkers in Nile tilapia (*Oreochromis niloticus*)," *Bulletin of Environmental Contamination and Toxicology*, vol. 97, no. 4, pp. 528–535, 2016.

[26] J. A. R. Paschoal, S. Rath, F. P. S. Airoldi, and F. G. R. Reyes, "Validação de métodos cromatográficos para a determinação de resíduos de medicamentos veterinários em alimentos," *Química Nova*, vol. 31, no. 5, pp. 1190–1198, 2008.

[27] A. L. Capriotti, C. Cavaliere, A. Lagana, S. Piovesana, and R. Samperi, "Recent trends in matrix solid-phase dispersion," *Trends in Analytical Chemistry*, vol. 43, pp. 53–66, 2013.

[28] J. C. Hashimoto, J. A. Paschoal, S. C. N. Queiroz, V. L. Ferracini, M. R. Assalin, and F. G. R. Reyes, "A simple method for the determination of malachite green and leucomalachite green residues in fish by a modified QuEChERS extraction and HPLC-MS/MS," *Journal of AOAC International*, vol. 95, no. 3, pp. 913–922, 2012.

[29] A. Kruve, A. Kunnapas, K. Herodes, and I. Leito, "Matrix effects in pesticide multiresidue analysis by liquid chromatography-mass spectrometry," *Journal of Chromatography A*, vol. 1187, no. 1-2, pp. 58–66, 2008.

[30] A. Kruve, I. Leito, and K. Herodes, "Combating matrix effects in LC/ESI/MS: the extrapolative dilution approach," *Analytica Chimica Acta*, vol. 651, no. 1, pp. 75–80, 2009.

[31] S. C. N. Queiroz, V. L. Ferracini, and M. A. Rosa, "Multi-residue method validation for determination of pesticides in food using QuEChERS and UPLC-MS/MS," *Quimica Nova*, vol. 35, pp. 185–192, 2012.

[32] C. C. Pericás, A. Maquieira, R. Puchades, B. Company, and J. Miralles, "Multiresidue determination of antibiotics in aquaculture fish by HPLC-MS/MS," *Aquaculture Research*, vol. 41, pp. 217–225, 2010.

[33] G. Stubbings and T. Bigwood, "The development and validation of a multiclass liquid chromatography tandem mass spectrometry (LC-MS/MS) procedure for the determination of veterinary drug residues in animal tissue using a QuEChERS (quick, easy, cheap, effective, rugged and safe) approach," *Analytica Chimica Acta*, vol. 637, no. 1-2, pp. 68–78, 2009.

[34] K. Zhang, J. W. Wong, P. Yang et al., "Multiresidue pesticide analysis of agricultural commodities using acetonitrile salt-out extraction, dispersive solid-phase sample clean-up, and high-performance liquid chromatography tandem mass spectrometry," *Journal of Agricultural and Food Chemistry*, vol. 59, no. 14, pp. 7636–7646, 2011.

[35] S. C. Nanita and N. L. T. Padivitage, "Ammonium chloride salting out extraction/cleanup for trace-level quantitative analysis in food and biological matrices by flow injection tandem mass spectrometry," *Analytica Chimica Acta*, vol. 768, pp. 1–11, 2013.

[36] S. P. Quesada, J. A. R. Paschoal, and F. G. Reyes, "A simple method for the determination of fluoroquinolone residues in tilapia (*Oreochromis niloticus*) and pacu (*Piaractus mesopotamicus*) employing LC-MS/MS QtoF," *Food Additives and Contaminants A*, vol. 30, no. 5, pp. 813–825, 2013.

[37] S. Y. Won, C. H. Lee, H. S. Chang, S. O. Kim, S. H. Lee, and D. S. Kim, "Monitoring of 14 sulfonamide antibiotic residues in marine products using HPLC-PDA and LC-MS/MS," *Food Control*, vol. 22, no. 7, pp. 1101–1107, 2011.

[38] S. Summa, S. L. Magro, A. Armentano, and M. Muscarella, "Development and validation of an HPLC/DAD method for the determination of 13 sulphonamides in eggs," *Food Chemistry*, vol. 187, pp. 477–484, 2015.

[39] S. Wang, H. Y. Zhang, L. Wang, Z. J. Duan, and I. Kennedy, "Analysis of sulphonamide residues in edible animal products: a review," *Food Additives and Contaminants*, vol. 23, no. 4, pp. 362–384, 2006.

[40] JPL, "The Japanese positive list system for agricultural chemical residues in foods. MRLs of agricultural chemicals, feed additives and veterinary drugs in foods. Ministry of Health Labour and Welfare Japan," 2006, http://www.ffcr.or.jp/zaidan/ffcrhome.nsf/pages/mrls-p/.

6

Rapid and Simultaneous Prediction of Eight Diesel Quality Parameters through ATR-FTIR Analysis

Maurilio Gustavo Nespeca (iD),[1] **Rafael Rodrigues Hatanaka,**[1] **Danilo Luiz Flumignan,**[2] **and José Eduardo de Oliveira**[1]

[1]*Centro de Monitoramento e Pesquisa da Qualidade de Combustíveis, Biocombustíveis, Petróleo e Derivados (Cempeqc), São Paulo State University (UNESP), R. Prof. Francisco Degni 55 Quitandinha, 14800-900 Araraquara, SP, Brazil*
[2]*Instituto Federal de Educação, Ciência e Tecnologia de São Paulo (IFSP), Campus Matão, Rua Estéfano D'avassi, 625 Nova Cidade, 15991-502 Matão, SP, Brazil*

Correspondence should be addressed to Maurilio Gustavo Nespeca; mauriliogn@gmail.com

Academic Editor: Karoly Heberger

Quality assessment of diesel fuel is highly necessary for society, but the costs and time spent are very high while using standard methods. Therefore, this study aimed to develop an analytical method capable of simultaneously determining eight diesel quality parameters (density; flash point; total sulfur content; distillation temperatures at 10% (T10), 50% (T50), and 85% (T85) recovery; cetane index; and biodiesel content) through attenuated total reflection Fourier transform infrared (ATR-FTIR) spectroscopy and the multivariate regression method, partial least square (PLS). For this purpose, the quality parameters of 409 samples were determined using standard methods, and their spectra were acquired in ranges of 4000–650 cm^{-1}. The use of the multivariate filters, generalized least squares weighting (GLSW) and orthogonal signal correction (OSC), was evaluated to improve the signal-to-noise ratio of the models. Likewise, four variable selection approaches were tested: manual exclusion, forward interval PLS (FiPLS), backward interval PLS (BiPLS), and genetic algorithm (GA). The multivariate filters and variables selection algorithms generated more fitted and accurate PLS models. According to the validation, the FTIR/PLS models presented accuracy comparable to the reference methods and, therefore, the proposed method can be applied in the diesel routine monitoring to significantly reduce costs and analysis time.

1. Introduction

Diesel fuel is a petroleum-derived product of great importance for a country's economy since most of the transportation of industrial and agricultural products depends on diesel vehicles [1, 2]. This fuel is a complex mixture composed mainly of paraffinic, olefinic, and aromatic hydrocarbons ranging from 8 to 28 carbon atoms and, in a lower concentration, substances containing oxygen, nitrogen, sulfur, and metals [3–5]. The diesel composition is influenced by several factors, such as the origin of crude oil, operating variables of the refinery, the addition of fractions from cracking process, and the insertion of additives to increase engine performance [3]. Therefore, the fuel quality is susceptible to many variables until the fuel reaches the consumer. In this perspective, the monitoring of diesel quality parameters is extremely important for commercialization, engine performance, consumer rights, business competition, and environmental risks [5, 6].

The assays performed to ensure the diesel quality are based on standardized procedures that require specific equipment to determine each physicochemical parameter. According to the standard methods, the quality assessment requires considerable sample volume and analysis time, besides the great expense of equipment maintenance and several specialized analysts [7–12]. Therefore, the development of methods to monitor diesel quality accurately, quickly, and environmentally friendly is highly necessary [13]. This becomes possible by attenuated total reflection Fourier transform infrared (ATR-FTIR) spectroscopy associated with

TABLE 1: Standard methods and equipment used to determine the quality parameters of diesel samples.

Diesel property	Method	Equipment
Density	ASTM D4052	DMA 4500 automatic densimeter (Anton Paar)
Flash point	ASTM D93	PMA-4 flash point automatic analyzer (Petrotest)
Total sulfur	ASTM D4294	EDX-800 spectrometer (Shimadzu)
Distillation	ASTM D86	AD-6 automatic atmospheric distiller (Tanaka)
Cetane index	ASTM D4737	—
Biodiesel content	EN 14078	Nicolet IR200 spectrometer (Thermo Scientific)

multivariate regression methods such as partial least square (PLS). Studies demonstrated the possibility to predict some diesel properties using midinfrared spectroscopy combined with chemometric tools [14–18], some aimed at the prediction of biodiesel content [16], and others were devoted to the identification of diesel adulteration with waste vegetable oils [17, 18].

In USA, European Community, and Japan, the regulations of diesel properties for consumption are established, respectively, by ASTM D975, EN 590, and JIS K2204 [19–21]. In Brazil, the regulation and supervision of fuels are performed according to ANP (National Agency of Petroleum, Natural Gas, and Biofuels) Resolution no. 30/2016, which requires that assays must be conducted according to ASTM, EN, or NBR standards [22]. According to this resolution, at least eight quality parameters of diesel are analyzed in official monitoring laboratories: aspect; color; density; flash point; total sulfur; volatility (distillation temperatures at 10% (T10), 50% (T50), and 85% (T85) recovery); cetane index; and biodiesel content [23].

The development of an alternative method for determining the physicochemical parameters of diesel through ATR-FTIR has several advantages for routine quality monitoring. The use of ATR-FTIR can reduce costs, increase analytical frequency, use smaller sample volume, and provide the determination of all required parameters using only one equipment. Moreover, infrared spectrometers are already purchased by monitoring laboratories for determination of biodiesel content in diesel according to EN 14078.

In view of the high costs and long time required to assess diesel quality by standard methods, this work aimed at the development of a simple and fast analytical method based on ATR-FTIR analysis and PLS regression method to determinate eight diesel quality parameters simultaneously. In this study, multivariate filters and variable selection techniques, such as genetic algorithm (GA), forward interval PLS (FiPLS), and backward interval PLS (BiPLS), were evaluated for the best model predictive ability.

2. Materials and Methods

2.1. Samples. For eight months, the quality parameters of 3549 samples of diesel fuel were analyzed by Cempeqc (Center for Monitoring and Research of the Quality of Fuels, Biofuels, Crude Oil and Derivatives) according to ASTM and EN standards. The samples were stored at 10°C for further spectroscopic analysis. The standards and equipment used in the determination of quality parameters are presented in Table 1.

Although an extensive sample set can provide greater robustness to a prediction model, this work aimed at the development of a simple method that can be easily reproduced by other laboratories. Therefore, we selected about 10% of the diesel samples for spectroscopic and chemometric analysis. The 3549 diesel samples were divided into groups using hierarchical cluster analysis (HCA) to select the most representative samples. An HCA was executed for each month, and the physicochemical parameters were used as variables. The clusters were performed using 60% of similarity, complete linkage method, and autoscale preprocessing to give the same influence for all variables. The software used for HCA was Pirouette (Infometrix), version 3.11. At the end of the eight months, 409 diesel samples were selected.

2.2. Spectroscopic Analysis. The infrared spectra of the 409 samples were obtained by a Nicolet 6700 FTIR spectrometer (Thermo Scientific, Waltham, USA) using 32 scans and 4 cm^{-1} resolution. A Smart ARK ATR sampling accessory of ZnSe crystal and angle of incidence 45° were used to acquire the infrared spectra. The ATR accessory required one milliliter for each sample, and a new background spectrum was acquired every hour to reduce the baseline shifting and ambient variations. The conditions of temperature and relative humidity during the analysis were 20.7 ± 2.0°C and 40 ± 9%, respectively.

2.3. Chemometric Analysis

2.3.1. Model Development. The chemometric analysis was executed using Matlab 2013a (MathWorks) with PLS toolbox 7.3.1 (Eigenvector Research Inc.). The FTIR spectra were converted into vectors of 1738 variables and the combination of the vectors resulted in the matrix **X** of dimension 409 by 1738. Prior to the development of PLS models, the sample set was separated into two-thirds for calibration (273 samples) and one-third for validation (136 samples). The Onion algorithm was used to select the samples with less covariance (based on distance from the mean) for each set and, consequently, to obtain greater sample representativeness in both sets [24, 25]. The algorithm was performed for each parameter to ensure that the calibration set had the largest range of reference values.

Initially, the PLS models were developed using the full spectra (full X-block) preprocessed by the mean center or autoscale, depending on the best fit. The number of latent variables (LV) was chosen based on the root mean square errors of calibration (RMSEC), cross-validation (RMSECV), and prediction (RMSEP) in order to minimize the prediction errors and avoid model overfitting [26, 27]. The cross-validation was performed using venetian blinds mode with 10 splits.

Then, statistical tests were applied according to ASTM E1655 [28] to detect the presence of outliers in the calibration and validation sets. Outliers include high leverage samples and samples whose reference values are inconsistent with the model. Therefore, samples with high leverage and studentized residuals were excluded from the sample sets.

2.3.2. Preprocessing Evaluation.

Spectral data usually present baseline shifting due to instrumental variations and reflectance deviations [29]. The baseline shifting is typically corrected by applying the first or second derivative, or by polynomials that correct the displacement based on a standard spectrum, for example, multiplicative scatter correction (MSC) and standard normal variate (SNV). In addition, digital filters such as smoothing are also used to improve the signal-to-noise ratio of spectral data [28]. Multivariate filters, such as generalized least squares weighting (GLSW) and orthogonal signal correction (OSC), are less usual preprocesses, but these filters are very useful to eliminate baseline shifting and increase signal-to-noise ratio [30–33]. Therefore, the following preprocessing was evaluated in modeling: mean center, autoscale, Savitzky–Golay smoothing and derivatives, SNV, MSC, GLSW, and OSC.

2.3.3. Variable Selection Methods.

Many studies have shown that variable selection is an efficient way to increase the signal-to-noise ratio and, as a consequence, improve the predictive ability of the model [34, 35]. When the noise dominates over the information related to the property of interest, the removal of variables often leads to better accuracy and performance of the analytical method [35, 36]. The selection of variables can be performed based on the spectral knowledge (manual approach) or through algorithms that search for variables that provide the minimum prediction error to the model. Some of the most popular methods for selecting variables are the interval selection method, such as the forward interval PLS (FiPLS), the backward interval PLS (BiPLS), and the genetic algorithm (GA), a technique that employs a probabilistic and nonlocal search process which manipulates binary strings with the coded experimental variables. Details on these variable selection methods can be found in [35].

In this study, four different approaches were evaluated to select variables: manual exclusion, FiPLS, BiPLS, and GA. The manual exclusion was carried out evaluating the spectral residues and loadings plots. Spectral regions with no absorbance or high relative standard deviation (RSD) were excluded from the data and compared with results obtained using the full spectra. Both iPLS methods were executed using interval size of 25 variables, and the number of intervals was determined by the algorithm to obtain the lowest

value of RMSECV. The GA was performed with a population size of 128 models, one variable by window, initial terms of 30%, the mutation rate of 0.5%, double crossover, 200 generations, and PLS regression method. All approaches were performed using only the calibration set to avoid overestimated results.

2.3.4. Model Validation.

The PLS models were statistically evaluated by figures of merit (FOM) according to ASTM E1655 and Valderrama et al. [28, 37]. The accuracy of the models, defined as the degree of agreement between a measured value and reference value, was assessed by the values of RMSECV, RMSEP, correlation coefficients (r), average relative errors (ARE), and relative percent difference (RPD). The RMSECV was obtained by cross-validation using the venetian blinds mode with 10 splits, and the RMSEP was obtained by the validation samples that were measured independently from the calibration samples. Then, the RMSECV and RMSEP were compared with the reproducibility of the reference method. The ARE was used as a parameter to evaluate the magnitude of the prediction errors in relation to the reference values [38]. The ARE value was calculated by

$$\text{ARE}(\%) = \frac{\sum_{i=1}^{n_v} (\widehat{y}_i - y_i/y_i)^2}{n_v} \times 100, \quad (1)$$

where y_i and $\widehat{y}_i$ correspond, respectively, to the reference value and predicted value by the model and n_v is the number of validation samples. The relative percent difference (RPD) was obtained by the ratio of the standard deviation of the validation set reference values to the RMSEP value. RPD values above 2.5 indicate that the model has acceptable accuracy over the measurement range, while values above 10 are considered excellent for alternative methods [39].

Linearity is an important parameter to evaluate the performance of the model since the PLS regression method is not suitable for nonlinear relationships between the variables x and the property of interest [40]. The linearity corresponds to the ability of the model to provide results directly proportional to the property of interest. One way to evaluate this parameter in multivariate models is through the residues of calibration and validation samples plots. If the distribution of residues is random, it can be said that the model shows a linear behavior. In addition to the residue plots, the linearity was also evaluated by the values of determination coefficients (R^2) and bias. This last FOM indicates the presence of systematic errors in the model. Bias can be assessed by a t-test for the validation samples at a confidence interval of 95%. The average bias was calculated by summing the differences between the reference value and the predicted value divided by the number of validation samples [28]:

$$\text{bias} = \frac{\sum_{i=1}^{n_v} (y_i - \widehat{y}_i)^2}{n_v}. \quad (2)$$

Then, the standard deviation of validation errors (SDV) was calculated as

$$\text{SDV} = \sqrt{\frac{\sum [(y_i - \widehat{y}_i) - bias]^2}{n_v - 1}}, \quad (3)$$

TABLE 2: Results of quality parameter assays using the reference methods in accordance with ANP Resolution no. 65 [41].

Quality parameter	Unit	Reproducibility	Repeatability	Range of conformity	Range of measured values	Number of nonconforming samples
Density	kg·m^{-3}	±0.5	±0.1	820.0 to 865.0	830.6 to 860.4	0
Flash point	°C	±3	±1	38	8 to 70	16
Total sulfur	% (w/w)	±0.01	±0.002	<0.05[a] or < 0.18[b]	0.01 to 0.19	4
T10	°C	±4.2	±1.8	>180.0	158.9 to 234.1	75
T50	°C	±3.0	±0.9	245.0 to 310.0	250.9 to 305.8	0
T85	°C	±5.2	±1.4	>360.0	247.3 to 369.5	40
Cetane index	—	±2.0	—	>42.0	41.1 to 54.2	1
Biodiesel content	% (v/v)	±0.2	±0.1	4.5–5.5	0.6 to 7.4	41

[a]For S50 diesel samples, maximum sulfur content is 0.05% (w/w); [b]for S1800 diesel samples, maximum sulfur content is 0.18% (w/w).

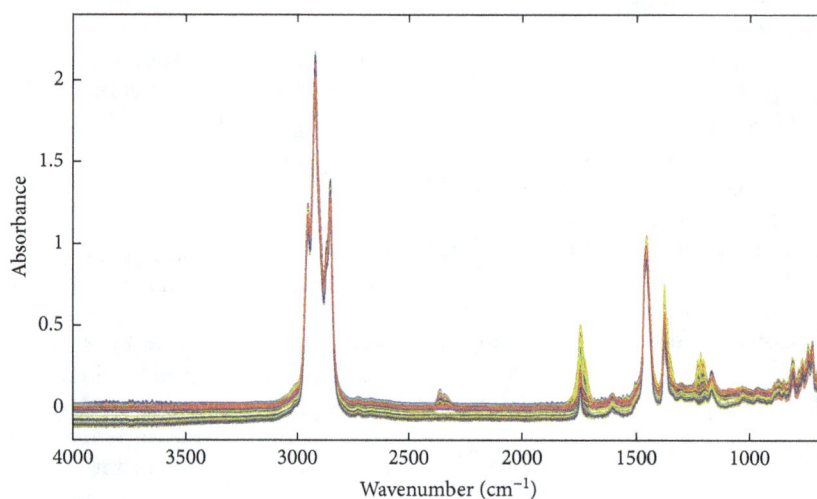

FIGURE 1: Infrared spectra of all diesel samples.

TABLE 3: Infrared vibrational groups of the diesel samples [42, 43].

Attribution	Wavenumber (cm^{-1})
CH$_3$ asymmetrical stretch	2953
CH$_3$ symmetric stretch	2870
CH$_3$ angular deformation	1379
CH$_2$ asymmetric stretch	2922
CH$_2$ symmetrical stretch	2853
CH$_2$ angular deformation	1464
CO$_2$ asymmetrical stretch	2350
CO$_2$ angular deformation	667
C=O carbonyl stretch	1750–1735
C–O stretch (aliphatic ester)	1300–1000
C=C stretch (alkenes)	1660–1600
C=C stretch (aromatic)	1600 and 1475
=C–H stretch (aromatic)	900–690
S–H stretch	2600–2550
C–S stretch	700–600

and finally, the value of t_{bias} was given by

$$t_{bias} = \frac{|\text{bias}|\sqrt{n_v}}{\text{SDV}}. \tag{4}$$

If the value obtained for t_{bias} was greater than the critical value for $n_v - 1$ degrees of freedom, then the multivariate model presented significant systematic errors.

The precision of the models was evaluated by the analysis of 14 replicates of 30 diesel samples performed on different days. The average of relative standard deviations (RSD) and the intermediate precision—calculated through (5), where n is the number of samples and m the number of replicates—were used as parameters [37]. Then, the intermediate precision was compared to the repeatability value of the reference method:

$$\text{intermediate precision} = \sqrt{\frac{\sum_{i=1}^{n}\sum_{j=1}^{m}\left(y_i - \widehat{y}_i\right)^2}{n(m-1)}}. \tag{5}$$

3. Results and Discussion

3.1. Physicochemical Assays. The values of reproducibility and repeatability of the reference methods, the range of

TABLE 4: The number of outliers removed from the calibration and validation sets.

	Density	Flash point	Total sulfur	T10	T50	T85	Cetane index	Biodiesel content
Calibration (273 samples)	2 (1%)	1 (1%)	2 (1%)	3 (1%)	3 (1%)	4 (2%)	3 (1%)	0
Validation (136 samples)	3 (2%)	1 (1%)	2 (2%)	4 (3%)	2 (2%)	1 (1%)	3 (2%)	0

TABLE 5: Comparison between mean center/autoscale and multivariate filters.

	Mean center/autoscale*				Multivariate filter			
	Number of LVs	RMSEP	r (val)	X-block variance (%)	y-block variance (%)	Number of LVs	RMSEP	r (val)	X-block variance (%)	y-block variance (%)
Density	12	0.5	0.9959	99.76	99.17	9	0.5	0.9957	99.99	99.14
Flash point	12	2.3	0.8735	99.72	83.78	8	2.3	0.8761	99.99	83.20
Total sulfur	7	0.013	0.9773	99.31	94.52	6	0.013	0.9806	99.99	95.05
T10	11	4.3	0.9192	99.69	86.90	9	4.2	0.9224	99.99	88.04
T50	7	3.5	0.9432	99.43	86.18	6	3.2	0.9508	89.92	89.47
T85	11	4.8	0.7909	99.67	70.31	8	4.7	0.7991	99.98	66.18
Cetane index	10	0.6	0.9084	99.63	83.02	9	0.6	0.9103	99.99	83.78
Biodiesel content	10	0.2	0.9337	99.71	88.50	5	0.2	0.9302	97.13	89.85

*The models for density, cetane index, and biodiesel content were preprocessed using mean centering.

measured values of each quality parameter, and the number of samples in nonconformity with ANP Resolution no. 65 [41] are shown in Table 2. The quality parameter that presented the highest number of nonconforming samples was T10, followed by T85 and biodiesel content. As ANP Resolution no. 65 allows only a variation of 0.5% (v/v) of biodiesel content, most of the samples were in a narrow range of concentration. The same occurred with the total sulfur but in two different ranges of concentration due to the availability of two types of commercial diesel with distinct sulfur content.

3.2. Spectroscopic Analysis.
The FTIR spectra of all diesel samples are represented in Figure 1. Functional groups of the constituents of samples could be observed by characteristic absorption bands of each group of atoms through the infrared spectra.

The most intense bands were caused by C–H groups stretch (3000–2800 cm^{-1}) and angular deformations (1464 cm^{-1} and 1379 cm^{-1}) [42]. The bands at 2350 cm^{-1} and 667 cm^{-1} were, respectively, results of asymmetrical stretch and angular deformation of CO_2 molecules present in the atmosphere [43]. The presence of biodiesel in the samples was observed by carbonyl absorption band (1750–1735 cm^{-1}) and aliphatic ester absorption band (1300–1000 cm^{-1}). Aromatic compounds had characteristic bands of low intensity in 900–675 cm^{-1} from the C–H out-of-plane angular deformation. The sulfur is present in diesel as mercaptans and sulfides, and it was observed by S–H axial stretch at 2600–2550 cm^{-1} and C–S axial stretch at 700–650 cm^{-1} [43, 44]. The S–H stretch was very weak; however, few groups have absorption in this region, so it was useful for the total sulfur parameter. The vibrational group attribution to each band is present in Table 3.

3.3. Chemometric Analysis

3.3.1. Outlier Detection.
During calibration, outlier statistics were applied to identify samples that had unusual leverage and studentized residuals. The outlier detection was performed prior to the variable selection because the exclusion of variables may reduce outlier detection capabilities of the model [28]. The number of outliers from each sample set is shown in Table 4. Considering the calibration and validation sample set with, respectively, 273 and 136 samples, the number of outliers (3% maximum) was not significant for the prediction models.

High studentized residual values may be the result of errors in the reference measurement, spectral acquisition error, reference value transcription, or even a failure of the model. Error in the spectral acquisition would lead to the presence of the same outlier in all models of prediction; however, different outliers were detected for each model. The absence of new outliers in the model after removal of the anomalous samples indicated that there was no failure in the model. Therefore, errors in the reference values were most likely responsible for the presence of outliers.

3.3.2. Preprocessing Evaluation.
The baseline shifting in the raw spectra was observed in Figure 1. The shifting may be the result of variations in the position of the ZnSe crystal since it was removed from the spectrometer for cleaning before each analysis. All the evaluated preprocessing—derivatives, MSC, SNV, GLSW, and OSC—provided baseline correction and higher correlation coefficients than mean center or autoscale preprocessing. Moreover, multivariate filters (GLSW and OSC) provided models with greater explained variance using fewer latent variables (Table 5). Therefore, all models were

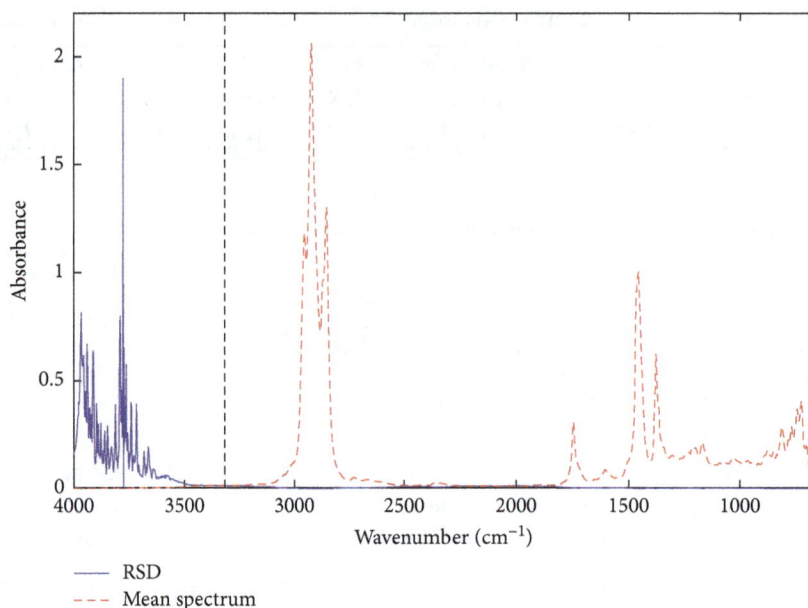

FIGURE 2: The mean spectrum (red line) and relative standard deviation (blue line) of 14 replicates of a diesel sample.

TABLE 6: PLS models using different variable selection approaches.

Variable selection	Number of variables	Number of LVs Density	RMSEP	r (val)	Number of variables	Number of LVs Flash point	RMSEP	r (val)
None	1738	9	0.50	0.996	1738	8	2.30	0.876
Manual	1124	10	0.55	0.995	1124	7	2.20	0.888
FiPLS	225	8	0.46	0.997	300	6	2.23	0.885
BiPLS	1363	8	0.49	0.996	1663	8	2.19	0.889
GA	439	9	0.43	0.997	380	7	2.17	0.889
		Sulfur content				*T10*		
None	1738	6	0.013	0.981	1738	9	4.19	0.922
Manual	1124	6	0.014	0.976	1124	10	4.16	0.923
FiPLS	175	5	0.011	0.987	350	8	4.24	0.920
BiPLS	1363	7	0.011	0.985	1563	10	4.13	0.925
GA	434	7	0.011	0.986	370	9	4.24	0.921
		T50				*T85*		
None	1738	6	3.20	0.951	1738	8	4.70	0.799
Manual	1124	5	3.24	0.951	1124	8	4.78	0.793
FiPLS	75	8	3.26	0.949	75	6	4.90	0.775
BiPLS	1638	6	3.10	0.954	1088	10	4.75	0.797
GA	410	5	2.97	0.957	395	6	4.81	0.788
		Cetane index				*Biodiesel content*		
None	1738	9	0.64	0.910	1738	5	0.21	0.930
Manual	1124	8	0.63	0.912	1124	9	0.20	0.936
FiPLS	225	7	0.62	0.914	150	8	0.21	0.932
BiPLS	1363	7	0.61	0.917	873	8	0.20	0.935
GA	405	6	0.62	0.914	452	6	0.20	0.937

preprocessed using OSC, except the model for T85, which presented better fit with GLSW preprocessing.

3.3.3. Variable Selection. The exclusion of regions without information of sample constituents or low signal-to-noise ratio may improve the performance of the models [35]. In Figure 2, the noisy spectral regions can be observed through the relative standard deviation (RSD), represented by the blue line, and calculated from the mean of 14 replicates, represented by the red line. In addition, there was no

TABLE 7: Validation results of prediction models for quality parameters of diesel.

	Density	Flash point	Sulfur content	T10	T50	T85	Cetane index	Biodiesel content
Unit	kg·m^{-3}	°C	% (w/w)	°C	°C	°C	—	% (v/v)
Measured interval	831.6 to 860.4	8 to 70	0.01 to 0.19	158.9 to 234.1	256.3 to 305.8	330 to 369.5	42.2 to 51.0	0.6 to 7.4
Variable selection	GA	GA	FiPLS	BiPLS	GA	None	BiPLS	GA
Number of variables	439	380	175	1563	410	1738	1363	452
Number of LVs	9	7	5	10	5	8	7	6
Accuracy								
RMSEC	0.4	2	0.01	3.9	3.0	4.6	0.6	0.2
RMSECV	0.5	2	0.01	4.4	3.4	5.1	0.6	0.2
RMSEP	0.4	2	0.01	4.1	3.0	4.7	0.6	0.2
r (cal)	0.997	0.930	0.987	0.940	0.954	0.814	0.919	0.946
r (CV)	0.996	0.900	0.984	0.922	0.941	0.770	0.903	0.936
r (val)	0.997	0.889	0.987	0.925	0.957	0.799	0.917	0.937
ARE (%)	0.04	3.72	14.10	1.68	0.77	1.06	0.95	3.18
RPD	12.89	2.19	5.93	2.63	3.44	1.64	2.49	2.83
Linearity								
R^2 cal	0.994	0.866	0.974	0.883	0.909	0.662	0.845	0.894
R^2 CV	0.993	0.809	0.969	0.851	0.886	0.593	0.815	0.877
R^2 val	0.994	0.791	0.974	0.855	0.916	0.639	0.841	0.878
Bias	0.002	0.049	−0.002	0.325	0.340	0.822	−0.086	−0.037
t_{bias}	0.05	0.26	2.20	0.91	1.33	2.06	1.66	2.16
Intermediary precision	1.5	2	0.01	1.9	1.5	2.1	0.4	0.1
RSD (%)	0.04	5.91	9.00	0.93	0.44	0.47	0.40	0.95

*t critical (95% confidence level; 135 degrees of freedom) = 1.98.

absorption by the components of diesel in the ranges 4000–3100 cm^{-1} and 2450–1950 cm^{-1}; thus, these spectral regions were excluded, and new models were developed.

The RMSEP and correlation coefficient of validation (r_{val}) obtained by the different variable selection approaches are presented in Table 6. The manual exclusion of variables provided better results only for the prediction models of flash point, T10, cetane index, and biodiesel content. The manual selection of variables had the risk of inadvertent exclusion of important variables for the modeling, impairing the performance of the model.

The selection of variables by interval selection methods reduces the values of RMSEC and RMSECV but might decrease the predictive ability of the model. The FiPLS method usually uses few intervals to correlate the spectral variables with the property of interest and, as consequence, the calibration model is more susceptible to overfitting and the prediction of unknown samples is impaired, especially properties that are correlated to several spectral variables. As the sulfur content is correlated only to the S–H and C–H bond variables, the FiPLS method provided the best fit to the model.

The distillation temperatures and the cetane index depend on the size and structure of the hydrocarbon chains of the diesel components; therefore, these are properties related to several functional groups with response in the midinfrared region. Thus, the selection methods such as BiPLS, which seek to exclude noisy variables rather than including variables more correlated to the property of interest, tend to be more suitable for optimization of these diesel parameters.

The analytical signals in the midinfrared region result in many correlated variables; that is, FTIR data present many collinearities. Normally, the problem of collinearity can be attenuated by the application of the genetic algorithm, since the spectral variables are manipulated in binary strings and the search for variables that provide a minor error of prediction is performed by a probabilistic and nonlocal process [35]. GA was the best variable selection approach for prediction models of density, flash point, T50, and biodiesel content.

In general, the selection of variables by iPLS and GA provided improvements in the predictive ability of the calibration models, except for T85, and the difference between the results obtained by both algorithms was not significant. The selected variables used in the best-fitted models are presented in the supplementary material (available here) attached to the article.

3.3.4. Model Performance. After defining the most appropriate variable selection method for each parameter, the figures of merit were determined for the prediction models

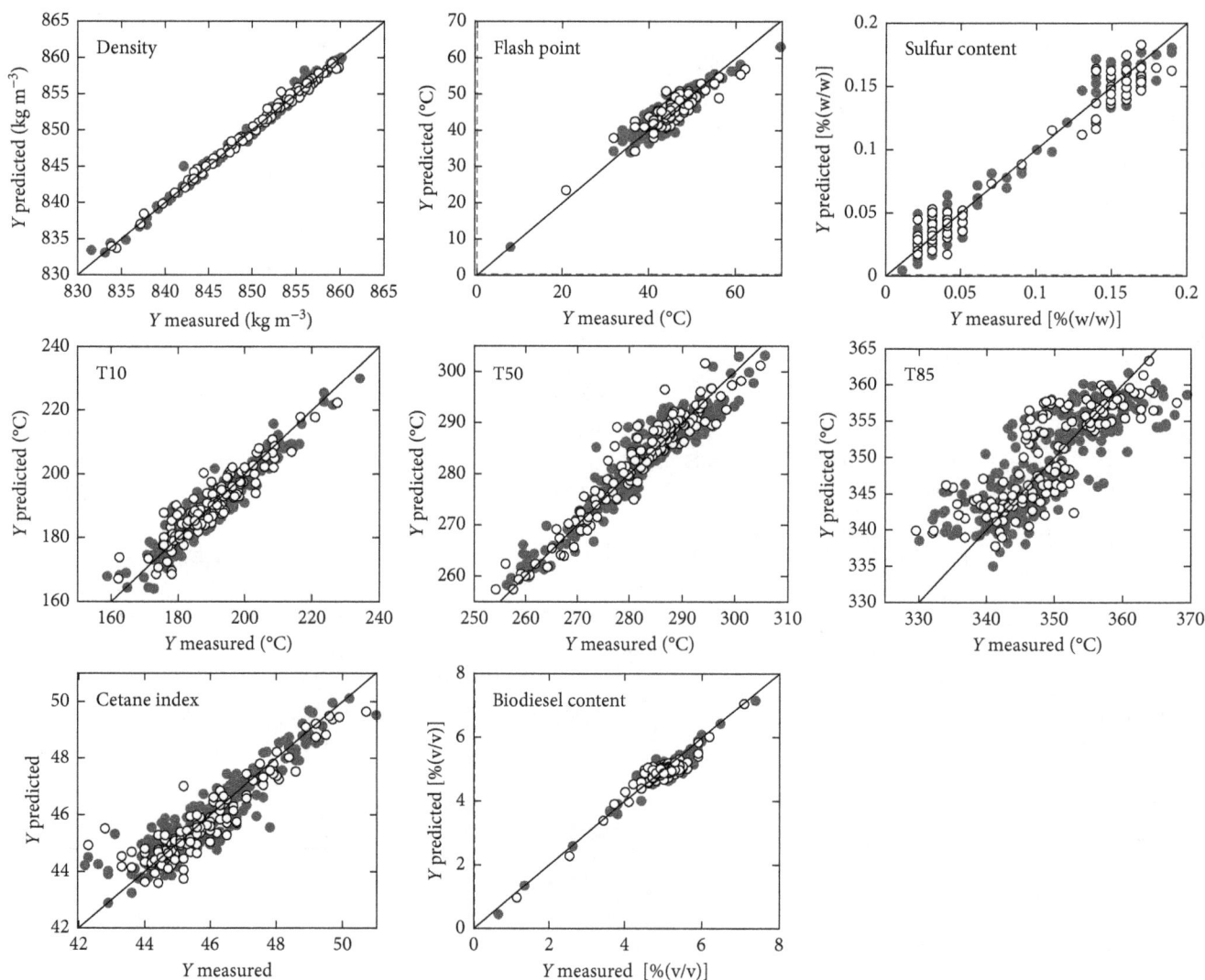

FIGURE 3: Predicted versus measured value plots of the PLS models.

(Table 7). The complexity of the diesel composition, consisting of hundreds of compounds, generates a large amount of information in the FTIR spectra and, therefore, the correlation between the matrix **X** and the property of interest requires a considerable number of LVs. Although the use of OSC reduces the collinearity problem and increases the captured variance of the **X** and **y** blocks, several analytical signals were correlated to the properties of diesel, so several LVs were required.

The accuracy of the models was evaluated by comparing the RMSEP values (Table 7) with the reproducibility values of the reference methods (Table 2). Since all models presented RMSEP values below or equivalent to the reproducibility value, the FTIR/PLS method could be considered accurate for predicting diesel parameters. In addition, the correlation coefficients were above 0.89, except for T85; thus, the predicted values were well correlated with the reference values (Figure 3).

Although the prediction model for sulfur content presented r_{val} equal to 0.987, the obtained ARE value was high when compared to the others. The high relative errors that

resulted in ARE equal to 14.10% were caused by the low sensitivity of the model for prediction of S500 diesel samples. However, the RPD value indicated that the model was accurate when the RMSEP value is compared to the sulfur content range of the validation sample set.

The determination coefficients (R^2) indicated that the prediction models of flash point and T85 presented lower linearity than the other parameters. Figure 3 shows that, for these parameters, the residues tend to be negative values with the increase of the reference value. Although the models have low bias values, the t-test revealed that there were systematic errors in the prediction models for sulfur content, T85, and biodiesel content. Since the models for sulfur and biodiesel content presented good linearity ($R^2_{val} > 0.88$), the presence of systematic errors can be reduced by the addition of more samples to the model.

The precision of the models was evaluated by the analysis of 30 diesel samples on 14 consecutive days. Although the samples were stored at 10°C between the analyses, the diesel fuel consists of semivolatile compounds and, therefore, changes in sample composition during the replicates acquisition imply an increase in measurement uncertainty. The intermediate precision values of the models were above the repeatability values of the reference methods, except for the biodiesel content prediction. However, the RSD values showed that almost all models had good precision (RSD below 1%) and only the models for prediction of flash point and sulfur content presented low precision.

Although the prediction models for flash point, sulfur content, and T85 have the limitations mentioned above, the conformity ranges of these parameters (Table 2) can be met by FTIR/PLS models with reliability since the accuracy and precision of the method are known. If an unknown sample is analyzed by FTIR and the result obtained is in the nonconformity range, it is recommended that the result is confirmed by the standard method. Since only about 2% of the diesel samples in Brazil presented nonconformities in 2017 [45], the FTIR/PLS method can be applied in routine monitoring of diesel quality to reduce the costs and time of analysis.

4. Conclusions

This study showed the possibility of applying ATR-FTIR spectroscopy with PLS regression method to predict the quality parameters (density; flash point; total sulfur content; distillation temperatures at 10% (T10), 50% (T50), and 85% (T85) recovery; cetane index; and biodiesel content) in commercial diesel samples.

All the evaluated preprocessing (derivatives, MSC, SNV, GLSW, and OSC) provided baseline correction and higher correlation coefficients. In addition, the GLSW and OSC preprocessing provided greater explained variance to the model using fewer latent variables. The selection of variables by iPLS or GA provided better predictive ability to the calibration models, except for T85. However, the difference between the results obtained by both algorithms was not significant.

According to the model validation, all PLS models presented acceptable accuracy when compared to the values of reproducibility and had good precision, except for sulfur content prediction of S500 diesel samples. Since the application of the ATR-FTIR/PLS method is able to reduce costs and increase considerably the analytical frequency, the diesel quality monitoring programs, as well as the final consumer, can benefit greatly from the application of the proposed method.

Conflicts of Interest

The authors declare that they have no conflicts of interest.

Acknowledgments

The authors would like to thank Capes, CNPq, and Fundunesp for providing academic scholarships and Cempeqc for providing financial support and the samples.

Supplementary Materials

In the supplementary material can be visualized the variables used in the PLS models that presented better predictive abilities. The variable selection method used for each diesel property is also presented in the supplementary material. Supplementary Figure 1: spectra variables of GA-PLS model for density prediction. Supplementary Figure 2: spectra variables of GA-PLS model for flash point prediction. Supplementary Figure 3: spectra variables of FiPLS model for sulfur content prediction. Supplementary Figure 4: spectra variables of BiPLS model for T10 prediction. Supplementary Figure 5: spectra variables of GA-PLS model for T50 prediction. Supplementary Figure 6: spectra variables of PLS model for T85 prediction. Supplementary Figure 7: spectra variables of BiPLS model for cetane index prediction. Supplementary Figure 8: spectra variables of GA-PLS model for biodiesel content prediction. (*Supplementary Materials*)

References

[1] BP, *BP Statistical Review of World Energy*, British Petroleum Company, London, UK, 2016.

[2] M. Colman, P. A. Sorichetti, and S. D. Romano, "Refractive index of biodiesel-diesel blends from effective polarizability and density," *Fuel*, vol. 211, pp. 130–139, 2018.

[3] J. G. Speight, *The Chemistry and Technology of Petroleum*, CRC Press, Boca Raton, FL, USA, 4th edition, 2006.

[4] G. D. Todd, R. L. Chessin, and J. Colman, *Toxicological Profile for Total Petroleum Hydrocarbons (TPH)*, Agency for Toxic Substances and Disease Registry, Atlanta, GA, USA, 1999.

[5] B. P. Vempatapu and P. K. Kanaujia, "Monitoring petroleum fuel adulteration: a review of analytical methods," *TrAC Trends in Analytical Chemistry*, vol. 92, pp. 1–11, 2017.

[6] D. A. Cunha, L. F. Montes, E. V. R. Castro, and L. L. Barbosa, "NMR in the time domain: a new methodology to detect adulteration of diesel oil with kerosene," *Fuel*, vol. 166, pp. 79–85, 2016.

[7] ASTM, *ASTM D4052: Standard Test Method for Density, Relative Density, and API Gravity of Liquids by Digital Density Meter*, ASTM International, West Conshohocken, PA, USA, 2013.

[8] ASTM, *ASTM D93: Standard Test Methods for Flash Point by Pensky-Martens Closed Cup Tester*, ASTM International, West Conshohocken, PA, USA, 2004.

[9] ASTM, *ASTM D4294: Standard Test Method for Sulfur in Petroleum and Petroleum Products by Energy Dispersive X-ray Fluorescence Spectrometry*, ASTM International, West Conshohocken, PA, USA, 2010.

[10] ASTM, *ASTM D86: Standard Test Method for Distillation of Petroleum Products and Liquid Fuels at Atmospheric Pressure*, 2008.

[11] D4737 A, *Standard Test Method for Calculated Cetane Index by Four Variable Equation*, ASTM International, vol. 5West Conshohocken, PA, USA, 2010.

[12] EN, *EN 14078: Determination of fatty acid methyl ester (FAME) content in middle distillates*, Infrared spectrometry method, European Committee for Standardization, Brussels, Belgium, 2009.

[13] P. M. Santos, R. S. Amais, L. A. Colnago, Å. Rinnan, and M. R. Monteiro, "Time domain-NMR combined with chemometrics analysis: an alternative tool for monitoring diesel fuel quality," *Energy and Fuels*, vol. 29, pp. 2299–2303, 2015.

[14] L. A. Pinto, R. K. H. Galvão, and M. C. U. Araújo, "Influence of wavelet transform settings on NIR and MIR spectrometric

analyses of diesel, gasoline, corn and wheat," *Journal of the Brazilian Chemical Society*, vol. 22, pp. 179–186, 2011.

[15] V. O. Santos, F. C. C. Oliveira, D. G. Lima et al., "A comparative study of diesel analysis by FTIR, FTNIR and FT-Raman spectroscopy using PLS and artificial neural network analysis," *Analytica Chimica Acta*, vol. 547, no. 2, pp. 188–196, 2005.

[16] M. Fernanda Pimentel, G. M. G. S. Ribeiro, R. S. Da Cruz, L. Stragevitch, J. G. A. Pacheco Filho, and L. S. G. Teixeira, "Determination of biodiesel content when blended with mineral diesel fuel using infrared spectroscopy and multivariate calibration," *Microchemical Journal*, vol. 82, no. 2, pp. 201–206, 2006.

[17] N. N. Mahamuni and Y. G. Adewuyi, "Fourier transform infrared spectroscopy (FTIR) method to monitor soy biodiesel and soybean oil in transesterification reactions, petrodiesel–biodiesel blends, and blend adulteration with soy oil," *Energy and Fuels*, vol. 23, no. 7, pp. 3773–3782, 2009.

[18] I. P. Soares, T. F. Rezende, and I. C. P. Fortes, "Study of the behavior changes in physical-chemistry properties of diesel/biodiesel (B2) mixtures with residual oil and its quantification by partial least-squares attenuated total reflection-Fourier transformed infrared spectroscopy (PLS/ATR-FTIR)," *Energy and Fuels*, vol. 23, no. 8, pp. 4143–4148, 2009.

[19] ASTM, *ASTM D975-Standard Specification for Diesel Fuel Oils*, ASTM International, West Conshohocken, PA, USA, 2015.

[20] EN, *EN 590: Automotive fuels-Diesel-Requirements and test methods*, European Committee for Standardization, Brussels, Belgium, 2009.

[21] PAJ-Petroleum Association of Japan, *JIS K2204-Petroleum Industry in Japan*, Petroleum Association of Japan, Tokyo, Japan, 2014.

[22] ANP, *Resolução ANP No. 30: Automotive fuels-Diesel-Specification of BX to B30 Diesel Fuel*, Brasília, DF, Brazil, 2016.

[23] ANP, *Panorama da Qualidade dos Combustíveis*, Brasília, DF, Brazil, 2011.

[24] A. G. Sousa, L. I. Ahl, H. L. Pedersen, J. U. Fangel, S. O. Sørensen, and W. G. T. Willats, "A multivariate approach for high throughput pectin profiling by combining glycan microarrays with monoclonal antibodies," *Carbohydrate Research*, vol. 409, pp. 41–47, 2015.

[25] S. Shrestha, L. Deleuran, and R. Gislum, "Classification of different tomato seed cultivars by multispectral visible-near infrared spectroscopy and chemometrics," *Journal of Spectral Imaging*, vol. 1, p. a1, 2016.

[26] D. M. Hawkins, "The problem of overfitting," *Journal of Chemical Information and Computer Sciences*, vol. 44, no. 1, pp. 1–12, 2004.

[27] C. V. Di Anibal, M. P. Callao, and I. Ruisánchez, "1H NMR variable selection approaches for classification. A case study: the determination of adulterated foodstuffs," *Talanta*, vol. 86, pp. 316–323, 2011.

[28] American Society for Testing and Materials, "ASTM E1655-standard practices for infrared multivariate quantitative analysis," *ASTM International*, vol. 5, p. 29, 2012.

[29] P. Gemperline, *Practical Guide to Chemometrics*, Koros Press Limited, London, UK, 2nd edition, 2006.

[30] Eigenvector Research, "Advanced preprocessing: multivariate filtering," 2013. http://wiki.eigenvector.com/index.php?title=Advanced_Preprocessing:_Multivariate_Filtering.

[31] M. L. Zhang, G. P. Sheng, Y. Mu et al., "Rapid and accurate determination of VFAs and ethanol in the effluent of an anaerobic H2-producing bioreactor using near-infrared spectroscopy," *Water Research*, vol. 43, no. 7, pp. 1823–1830, 2009.

[32] L. Laghi, A. Versari, G. P. Parpinello, D. Y. Nakaji, and R. B. Boulton, "FTIR spectroscopy and direct orthogonal signal correction preprocessing applied to selected phenolic compounds in red wines," *Food Analytical Methods*, vol. 4, no. 4, pp. 619–625, 2011.

[33] P. Roudier, C. B. Hedley, C. R. Lobsey, R. A. Viscarra Rossel, and C. Leroux, "Evaluation of two methods to eliminate the effect of water from soil vis–NIR spectra for predictions of organic carbon," *Geoderma*, vol. 296, pp. 98–107, 2017.

[34] M. Vohland, M. Ludwig, S. Thiele-Bruhn, and B. Ludwig, "Determination of soil properties with visible to near- and mid-infrared spectroscopy: effects of spectral variable selection," *Geoderma*, vol. 223–225, pp. 88–96, 2014.

[35] Z. Xiaobo, Z. Jiewen, M. J. W. Povey, M. Holmes, and M. Hanpin, "Variables selection methods in near-infrared spectroscopy," *Analytica Chimica Acta*, vol. 667, no. 1-2, pp. 14–32, 2010.

[36] T. Mehmood, K. H. Liland, L. Snipen, and S. Sæbø, "A review of variable selection methods in partial least squares regression," *Chemometrics and Intelligent Laboratory Systems*, vol. 118, pp. 62–69, 2012.

[37] P. Valderrama and J. W. B. Braga, "Estado da arte de figuras de mérito de calibraçao multivariada," *Química Nova*, vol. 32, no. 5, pp. 1278–1287, 2009.

[38] D. L. Flumignan, F. de Oliveira Ferreira, A. G. Tininis, and J. E. de Oliveira, "Multivariate calibrations in gas chromatographic profiles for prediction of several physicochemical parameters of Brazilian commercial gasoline," *Chemometrics and Intelligent Laboratory Systems*, vol. 92, no. 1, pp. 53–60, 2008.

[39] C. J. Lomborg, J. B. Holm-Nielsen, P. Oleskowicz-Popiel, and K. H. Esbensen, "Near infrared and acoustic chemometrics monitoring of volatile fatty acids and dry matter during co-digestion of manure and maize silage," *Bioresource Technology*, vol. 100, no. 5, pp. 1711–1719, 2009.

[40] B. K. Lavine and J. Workman, "Chemometrics," *Analytical Chemistry*, vol. 80, no. 12, pp. 4519–4531, 2012.

[41] ANP, *Resolução ANP No. 65: Automotive fuels-Diesel-Specification of Diesel Fuel*, Brasília, DF, Brazil, 2011.

[42] R. M. Silverstein, F. X. Webster, and D. J. Kiemle, "Spectrometric identification of organic compounds," *Microchemical Journal*, vol. 21, no. 4, p. 496, 2005.

[43] N. B. Colthup, L. H. Daly, and S. E. Wiberley, *Introduction to Infrared and Raman Spectroscopy*, Academic Press, Cambridge, MA, USA, 1990.

[44] M. C. Breitkreitz, I. M. Raimundo, J. J. R. Rohwedder et al., "Determination of total sulfur in diesel fuel employing NIR spectroscopy and multivariate calibration," *Analyst*, vol. 128, no. 9, pp. 1204–1207, 2003.

[45] ANP, "Boletim de monitoramento da qualidade dos combustíveis," 2017, http://www.anp.gov.br/wwwanp/publicacoes/boletins-anp/2388-pmqc-edicoes-anteriores.

7

The Bone Black Pigment Identification by Noninvasive, In Situ Infrared Reflection Spectroscopy

Alessia Daveri (iD),[1] **Marco Malagodi,**[2] **and Manuela Vagnini**[1]

[1]*Laboratorio di Diagnostica per i Beni Culturali di Spoleto, Rocca Albornoziana, Piazza B. Campello 2, 06049 Spoleto, Italy*
[2]*Laboratorio Arvedi di Diagnostica Non Invasiva, Università di Pavia, via Bell'Aspa 3, 26100 Cremona, Italy*

Correspondence should be addressed to Alessia Daveri; a.daveri@diagnosticabeniculturali.it

Academic Editor: Gabriele Giancane

Two real case studies, an oil painting on woven paper and a cycle of mural paintings, have been presented to validate the use of infrared reflection spectroscopy as suitable technique for the identification of bone black pigment. By the use of the sharp weak band at $2013\,cm^{-1}$, it has been possible to distinguish animal carbon-based blacks by a noninvasive method. Finally, an attempt for an eventual assignment for the widely used sharp band at $2013\,cm^{-1}$ is discussed.

1. Introduction

Over the centuries, carbon-based black pigments have represented a considerable group of pigments used in different types of works of art since ancient times. According to the convention proposed by Winter [1], the carbon-based pigments can be classified according to the starting materials (plants, animals, and minerals) and to the manufacturing processes (flame carbons, chars, and cokes) [2].

Several studies have been devoted to the molecular characterization of carbon-based black pigments by means of different destructive and microdestructive techniques such as Raman, SEM-EDX, XRD, ICP-AES, and FTIR [3–7]. Generally, the diagnostic difficulties which lie in the characterization of this pigment are due to the various origins and/or manufacturing processes related to (i) characterization of carbonaceous phase [8] and (ii) identification of noncarbon constituents. Regarding this last point, the most obvious example of noncarbon constituents is the inorganic materials in animal bone pigments. According to Winter's classification, these pigments belong to the coke group, being produced from plastic precursor like the collagen [1]. Indeed, in the bone black, the collagen forms a coke that is intimately mixed with hydroxyapatite $Ca_5(OH)(PO_4)_3$, which consequently could be considered a clue for its identification.

Numerous studies have been published concerning bone black identification through the detection of phosphates by microdestructive techniques (SEM-EDS, micro-Raman, micro-FTIR, and micro-XRD) [3, 9, 10]. From these, it can be deduced that micro-Raman technique is not always able to distinguish the origin of those pigments because the detectable band at $960\,cm^{-1}$ assignable to the symmetric stretching of phosphate is not always visible in bone blacks [9–11].

SEM-EDS study may reveal the nature of the bone black pigment both considering the elemental composition, namely, the combined detection of Ca and P, and the different morphologies that diverse sources and origin of carbon-based material show [1, 2]. In addition, elemental composition analysis may also distinguish between ivory and bone black on the basis of the higher content of magnesium in the ivory black-based pigment [3, 12].

XRD has proven to be a valuable tool for identifying the crystalline noncarbon constituent hydroxyapatite in the ivory and bone blacks [3].

Finally, by infrared spectroscopy, it is possible to distinguish animal carbon-based blacks thank to the presence of a sharp band at $2013\,cm^{-1}$ in the infrared spectra of bone pigments [13, 14]. Although this band is generally used for identifying and distinguishing the bone black pigments from the others, its assignment is still uncertain.

In all the previous cited studies, the investigation was conducted by both destructive and microdestructive techniques and nondestructive techniques which generally imply the need for sampling. In recent years, the noninvasive approach has been preferred in order to preserve the integrity of artworks and to avoid their transportation. Among the portable techniques, FTIR in reflection mode has been already demonstrated a valuable tool for the noninvasive identification of both organic and inorganic materials (including bone black pigments) in work of arts [15, 16]. The main limitation of the technique is related to the possible distortion arising from the specular and diffuse components of the reflected light that may strongly alter the conventional appearance of the infrared spectra, thus making any interpretation difficult.

Therefore, powdered commercial bone black pigments were analyzed by infrared spectroscopy in transmission and reflection mode. An oil painting on woven paper and a mural painting were investigated by infrared reflection spectroscopy for the noninvasive identification of animal carbon black pigments. Finally, the bone black and hydroxyapatite spectra were compared in order to find a possible molecular assignment of the widely used infrared band at $2013\,cm^{-1}$.

2. Materials and Methods

2.1. Bone Black Pigments and Model Painting Replicas. Ivory black pigments by Maimeri and Bresciani, lamp black by Zecchi, and hydroxyapatite by Carlo Erba were selected. The test panels were prepared applying the two commercial black pigments with an acrylic binder (primal), onto a commercial panel support primed with calcium carbonate and an acrylic medium. The pigment/binding medium ratio was $1:1$ wt/wt.

2.2. Real Artworks. An oil painting on woven paper depicting a "drummer boy," signed HD in the lower left and ascribable to the style of the French caricaturists of the late XIXs, belonging to a private collection, previously a restoration procedure, has been investigated by means of noninvasive technique.

A cycle of mural paintings, executed by Lo Spagna in 1526-1528 has been analyzed during a wide diagnostic campaign at the presbytery of St. Giacomo Church in Spoleto, Italy.

2.3. Portable Mid-Infrared Spectrophotometer. The portable mid-infrared spectrophotometer ALPHA-R Bruker Optics is equipped with a Globar radiation source, a modified Michelson interferometer (RockSolid™) and a DLaTGS detector. Its weight is about $7\,kg$, and its dimensions are $20 \times 30 \times 12\,cm^3$. Exchangeable QuickSnap™ sampling modules have allowed us to perform transmission and external reflection measurements.

2.4. Sample Compartment Module. The universal sampling module is equipped with a transmission compartment with standard sample holder. The spectra have been recorded with 100 scans, a resolution of $4\,cm^{-1}$ and an effective range of $7500-375\,cm^{-1}$ in KBr pressed disks.

2.5. Module for External Reflection. The module for external reflection enables us to work with an angle of incidence of about 20°. It is equipped with an integrated video camera for the control and monitoring of the sampling area. The investigated sample area is about $28\,mm^2$. The spectrum intensity has been defined as the pseudo absorbance as log $(1/R)$. The background correction has been measured using the spectrum from a gold mirror surface. The spectra have been collected in the $7500-375\,cm^{-1}$ range, with a resolution of $4\,cm^{-1}$ and 186 scans.

2.6. Optical Microscopy. A preliminary evaluation of cross section and a study of the different layers were performed by a light-polarized microscope Olympus BX51TF, equipped with the Olympus TH4-200 lamp (visible light) and the Olympus U-RFL-T (UV light). The sessions were carried out in reflection mode at different magnifications, (10x, 20x, and 50x).

2.7. Scanning Electron Microscopy (SEM-EDS). Scanning electron microscopy (SEM) images and energy-dispersive X-ray spectra (EDS) were collected by using a Tescan FE-SEM, MIRA XMU series equipped with EDAX spectrometer, at an accelerating voltage of 15–20 kV, and high vacuum. The sample surfaces were metalized with a coating of graphite by Cressington 208HR sputter.

3. Results and Discussion

The comparison of FTIR spectra in transmission and in reflection mode of the animal commercial black pigments as pure powder and applied onto panel is shown in Figure 1. The transmission infrared spectra show the characteristic phosphate group bands: the ν_3 $(PO_4)^{3-}$ at 1087 and $1038\,cm^{-1}$, the ν_1 $(PO_4)^{3-}$ at 875 and $962\,cm^{-1}$, the ν_4 $(PO_4)^{3-}$ at 630, 604, and $567\,cm^{-1}$, and the ν_2 $(PO_4)^{3-}$ at $469\,cm^{-1}$ (Figure 1(b)). The identical features, together with the acrylic binder signals (highlighted with B, in Figure 1(a)), are recognizable in the infrared spectra in reflection mode as derivative and/or reststrhalen bands that are strongly distorted by the surface reflection and difficult to use for diagnostic purposes [17]. All the FTIR spectra display also a weak-sharp band at $2013\,cm^{-1}$. This not yet unsigned feature is not affected by the surface reflection, and it is diagnostic for the identification and distinction of the animal black pigments [13, 14].

This peculiar feature is easily detected in real artworks by means of infrared spectroscopy working in transmission, as a microdestructive technique [13, 14], and in reflection mode as noninvasive method directly on the painting surfaces. Some papers have already been published highlighting the effectiveness of this technique for detecting the presence of

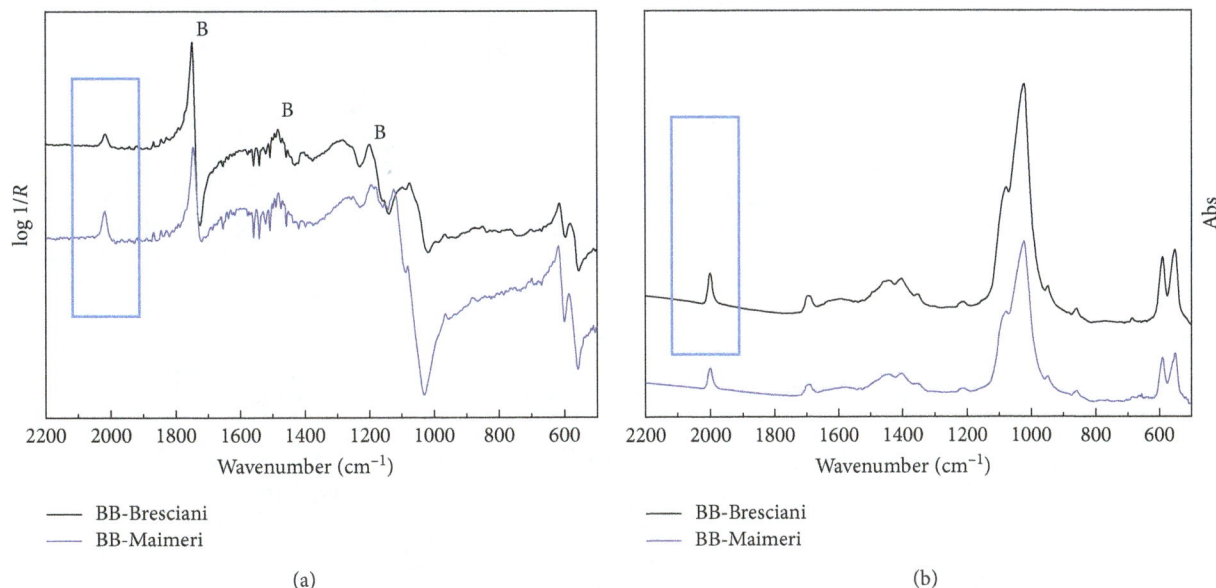

FIGURE 1: FTIR spectra in (a) reflection mode and (b) transmission mode of commercial animal black pigments and mockup, respectively. The acrylic binder (B) signals are highlighted.

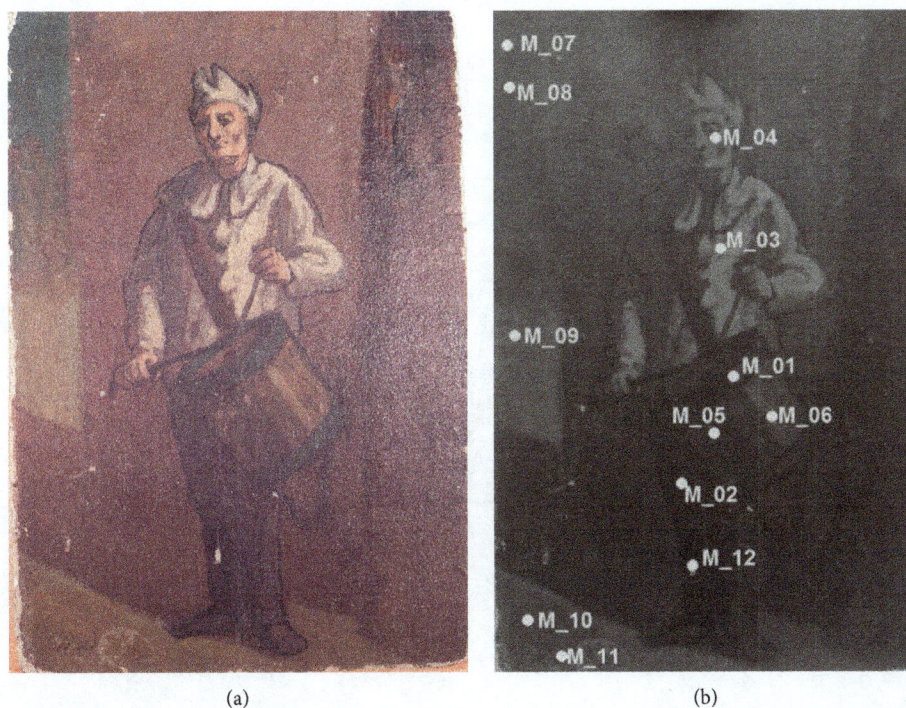

FIGURE 2: The visible image of the oil painting and the sampling by infrared reflection spectroscopy.

this band for the noninvasive and in situ identification of animal black on real works of art [15, 16].

Below are details concerning two studies carried out on different material types of artworks are reported: an oil painting on woven paper probably ascribable to *Honorè Daumier* and a mural painting by *Lo Spagna*.

3.1. Painting Oil on Woven Paper. The oil painting on woven paper depicting a "drummer boy" examined is shown in Figure 2.

The completely noninvasive study was focused on the characterization of the original materials. Firstly, multispectral imaging techniques, then X-ray fluorescence point by point analyses, and finally reflection infrared spectroscopy were performed.

The infrared spectrum profiles recorded from two blue areas (M_01 and M_08) and from a brown area of the musician's trousers (Figure 3) highlight the diagnostic signal of the bone black at the 2013 cm^{-1} in both investigated

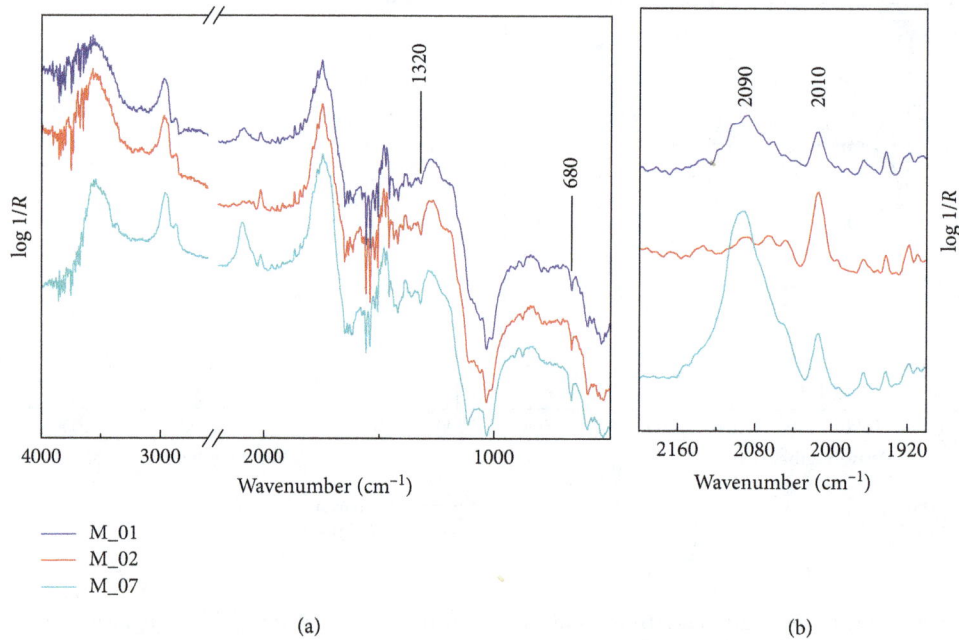

FIGURE 3: Reflection infrared spectra recorded from the blue hues of the drummer (M_01), from the background (M_08), and from the dark brown area of the musician pant (M_02).

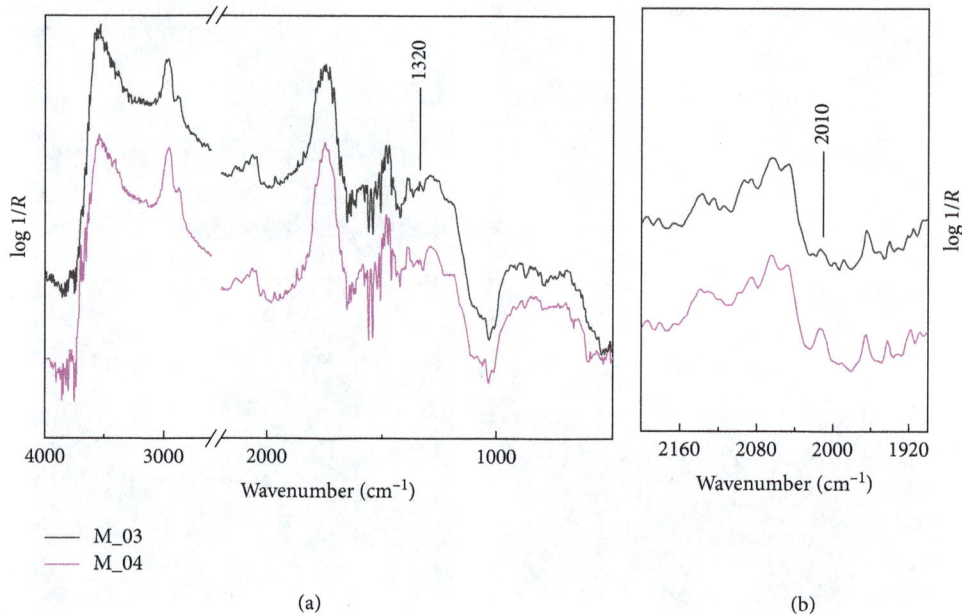

FIGURE 4: Reflection infrared spectra collected on the grey (M_03) area of the musician dress and on the flesh (M_04).

colors. In blue hue, the bone black pigment is blended with Prussian blue, such as the infrared spectra display the characteristic CN stretching at 2094 cm^{-1} [18]. As mentioned above, the phosphate fundamental bands are strongly distorted by the surface reflection and of little use for diagnostic purposes. The spectra display intense derivative bands at 2920 and 2850 cm^{-1}, derivative-like bands at about 1740 and 1460 cm^{-1} ascribable to the lipidic binder, and a signal at 1320 cm^{-1} associated to calcium oxalate [19]. The presence of lead white, as preparation layer, recognizable by

the ν_4 at 680 cm^{-1}, makes the detection of phosphate bands more challenging.

However, it is important to emphasize that the bone black is still visible even in the presence of barium sulphate (Figure 4) used by the artist as white pigment. Indeed, the infrared spectra recorded on a grey area of the musician's costume and in the flesh clearly exhibit the diagnostic feature at 2013 cm^{-1} of the animal black pigment in spite of the evident combination bands of the sulphate anion in BaSO$_4$ (Figure 3(b)) [20].

(a)

(b)

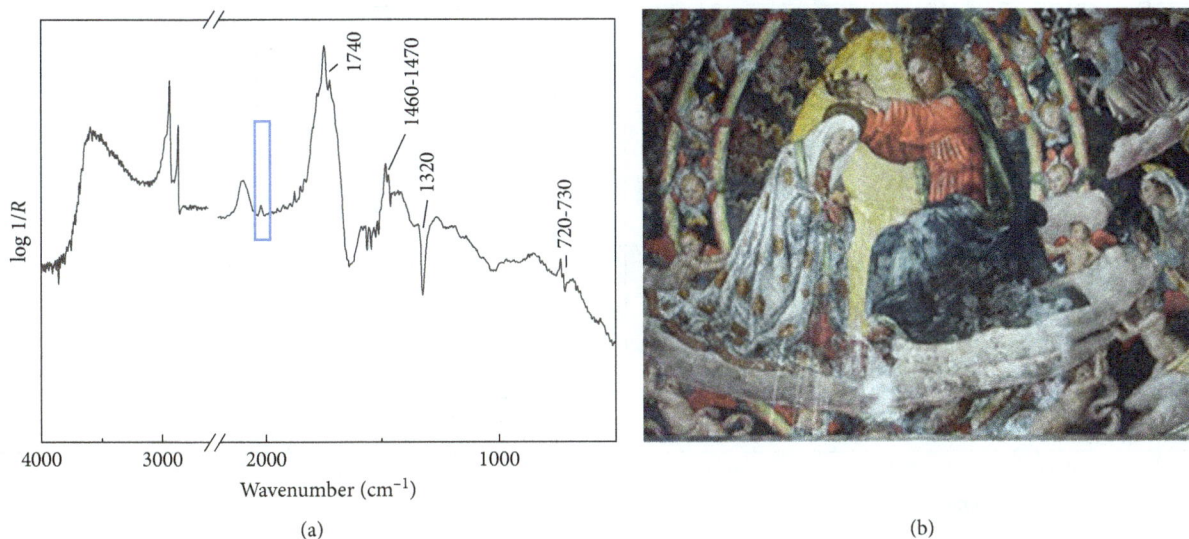

FIGURE 5: (a) Reflection infrared spectrum collected on a blue sky area of the apse and (b) visible image of the St. Giacomo Church apse.

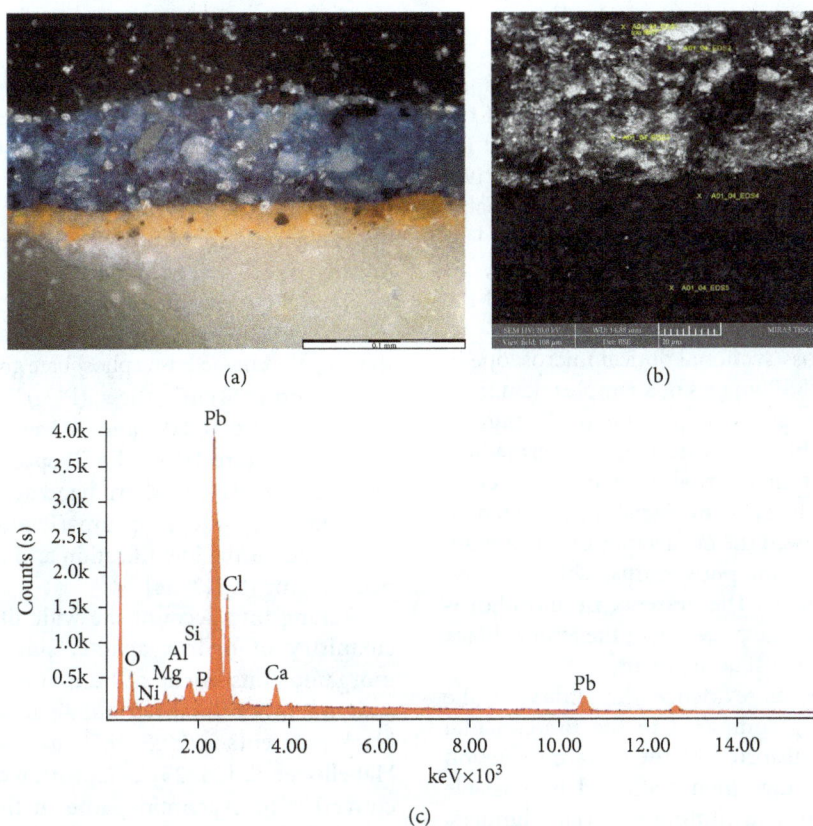

(a)

(b)

(c)

FIGURE 6: Stratigraphic section of the sample taken on the mural painting. (a) Optical microscope in visible light and (b) the scanning electron microscope showed the points of analysis. (c) EDS spectrum of the blue layer.

3.2. Mural Painting by Lo Spagna.

During the restoration project of the cycle of mural paintings by Lo Spagna in the St. Giacomo Church in Spoleto, a careful diagnostic campaign was planned. The diagnostic support has become necessary for the restorer, especially during the removal of retouching in blue sky and mantle, probably belonging to the three important documented restorations during the nineteenth century.

The noninvasive measurements carried out by FTIR spectrometer in the blue areas revealed the presence of azurite, as the original pigment, and Prussian blue as retouching pigment. In Figure 5, the infrared spectrum collected in the apse on a blue hue clearly display the characteristic CN stretching at $2094 \, cm^{-1}$ of the Prussian blue [18] and the diagnostic signal of the bone black at the

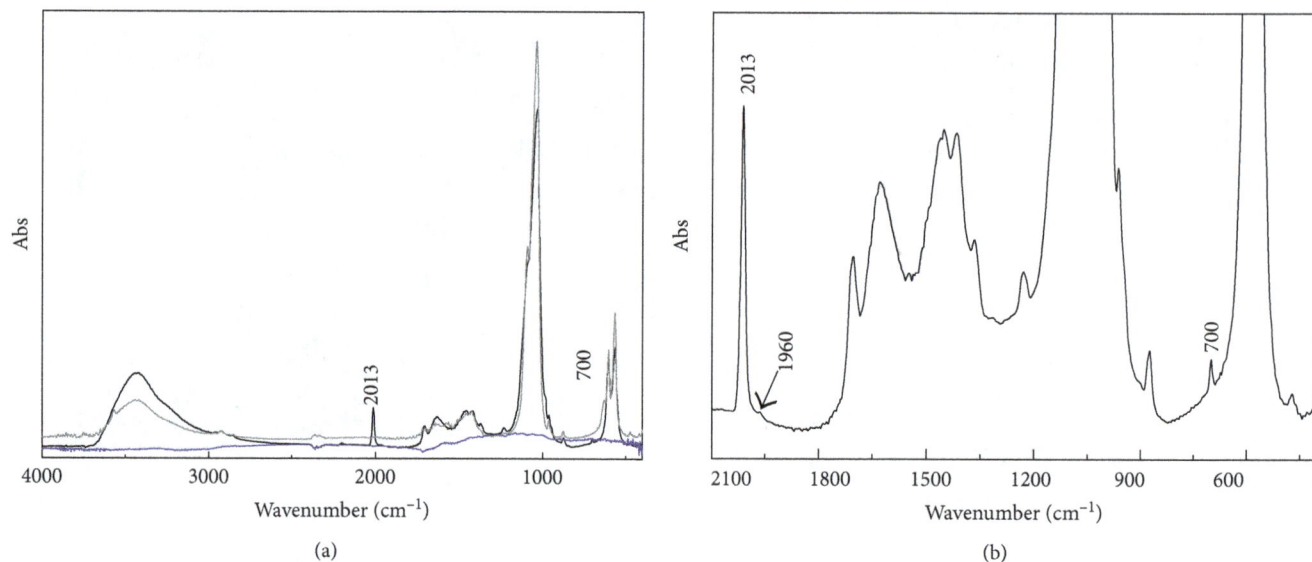

FIGURE 7: (a) Transmission spectra of vegetal (blue line) and animal black pigments (black line) compared to hydroxyapatite (grey line) standard; (b) enlarged view of the bone black spectrum highlighting the cyanamidapatite characteristic features.

2013 cm^{-1} band. The spectrum furthermore shows an intense derivative signal at 1320 cm^{-1} characteristic of the CO symmetric stretching of calcium oxalate [19] and intense derivative bands at 2855, 2920, 1740, and derivative-shaped doublets at 1470/1460 and 730/720 cm^{-1} ascribable to a natural wax [21]. The occurrence of oxalate, as degradation products, and of a lipidic component, as retouching materials, hides the fundamental modes of the phosphate group of the bone black pigment.

Figure 6 shows the cross-sectional optical microscope in visible light and the SEM-BSE images of a sample taken from a blue area in the sky of apse. It is possible to distinguish three different layers: the blue one external, an intermediate one of orange color, and an internal layer of white color ascribable to the plaster. The SEM-EDS analysis collected on the external blue layer showed the occurrence of lead a main element and silicon, calcium, phosphorus, chlorine, and copper as minor components. The presence of phosphorus in the blue layer confirmed the presence of the animal black pigment in mixture with the blue pigment.

These examples show in evidence the utility of the weaker band at 2013 cm^{-1} compared to the fundamental modes that are generally distorted by the surface reflection [17] and fall in spectral range highly affected by possible overlapping with the other constituting materials (binders, pigments, grounds, and degradation products). Despite its usefulness for the identification of an animal black pigment by noninvasive technique, its assignment is still uncertain.

3.3. Band Assignment. Vila et al. [13] associate this band probably to a degradation product of the bone compounds produced after the thermal treatment in the synthesis of the black pigment [1, 2]. Tomasini et al. attribute the signal to the cyano group of isocyanate, thiocyanate, and isothyocianate related to degradation products of protein [14]. As reported

in literature, the fresh bone infrared spectrum shows only the signals related to organic matter [22, 23].

The two commercial animal black pigments, characterized by portable X-ray diffraction, are mainly constituted of hydroxyapatite with impurities of quartz. In Figure 7, the bone black pigment and the hydroxyapatite transmission infrared spectra are compared.

The hydroxyapatite component is clearly underlined in the infrared spectra of bone black pigment commercial sample showing the characteristic phosphate group bands: the ν_3 (PO$_4$)$^{3-}$ at 1087 and 1038 cm^{-1}, the ν_1 (PO$_4$)$^{3-}$ at 875 and 962 cm^{-1}, the ν_4 (PO$_4$)$^{3-}$ at 630, 604, and 567 cm^{-1}, and the ν_2 (PO$_4$)$^{3-}$ at 469 cm^{-1} (Figure 7(b)). The IR spectra from the commercial pigments show also a sharp band at 2013 cm^{-1} not present in the hydroxyapatite spectrum (Figure 7(b) and insert) and diagnostic for the identification and distinction of the animal black pigments [13, 14].

Taking into account the wide literature concerning the chemistry of hydroxyapatite, since its important role as inorganic constituent of human bone and teeth, a further assignment for the sharp and distinctive IR band of the bone black pigments at 2013 cm^{-1} may be considered. In detail, Habelitz et al. [24, 25] demonstrated that hydroxyapatite is converted to cyanamidapatite in the presence of dry ammonia and graphite at temperature between 800 and 1300°C. The authors taking into account the results from IR, XRD, and ^{31}P NMR suggest that nitrogen enters into the structure as [CN$_2$]$^{2-}$ ion, substituting the highly mobility [OH]$^-$ groups of hydroxyapatite and forming cyanamidapatite. Their published infrared spectra of the ammonia-treated hydroxyapatite show a sharp band at 2015 cm^{-1} assigned to the ν_3 (CN$_2$)$^{-2}$ and a signal at about 700 cm^{-1} associated to the bending mode ν_2. As shown in Figure 7(b), these features are clearly detectable in the IR transmission spectra of the commercial bone blacks. These findings may suggest that a similar reaction could occur when burning the animal bone to produce the black pigment. In fact, the protein constituting the collagen may be the source of nitrogen, whilst the combustion process would provide for carbon favouring the formation of cyano ion substituting the OH groups in the hydroxyapatite structure. To conclude, the sharp band at 2013 cm^{-1}, diagnostic for the animal black pigments, is associated with cyanamidapatite, a degradation product of the synthesis of the black pigment.

4. Conclusion

Infrared reflection spectroscopy has proven to be a powerful tool for the noninvasive identification of animal carbon black pigments. Two real case studies carried out on different painting materials have been presented in order to highlight the band at 2013 cm^{-1} as marker for the noninvasive identification of animal carbon-based pigments. Previous published data allowed us to suggest a possible assignment of this diagnostic feature. Indeed, the band at 2013 cm^{-1} is associated with the formation of cyanamideapatite during the synthesis of the pigment, involving the bone combustion in the absence of air.

Conflicts of Interest

The authors declare that they have no conflicts of interest.

Acknowledgments

This research was carried out as part of the Regione Umbria project "Sviluppo delle attività di ricerca, valutazione e tutela conservativa," Progetto 1 del Primo atto integrativo all'APQ "Tutela e prevenzione dei beni culturali." The authors thank Professor Carla Mancini for the technical support during the noninvasive analysis of the "drummer boy" oil painting. Thanks are due to Soprintendenza Archeologica, belle arti e paesaggio dell'Umbria and Arcidiocesi di Spoleto-Norcia for the access to the mural painting by Lo Spagna in the San Giacomo Church in Spoleto.

References

[1] J. Winter, "The characterization of pigments based on carbon," *Studies in Conservation*, vol. 28, no. 2, pp. 49–66, 1983.

[2] B. H. Berrie, "Artists' pigments," in *Pigment Compendium*, N. Eastaugh, V. Walsh, T. Chaplin, and R. Siddall, Eds., vol. 4, Butterworth-Heinemann, Oxford, UK, 2004.

[3] A. van Loon and J. J. Boon, "Characterization of the deterioration of bone black in the 17th century *Oranjezaal* paintings using electron-microscopic and micro-spectroscopic imaging techniques," *Spectrochimica Acta Part B: Atomic Spectroscopy*, vol. 59, no. 10-11, pp. 1601–1609, 2004.

[4] A.-M. Bakr, T. Kawiak, M. Pawlikowski, and Z. Sawlowicz, "Characterisation of 15th century red and black pastes used for wall decoration in the Qijmas El-Eshaqi mosque (Cairo, Egypt)," *Journal of Cultural Heritage*, vol. 6, no. 4, pp. 351–356, 2005.

[5] J. van der Weerd, G. D. Smith, S. Firth, and R. J. H. Clark, "Identification of black pigments on prehistoric Southwest American potsherds by infrared and Raman microscopy," *Journal of Archaeological Science*, vol. 31, no. 10, pp. 1429–1437, 2004.

[6] E. P. Tomasini, B. Gómez, E. B. Halac et al., "Identification of carbon-based black pigments in four South American polychrome wooden sculptures by Raman microscopy," *Heritage Science*, vol. 3, p. 19, 2015.

[7] E. P. Tomasini, C. M. Favier Dubois, N. C. Little, S. A. Centeno, and M. S. Maier, "Identification of pyroxene minerals used as black pigments in painted human bones excavated in Northern Patagonia by Raman spectroscopy and XRD," *Microchemical Journal Volume*, vol. 121, pp. 157–162, 2015.

[8] T. Gatta, L. Campanella, C. Coluzza et al., "Characterization of black pigment used in 30 BC fresco wall paint using instrumental methods and chemometry," *Chemistry Central Journal*, vol. 6, no. 2, 2012.

[9] A. M. Correia, R. J. H. Clark, M. I. M. Ribeiro, and M. L. T. S. Duarte, "Pigment study by Raman microscopy of 23 paintings by the Portuguese artist Henrique Pousão (1859–1884)," *Journal of Raman Spectroscopy*, vol. 38, no. 11, pp. 1390–1405, 2007.

[10] A. M. Correia, M. J. V. Oliveira, R. J. H. Clark, M. I. M. Ribeiro, and M. L. T. S. Duarte, "Characterization of Pousão pigments and extenders by micro-X-ray diffractometry and infrared and Raman microspectroscopy," *Analytical Chemistry*, vol. 80, no. 5, pp. 1482–1492, 2008.

[11] K. Castro, M. Pérez-Alonso, M. D. Rodríguez-Laso, N. Etxebarria, and J. M. Madariaga, "Non-invasive and non-destructive micro-XRF and micro-Raman analysis of a decorative wallpaper from the beginning of the 19th century," *Analytical and Bioanalytical Chemistry*, vol. 387, no. 3, pp. 847–860, 2007.

[12] A. Freund, G. Eggert, H. Kutzke, and B. Barbier, "On the occurrence of magnesium phosphates on ivory," *Studies in Conservation*, vol. 47, no. 3, pp. 155–160, 2002.

[13] A. Vila, N. Ferrer, and J. F. García, "Chemical composition of contemporary black printing inks based on infrared spectroscopy: Basic information for the characterization and discrimination of artistic prints," *Analytica Chimica Acta*, vol. 591, no. 1, pp. 97–105, 2007.

[14] E. Tomasini, G. Siracusano, and M. S. Maier, "Spectroscopic, morphological and chemical characterization of historic pigments based on carbon. Paths for the identification of an artistic pigment," *Microchemical Journal*, vol. 102, pp. 28–37, 2012.

[15] C. Miliani, F. Rosi, A. Burnstock, B. G. Brunetti, and A. Sgamellotti, "Non-invasive in-situ investigations versus micro-sampling: a comparative study on a Renoirs painting," *Applied Physics A*, vol. 89, no. 4, pp. 849–856, 2007.

[16] G. Van der Snickt, C. Miliani, K. Janssens et al., "Material analyses of 'Christ with singing and music-making Angels', a late 15th-C panel painting attributed to Hans Memling and assistants: part I. non-invasive *in situ* investigations," *Journal of Analytical Atomic Spectrometry*, vol. 26, pp. 2216–2229, 2011.

[17] C. Miliani, F. Rosi, A. Daveri, and B. G. Brunetti, "Reflection infrared spectroscopy for the non-invasive in situ study of artists' pigments," *Applied Physics A*, vol. 106, no. 2, pp. 295–307, 2012.

[18] F. Rosi, A. Burnstock, K. Jan Van den Berg, C. Miliani, B. G. Brunetti, and A. Sgamellotti, "A non-invasive XRF study supported by multivariate statistical analysis and reflectance FTIR to assess the composition of modern painting materials," *Spectrochimica Acta Part A: Molecular and Biomolecular Spectroscopy*, vol. 71, no. 5, pp. 1655–1662, 2009.

[19] L. Monico, F. Rosi, C. Miliani, A. Daveri, and B.G. Brunetti, "Non-invasive identification of metal-oxalate complexes on polychrome artwork surfaces by reflection mid-infrared spectroscopy," *Spectrochimica Acta Part A: Molecular and Biomolecular Spectroscopy*, vol. 116, pp. 270–280, 2013.

[20] F. Rosi, A. Daveri, B. Doherty et al., "On the use of overtone and combination bands for the analysis of the $CaSO_4$—H_2O system by mid-infrared reflection spectroscopy," *Applied Spectroscopy*, vol. 64, no. 8, pp. 956–963, 2010.

[21] C. Invernizzi, A. Daveri, M. Vagnini, and M. Malagodi, "Non-invasive identification of organic materials in historical stringed musical instruments by reflection infrared spectroscopy: a methodological approach," *Analytical and Bioanalytical Chemistry*, vol. 409, no. 13, pp. 3281–3288, 2017.

[22] M. Petraa, J. Anastassopouloua, T. Theologis, and T. Theophanides, "Synchrotron micro-FT-IR spectroscopic evaluation of normal paediatric human bone," *Journal of Molecular Structure*, vol. 733, no. 1–3, pp. 101–110, 2005.

[23] I. E. Chesnick, F. A. Avallone, R. D. Leapman, W. J. Landis, N. Eidelman, and K. Potter, "Evaluation of bioreactor-cultivated bone by magnetic resonance microscopy and FTIR microspectroscopy," *Bone*, vol. 40, no. 4, pp. 904–912, 2007.

[24] S. Habelitz, L. Pascual, and A. Durán, "Transformation of tricalcium phosphate into apatite by ammonia treatment," *Journal of Materials Science*, vol. 36, no. 17, pp. 4131–4135, 2001.

[25] S. Habelitz, L. Pascual, and A. Durán, "Nitrogen-containing apatite," *Journal of the European Ceramic Society*, vol. 19, no. 15, pp. 2685–2694, 1999.

8

Fluorescence Spectroscopy Applied to Monitoring Biodiesel Degradation: Correlation with Acid Value and UV Absorption Analyses

Maydla dos Santos Vasconcelos,[1] **Wilson Espíndola Passos,**[1] **Caroline Honaiser Lescanos,**[2] **Ivan Pires de Oliveira** ⓘ**,**[3] **Magno Aparecido Gonçalves Trindade,**[1] **Anderson Rodrigues Lima Caires,**[4] **and Rozanna Marques Muzzi**[1]

[1]*Faculty of Exact Sciences and Technology, Federal University of Grande Dourados, Dourados, MS, Brazil*
[2]*Faculty of Medical Sciences, State University of Campinas, Campinas, SP, Brazil*
[3]*Institute of Chemistry, State University of Campinas, Campinas, SP, Brazil*
[4]*Institute of Physics, Federal University of Mato Grosso do Sul, Campo Grande, MS, Brazil*

Correspondence should be addressed to Ivan Pires de Oliveira; ivan.pires.oliveira@hotmail.com

Academic Editor: Boryana M. Nikolova-Damyanova

The techniques used to monitor the quality of the biodiesel are intensely discussed in the literature, partly because of the different oil sources and their intrinsic physicochemical characteristics. This study aimed to monitor the thermal degradation of the fatty acid methyl esters of *Sesamum indicum* L. and *Raphanus sativus* L. biodiesels (SILB and RSLB, resp.). The results showed that both biodiesels present a high content of unsaturated fatty acids, ~84% (SILB) and ~90% (RSLB). The SILB had a high content of polyunsaturated linoleic fatty acid (18 : 2), about 49%, and the oleic monounsaturated (18 : 1), ~34%. On the other hand, RSLB presented a considerable content of linolenic fatty acid (18 : 3), ~11%. The biodiesel samples were thermal degraded at 110°C for 48 hours, and acid value, UV absorption, and fluorescence spectroscopy analysis were carried out. The results revealed that both absorption and fluorescence presented a correlation with acid value as a function of degradation time by monitoring absorptions at 232 and 270 nm as well as the emission at 424 nm. Although the obtained correlation is not completely linear, a direct correlation was observed in both cases, revealing that both properties can be potentially used for monitoring the biodiesel degradation.

1. Introduction

The human interest in using the biomass available on the planet has been increased due to the global population growth, accelerated industrial development, reduced global petroleum reserves, and concerns about environmental impacts [1]. In this context, the search for alternative sources of renewable energy has growing interest. Besides renewable, the alternative energy sources should also be ecologically correct, socially sustainable, and economically viable. Clearly, biofuels compose part of the alternative ecologically friendly energy because they are produced from renewable sources and also more friendly to the environment than fossil-derived fuels [2].

Typically, oleaginous species are used as the main source of raw materials to produce biofuel [3]. Biodiesels can be produced by different oleaginous sources depending on the raw material available in each country and/or region. Consequently, it have been reported that biodiesels can be synthesized from soybean, corn, canola, palm, animal fats, recycled greases, and others [4, 5]. For instance, in 2016/2017 soybean oil was the main feedstock for biodiesel production worldwide (341 million tons), for a total of 558 million tons of total oilseeds [6].

The desirable characteristics of the raw materials for biodiesel production include (i) adaptability to local growing conditions; (ii) regional availability; (iii) high oil content;

(iv) favorable fatty acid composition; (v) compatibility with existing agricultural practices; (vi) agricultural inputs; (vii) by-product markets; (viii) land compatibility; and (ix) rotational adaptability with other cultures [7].

Nevertheless, depending on the region, different vegetable oils are available to biodiesel production according to the agricultural, social, and commercial factors as previously mentioned. As consequence, different biodiesels are produced with particular chemical compositions, such as secondary compounds (polyphenols, carotenoids, and chlorophylls) and triacylglycerols (TAGs) with different lengths and unsaturation contents of the fatty acid carbon chain. The secondary compounds and TAG's compositions are directly associated with the susceptibility of biodiesel degradation: (i) secondary compounds (antioxidants) can promote biodiesel stability, and (ii) unsaturated carbon chain content makes biodiesel more susceptible to the oxidation [8].

The biodiesel commercialization is dependent of several quality parameters [9]. Among them, acid value (AV) is an important parameter to evaluate the biodiesel characteristics. This parameter indicates the formation of organic acids from TAG's oxidation by molecular oxygen [10]. High AV indicates that biodiesel is very degraded, and its commercialization is impaired [11].

In the course of biodiesel degradation process, several small molecules are generated, such as peroxides, aldehydes, ketones, and carboxylic acids, among others [9]. These molecules are formed by reaction between carbon chains of biodiesel with singlet oxygen, forming oxidized compounds [12]. Recent studies have demonstrated the potential of the spectroscopic technique in determining some important transformations in the molecular structure of the alkyl esters from biodiesel during degradation process [13, 14]. Despite the fact that the formed degradation, specifically carboxylic acids, be satisfactory quantified by the titration method (AV parameter), it remains unclear the relation of generating these compounds with biodiesel light absorption/emission.

Our previous works have shown that the acidification of vegetable oils reflects on the changes in the absorbances at 232 and 270 nm with direct impact on the emission profile [15, 16]. However, the optical behavior is still not completely understood for biodiesel, especially for unconventional vegetable sources studied here (from *Raphanus sativus* L. and *Sesamum indicum* L. seeds). In this sense, the present study aimed to apply fluorescence spectroscopy for monitoring biodiesel degradation and determine its correlation with acid value and UV absorbance changes.

2. Materials and Methods

2.1. Chemicals, Materials, and Equipment. Methanol (Impex; 99.8%); KOH (Dinâmica; 95%); MgSO$_4$ (Impex; 98%); ethyl ether (Impex; 98%); ethyl alcohol (Dinâmica; 96%); hexane (Panreac; 95%); laboratory oven (Sterilifer SXCR42; 42 liters); hydraulic press (Ecirtec-mpe40); gas chromatograph (GC-2012-plus-Shimadzu); 743-Rancimat Metrohm; UV-Vis

spectrometer (Cary 50 Varian); and spectrofluorimeter (Cary Eclipse Varian) were used.

2.2. Oil Extraction. *Raphanus sativus* L. (RSL) and *Sesamum indicum* L. (SIL) seeds were commercially obtained from local companies. The seeds were dehydrated in laboratory oven under temperature of 60°C for 14 hours. Then, the seeds were submitted to hydraulic press extraction obtaining a yield of extracted oil of 32% (w/w) and 45% (w/w) for RSL and SIL oils, respectively. The extracted oils were kept in a dark environment (refrigerator) at −4°C until methyl esters synthesis. Unrefined crude oils were used in all analyses.

2.3. Biodiesel Production. Primary, the acid values for RSL and SIL oils were titrated, obtaining the values of 0.35 and 0.82 mg KOH/g, respectively. The biodiesel was obtained by the transesterification of the extracted crude oils, using the potassium methoxide as a catalyst. Initially, the potassium methoxide was obtained by reacting 40 mL of methanol with 1.8 g of KOH under stirring at 60°C for 30 min. After that 90 g of oil was added in the solution and remained under stirring at 60°C during approximately 2 h for the transesterification reaction. This reaction was monitored by thin layer chromatography as described in our recent study [17, 18]. After the transesterification process, the mixture was decanted to separate the biodiesel from the by-product glycerol. The phase containing the esters of interest was washed with distilled water and then dried with anhydrous MgSO$_4$, filtered and concentrated in a rotary evaporator at 55°C for approximately 3 h [15].

2.4. Sample Degradation. The degradation was carried out as follows: (i) the biodiesel samples were weighed in amber flasks, approximately 3.5 g of sample (in each flask) in which the samples were prepared in triplicate; (ii) all samples were subjected to heating in an air circulation laboratory oven at 110°C for 48 hours; (iii) initially, flasks were removed and stored ($n = 3$) every 2 hours up to the 24 hour period, and finally, flasks were sampled in 36 and 48 hours. The samples were stored at approximately −4°C until analyses. The biodiesel oxidation was governed by the biodiesel-atmospheric air interaction without any additional air flow rate. The biodiesel-air contact was defined simply by the size of the flask aperture, a radius of approximately 0.8 cm with 2.0 cm^2 of area with constant air renovation. All analyses were performed in triplicates considering three independent flasks containing degraded biodiesel samples.

2.5. Fatty Acid Composition. The analyses of the fatty acid composition were performed by using a gas chromatography system according to the method Ce 2-66 [19]. The fused silica SP-2560 column (100 m and 0.25 mm) was used in the separation process. The isothermal temperature was programmed at 140°C for 5 min followed by heating at 4°C min^{-1} up to 240°C, remaining at this level for 30 min. The temperature of the vaporizer was 250°C and detector was 260°C, using helium as carrier gas.

2.6. Oxidative Stability Study. The oxidative stability, expressed by the induction period (IP) in hours, was determined in duplicate by a Rancimat apparatus according to the EN 14112 method. Samples weighing 3.0 g (±0.1) were added into a sealed glass vessel reaction, heated at constant temperature of 110°C, and analyzed under a constant air flow rate (10 L·h^{-1}) passing through the samples and then into measuring vessel containing 50 mL ultrapure water [20].

2.7. Acid Value Analysis. The variation of total acid value during the degradation period was evaluated by using the classical titrations. In three Erlenmeyers were added 2.0 g of sample, 25 mL of the solution ethyl ether : ethyl alcohol (2 : 1), and two drops of the phenolphthalein indicator. The samples were then titrated with 0.1 mol·L^{-1} potassium hydroxide, duly standardized, until the appearance of the pink coloration observed for at least 30 seconds [21].

2.8. UV Absorption and Fluorescence Spectroscopy Analysis. All samples were diluted in a hexane HPLC grade. The concentration was 0.025% (w/v) and 0.030% (w/v) for the biodiesels produced from *S. indicum* L. (SILB) and *R. sativus* L. (RSLB), respectively. Molecular absorption and emission analyses were performed at room temperature using a 10 mm thick quartz cell.

2.9. Statistical Analyses. The statistical analyses were performed by two-way ANOVA followed by the Tukey's ($P < 0.05$) test using GraphPad Prism (GraphPad 164 Software, Version 6.0, San Diego, CA).

3. Results and Discussion

3.1. Fatty Acid Composition. Table 1 shows the composition of fatty acids present in each oilseed. The results show that both feedstocks have a high content of unsaturated fatty acids, ~84% for *S. indicum* L. and ~90% for *R. sativus* L. As can be clearly seen, SIL has a higher content of linoleic polyunsaturated fatty acid (18 : 2) and oleic monounsaturated fatty acid (18 : 1), which were approximately 49% and 34%, respectively. On the other hand, RSL presented a considerable content of linolenic fatty acid (18 : 3), ~11%. It was also observed that RSL presented two monounsaturated fatty acids, which were not found in SIL (eicosenoic acid (20 : 1) and erucic acid (22 : 1)). Moreover, significant quantities of oleic (18 : 1), >32%, for both feedstocks were quantified. The fatty acid composition in the feedstock is very important because the biodiesel degradation can be estimated by knowing the type and quantity of fatty acids. This estimation is possible considering the unsaturations present in the carbon chains from fatty acids because these regions are susceptible to atmospheric oxygen attack. The oxygen reaction leads to the breaking of these chains to form unwanted oxidized compounds that change the initial physicochemical properties of the samples [22]. However, similar quantities of unsaturated fatty acids (~84% for SIL and ~90% for RSL) may not reflect an identical

TABLE 1: Fatty acid methyl ester composition in biodiesels from *Sesamum indicum* L. and *Raphanus sativus* L. [16].

Fatty acids (%)	Description	*S. indicum* L.	*R. sativus* L.
Palmitic	C 16 : 0	8.54 ± 0.03	6.38 ± 0.66
Palmitoleic	C 16 : 1	0.00 ± 0.00^A	0.33 ± 0.01^A
Stearic	C 18 : 0	6.83 ± 0.01	2.32 ± 0.14
Oleic	C 18 : 1(n-9)	34.12 ± 0.09	31.72 ± 2.14
Vaccenic	C 18 : 1(n-7)	0.00 ± 0.00^B	0.00 ± 0.00^B
Linoleic	C 18 : 2(n-6)	48.86 ± 0.08	18.35 ± 1.27
Arachidic	C 20 : 0	0.87 ± 0.03^C	0.93 ± 0.06^C
Eicosenoic	C 20 : 1	0.18 ± 0.01	8.80 ± 0.59
Linolenic	C 18 : 3(n-3)	0.41 ± 0.00	11.38 ± 0.67
Docosanoic	C 22 : 0	0.19 ± 0.01^D	0.53 ± 0.03^D
Erucic	C 22 : 1	0.00 ± 0.00	17.89 ± 1.13
Lignoceric	C 24 : 0	0.00 ± 0.00^E	0.63 ± 0.04^E
Nervonic	C 24 : 1	0.00 ± 0.00^F	1.80 ± 0.29^F
Σ saturated	—	$\mathbf{16.43 \pm 0.02}$	$\mathbf{10.79 \pm 0.93}$
Σ monounsaturated	—	$\mathbf{34.30 \pm 0.09}$	$\mathbf{60.54 \pm 4.16}$
Σ polyunsaturated	—	$\mathbf{49.27 \pm 0.08}$	$\mathbf{29.73 \pm 1.94}$

Values are mean ± sd (standard deviation) ($n = 3$) expressed as a percentage in relation to total fatty acids quantified. Values on the same line highlighted by equal superscript letters indicate no significant differences ($P < 0.05$).

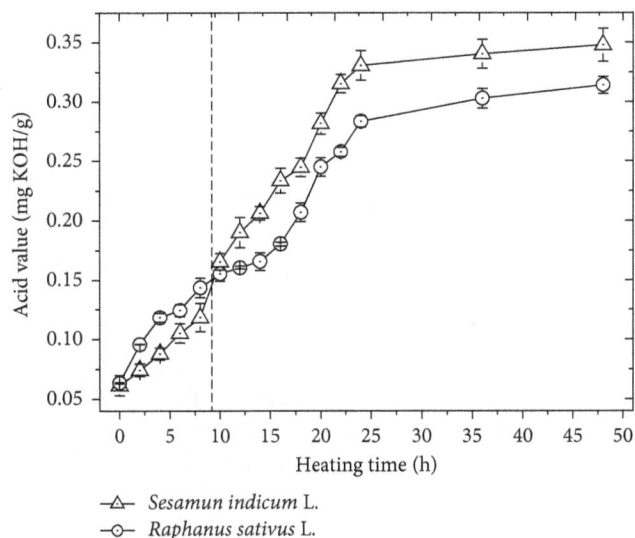

FIGURE 1: Acid values for biodiesels from *S. indicum* L. and *R. sativus* L. during the thermal degradation process. Error bars represent standard deviations.

profile of degradation due to the presence of secondary compounds (intrinsic antioxidants), depending on its chemical feature and content in the oilseeds [23, 24].

3.2. Acid Value Determination. The monitoring of the degradation process was carried out by the quantification of the acidic compounds (Figure 1). In general, a similar profile for the formation of acidic compounds in both biodiesels was observed. There is an increase in the formation of these compounds in the first 24 hours of thermodegradation,

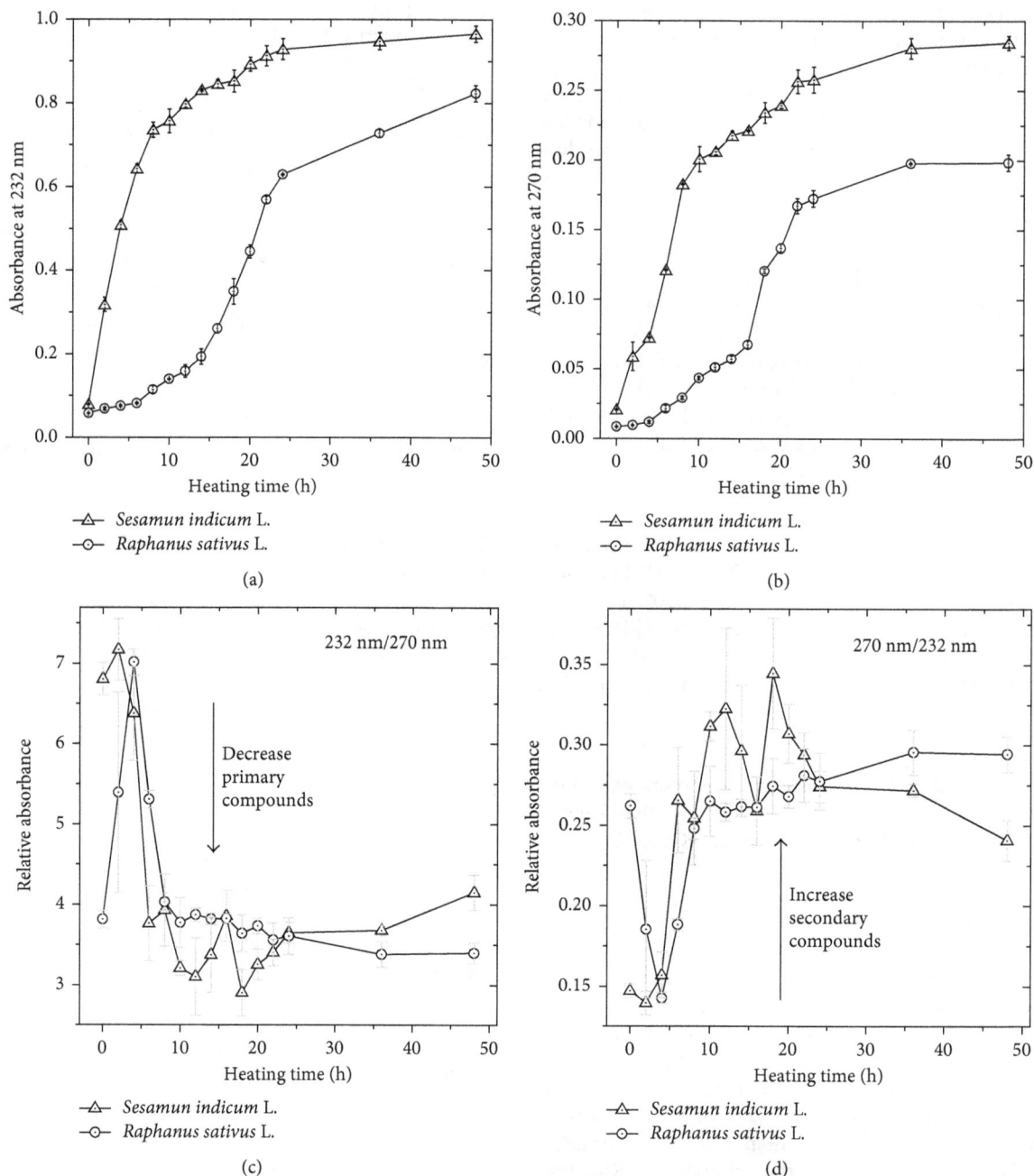

FIGURE 2: Absorbance values at (a) 232 nm and (b) 270 nm for biodiesels from *S. indicum* L. and *R. sativus* L. during the thermal degradation process. (c) and (d) show the absorbance ratios with formation of secondary compounds after ~10 h of thermal degradation. Error bars represent standard deviations.

tending to a plateau with a AV of ~0.35 and ~0.32 mg KOH/g for SILB and RSLB, respectively, until 48 hours. It is interesting to note that there is a change in the acidity profile between these two biodiesels in approximately 9 hours, presenting a relative inversion. This means that the RSLB presents lower susceptibility to oxidation and formation of acidic compounds than SILB after 9 h of induced degradation. The difference in the acid values from SILB to RSLB may be related to the major content of linoleic acid (18 : 2), approximately 49% present in the SIL fatty acid composition. This trend makes the biodiesel derived from SIL more susceptible to degrade than biodiesel from RSL. In this process, the small organic acids can be formed from

radical reactions; in which, the oxygen reacts with the unsaturations forming organic hydroperoxides, further the formation of aldehydes and finally oxidized to a stable product such as carboxylic acid [22, 25]. In summary, Figure 1 shows that although the degradation profiles are not equal, a similar response of acid value analyses was observed for both SILB and RSLB.

3.3. *UV Absorption Measurement.* The carbonic chains of the methyl esters undergo oxidation exhibit absorption of light at two characteristic wavelengths, at 232 nm and 270 nm.

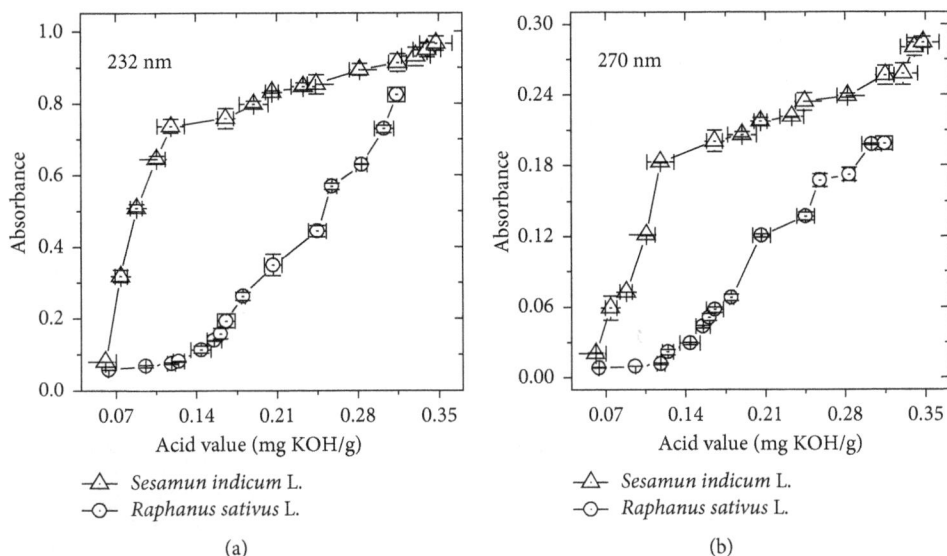

FIGURE 3: Correlation between the acid values and the absorbance values at (a) 232 nm and (b) 270 nm for the biodiesels from *S. indicum* L. and *R. sativus* L. Error bars represent standard deviations.

These wavelengths refer to modifications in the structures of the carbon chains and/or the formation of new molecules (degradation products). At the initial stage of degradation, there is a more prominent increase in absorption at 232 nm (primary oxidation), which occurs due to the formation of conjugated dienes. At the end of the degradation process, there is a more prominent increase in absorption at 270 nm (secondary oxidation), associated with aldehydes, ketones, and carboxylates [26, 27].

In Figure 2(a), it is possible to see the formation of the primary compounds in the biodiesels. There was a higher formation of these compounds in SILB in the first 10 hours of induced degradation. On the other hand, RSLB presents lower production of these primary compounds in the first 15 hours, being formed more intensely after 24 hours. It is interesting to observe a similar trend for the response of the measured absorbance at 270 nm (Figure 2(b)), comparing with the absorbance at 232 nm. This means that the primary and secondary compounds are generated proportionally in both biodiesels [28]. Moreover, the higher absorption at 232 and 270 nm for SILB can be attributed to its composition in terms of linoleic acids.

Previous studies have reported that an estimate of the level of the biodiesel oxidation can be obtained by analyzing the absorbance ratio of 232 to 270 nm [26, 27, 29]. Figures 2(c) and 2(d) show that both biodiesels form large amounts of more oxidized compounds after ~10 hours during the thermal degradation. This response can be understood by the similarity in the unsaturated fatty acids for both biodiesels (~84% for SIL and ~90% for RSL).

In addition, the correlation behavior between the quantified acid values and the absorbance responses of the primary and secondary compounds generated in the oxidative process was performed. Figure 3 shows that, in general, considering at both wavelengths, there is a direct, but not linear correlation in the values for both biodiesels. Moreover, an opposite behavior was observed for each

biodiesel. The inclination of the correlation curve is greater for SILB than for RSLB, suggesting a higher sensitivity of absorbance measurements comparatively to acid value titration for the SILB. In this sense, the results suggest that the oxidative stage of the SILB can be better studied by monitoring the alterations in the UV absorbances. Differently, RSLB degradation can be evaluated with similar sensitivity applying AV analysis or UV absorption at 232 or 270 nm.

3.4. Fluorescence Profiles of Biodiesels.
Alternatively, the excitation/emission profile was investigated through the contour maps presented in Figures 4 and 5 for SILB and RSLB, respectively. Figure 4 shows changes in fluorescence intensities from 280 to 720 nm with excitation of 220 to 400 nm. Figures 4(a) and 4(d) reveal that the oxidation process induced a strong reduction in the emission intensity at ~320 nm (excitations at ~235 and ~285 nm). Similarly, a decrease in the emission intensity at ~630 nm, when excited at 235 and 285 nm, was also observed as presented in Figures 4(c) and 4(f). This last emission is attributed to chlorophylls [27, 30], clearly degraded under thermal conditions. Additionally, Figures 4(b) and 4(e) show that an increase in the emission intensity in the 415–430 nm range was determined, with excitation from 345 to 380 nm, revealing the production of fluorescent compounds during the degradation process. This behavior is similar to that observed by other studies and corroborates with our previous studies which demonstrated that conjugated tetraenes are produced during the biodiesel degradation [31, 32].

Figure 5 shows changes in fluorescence intensities from 310 to 500 nm, with excitation of 220 to 400 nm for RSLB. Figures 5(a) and 5(d) show the reduction in the emission intensities at ~320 nm. According to the literature, the emission at 322 nm region is related to the natural antioxidant tocopherols [33]. The results also revealed a decrease in the emission intensity from ~480 to ~500 nm as presented

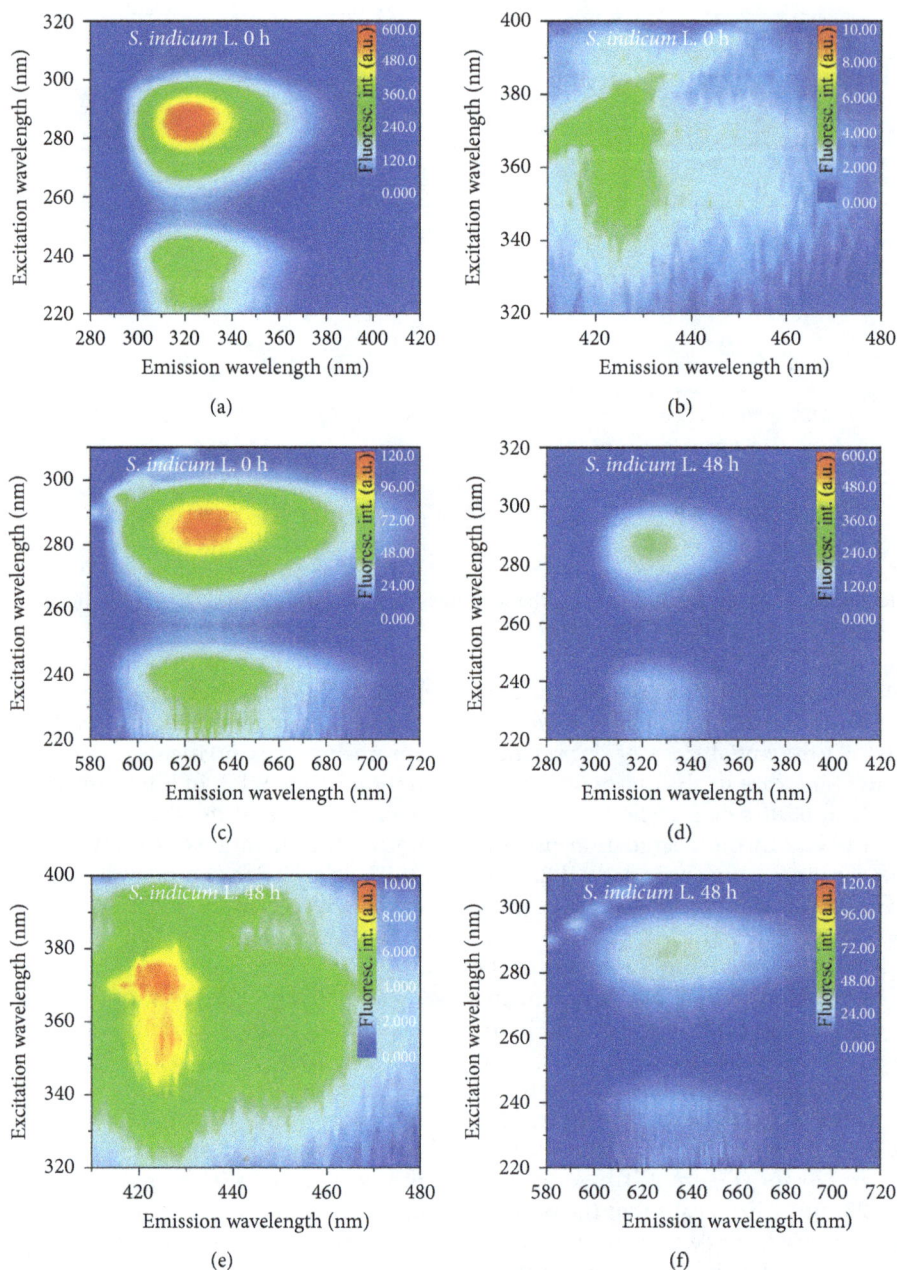

FIGURE 4: Fluorescence intensity maps of *S. indicum* L. biodiesel (a, b, c) before the oxidative process and (d, e, f) after 48 h of degradation. Scattering not omitted.

in Figures 5(c) and 5(f). This emission region was recently reported to be associated with the carotenoids as reported by Silva et al. [30]. Similarly to that observed for SILB, it was possible to observe the appearance of emissions at ~425 nm, directly associated with the degradation compounds. As previously discussed, this emission is associated with the conjugated tetraenes which are formed during the degradation of the methyl esters [32, 34].

From the emission/excitation contour maps, some candidate wavelengths were selected to monitor the oxidative process of SILB and RSLB, as presented in Figure 6. The fluorescence profile of SILB shows that the emission intensity at 322 nm decays rapidly to ~300 a.u. in the first 10 hours (Figure 6(a)). For RSLB, it was observed that the intensity decrease is less abrupt (Figure 6(b)), but equally significant. In addition, it is interesting to note the convergence of fluorescence intensity to specific values. Moreover, these behaviors are very similar to that observed for AV analysis and UV absorbances profiles, with change on the response at approximately 10 hours of degradation time. Similar decrease of emission may be observed at wavelengths close to 630 nm for SILB limiting to the initial 15 hours of the degradation process (Figure 6(e)). For RSLB, the decrease in emission at 488 nm occurs similarly to that observed for emission at 321 nm with satisfactory response in the initial 24 hours of the oxidative process (Figure 6(f)).

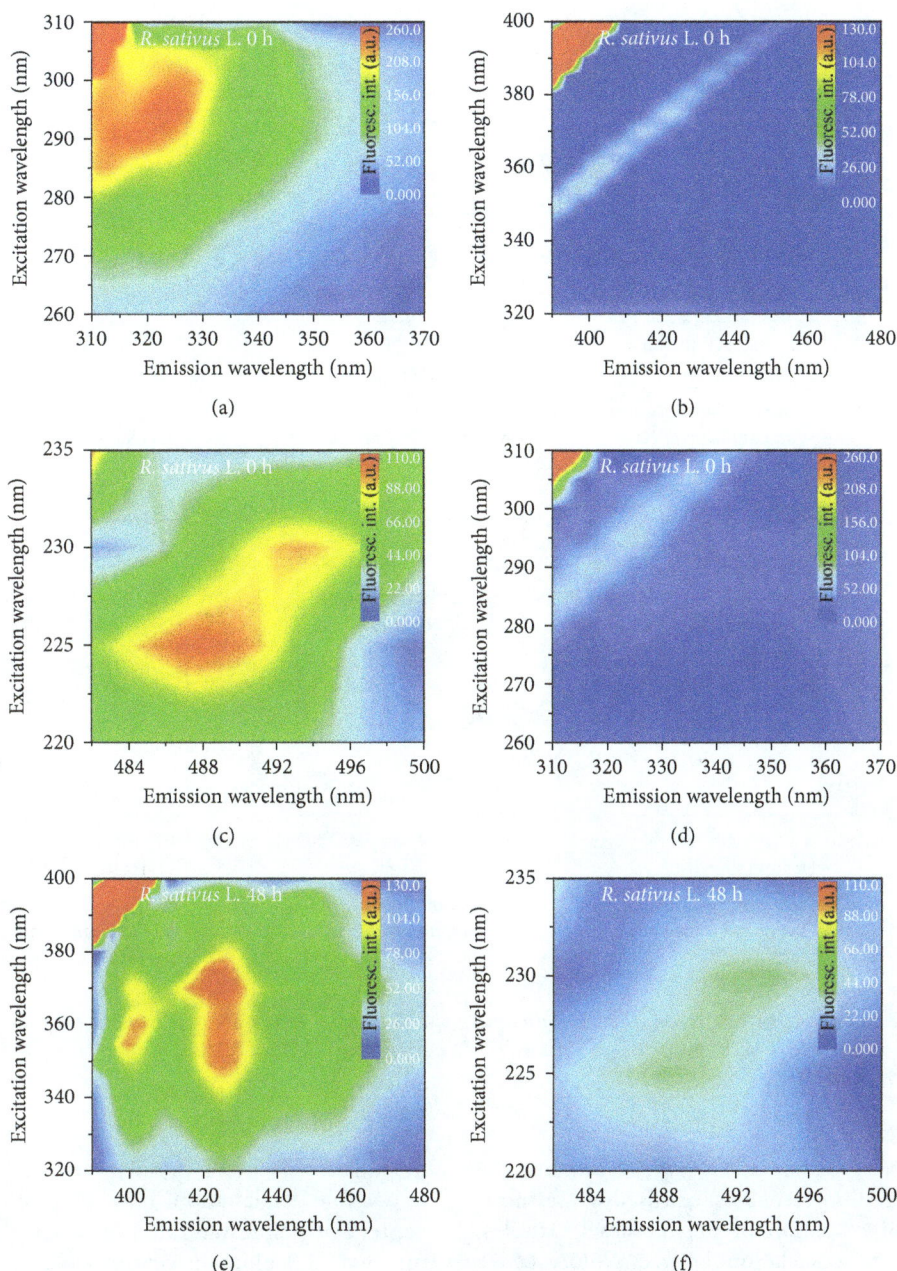

FIGURE 5: Fluorescence intensity maps of *R. sativus* L. biodiesel before the oxidative process and after 48 h of degradation. Scattering not omitted.

On the other hand, the oxidation process may be monitored by increase of the emission at 424 nm, being a very promising parameter to monitor the oxidative process of both biodiesels (Figures 6(c) and 6(d)).

3.5. Fluorescence, Absorbance, and Acid Value Correlations.

The emission profile at 424 nm may be associated with classical technique acid value and more recently with absorptions at 232 and 270 nm [16]. Figure 7(a) shows the correlation of fluorescence intensity at 424 nm with acid values for RSLB and SILB, respectively. These results show that there is

a direct correlation in both cases, indicating that the emission at 424 nm is a potential tool for monitoring the oxidative stability for both SILB and RSLB. These correlations will be better addressed in Section 3.6.

Figures 7(b) and 7(c) show the obtained correlations of emission at 424 nm with absorbance at 232 and 270 nm, respectively. The results present a direct but not linear correlation. Furthermore, interesting to note different correlation plotted for each biodiesel. These results can be associated with specific functions, such previously demonstrated [16]; however, it is clear that there is a nonlinear correlation.

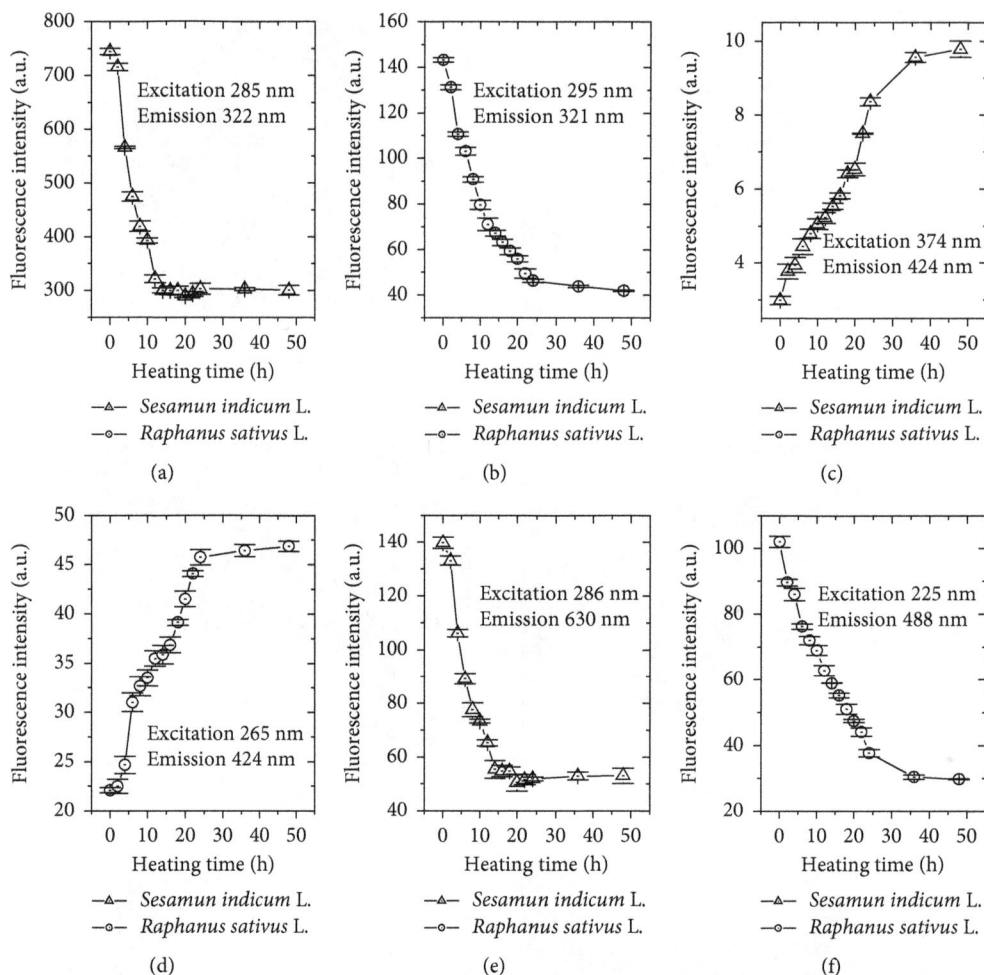

FIGURE 6: Fluorescence intensities of biodiesels from *S. indicum* L. and *R. sativus* L. during the thermal degradation process at specific wavelengths: (a) emission at ~322 and (b) emission at 321 nm; (c, d) emission at 424 nm; (e) emission at 630 nm; (f) emission at 488 nm. Error bars represent standard deviations.

3.6. AV Titration versus Emission at 424 nm. To evaluate the potential of molecular fluorescence spectroscopy for monitoring the degree of oxidation of the biodiesel samples, a close analysis of the correlation between fluorescence intensity at 424 nm and acid value was performed as shown in Figure 8. Clearly, fluorescence emission at 424 nm can be used to estimate the acid values using linear equations (1) and (2), respectively, which were obtained from the data fitting presented in Figure 8:

$$Em_{424} = 15.6 + 108.3*AV, \qquad (1)$$

$$Em_{424} = 2.1 + 17.1*AV. \qquad (2)$$

3.7. Consideration about Oxidative Stability of Biodiesels. The induction periods determined in this work for biodiesels from SILB and RSLB were (0.07 ± 0.01) h and (1.60 ± 0.10) h, respectively. These results show that RSLB is more stable than SILB. In fact, in Figure 2, it was possible to see that the formation of primary and secondary compounds of

degradation is more rapidly formed for SILB than RSLB; a possible explanation may be centered on the higher content of polyunsaturated fatty acids present in the SILB, making SILB more susceptible to degradation.

The classical method for evaluating the stability of the biodiesels is the Rancimat method [35]. However, several alternatives methods were proposed to increase the knowledge about a particular sample, adding with information obtained from more classical analyses such as titration of peroxide, iodine, and acid values [36]. Other methods, such as infrared spectroscopy, PetroOXY, ultrasonic-accelerated oxidation, differential scanning calorimetry (DSC), and thermogravimetric analysis (TGA) [8, 37], have been also recently applied to monitor the biodiesel quality. Each of these techniques can provide specific sample diagnostic which they should be crossed and interpreted to obtain a detailed picture of the biodiesel quality condition. In this context, the present study shows that due to the close correlation between biodiesel emission and acid values, fluorescence spectroscopy can be potentially applied to monitoring the biodiesel degradation.

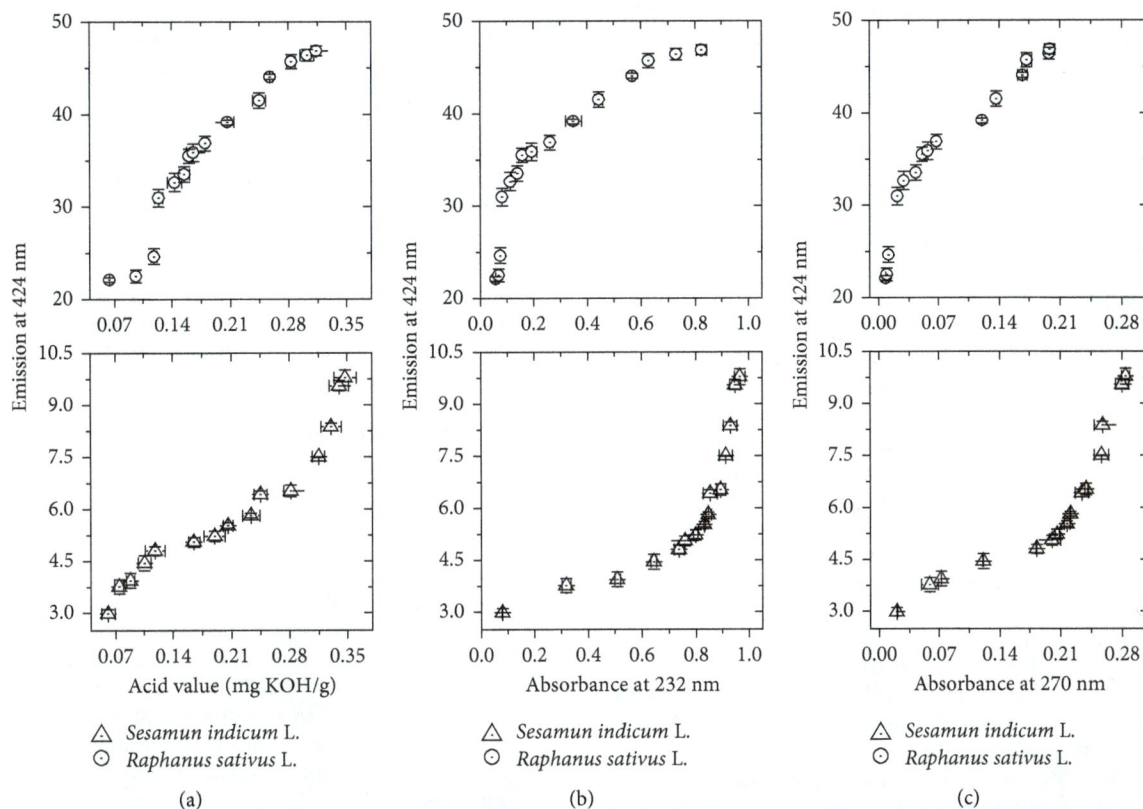

FIGURE 7: Correlation between emission at 424 nm (excitation at 374 and 265 nm for *S. indicum* L. and *R. sativus* L., resp.) and (a) acid value, (b) absorbance at 232 nm, and (c) absorbance at 270 nm for SILB and RSLB, respectively. Error bars represent standard deviations.

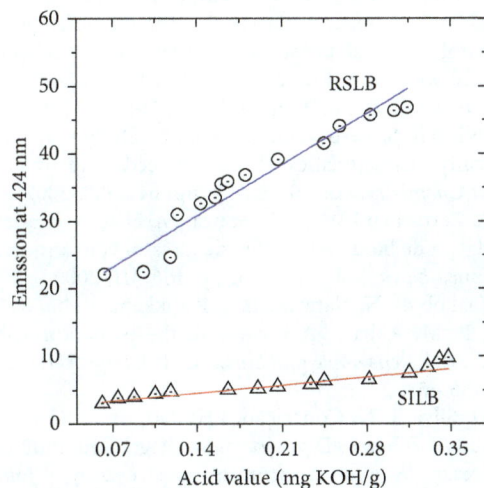

FIGURE 8: Fitting for the linear correlation between emission at 424 nm and acid value for RSLB and SILB.

4. Conclusion

This work showed that the composition of fatty acids was feedstock dependent, with high content of unsaturated fatty acids, principally polyunsaturated in SILB. The high content of unsaturated fatty acids reflects in the transformations undergone by RSLB and SILB carbon chains accused by acid value analysis, UV absorption, and fluorescence analysis.

The oxidation stages of the biodiesels were monitored by the formation of acidic compounds through the classic determination of the acid value, obtaining a similar change for both biodiesels. It was also found that the absorption profiles at 232 and 270 nm were more sensible to discriminate the biodiesels from different feedstocks under degradation process, evidenced by the correlation between the absorbance and acid value analysis. Alternatively, our results showed that molecular fluorescence spectroscopy can be used for monitoring the degradation stages of the RSLB and SILB according to the linear correlation between the emission at 424 nm and AV. Finally, it was demonstrated that the acid value for both biodiesels can be predicted by analyzing the emission at 424 nm. We summarize that the present findings revealed that fluorescence spectroscopy can be potentially used to monitor the biodiesel degradation.

Conflicts of Interest

The authors declare that there are no conflicts of interest regarding the publication of this paper.

Acknowledgments

The authors would like to thank Conselho Nacional de Desenvolvimento Científico e Tecnológico (CNPq),

Coordenação de Aperfeiçoamento de Pessoal de Nível Superior (CAPES), and Fundação de Apoio ao Desenvolvimento do Ensino, Ciência e Tecnologia do Estado de Mato Grosso do Sul (FUNDECT), for their assistance in completing this work.

References

[1] F. Ma and M. A. Hanna, "Biodiesel production: a review," *Bioresource Technology*, vol. 70, no. 1, pp. 1–15, 1999.

[2] A. E. Atabani, A. S. Silitonga, I. A. Badruddin, T. M. I. Mahlia, H. H. Masjuki, and S. Mekhilef, "A comprehensive review on biodiesel as an alternative energy resource and its characteristics," *Renewable and Sustainable Energy Reviews*, vol. 16, no. 4, pp. 2070–2093, 2012.

[3] G. Santori, G. Di Nicola, M. Moglie, and F. Polonara, "A review analyzing the industrial biodiesel production practice starting from vegetable oil refining," *Applied Energy*, vol. 92, pp. 109–132, 2012.

[4] Associação Brasileira das Indústrias de Óleos Vegetais, 2017, http://www.abiove.org.br/site/index.php?page=estatistica&area= NC0yLTE=.

[5] U.S. Energy Information Adminstration, 2017, https://www.eia.gov/biofuels/biodiesel/production/table2.pdf.

[6] W. Vogel, "European and world demand for biomass for the purpose of biofuel production in relation to supply in the food and feedstuff markets," 2016.

[7] X. Yuan, J. Liu, G. Zeng, J. Shi, J. Tong, and G. Huang, "Optimization of conversion of waste rapeseed oil with high FFA to biodiesel using response surface methodology," *Renewable Energy*, vol. 33, no. 7, pp. 1678–1684, 2008.

[8] R. K. Saluja, V. Kumar, and R. Sham, "Stability of biodiesel-a review," *Renewable and Sustainable Energy Reviews*, vol. 62, pp. 866–881, 2016.

[9] I. P. Lôbo, S. L. C. Ferreira, and R. S. da Cruz, "Biodiesel: parâmetros de qualidade e métodos analíticos," *Química Nova*, vol. 32, no. 6, pp. 1596–1608, 2009.

[10] R. L. McCormick and S. R. Westbrook, "Storage stability of biodiesel and biodiesel blends," *Energy and Fuels*, vol. 24, no. 1, pp. 690–698, 2010.

[11] D. P. C. De Quadros, E. S. Chaves, J. S. A. Silva, L. S. G. Teixeira, A. J. Curtius, and P. A. P. Pereira, "Contaminantes em biodiesel e controle de qualidade," *Revista Virtual de Quimica*, vol. 3, no. 5, pp. 376–384, 2011.

[12] E. Choe and D. B. Min, "Mechanisms and factors for edible oil oxidation," *Comprehensive Reviews in Food Science and Food Safety*, vol. 5, no. 4, pp. 169–186, 2006.

[13] J. C. Da Silva, A. Queiroz, A. Oliveira, and V. Kartnaller, "Advances in the application of spectroscopic techniques in the biofuel area over the last few decades," *Frontiers in Bioenergy and Biofuels*, pp. 25–58, 2017.

[14] I. Demshemino, M. Yahaya IsiomaNwadike, and L. Okoro, "A review on the spectroscopic analyses of biodiesel," *European International Journal of Science and Technology*, vol. 2, no. 7, pp. 137–149, 2013.

[15] I. P. de Oliveira, W. Correa, P. Neves et al., "Optical analysis of the oils obtained from *Acrocomia aculeata* (jacq.) lodd: mapping absorption-emission profiles in an induced oxidation process," *Photonics*, vol. 4, no. 1, p. 3, 2017.

[16] I. P. Oliveira, A. F. Souza, C. H. Lescano, A. R. L. Caires, and R. M. Muzzi, "Thermal oxidation analysis of forage turnip (*Raphanus sativus* L. var. oleiferus Metzg.) oil," *Journal of the American Oil Chemists' Society*, vol. 92, no. 3, pp. 403–408, 2015.

[17] M. Comin, A. C. D. de Souza, A. C. Roveda et al., "Alternatives binary and ternary blends and its effects on stability of soybean biodiesel contaminated with metals," *Fuel*, vol. 191, pp. 275–282, 2017.

[18] A. C. D. de Souza, M. Comin, L. H. de Oliveira et al., "Application of solvent dye in the field of biodiesel preservation," *Coloration Technology*, vol. 133, no. 2, pp. 165–169, 2017.

[19] AOAC, *Official methods of analysis of AOAC international*, Association of Analytical Communities, Gaithersburg, MD, USA, 17th edition, 2000.

[20] F. Lacoste and L. Lagardere, "Quality parameters evolution during biodiesel oxidation using Rancimat test," *European Journal of Lipid Science and Technology*, vol. 105, no. 34, pp. 149–155, 2003.

[21] A. F. F. de Vasconcelos and O. E. S. Godinho, "Uso de métodos analíticos convencionados no estudo da autenticidade do óleo de copaíba," *Química Nova*, vol. 25, no. 6b, pp. 1057–1060, 2002.

[22] M. B. Dantas, A. R. Albuquerque, A. K. Barros et al., "Evaluation of the oxidative stability of corn biodiesel," *Fuel*, vol. 90, no. 2, pp. 773–778, 2011.

[23] K. P. Suja, A. Jayalekshmy, and C. Arumughan, "Free radical scavenging behavior of antioxidant compounds of sesame (*Sesamum indicum* L.) in DPPH•system," *Journal of Agricultural and Food Chemistry*, vol. 52, no. 4, pp. 912–915, 2004.

[24] S. S. Beevi, L. N. Mangamoori, and B. B. Gowda, "Polyphenolics profile and antioxidant properties of *Raphanus sativus* L.," *Natural Product Research*, vol. 26, no. 6, pp. 557–563, 2012.

[25] I. A. Nehdi, "*Cupressus sempervirens* var. horizentalis seed oil: Chemical composition, physicochemical characteristics, and utilizations," *Industrial Crops and Products*, vol. 41, no. 1, pp. 381–385, 2013.

[26] S. N. Shah, O. K. Iha, F. C. S. C. Alves, B. K. Sharma, S. Z. Erhan, and P. A. Z. Suarez, "Potential application of turnip oil (*Raphanus sativus* L.) for biodiesel production: physical-chemical properties of neat oil, biofuels and their blends with ultra-low sulphur diesel (ULSD)," *BioEnergy Research*, vol. 6, no. 2, pp. 841–850, 2013.

[27] M. Elleuch, S. Besbes, O. Roiseux, C. Blecker, and H. Attia, "Quality characteristics of sesame seeds and by-products," *Food Chemistry*, vol. 103, no. 2, pp. 641–650, 2007.

[28] R. A. Ferrari and W. L. De Souza, "Avaliação da estabilidade oxidativa de biodiesel de óleo de girassol com antioxidantes," *Química Nova*, vol. 32, no. 1, pp. 106–111, 2009.

[29] Z. Yaakob, B. N. Narayanan, S. Padikkaparambil, K. S. Unni, and P. M. Akbar, "A review on the oxidation stability of biodiesel," *Renewable and Sustainable Energy Reviews*, vol. 35, pp. 136–153, 2014.

[30] V. D. Silva, J. N. Conceição, I. P. Oliveira et al., "Oxidative stability of baru (*Dipteryx alata* vogel) oil monitored by fluorescence and absorption spectroscopy," *Journal of Spectroscopy*, vol. 2015, Article ID 803705, 6 pages, 2015.

[31] N. Tena, D. L. García-gonzález, and R. Aparicio, "Evaluation of virgin olive oil thermal deterioration by fluorescence spectroscopy," *Journal of Agricultural and Food Chemistry*, vol. 57, no. 22, pp. 10505–10511, 2009.

[32] Y. G. M. Kongbonga, H. Ghalila, M. B. Onana et al., "Characterization of vegetable oils by fluorescence spectroscopy," *Food and Nutrition Sciences*, vol. 2, no. 7, pp. 692–699, 2011.

[33] R. Cheikhousman, M. Zude, D. J. R. Bouveresse, C. L. Léger, D. N. Rutledge, and I. Birlouez-Aragon, "Fluorescence spectroscopy for monitoring deterioration of extra virgin

olive oil during heating," *Analytical and Bioanalytical Chemistry*, vol. 382, no. 6, pp. 1438–1443, 2005.

[34] K. F. Magalhães, A. R. L. Caires, M. S. Silva, G. B. Alcantara, and S. L. Oliveira, "Endogenous fluorescence of biodiesel and products thereof: investigation of the molecules responsible for this effect," *Fuel*, vol. 119, pp. 120–128, 2014.

[35] L. S. De Sousa, C. V. R. De Moura, J. E. De Oliveira, and E. M. De Moura, "Use of natural antioxidants in soybean biodiesel," *Fuel*, vol. 134, pp. 420–428, 2014.

[36] L. Geng, "The impact of antioxidants on the oxidation stability of biodiesel," in *Proceedings of the IEEE 2nd International Conference on Computing, Control and Industrial Engineering*, vol. 1, pp. 62–65, Wuhan, China, August 2011.

[37] R. S. Leonardo, M. L. Murta Valle, and J. Dweck, "An alternative method by pressurized DSC to evaluate biodiesel antioxidants efficiency," *Journal of Thermal Analysis and Calorimetry*, vol. 108, no. 2, pp. 751–759, 2012.

9

Quantitative Analysis of Phenolic Acids and Flavonoids in *Cuscuta chinensis* Lam. by Synchronous Ultrasonic-Assisted Extraction with Response Surface Methodology

Kun-ze Du,[1,2] **Jin Li,**[1] **Xinrong Guo,**[1,2] **Yuhong Li** iD**,**[1] **and Yan-xu Chang** iD[1,2]

[1]*Tianjin State Key Laboratory of Modern Chinese Medicine, Tianjin University of Traditional Chinese Medicine, Tianjin 300193, China*
[2]*Tianjin Key Laboratory of Phytochemistry and Pharmaceutical Analysis, Tianjin University of Traditional Chinese Medicine, Tianjin 300193, China*

Correspondence should be addressed to Yuhong Li; liyuhong@tjutcm.edu.cn and Yan-xu Chang; tcmcyx@126.com

Academic Editor: Anna Vallverdu-Queralt

An effective ultrasonic-assisted extraction method for the separation of phenolic acids and flavonoids in *Cuscuta chinensis* Lam. was conducted by combining uniform design (UD) coupled with response surface methodology (RSM) and orthogonal design (OD) experiment. A sensitive and selective high-performance liquid chromatography-electrospray ionization tandem triple quadrupole mass spectrometry (HPLC-ESI-MS/MS) method was applied to quantify the sixteen active ingredients (chlorogenic acid, cryptochlorogenic acid, neochlorogenic acid, isochlorogenic acid A, isochlorogenic acid B, isochlorogenic acid C, caffeic acid, hyperin, isoquercitrin, quercetin, campherol, *p*-coumaric acid, isorhamnetin, rutin, astragalin, and apigenin). The extraction method was optimized with respect to concentration of extraction solvent, extraction time, and ratio of liquid to solid as a consequence of getting a high sensitive and feasible method for simultaneous determination of contents of multiple components and evaluation of quality control of *Cuscuta chinensis* Lam. from different origins. It was also considered useful and valuable in the further study for quality control of *Cuscuta chinensis* Lam.

1. Introduction

Traditional Chinese medicines (TCMs) occupy incomparable position in the pharmaceutical industry due to their extensive activity for preventing and treating diseases. Because of their clinic application and the contribution for drug discovery, TCMs have been drawn widespread attention in the world [1]. Only one or very limited kind of constituent was stipulated as a marker for the quality control of herb according to the authoritative Chinese Pharmacopoeia 2015. However, hundreds of constituents could be extracted from a single herb, which may exert various pharmacological functions and diverse bioactivities [2]. On account of the intimate connection with pharmacological activity, the quality of TCMs is increasingly strict and more components in TCMs should be identified and quantified grimly.

As a commonly used TCM, cuscutae semen, dried fruits of *Cuscuta chinensis* Lam., is widely distributed in China. This annual parasitic herb is parasitic on the legume, compositae, chenopodiaceae, and other herbs commonly, such as *Artemisia lavandulaefolia* DC, *Lespedeza chinensis* G. Don, and *Vicia cracca*. It has been employed in various clinical applications, including female infertility, preventing abortion, male reproductive system disease, chyluria, and chloasma faciei. [3] The extensive modern pharmacological studies have indicated that *Cuscuta chinensis* Lam. could decrease the apoptosis of cardiomyocytes [4], exhibit antifibrotic effect [5], suppress the inflammatory response [6], improve sexual potency, prevent abortion, and enhance liver and kidney conditions [3]. Phytochemical compounds of *Cuscuta chinensis* Lam. incorporate flavonoids, phenolic acids, volatile oils, hydroquinones, lignans, fatty acids, resin glycosides, steroids, polysaccharides, and alkaloids. By reason of the highest proportion in the chemical components of *Cuscuta chinensis* Lam., flavonoids and phenolic acids are two kinds of major bioactive compounds isolated from this TCM. A few results of studies suggested that some *Cuscuta chinensis* Lam. extracts could exert various pharmacological activities, for instance, flavonoids, phenolic acids, and phenolic compounds acting as free-radical scavengers could be responsible for antioxidative activity [3, 7].

A few analytical methods of quality control for *Cuscuta chinensis* Lam. were applied, such as HPLC-UV [8–10], high-performance capillary electrophoresis (HPCE) [11], and thin-layer chromatography (TLC) [12]. However, these practical methods either have very low sensitivities or were simply applied for evaluating limited kinds of active constituents such as flavonoids and polysaccharides. Due to the characteristics of multicomponent and multitarget of TCMs, it is difficult to comprehensively reveal the quality of TCMs via determining the few compounds merely. Therefore, it is necessary to establish a more precise analytical method of determining multiple compounds for assessing the quality of TCMs.

More useful information and optimum experimental conditions could be achieved by a good design and suitable model of experiment. Recently, uniform design (UD), which was shown to be a promising experimental design method and more effective than orthogonal design (OD) relying on numerical experiments, was reportedly used in the field of chemometrics, sciences, pharmaceutics, engineering, and manufacturing [13–18]. Response surface methodology (RSM) could be used to improve and optimize the complex experimental processes as a collection of statistical and mathematical techniques [19]. The various parameters and their interactions could be evaluated efficiently by this data analysis technology reducing the experimental group number [20, 21]. Uniform design underlined the uniformity of space filling in the experimental domain and the largest possible number of levels for each factor among all experimental designs [22]. UD can be simply used as a criterion to obtain better orthogonal designs.

The method in this study was developed to simultaneously determine sixteen flavonoids and phenolic acids of *Cuscuta chinensis* Lam. in different origins using liquid chromatography tandem mass-mass spectrum (HPLC-MS/MS) in the multiple reaction monitoring (MRM) acquisition mode for comprehensive quality control of *Cuscuta chinensis* Lam. To our knowledge, this is the first ever method identified to detect and quantify sixteen active components (Figure S1 in Supplementary Materials) of *Cuscuta chinensis* Lam. extracted via optimum conditions combined UD coupled with RSM and OD experiment. Moreover, the proposed method will be possible to lay the material basis for the evaluation and control of the quality of TCMs.

2. Experimental

2.1. Chemicals and Reagents. Four reference compounds including chlorogenic acid, caffeic acid, quercetin, and catechin were purchased from the National Institute for the Control of Pharmaceutical and Biological Products.

Neochlorogenic acid, cryptochlorogenic acid, isochlorogenic acid A, isochlorogenic acid B, isochlorogenic acid C, *p*-coumaric acid, gallic acid, hyperin, isoquercitrin, campherol, rutin, isorhamnetin, astragalin, and apigenin were purchased from Chengdu Must Bio. Sci. and Chengdu Desite Bio-Technology Co., Ltd (Chengdu, China). Deionized water used for sample preparations and buffer solutions was purified by a Milli-Q Academic ultrapure water system (Millipore, Milford, MA, USA). Acetonitrile and methanol were purchased from Merck (Germany). Formic acid was purchased from Anaqua Chemicals Supply (ACS). All other chemicals were of analytical grade.

2.2. Herbal Plant. Various batches of *Cuscuta chinensis* Lam. were purchased from various provinces in China. The authenticity of *Cuscuta chinensis* Lam. species were identified by Professor Lin Ma (Tianjin University of Traditional Chinese Medicine), and the voucher specimens were deposited at Tianjin University of Traditional Chinese Medicine. The *Cuscuta chinensis* Lam. was smashed into powder using a pulverizer (Zhongcheng Pharmaceutical Machinery) and dried at 40°C. Then, the powders were passed over 50 meshes, which were prepared for the following tests.

2.3. Preparation of Standard Solutions and Samples. All standard solutions were individually dissolved in methanol at a stock concentration of $1\,mg\cdot mL^{-1}$. A series of mixed standard solutions were diluted with methanol in different concentrations. The stock solutions of catechin and gallic acid were also dissolved and diluted with methanol to a final concentration of $25\,\mu g\cdot mL^{-1}$ and $1\,\mu g\cdot mL^{-1}$, respectively. All of the solutions were stored at 4°C until analysis.

Powdered sample (0.1 g) adding $10\,\mu L$ of $25\,\mu g\cdot mL^{-1}$ catechin and $1\,\mu g\cdot mL^{-1}$ gallic acid as internal standards were suspended in 45% ethanol-water (10 mL) and extracted in an ultrasonic bath for 120 min. Then, the sample was centrifuged at 14,000 ×g for 10 min. The supernatant was transferred and filtrated through a $0.22\,\mu m$ membrane prior to injection. $1\,\mu L$ of the solution was injected for analysis. The schematic diagram of the extraction method is shown in Figure 1.

2.4. Preparation of Quality Control Samples. Quality control (QC) samples of chlorogenic acid, caffeic acid, quercetin, neochlorogenic acid, cryptochlorogenic acid, isochlorogenic acid A, isochlorogenic acid B, isochlorogenic acid C, *p*-coumaric acid, hyperin, isoquercitrin, campherol, rutin, isorhamnetin, astragalin, and apigenin were prepared at low, medium, and high concentration levels by dissolving appropriate mixed standard solutions in methanol, respectively.

2.5. HPLC Condition and MS Condition. Qualitative analysis of the samples was performed using an Agilent HPLC 1200 system (Agilent Technologies, USA) coupled to an API 3200 triple quadrupole instrument (Agilent Corporation, CA, USA) with an electrospray ionization (ESI) source (Concord, Ontario, Canada). An Agilent Eclipse Plus C18 column

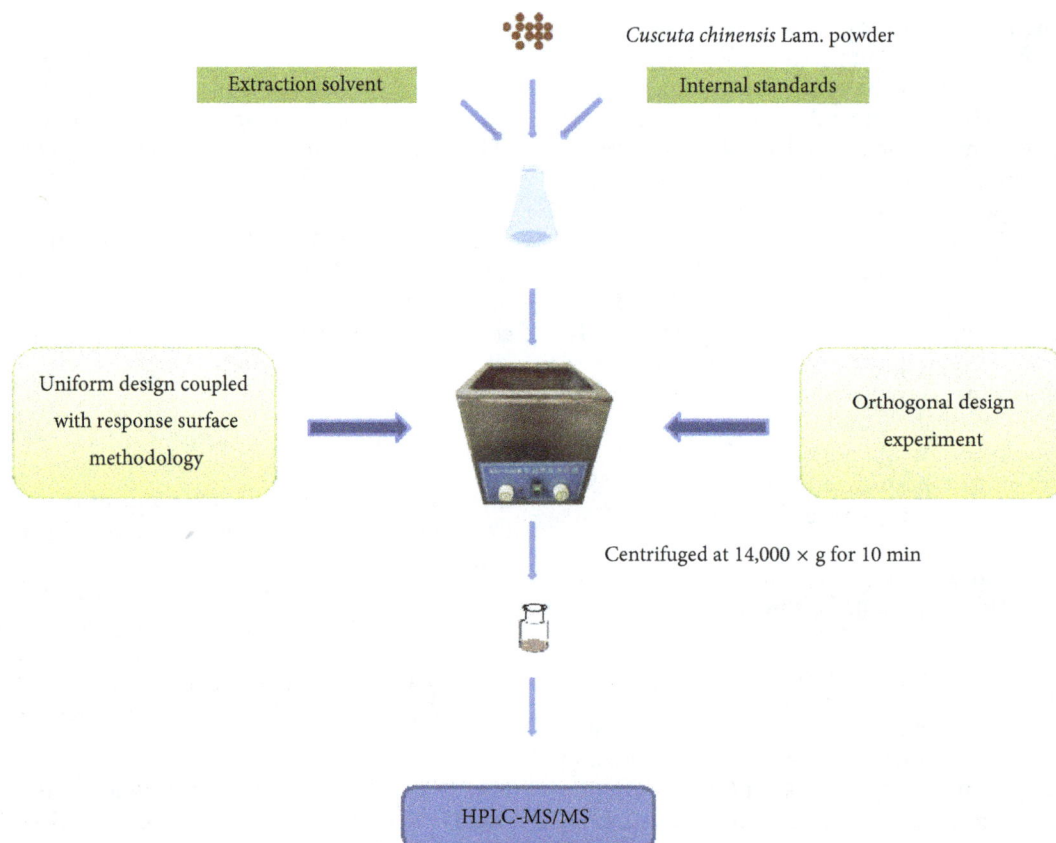

FIGURE 1: The schematic diagram of ultrasonic-assisted method coupled with HPLC-ESI-MS/MS.

(1.8 μm, 4.6 mm × 150 mm) was equipped with a security guard Agilent C18 column (5 μm, 2.1 mm × 12.5 mm). The mobile phase for the developed method consisted of acetonitrile (solvent A) and 0.05% aqueous formic acid in water (solvent B). The method employed a stepwise linear gradient as follows: 15%–19% solvent A at 0–3 min, 19%–20% solvent A at 3–9 min, 20%–30% solvent A at 9–12 min, 30%–48% solvent A at 12–12.5 min, 48%–52% solvent A at 12.5–14 min, 52%–54% solvent A at 14–17 min, 54%–60% solvent A at 17–18.5 min, and then 60%–81% solvent A at 18.5–20 min. The column was set at 35°C. In addition, the injection volume and low rate were 3 μL and 3 mL·min^{-1}, respectively.

The mass spectrometer was operated in the negative ion mode with curtain gas (CUR) of 45 psi, collision gas (CAD) of 5 psi, ion spray voltage (IS) of −4500 V, capillary temperature of 700°C, ion source gas 1 (GS1) of 40 psi, and ion source gas 2 (GS2) of 60 psi. The instrument was used in the tandem MS mode, by using the experiment of multiple reaction monitoring (MRM). A tandem mass spectrometry experiment that allows the selective isolation of the precursor ion in Q1, its subsequent fragmentation in a collision cell, and the final monitoring of a selected product ion in Q3 was done for the analysis of sixteen analytes and internal standards as shown in Table 1. The other parameters of eighteen compounds including declustering potential (DP), entrance potential (EP), collision energy (CE), collision cell exit potential (CXP), dwell time (DT), and retention time (RT) are also listed in Table 1.

3. Results and Discussion

3.1. Optimization of Extraction Procedure

3.1.1. Optimization of Extraction by Uniform Design Coupled with Response Surface Methodology. As for the efficient extraction of active compounds including phenolic acids and flavonoids from Cuscuta chinensis Lam., some parameters which influenced the extraction efficiency were optimized. In this study, ethanol was chosen as the extraction solvent. In order to obtain the optimal extraction condition, the relationship among concentration of extraction solvent (X_1), extraction time (X_2), and ratio of liquid to solid (X_3) was researched by uniform design (U12 (12 × 6 × 6)) (Table S1 in Supplementary Materials). The quadratic polynomial step-by-step regression method and data were analyzed by Uniform Design Version 3.00 software. A model was given as below to predict the response variables:

$$Y = b_0 + b_1 X_1 + b_2 X_2 + b_3 X_3 + b_1^2 X_1^2 + b_2^2 X_2^2 + b_2^3 X_2^3 \\ + b_1 b_2 X_1 X_2 + b_1 b_3 X_1 X_3 + b_2 b_3 X_2 X_3, \quad (1)$$

where Y is the predicted dependent variable; b_0 is a constant that fixes the response at the central point of the experiment; b_1, b_2, and b_3 are the regression coefficients for the linear effect terms; $b_1 b_2$, $b_1 b_3$, and $b_2 b_3$ are the interaction effect terms; and b_1^2, b_2^2, and b_3^2 are the quadratic effect terms, respectively [23]. Firstly, the extraction options for flavonoids (Y_1) were statistically analyzed, and the predicted model developed for flavonoids (Y_1) was as follows:

TABLE 1: Mass spectrometric parameters of analytical compounds and internal standards.

No.		Q1	Q3	DP (V)	EP (V)	CE (V)	CXP (V)	Acquisition time (second)	RT (min)
1	Chlorogenic acid	353.1	190.8	−24	−6	−33	−13	100	4.7
2	Cryptochlorogenic acid	353.2	173.0	−33	−5	−22	−2	100	4.96
3	Neochlorogenic acid	353.1	135.1	−24	−9	−46	−2	100	3.29
4	Isochlorogenic acid A	514.9	353.2	−64	−5	−28	−28	100	15.36
5	Isochlorogenic acid B	515.3	353.3	−67	−9	−28	−29	100	14.58
6	Isochlorogenic acid C	515.4	353.2	−51	−5	−21	−29	100	16.16
7	Caffeic acid	178.9	135.1	−33	−7	−23	−1	100	6.95
8	Hyperin	463.1	300.2	−71	−9	−42	−22	100	11.73
9	Isoquercitrin	463.2	300.0	−65	−8	−37	−25	100	12.28
10	Quercetin	301.2	151.2	−67	−8	−30	−2	100	17.25
11	Campherol	285.2	117.3	−80	−8	−56	−1	100	18.18
12	p-Coumaric acid	163.2	118.7	−36	−8	−23	−1	100	10.53
13	Isorhamnetin	315.3	300.2	−65	−10	−35	−25	100	18.34
14	Rutin	609.3	300.2	−28	−5	−46	−25	100	16.57
15	Astragalin	447.1	284.1	−65	5	−41	−22	100	15.29
16	Apigenin	269.2	117.1	−59	−9	−54	−1	100	17.9
IS1	Gallic acid	169.0	124.9	−32	−6	−24	−1	100	2.64
IS2	Catechin	289.1	123.0	−55	−9	−38	−2	100	4.94

$$Y = 2.07 - 2.82e^{-2}X_1 - 2.92e^{-2}X_2 - 4.27e^{-3}X_3$$
$$+ 1.04e^{-4}X_1^2 + 1.93e^{-4}X_1X_2 + 2.48e^{-5}X_1X_3 \quad (2)$$
$$+ 1.16e^{-4}X_2^2 + 2.19e^{-5}X_3^2.$$

According to the analysis of variance, the model with a good coefficient of determination ($R = 0.95$) was not significant (F-value 8.845, $P > 0.05$) which implied that the three factors (X_1, X_2, and X_3) had no influence to the extraction efficiency of flavonoids. The extraction efficiency for phenolic acids was significantly influenced by the concentration of the extraction solvent ($P < 0.05$). On increasing the temperature from 20 to 120°C (45% ethanol), the phenolic acids content increased by about 0.106 g·g⁻¹. In addition, on increasing the concentration of ethanol from 40% to 45%, the phenolic acids content increased by about 0.136 g·g⁻¹. Further increasing the concentration of ethanol, the phenolic acids content was decreasing (Figure 2(a)). The extraction conditions for phenolic acids (Y_2) were statistically analyzed and the predicted model established for phenolic acids (Y_2) is shown as follows:

$$Y_2 = -0.415 + 1.33e^{-2}X_1 + 2.58e^{-3}X_2 + 1.34e^{-3}X_3$$
$$- 7.57e^{-5}X_1^2 - 3.54e^{-5}X_1X_2 - 3.00e^{-5}X_1X_3 \quad (3)$$
$$- 9.38e^{-6}X_2^2 + 1.90e^{-5}X_2X_3 + 1.40e^{-6}X_3^2.$$

After the three revisions of the regression equation and eliminating nonsignificant items (F-value< F-critical value, $P > 0.05$), the predicted model established for phenolic acids (Y_2) was modified as follows:

$$Y = -0.415 + 1.33e^{-2}X_1 + 2.58e^{-3}X_3 + 1.34e^{-3}x_1^2$$
$$- 7.57e^{-5}X_1X_2 - 3.54e^{-5}X_1X_3 - 3.00e^{-5}X_2X_3. \quad (4)$$

On the basis of the elimination consequence (X_2^2), X_2, and (X_3^2) were eliminated which indicated that the effect of extraction time (X_2) and ratio of liquid to solid (X_3) was not significant for the extraction for phenolic acids. Last but not the least, the effect of extraction parameters for total

flavonoids and phenolic acids was conducted by RSM and developed a predicted model. The extraction efficiency for the total contents was influenced by the concentration of extraction solvent ($P < 0.05$). The total contents of flavonoids and phenolic acids ranged from 0.180 g·g⁻¹ to 0.420 g·g⁻¹, when the temperature increased from 20°C to 120°C (45% ethanol). In addition, there is a rising trend for the total contents of flavonoids and phenolic acids from 40% ethanol to 45% ethanol. Otherwise, the trend was changed when the concentration of ethanol declined continuously (Figure 2(b)). The predicted model established for total flavonoids and phenolic acids (Y_3) is shown as follows:

$$Y = -1.96 + 4.49e^{-2}X_1 + 2.00e^{-2}X_2 + 3.28e^{-3}X_3$$
$$- 2.07e^{-4}X_1^2 - 1.69e^{-4}X_1X_2 - 7.81e^{-5}X_1X_3 \quad (5)$$
$$- 8.18e^{-5}X_2^2 + 4.70e^{-5}X_2X_3.$$

After quadruplicate revision of the regression equation and eliminating nonsignificant items (F-value< F-critical value, $P > 0.05$), the predicted model established for total flavonoids and phenolic acids (Y_3) was modified as follows:

$$Y = -0.218 + 1.00e^{-2}X_1 + 1.24e^{-3}X_3 - 6.45e^{-5}X_1^2$$
$$- 1.54e^{-5}X_1X_2 - 2.43e^{-5}X_1X_3 + 1.77e^{-5}X_2X_3. \quad (6)$$

The items including (X_2^2), X_2, and X_3 were eliminated (F-value< F-critical value, $P > 0.05$), which revealed that extraction time (X_2) and ratio of liquid to solid (X_3) were not significant factors in the extraction procedure.

Synthesizing the RSM analysis and the predicted model, 10 mL 45% ethanol-water and ultrasonic extraction for 120 min at room temperature were selected as the optimum extraction method.

3.1.2. Optimization of Extraction by Orthogonal Design.
After applying uniform design, which narrowed effectively the range of extraction conditions, some of the sophisticated

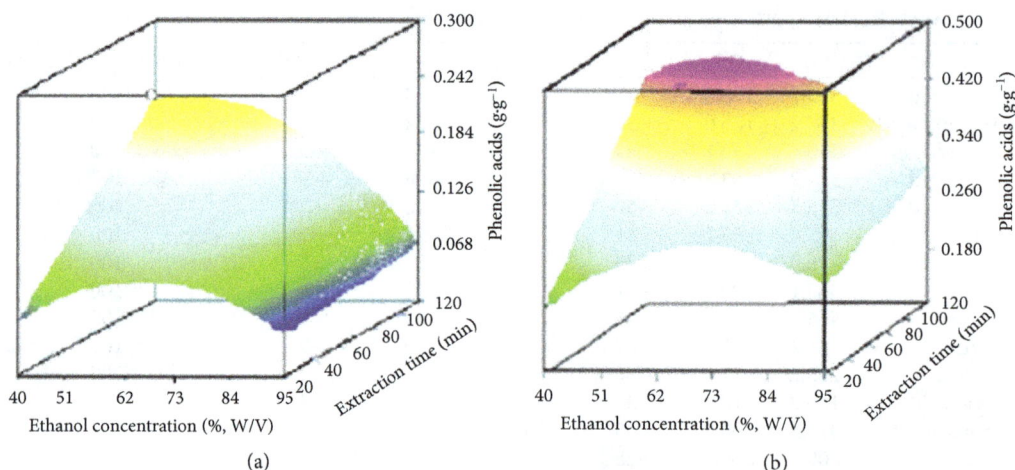

FIGURE 2: Effect of ethanol concentration and extraction time on the extraction efficiency of phenolic acids (a) and total analytes (b) in *Cuscuta chinensis* Lam.

tests should be investigated in succession by orthogonal design in order to obtain more efficient consequence. OD experiment was used to further optimize the extraction condition as an effective design method with small experimental range and less examined factors and levels. During this OD procedure, the experiment points for both training set and prediction set were concentrated in only a certain region derived from the results of UD coupled with RSM.

Firstly, the OD was designed to confirm the effects of various factors for extraction efficiency of flavonoids, including concentration of ethanol (A: 35% (A_1), 40% (A_2), 45% (A_3)), extraction time (B: 80 min (B_1), 100 min (B_2), 120 min (B_3)), ratio of sample to solvent (C: 100 g·mL^{-1} (C_1), 120 g·mL^{-1} (C_2), 125 g·mL^{-1}(C_3)), and D as the void item. When concentration of ethanol, extraction time, and ratio of sample to solvent was set to 45%, 120 min, and 125 g·mL^{-1}, respectively, the best extraction efficiency could be obtained for analyzing flavonoids (Table S2 in Supplementary Materials). According to the result of the ANOVA test, F-value of extraction time exceeded F-critical values which meant this factor had significance in this extraction procedure. However, there was no significance in the other factors because their F-values were under F-critical value. Considering the above results and economy, 45%, 120 min, and 100 g·mL^{-1}were chosen as concentration of ethanol, extraction time and ratio of sample to solvent for extracting flavonoids, respectively. Then, the following OD was carried out to investigate the effect of factors for extracting phenolic acids solely. The results displayed that the best extraction efficiency could be acquired for phenolic acids while concentration of ethanol, extraction time, and ratio of sample to solvent were set to 45%, 120 min, and 120 g·mL^{-1}, respectively (Table S3 in Supplementary Materials). But, the ANOVA test of this OD showed that the ratio of sample to solvent in this part had no significance. Thus, best extraction factors for phenolic acids were chosen as 45% ethanol, 120 min of extraction time, and 120 g·mL^{-1} of sample to solvent. The last but not the least, further OD was applied to

verify the optimum conditions for total phenolic acids and flavonoids. The results revealed that the extraction efficiency was best with 45% ethanol, 120 min of extraction time, and 100 g·mL^{-1} of sample to solvent (Table S4 in Supplementary Materials). The following ANOVA test also indicated that the concentration of ethanol and extraction time were significant factors, whereas the ratio of sample to solvent was not significant as the F-value was under the F-critical value. By comprehensive consideration of the cost of production and the extraction efficiency of phenolic acids and flavonoids, 100 g·mL^{-1} of sample to solvent, 45% ethanol-water, and ultrasonic extraction for 120 min at room temperature was selected as the optimum extraction method. This consequence revealed that the best pronounced condition could be obtained by OD experiment particularly and the optimum condition conducted by OD was the same with that obtained by UD.

3.2. Optimization of Chromatographic Conditions

3.2.1. Optimization of HPLC Condition. To obtain good separation and stable peak retention value, the factors of the HPLC-ESI-MS/MS method was optimized in detail for analyzing the active compounds of *Cuscuta chinensis* Lam. The factors concerning the separation of sixteen components comprised the type of acid (formic acid, acetic acid, and ammonium formate), the concentration of acid (0.01%, 0.05%, and 0.1%), column temperature (25°C, 35°C, and 45°C), and flow rate (0.3 mL·min^{-1}, 0.4 mL·min^{-1}, and 0.5 mL·min^{-1}). The effect of each factor was studied by orthogonal experiment design L9 (3^4). Taking into account centered migration time and good peak response, hyperoside was selected as a reference to find the optimum HPLC condition, which is a main active component of *Cuscuta chinensis* Lam. The results (Table S5) showed that acetonitrile-0.05% formic acid aqueous solution in a step linear gradient, column temperature at 35°C, and 0.5 mL·min^{-1} flow rate should be finally set for the qualitative analysis of active compounds.

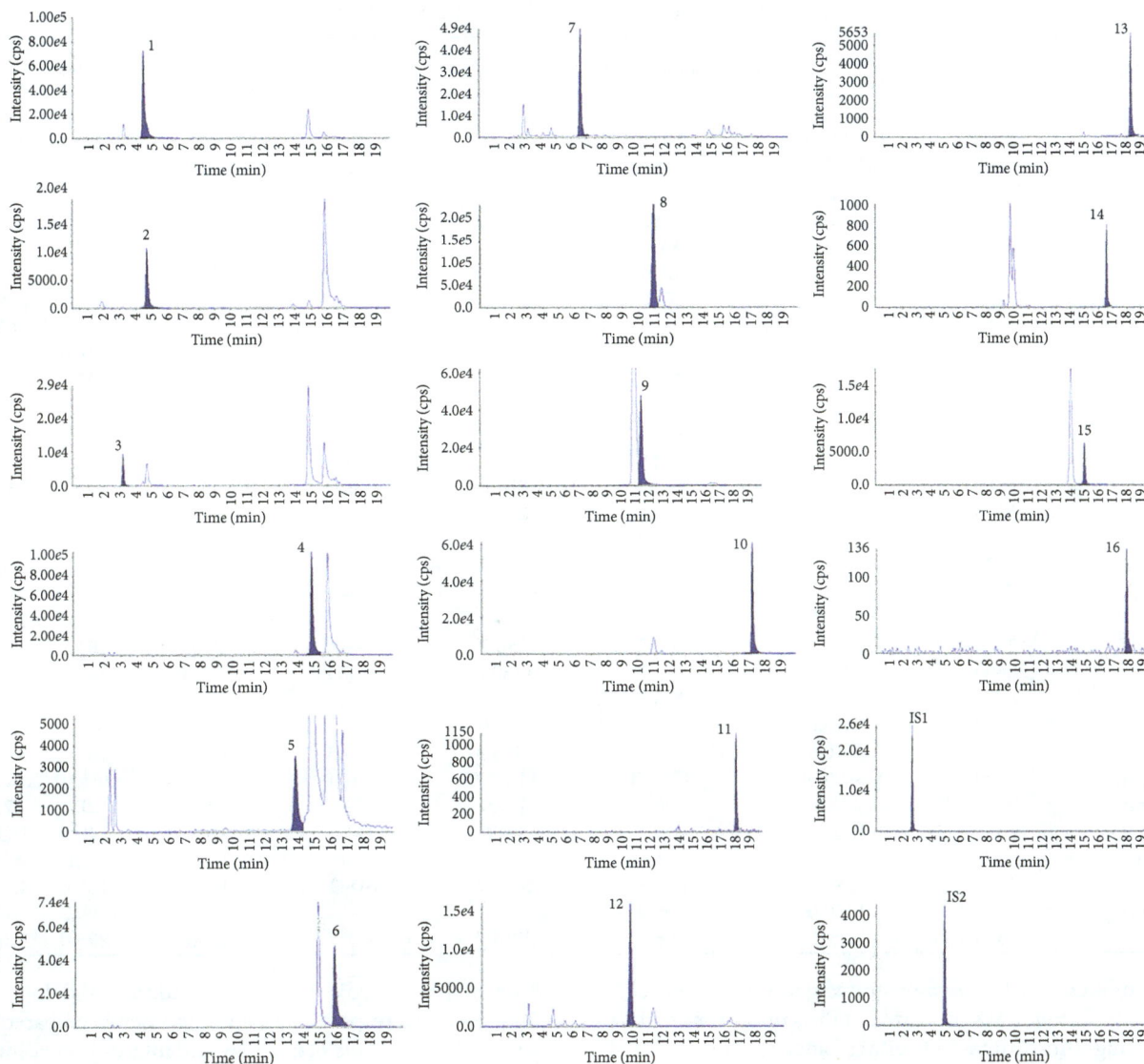

FIGURE 3: Typical chromatograms of LC–MS/MS for analytes and internal standards in *Cuscuta chinensis* Lam. 1, chlorogenic acid; 2, cryptochlorogenic acid; 3, neochlorogenic acid; 4, isochlorogenic acid A; 5, isochlorogenic acid B; 6, isochlorogenic acid C; 7, caffeic acid; 8, hyperin; 9, isoquercitrin; 10, quercetin; 11, campherol; 12, *p*-coumaric acid; 13, isorhamnetin; 14, rutin; 15, astragalin; 16, apigenin; IS1, catechin; IS 2, gallic acid.

3.2.2. Optimization of MS Condition.

A tandem mass spectrometry approach was employed to obtain a quantitative determination of sixteen active markers and two internal standards. In order to increase the sensitivity of detection and improve the response value, optimization experiments of ion source parameters including CUR (30 psi, 35 psi, 40 psi, 45 psi), CAD (3 psi, 5 psi, 7 psi, 9 psi), IS (−2500 V, −3000 V, −3500 V, −4000 V, −4500 V), GS1 (25 psi, 30 psi, 35 psi, 40 psi, 45 psi), GS2 (30 psi, 40 psi, 50 psi, 60 psi), and TEM (300°C, 400°C, 500°C, 600°C, 700°C) were carried out. Consequently, the optimum ion source parameters were achieved when CUR was 45 psi, CAD was 5 psi, IS was −4500 V, GS1 was 40 psi, GS2 was 60 psi, and TEM was 700°C.

3.3. Selection of Internal Standard.

In order to avoid the error caused by sample consistency and sample discrimination effect, internal standard should be added into the analytical sample solution. Considering the chemical property of analytical compounds and the interaction of retention time between analytes and internal standard, two kinds of compounds were selected as internal standard, covering catechin and gallic acid. Catechin and gallic acid were used as internal standards for the eight flavonoids and eight phenolic acids, respectively, because their chemical properties were similar with these two kinds of analytes and they would not impact the peaks of the targets in the chromatogram.

3.4. Method Validation

3.4.1. Specificity.

Figure 3 shows the chromatograms of sixteen active components of *Cuscuta chinensis* Lam. in the MRM acquisition mode. The results of HPLC-MS/MS

TABLE 2: Contents of phenolic acids in samples from different origins (ng·g^{-1}).

Collecting locations	Chlorogenic acid	Cryptochlorogenic acid	Neochlorogenic acid	Isochlorogenic acid A	Isochlorogenic acid B	Isochlorogenic acid C	Caffeic acid	Total phenolic acids
Qingdao	12000.0	1043.0	102.3	536.5	247.0	341.3	51.6	14321.7
Inner Mongolia	20700.0	2570.0	1021.3	3703.3	518.7	1416.7	88.1	30018.1
Ningxia	7060.0	652.3	97.4	260.7	—	417.7	82.0	8570.1
Beijing	567.0	361.0	138.7	—	—	—	—	1066.7
Suizhou	6966.7	736.7	91.6	119.0	—	358.7	134.7	8407.4
Zhangshu	343.3	175.0	65.0	63.6	—	285.7	—	932.6
Guizhou	20466.7	788.7	467.3	32633.3	1643.3	6716.7	213.3	62929.3
Yunnan	196.7	202.3	88.1	142.1	—	—	—	629.2
Henan	11900.0	501.7	141	187.0	375.0	689.0	119.0	13912.7
Anhui1	8710.0	423.5	101.1	675.0	288.7	398.7	127.5	10724.5
Tianjin	11633.3	—	—	149.5	266.0	410.5	—	12459.3
Xinjiang	9625.0	1640.0	178.5	212.0	321.3	781.7	141.0	12899.5
Jiangsu	603.3	—	143.0	—	—	—	—	746.3
Bozhou	21366.7	2060.0	211.0	857.0	411.0	757.0	125.7	25788.4
Anhui2	15233.3	1900	310.7	568.7	334.3	968.3	118.0	19433.3
Shanxi	5090.0	705.7	145.3	75.3	—	—	144.3	6160.6
Tangshan	143.5	—	522.5	—	—	—	—	666.0
Liaoning	5746.7	—	215.7	102.5	—	—	30.1	6095.0
Zhejiang	12500.0	1143.3	110.0	165.0	—	385.0	86.4	14389.7
Anhui3	13500.0	1466.7	208.0	256.7	—	330.3	105.7	15867.4
Jiangsu	20466.7	2010.0	312.7	1380.0	363.0	720.0	141.3	25393.7
Henan	2610.0	—	229.3	—	—	—	11.6	2850.9
Hebei	50500.0	5895.0	851.0	15150	1795.0	7005.0	2190.0	83386.0
Chifeng	23733.3	2730.0	1293.3	3953.3	585.3	1366.7	167.7	33829.6
Chongqing	30200.0	5100.0	1780.0	44633.3	3310.0	13300.0	3940.0	102263.3
Lingchuan	4995.0	633.3	120.3	453.0	—	487.5	70.1	6759.2
Shanxi	16033.3	1305.0	110.0	391.0	—	417.7	140.0	18397.0
Shanghai	4086.7	—	101.6	—	—	—	19.1	4207.4
Xuzhou	32266.7	3180.0	896.3	5056.7	618.3	1423.3	157.0	43598.3
Guangdong	15150.0	1550.0	165.9	—	—	—	100.2	16966.1
Qiyang	16333.3	605.3	256.3	2816.7	477.7	1090.0	295.0	21874.3

analysis of these markers demonstrated good shape of peaks, and no interfering peaks presented in the sample for analysis at the migration times of either analytes or internal standards.

3.4.2. Linearity, Limits of Detection, and Repeatability.

The calibration curve consisted of six concentration levels. The calibration graph was constructed by adding $10 \mu L$ of $25 \mu g \cdot mL^{-1}$ catechin and $1 \mu g \cdot mL^{-1}$ gallic acid as internal standard. Sixteen plotted calibration curves and correlation coefficients ($r > 0.999$) confirmed that the curves were linear in the concentration ranges of each component. The limit of detection (LOD) and limit of quantification (LOQ) were considered as the concentrations of the compound that produced a signal-to-noise (S/N) ratio of 3 and 10, respectively. LODs and LOQs of each component ranged from $0.001 \, ng \cdot mL^{-1}$ to $14 \, ng \cdot mL^{-1}$ and $0.005 \, ng \cdot mL^{-1}$ to $34.9 \, ng \cdot mL^{-1}$, respectively (Table S6). The repeatability was evaluated by preparing the same sample with the same optimized extraction method ($n = 6$). The repeatability RSDs of sixteen targets were not more than 2.9%.

3.4.3. Precision and Accuracy.

Precision and accuracy were obtained by evaluating quality control samples containing sixteen active ingredients at low, medium, and high concentrations with respect to a calibration graph prepared each day ($n = 6$). The results of intraday and interday precision and accuracy are illustrated in Table S7. The accuracies of sixteen active components for both intraday and interday were within the range of 87.8%–120.7%. The RSDs for both intraday and interday were below 13.9%. These results indicated that the present method was accurate, reliable, and reproducible.

3.4.4. Stability.

The stability of sixteen active components at ambient temperature was assessed by analyzing quality control samples at low, medium, and high concentrations over 48 h storage at 4°C. The data of stability studies presented in Table S7 showed that RSDs of stability of sixteen active components were below 15.4%, when they were determined at 0 h, 2 h, 4 h, 6 h, 8 h, 12 h, 24 h, and 48 h, respectively. The results demonstrated that all of the components have good stability within 48 h of storage at the concentrations studied and this analytical method could be used to determine these compounds.

3.4.5. Recovery.

The recoveries were determined for the purpose of evaluation of precision and accuracy of the extraction method. The extraction recoveries of sixteen components were evaluated after extraction from Cuscuta

TABLE 3: Contents of flavonoids in samples from different origins (ng·g^{-1}).

Collecting locations	Hyperin	Isoquercitrin	Quercetin	Campherol	p-Coumaric acid	Isorhamnetin	Rutin	Astragalin	Apigenin	Total flavonoids
Qingdao	19233.3	3240.0	49.3	3705.0	464.7	52.6	155.0	10266.7	1.1	37167.7
Inner Mongolia	34700.0	2276.7	60.1	451.0	479.3	49.6	119.0	926.3	0.8	39062.8
Ningxia	31833.3	1600.0	81.5	788.3	754.3	57.1	755.3	3270.0	0.9	39140.7
Beijing	3983.3	558.0	19.9	91.3	652.3	17.3	63.0	2270.0	1.0	7656.1
Suizhou	15933.3	3200.0	255.0	3316.7	779.3	123.0	51.1	7880.0	1.3	31539.7
Zhangshu	732.7	121.7	21.8	166.0	89.2	15.1	69.7	320.7	0.6	1537.5
Guizhou	3906.7	660.0	39.7	485.0	278.0	26.6	3583.3	1890.0	0.7	10870
Yunnan	2220.0	431.0	17.9	86.7	266.0	12.7	46.7	3020.0	0.7	6101.7
Henan	30166.7	5500.0	155.0	4113.3	1373.3	99.6	56.7	5416.7	1.5	46882.8
Anhui1	27000.0	3693.3	381.7	2356.7	906.0	129.7	201.3	4630.0	1.6	39300.3
Tianjin	20166.7	3533.3	119.0	3185.0	724.0	99.7	34.6	10295.0	1.0	38158.3
Xinjiang	12300.0	3380.0	78.8	5770.0	592.7	128.1	100.7	26900.0	1.0	49251.3
Jiangsu	6645.0	1226.7	20.0	186.0	690.7	16.1	96.0	6166.7	1.2	15048.4
Bozhou	23133.3	3723.3	44.2	4576.7	760.7	62.0	166.3	9186.7	1.2	41654.4
Anhui2	25433.3	4773.3	159.0	4066.7	504.7	143.3	48.0	19600.0	0.9	54729.2
Shanxi	13133.3	943.0	102.4	2273.3	637.0	89.3	431.0	4393.3	0.9	22003.5
Tangshan	880.0	209.0	21.9	196.7	959.0	15.3	29.9	2590.0	0.8	4902.6
Liaoning	13333.3	1400.0	19.6	112.5	571.0	13.8	286.7	3203.3	1.1	18941.3
Zhejiang	14133.3	2460.0	66.7	2800.0	551.7	65.2	56.6	6360.0	16.4	26509.9
Anhui3	20200.0	3676.7	54.5	4560.0	445.3	73.9	103.2	10346.7	1.2	39461.5
Jiangsu	29233.3	4030.0	106.5	2635.0	588.0	75.1	141.0	8493.3	1.2	45303.4
Henan	32250.0	6116.7	33.0	355.3	1153.3	17.3	322.7	29500.0	1.2	69749.5
Hebei	17666.7	807.0	18.9	431.3	1136.7	14.0	1316.7	1263.3	0.8	22655.4
Chifeng	24066.7	1510.0	58.1	98.6	359.0	30.1	117.0	134.3	0.7	26374.5
Chongqing	2326.7	472.0	20.1	426.3	856.7	18.5	4063.3	1500.0	2.5	9686.1
Lingchuan	6836.7	1130.0	38.1	1273.3	259.3	36.8	72.0	4990.0	0.7	14636.9
Shanxi	32066.7	4366.7	116.7	3610.0	718.0	81.9	52.1	5030.0	1.2	46043.3
Shanghai	25600.0	4180.0	19.5	256.0	458.7	21.7	68.1	5280.0	1.1	35885.1
Xuzhou	24400.0	1793.3	19.0	268.3	446.3	18.8	49.3	986.3	0.7	27982.0
Guangdong	7020.0	1206.7	14.0	1080.0	227.3	13.0	91.2	3116.7	0.7	12769.6
Qiyang	21666.7	3496.7	66.2	2190.0	590.3	44.5	531.3	4956.7	1.5	33543.9

chinensis Lam. to an equivalent amount of the standard solution including chlorogenic acid, cryptochlorogenic acid, neochlorogenic acid, isochlorogenic acid A, iso-chlorogenic acid B, isochlorogenic acid C, caffeic acid, hyperin, isoquercitrin, quercetin, campherol, p-coumaric acid, isorhamnetin, rutin, astragalin, and apigenin ($n = 6$). The mean recoveries of sixteen active components determined in the range of 89.9%–112.7% and RSDs were below 11.4% (Table S6). The results suggested that the extraction method showed good precision and accuracy.

3.5. Method Application. The proven ultrasonic-assisted extraction method coupled with HPLC-MS/MS method was applied to quantify sixteen active components in 31 batches of *Cuscuta chinensis* Lam. from different origins. The data of quantitative analysis illustrated that the total content of phenolic acids and flavonoids in *Cuscuta chinensis* Lam. from different origins varied significantly as shown in Tables 2 and 3. The total content of phenolic acids of *Cuscuta chinensis* Lam. from Chongqing province was relatively higher than others, which was the lowest from Yunnan. The highest total content of flavonoids of *Cuscuta chinensis* Lam. of these samples was determined from Henan province, and the lowest was assessed from Zhangshu.

4. Conclusion

In this research, an optical extraction condition was performed to get high yields of multiple active compositions by combining uniform design coupled with RSM and orthogonal design experiment. The HPLC-MS/MS procedure has been proposed to simultaneously determine sixteen active components present in different batches of *Cuscuta chinensis* Lam. with this effective extraction method. The developed method has been successfully applied to analyze the contents of phenolic acids and flavonoids using catechin and gallic acid as internal standards. The results demonstrated that this method offered excellent selectivity and sensitivity. Meanwhile, the use of HPLC-MS/MS could quantify phenolic acids and flavonoids and be successfully applied for quality control of *Cuscuta chinensis* Lam. in different origins. The contents of either phenolic acids or flavonoids of *Cuscuta chinensis* Lam. from the south of China were generally higher than those from the north of China. Consequently, the proposed HPLC-MS/MS method could be taken into consideration for future study of evaluation of quality control of TCMs.

Conflicts of Interest

The authors declare that there are no conflicts of interest regarding the publication of this paper.

Authors' Contributions

Jin Li and Kun-ze Du contributed equally to this study.

Acknowledgments

This research was supported by the National Natural Science Foundation of China (81374050) and Special Program of Talents Development for Excellent Youth Scholars in Tianjin.

Supplementary Materials

Supplementary material contains revised articles on the experiment design for optimization of extraction of analytes from *Cuscuta chinensis* Lam. and the validation results for the analytical method. Table S1: uniform design (U_{12} ($12 \times 6 \times 6$)) for optimization of extraction of phenolic acids and flavonoids from *Cuscuta chinensis* Lam. Table S2: the results of orthogonal design [L_9 (3^4)] for optimizing of extraction of flavonoids from *Cuscuta chinensis* Lam. Table S3: the results of orthogonal design [L_9 (3^4)] for optimizing of extraction of phenolic acids from *Cuscuta chinensis* Lam. Table S4: the results of orthogonal design [L_9 (3^4)] for optimizing the extraction of flavonoids and phenolic acids from *Cuscuta chinensis* Lam. Table S5: the results of orthogonal design [L_9 (3^4)] for optimizing the HPLC conditions based on hyperoside of *Cuscuta chinensis* Lam. Table S6: the calibration curves, linearity ranges, LODs, LOQs, and recoveries of sixteen compounds in samples ($n = 6$). Table S7: intraday and interday accuracy and precision and stability of sixteen compounds ($n = 6$). Figure S1: the chemical structure of sixteen compounds and two internal standards. (*Supplementary Materials*)

References

[1] J. D. McChesney, S. K. Venkataraman, and J. T. Henri, "Plant natural products: back to the future or into extinction?," *Phytochemistry*, vol. 68, no. 14, pp. 2015–2022, 2007.

[2] K. Li, Y. Fan, H. Wang, Q. Fu, Y. Jin, and X. Liang, "Qualitative and quantitative analysis of an alkaloid fraction from *Piper longum* L. using ultra-high performance liquid chromatography-diode array detector-electrospray ionization mass spectrometry," *Journal of Pharmaceutical and Biomedical Analysis*, vol. 109, pp. 28–35, 2015.

[3] S. Donnapee, J. Li, X. Yang et al., "*Cuscuta chinensis* Lam.: a systematic review on ethnopharmacology, phytochemistry and pharmacology of an important traditional herbal medicine," *Journal of Ethnopharmacology*, vol. 157, pp. 292–308, 2014.

[4] S.-L. Sun, L. Guo, Y.-C. Ren et al., "Anti-apoptosis effect of polysaccharide isolated from the seeds of *Cuscuta chinensis* Lam on cardiomyocytes in aging rats," *Molecular Biology Reports*, vol. 41, no. 9, pp. 6117–6124, 2014.

[5] J. S. Kim, S. Koppula, M. J. Yum et al., "Anti-fibrotic effects of *Cuscuta chinensis* with in vitro hepatic stellate cells and a thioacetamide-induced experimental rat model," *Pharmaceutical Biology*, vol. 55, no. 1, pp. 1909–1919, 2017.

[6] S. Y. Kang, H. W. Jung, M.-Y. Lee, H. W. Lee, S. W. Chae, and Y.-K. Park, "Effect of the semen extract of *Cuscuta chinensis* on inflammatory responses in LPS-stimulated BV-2 microglia," *Chinese Journal of Natural Medicines*, vol. 12, no. 8, pp. 573–581, 2014.

[7] J. Liu, J. Tian, J. Li et al., "The in-capillary DPPH-capillary electrophoresis-the diode array detector combined with reversed-electrode polarity stacking mode for screening and quantifying major antioxidants in *Cuscuta chinensis* Lam," *Electrophoresis*, vol. 37, no. 12, pp. 1632–1639, 2016.

[8] H. Hajimehdipoor, B. M. Kondori, G. R. Amin, N. Adib, H. Rastegar, and M. Shekarchi, "Development of a validated HPLC method for the simultaneous determination of flavonoids in *Cuscuta chinensis* Lam. by ultra-violet detection," *DARU Journal of Pharmaceutical Sciences*, vol. 20, no. 1, p. 57, 2012.

[9] S. Yang, H. Xu, B. Zhao et al., "The difference of chemical components and biological activities of the crude products and the salt-processed product from semen Cuscutae," *Evidence-Based Complementary and Alternative Medicine*, vol. 2016, Article ID 8656740, 9 pages, 2016.

[10] M. Ye, Y. Li, Y. Yan, H. Liu, and X. Ji, "Determination of flavonoids in semen Cuscutae by RP-HPLC," *Journal of Pharmaceutical and Biomedical Analysis*, vol. 28, no. 3-4, pp. 621–628, 2002.

[11] P. Hu, G. A. Luo, R. J. Wang et al., "Identification of semen Cuscutae by HPCE," *Yao Xue Xue Bao = Acta Pharmaceutica Sinica*, vol. 32, no. 7, pp. 549–552, 1997.

[12] M. Ye, P. Zhou, Y. Yan et al., "Identification of seeds of *Cuscuta australis* and *C. chinensis* by TLC and HPLC," *Zhong Yao Cai*, vol. 24, no. 2, pp. 97–99, 2001.

[13] Q. S. Xu, Y. D. Xu, L. Li, and K. T. Fang, "Uniform experimental design in chemometrics," *Journal of Chemometrics*, vol. 32, no. 11, p. e3020, 2018.

[14] Y. Tian and F. Gao, "The water-fertilizer coupling study of high protein soybean-based on uniform design and partial least squares regression modeling," *Journal of Nuclear Agricultural Sciences*, vol. 3, pp. 561–568, 2018.

[15] A. M. Elsawah and K. T. Fang, "New results on quaternary codes and their Gray map images for constructing uniform designs," *Metrika*, vol. 81, no. 3, pp. 307–336, 2018.

[16] X. U. Xiao-Hu, X. Chen, Y. E. Liang et al., "Ratio of effective components of compound honeysuckle traditional Chinese medicine mouthwash by uniform design," *Chinese Journal of Conservative Dentistry*, vol. 1, pp. 20–25, 2018.

[17] Y. Liu, S. Liu, Y. Feng et al., "Uniform-design-based optimization for fuel reactor of chemical looping combustio," *International Journal of Chemical Reactor Engineering*, vol. 15, no. 6, 2017.

[18] Y. Zhang, D. J. Sun, and A. Kondyli, "An empirical framework for intersection optimization based on uniform design," *Journal of Advanced Transportation*, vol. 2017, Article ID 7396250, 10 pages, 2017.

[19] F. Gu, F. Xu, L. Tan, H. Wu, Z. Chu, and Q. Wang, "Optimization of enzymatic process for vanillin extraction using response surface methodology," *Molecules*, vol. 17, no. 8, pp. 8753–8761, 2012.

[20] F.-l. Gu, S. Abbas, and X.-m. Zhang, "Optimization of maillard reaction products from casein-glucose using

response surface methodology," *LWT–Food Science and Technology*, vol. 42, no. 8, pp. 1374–1379, 2009.

[21] A. Simsek, E. S. Poyrazoglu, S. Karacan, and Y. Sedat Velioglu, "Response surface methodological study on HMF and fluorescent accumulation in red and white grape juices and concentrates," *Food Chemistry*, vol. 101, no. 3, pp. 987–994, 2007.

[22] B. Dejaegher and Y. Vander Heyden, "The use of experimental design in separation science," *Acta Chromatographica*, vol. 21, no. 2, pp. 161–201, 2009.

[23] A. Ghasemzadeh, H. Z. Jaafar, E. Karimi, and A. Rahmat, "Optimization of ultrasound-assisted extraction of flavonoid compounds and their pharmaceutical activity from curry leaf (*Murraya koenigii* L.) using response surface methodology," *BMC Complementary and Alternative Medicine*, vol. 14, no. 1, p. 318, 2014.

10

A Convenient and Sensitive LC-MS/MS Method for Simultaneous Determination of Carbadox- and Olaquindox-Related Residues in Swine Muscle and Liver Tissues

Heying Zhang,[1] Wei Qu,[1] Yanfei Tao,[2] Dongmei Chen,[2] Shuyu Xie,[1] Lingli Huang,[1,2] Zhenli Liu,[1,2] Yuanhu Pan (iD),[1,2] and Zonghui Yuan (iD)[1,2]

[1]MOA Laboratory for Risk Assessment of Quality and Safety of Livestock and Poultry Products, Huazhong Agricultural University, Wuhan, Hubei 430070, China
[2]National Reference Laboratory of Veterinary Drug Residues and MAO Key Laboratory for Detection of Veterinary Drug Residues, Huazhong Agricultural University, Wuhan, Hubei 430070, China

Correspondence should be addressed to Yuanhu Pan; panyuanhu@mail.hzau.edu.cn

Academic Editor: Erwin Rosenberg

This paper presents a convenient and sensitive LC-MS/MS method for the simultaneous determination of carbadox and olaquindox residues, including desoxyolaquindx (DOLQ), desoxycarbadox (DCBX), quinoxaline-2-carboxylic acid (QCA), 3-methyl-quinoxaline-2-carboxylic acid (MQCA), and the glycine conjugates of QCA and MQCA (namely, QCA-glycine and MQCA-glycine, resp.) in swine muscle and liver tissues. Tissue samples were extracted with 2% metaphosphoric acid in 20% methanol and cleaned up by solid-phase extraction (SPE) on a mixed-mode anion-exchange column (Oasis MAX). Analysis was performed on a C_{18} column by detection with mass spectrometry in the multiple reaction monitoring (MRM) mode. The limits of detection (LODs) of the six analytes were determined to be $0.01\,\mu g\cdot kg^{-1}$ to $0.25\,\mu g\cdot kg^{-1}$, and the limits of quantification (LOQs) were $0.02\,\mu g\cdot kg^{-1}$ to $0.5\,\mu g\cdot kg^{-1}$. The total recoveries of the six analytes in all tissues were higher than 79.1% with the RSD% less than 9.2%. The developed method can determine the real residue level of QCA and MQCA, whether they are present in free form or as glycine conjugates in tissues, together with the carcinogenic desoxy metabolites DCBX and DOLQ with high recovery. Therefore, this method was suitable for routine analysis of residue control programmes and the residue depletion study of CBX and OLQ on swine.

1. Introduction

Carbadox (CBX) and olaquindox (OLQ) have been used as antimicrobial drugs in the feed of swine for growth promotion and the increased rate of weight gain and to control swine dysentery and bacterial enteritis in young swine [1]. Since 1998, both drugs have been withdrawn from market in the European Union due to possible carcinogenic and mutagenic effects of the drugs and their desoxy metabolites [2–5]. Nowadays, CBX is still permitted to use in the USA and several countries, and OLQ is permitted in China as a premix used in swine feed for growth promotion. Metabolism studies have revealed that OLQ was *in vivo*

transformed into desoxyolaquindox (DOLQ) by the reduction of the *N*-oxide, and the latter could be further hydrolyzed to 3-methyl-quinoxaline-2-carboxylic acid (MQCA). More than 80% of the extracted residues in the liver and kidney tissues of swine were DOLQ, and the second metabolite that accounts for 13% of the total residue in the liver was identified as a carboxylic acid derivative (named 2-carboxymethylaminocarbonyl-3-methylquinoxaline, identical to the structure of the glycine conjugate of MQCA) [6]. The metabolism of CBX was characterized by the rapid reduction of the *N*-oxide groups to give desoxycarbadox (DCBX) and the cleavage of the methyl carbazate side chain to give the carboxaldehyde which was further oxidized to the

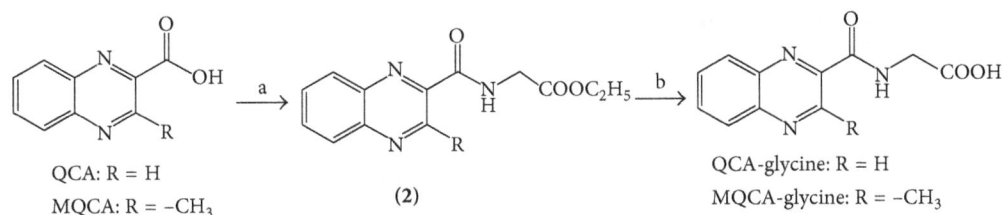

SCHEME 1: Synthesis route of QCA-glycine and MQCA-glycine. Reagents and conditions: (a) glycine ethyl ester, DCC, DMAP, and DCM at room temperature for 4 h; (b) (1) 1 M NaOH at room temperature for 2 h and (2) HCl.

corresponding carboxylic acid, quinoxaline-2-carboxylic acid (QCA). The major urinary metabolite was shown to be the QCA, which was present in a free form and as its glycine conjugate [7]. Since 1991, QCA and MQCA have been designated as the residue markers for CBX and OLQ, respectively, and maximum residue limits (MRLs) of 30 and $5 \mu g \cdot kg^{-1}$ in the liver and muscle tissues of pigs were recommended [8, 9]. However, at its 60th meeting in 2003, the Committee reported that QCA is not a suitable marker for monitoring carcinogenic metabolites of CBX, and QCA does not ensure the absence of carcinogenic residues in the liver and muscle tissues [10]. To ensure the safety of porcine products concerning the two drugs and especially to guarantee carcinogenic-free metabolites in the tissues of food-producing animals, many methods have been developed including HPLC-UV [11], GC-MS [12, 13], LC-MS [8, 14–16], and LC-MS/MS [8, 17, 18]. However, these methods usually focus on the determination of MQCA and QCA, sometimes including DCBX; moreover, the sample pretreatment involved either alkaline hydrolysis or protease digestion (or digestion with the enzymes) in order to release the conjugated or protein-binding "marker residue," followed by liquid-liquid extraction and cleanup with solid-phase extraction (SPE), which are relatively laborious and time-consuming and usually lead to low recovery especially for DCBX [8].

Since no analytical methods to quantify the main metabolites of CBX and OLQ were reported yet, the aim of this work was to develop a sensitive and reliable LC-MS/MS method for simultaneous determination of the carcinogenic desoxy metabolites DOLQ and DCBX and the marker residues QCA and MQCA (presented in free form or as glycine conjugates, namely, QCA-glycine and MQCA-glycine, resp.) in swine muscle and liver tissues.

2. Experimental

2.1. Chemicals and Reagents. Standards for DCBX, DOLQ, and MQCA were obtained from the Institute of Veterinary Pharmaceuticals (Huazhong Agricultural University, Wuhan, China). QCA was purchased from Sigma-Aldrich (St. Louis, MO, USA). The glycine conjugations of MQCA and QCA were prepared using a common method as illustrated in Scheme 1. Briefly, MQCA (or QCA), DCC, glycine ethyl ester, and a catalytic amount of DMAP were stirred at room temperature for 4 hours. The urea was filtered off, the solvent was evaporated in a vacuum, and the intermediate (2) was obtained. Then, (2) was hydrolyzed with 1 M sodium hydroxide to give the desired compound MQCA-glycine (or QCA-glycine). After

recrystallizing from methanol for 3 times, QCA-glycine and MQCA-glycine with purity above 98% were obtained. Distilled water was further purified by passing through a Milli-Q Integral water purification system (Millipore, Bedford, USA). HPLC-grade methanol and acetonitrile were purchased from Tedia (Fairfield, OH, USA).

The stock standard solutions of all analytes $(1 \text{ mg} \cdot mL^{-1})$ were prepared by exactly weighing the necessary quantities of the substances and dissolving each in methanol. All stock solutions of the individual substances were stored in amber vials at $-20°C$ and were stable for 6 months. Working combined mixed standard solutions $(1.0 \mu g \cdot g^{-1})$ were prepared by diluting the stock standards in methanol. The working standard mixture solutions were stable for at least 3 months for all substances when stored in amber vials below $4°C$.

2.2. LC-MS/MS Analysis. The LC-MS system (Shimadzu Co., Ltd.) was composed of an autosampler (SIL-20AC), a solvent delivery pump (LC-20AD), and a column oven (CTO-20AC) with an API5000 triple-quadrupole instrument (Applied Biosystems, Foster City, CA); Biosystems Sciex Analyst software version 1.5 was used to process data. Chromatographic separation was performed on a Thermo Scientific Hypersil Gold C_{18} column (150 mm × 2.1 mm i.d.; 5 μm particles) using a gradient elution consisting of mobile phase A (0.1% formic acid in water) and mobile phase B (acetonitrile). The flow rate was 0.2 mL/min with linear gradient at the following conditions: 0–12 min 85% to 20% A, 12–17 min 20% A, and 17–17.5 min return to 85% A. The injection volume was 10 μL. The source/gas conditions were as follows: the curtain gas was set at 6 psi, while the ion source gas 1 (GS1) and ion source gas 2 (GS2) were set at 60 psi and 50 psi, respectively. The compound conditions were entrance potential 10.0 V and collision cell exit potential 2.0 V. The mass spectrometer was operated in a multiple reaction monitoring (MRM) mode that selected one precursor ion and two suitable product ions for each target compound. The parameters of the m/z and collision energy of parent ions and quantitative product ions of DCBX, DOLQ, QCA, MQCA, QCA-glycine, and MQCA-glycine are shown in Table 1.

2.3. Sample Extraction and Cleanup. Aliquots of 2.0 ± 0.1 g of homogenized tissue samples were placed into a 50 mL centrifugal tube. 5 mL of 5% (w/v) metaphosphoric acid in 20% (v/v) methanol was added into the samples, and the mixtures were sonicated in an ultrasonic bath at room

TABLE 1: Optimum precursor and product ions with the respective collision energy (eV) for MS/MS.

Compound	Q1 (m/z)	Q3 (m/z)	CE (eV)	DP (V)
MQCA	189.1	145.0[a]	18	51
		101.9	35	75
QCA	175.3	129.2[a]	15	54
		101.9	38	77
DOLQ	232.2	143.2[a]	15	51
		102.0	38	85
DCBX	231.1	199.0[a]	35	95
		143.2	38	91
MQCA-glycine	246.3	200.1	10	45
		143.2[a]	35	90
QCA-glycine	232.1	186.1	10	48
		131.1[a]	32	80

[a]The most abundant ion (also used for quantification).

temperature for 10 min. After centrifugation at 8000 rpm for 10 min, the supernatants were transferred to a new tube, and the residue was extracted again as described above. The extracted solutions were merged, and $1 \, mol \cdot L^{-1}$ sodium acetate was added to adjust the pH to about 7. The extracts were ready for further SPE cleanup.

The Oasis MAX cartridge column (60 mg, 3 mL) (Waters Corp., Milford, MA, USA) was conditioned sequentially with 3 mL of methanol and 3 mL of water. The filtered extract was loaded onto the column at a flow rate of $1 \, mL \cdot min^{-1}$. The cartridge was washed with 30 mL sodium acetate/methanol (95 : 5, v/v) and then dried by purging air at a rate of $10 \, mL \cdot min^{-1}$ for 5 min. The analytes were eluted with 15 mL 2% trifluoroacetic acid in methanol at a flow rate of $1.0 \, mL \cdot min^{-1}$. The eluent was evaporated to dryness under a stream of nitrogen at 40°C and reconstituted in acetonitrile-water solution (15 : 85 (v/v), 200 μL) for LC-MS/MS analysis.

2.4. Validation Study. The analytical method developed for determination of the residues of CBX and OLQ and their main metabolites in swine muscle and liver tissues was validated according to the EU Decision 2002/657/EC [19]. The selectivity, matrix effects, linearity, CCα, CCβ, accuracy, and precision of the method were evaluated by spiking the six reference compounds into the blank matrices.

2.4.1. Selectivity. The selectivity of the method was evaluated by duplicate analysis of 10 blank samples of swine muscle and liver tissues. The blank samples were spiked with the six analytes (LOQ level) and processed by the proposed extraction method. The chromatograms of blank samples were compared and analyzed with those of spiked plasma samples.

2.4.2. Linearity and Matrix Effects. Linearity was evaluated from the calibration curves by triplicate analysis of blank tissue samples fortified with the analytes at five concentration levels (DCBX, DOLQ, QCA-glycine, and MQCA-glycine at

0.05, 0.10, 0.2, 0.5, and 5 μg·kg^{-1}; QCA and MQCA at 0.50, 1.0, 2.5, 5.0, and 10.0 μg·kg^{-1}). Linearity was expressed as the coefficient of linear correlation (r) and from the slope of the calibration curve. To evaluate the degree of ion suppression or signal enhancement, matrix-matched calibration curves were established. Matrix effects were assessed by comparing the slopes of these calibration curves using the following formula: matrix effect (ME) = $1 - (a_{matrix}/a_{standard}) \times 100$, where a_{matrix} and $a_{standard}$ are the slopes of calibration straight lines for standard and matrix-matched calibration graphs.

2.4.3. Detection Limits and Detection Capability. The limit of detection (LOD) was determined by successive analysis of spiked matrices with decreasing amounts of every standard until a signal-to-noise (S/N) ratio of 3 : 1 was reached. The limit of quantification (LOQ) was defined as the lowest concentration of an analyte that can be quantified with acceptable precision and accuracy.

The CCα and CCβ were determined by analysis of 20 blank tissue samples, and the signal-to-noise (S/N) ratio is calculated at the time window in which the analyte is expected. The CCα values were calculated as three times the S/N ratio. The CCβ was calculated by analyzing 20 blank samples spiked with concentration at CCα. Then, the CCβ value was added up to 1.64 times the corresponding standard deviation.

2.4.4. Accuracy and Precision. The accuracy and precision were determined in all matrices of the six analytes at three different concentrations (low, middle, and high levels) in six replicates using different analytical batches. The recovery was calculated by the following formula: (the measured level/the fortified level) × 100. The precision of the method was determined by intraday and interday results and expressed by the relative standard deviation (RSD). Intraday precision was conducted on the same day. Interday precision was determined by repeating the study on 3 consecutive days.

3. Results and Discussion

3.1. Extraction and Cleanup Procedure. In previous studies on quantitative determination of carbadox and olaquindox residues in the porcine muscle or liver, the samples were subjected either to alkaline hydrolysis at 105°C for 1 h or protease digestion overnight to release the QCA and MQCA [11, 12]. These methods are obviously tedious and time-consuming and usually lead to low recovery and sample throughput. In addition, it has been noted that the residues of the carcinogenic and mutagenic metabolites (i.e., DCBX and DOLQ) must be monitored for the regulation of these AGPs in food animal production. Boison et al. have demonstrated that even QCA was stable under this excessive pretreatment process, and DCBX was very rapidly degraded and undetectable [8]. So, the extraction and cleanup procedure should meet the request that QCA and MQCA, whether they exist in free form or as conjugate with glycine,

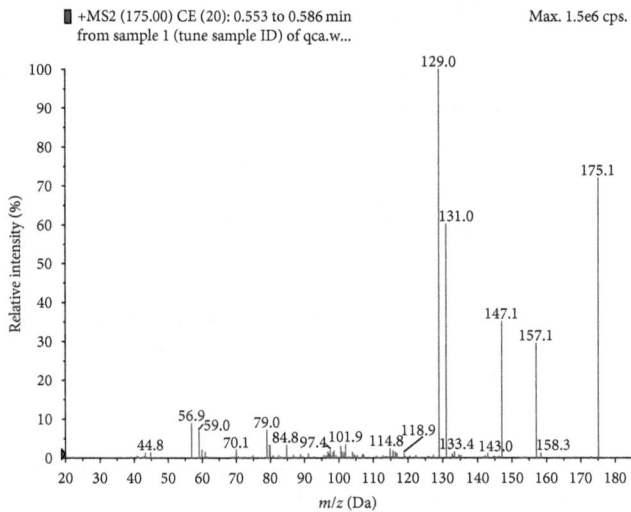

(a)

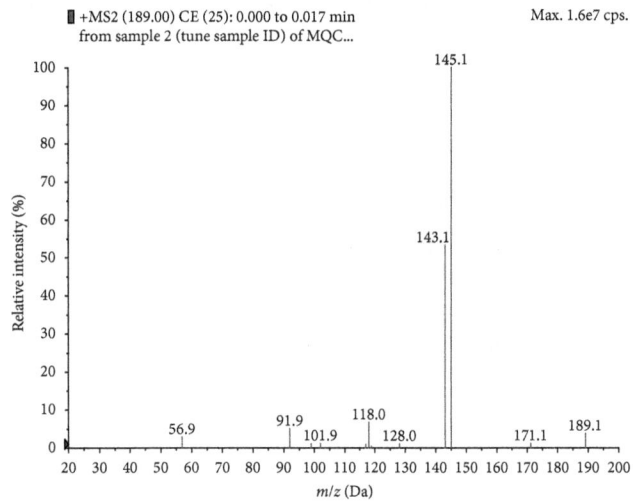

(b)

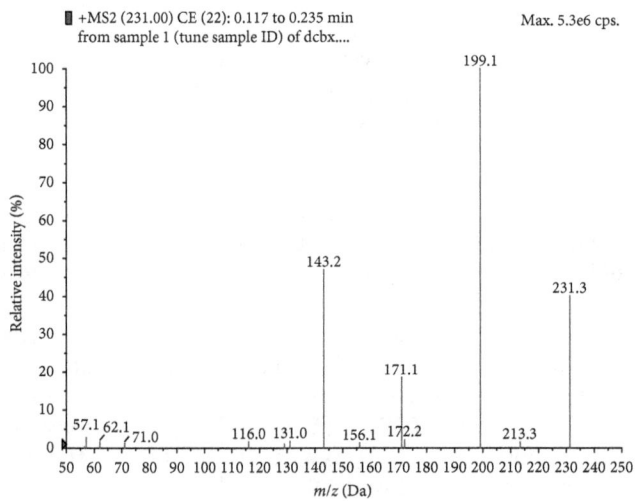

(c)

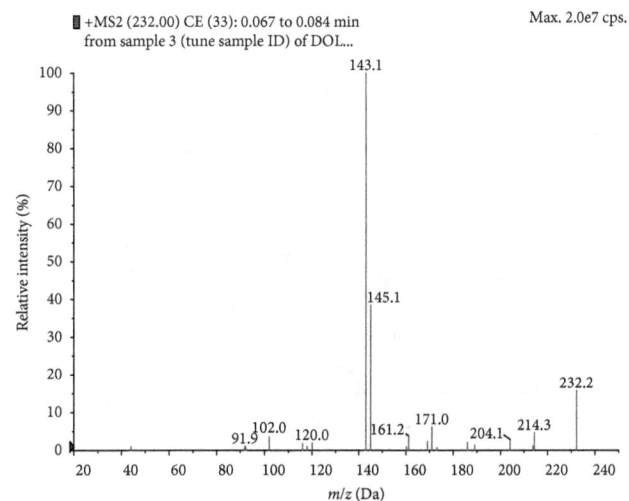

(d)

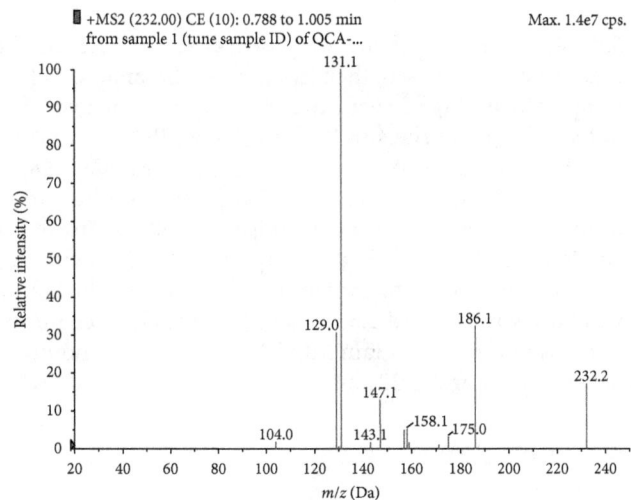

(e)

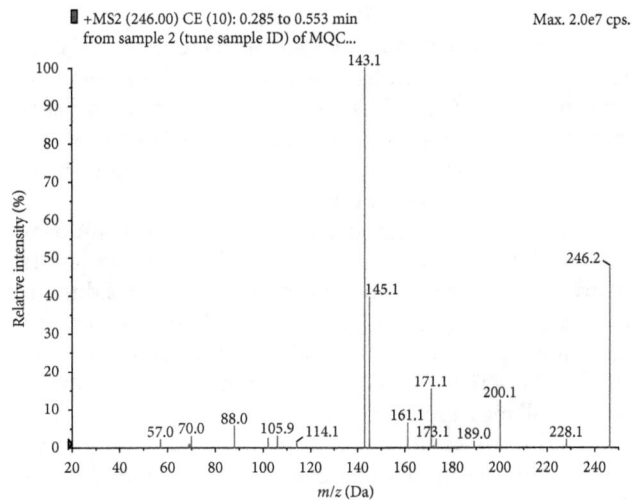

(f)

FIGURE 1: Mass spectra of (a) QCA, (b) MQCA, (c) DCBX, (d) DOLQ, (e) QCA-glycine, and (f) MQCA-glycine.

together with DCBX and DOLQ, could be determined with high recovery.

In this work, 2% metaphosphoric acid in 20% methanol was used as the extracting solvent and applied for deproteination to release the binding metabolites. Without further alkaline hydrolysis or protease digestion, the analytes were subjected to cleanup by SPE columns (Waters Oasis MAX). The analytes could be eluted smoothly with 2% trifluoroacetic acid in methanol. This mixed-mode anion-exchange sorbent for sample cleanup proved to be effective to decrease the matrix effect and to enhanced a good recovery, thus significantly improving the measurement precision.

3.2. LC-MS/MS of the Analytes. The MS/MS of QCA and MQCA exhibits a similar MS/MS fragmentation pattern which has been described previously [15, 16]. DCBX showed a molecular ion at m/z 231 amu and a prominent product ion at m/z 231 → 199 amu, resulting from α-fragmentation of the methyl ester to loss of OCH_3. The second prominent product ion at m/z 143 was attributed to the successive loss of carbon monoxide and N_2 that is supported by the observation of a moderate peak at m/z 199 → 171 amu. DOLQ showed a molecular ion at m/z 232 amu and two prominent product ions at m/z 143 and m/z 145. QCA-glycine and MQCA-glycine showed similar fragmentation patterns, which was characterized by the successive loss of water and carbon monoxide of glycine to produce the product ions at m/z 232 → 186 amu and m/z 246 → 200 amu, respectively. The second prominent product ions of them were at m/z 129 and m/z 143, just like those of QCA and MQCA, respectively. The mass spectra of MQCA, QCA, DCBX, DOLQ, QCA-glycine, and MQCA-glycine are shown in Figure 1.

3.3. Method Validation. Figure 2 shows the representative MRM chromatograms of blank liver samples spiked with the analytes at 0.5, 0.5, 0.05, 0.02, 0.05, and 0.05 $\mu g \cdot kg^{-1}$ for QCA, MQCA, DCBX, DOLQ, QCA-glycine, and MQCA-glycine, respectively. Under the optimized condition, no significant interfering peaks were observed in chromatograms of all tested matrices near the retention time of the analyte. All the peaks of the analytes were detected to have high resolution and good peak shape.

According to the European Union Commission Decision 657/2002 [19], a minimum of 4 identification points (IPs) are required to unambiguously confirm the residues in food of animal origin. The described method monitored 2 product ion scores and 3 IPs, plus 1 IP for the precursor ion, to fulfill the criteria for confirmation of the analytes. Moreover, all measured ion ratios in unknown samples must correspond to those in standards within predefined limits. The maximum permitted tolerance in the ion ratios varies with the relative intensity of the product ions to the base peak. In this method, the occurrence of MQCA, QCA, DCBX, DOLQ, QCA-glycine, and MQCA-glycine in real samples was confirmed by comparing the ion ratios of the two MRM transitions with those obtained from the matrix-matched calibration curve standards. All of the samples used

for the validation study met the relevant identification criteria (Table 2).

The linearity of the analytical response across the studied range for all the analytes was excellent, with correlation coefficients higher than 0.997. The linear dynamic range of the mass spectrometer was also estimated for all the analytes from a matrix-matched calibration curve. The matrix spike curve showed good linearity within the tested range 0.02–50 $\mu g \cdot kg^{-1}$ for DOLQ, 0.05–50 $\mu g \cdot kg^{-1}$ for DCBX, QCA-glycine, and MQCA-glycine, and 0.5–50 $\mu g \cdot kg^{-1}$ for QCA and MQCA, respectively, and the matrix effects were found to be in a range of 11.5–23.8%. Results for the calibration curve and coefficient in the swine liver are shown in Table 3. The LOQs of QCA and MQCA were determined to be 0.5 $\mu g \cdot kg^{-1}$, which were comparable to those reported in literature [8, 15]. When QCA and MQCA conjugated with glycine, the LOQs could be lowered by an order of magnitude (0.5 $\mu g \cdot kg^{-1}$ versus 0.05 $\mu g \cdot kg^{-1}$). The two desoxy metabolites, DCBX and DOLQ, showed high sensitivity with an LOQ of 0.05 $\mu g \cdot kg^{-1}$ and 0.02 $\mu g \cdot kg^{-1}$, respectively. CCα and CCβ of the analytes in swine tissues ranged from 0.02 $\mu g/kg$ to 0.3 $\mu g/kg$ and 0.05 $\mu g/kg$ to 0.5 $\mu g/kg$, respectively. The mean accuracy values obtained in the recovery tests were between 79.1% and 91.1%. In 2007, the European Union Reference Laboratory (Fougeres, France) proposed for DCBX, QCA, and MQCA in meat a recommended concentration of 10 $\mu g \cdot kg^{-1}$ as a minimum requirement for the analytical method [20]. The developed method well fulfilled the requirements for the monitoring of the residues related to CBX and OLQ. The relative standard deviation (RSD) of interday values of the six compounds analyzed by the present method was 5.5–9.2% and for the intraday test was 5.0–7.4% (Table 4). The results showed that the established method was accurate and precise and fit for the purpose of CBX and OLQ residue detection in swine muscle and liver tissues.

3.4. Applicability of the Proposed Method. The method established above was then applied to the analysis of real samples including 50 swine muscles and 50 swine livers from different supermarkets in Wuhan (Hubei Province, China). All the samples were analyzed for QCA, MQCA, DCBX, DOLQ, QCA-glycine, and MQCA-glycine residues to determine whether there exists misuse/or illegal use of the AGPs in the locality. Eight samples out of 50 swine liver samples were found to contain DOLQ residues (the amount was between 0.65 and 2.35 $\mu g \cdot kg^{-1}$), in which three samples were also found to contain MQCA residues (the amount was between 0.76 and 1.22 $\mu g \cdot kg^{-1}$).

4. Conclusions

This paper describes a convenient and sensitive LC-MS/MS method for the simultaneous determination of residues of the metabolites of CBX and OLQ in swine muscle and liver tissues. Since there existed a doubt whether QCA and MQCA are suitable residue markers for the regulation of these two AGPs in food animal production, the residues of

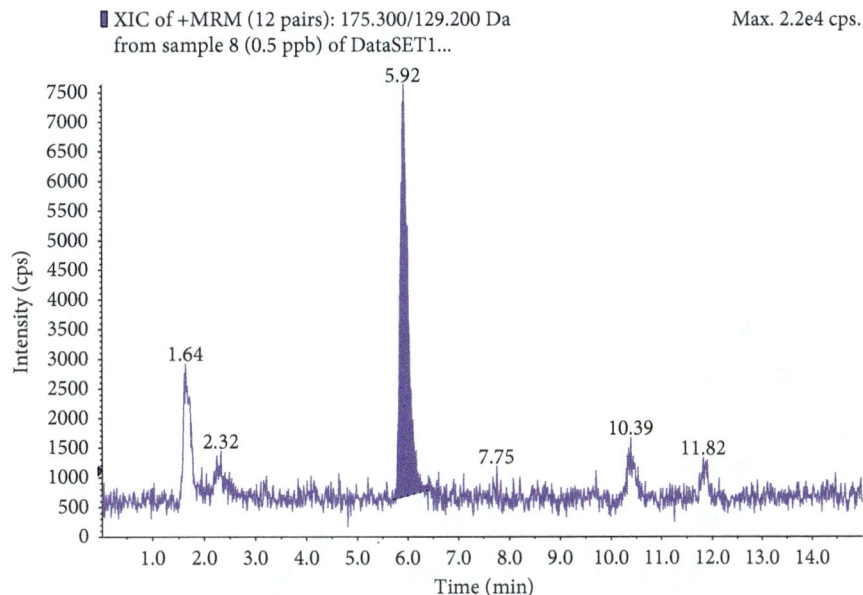

(a)

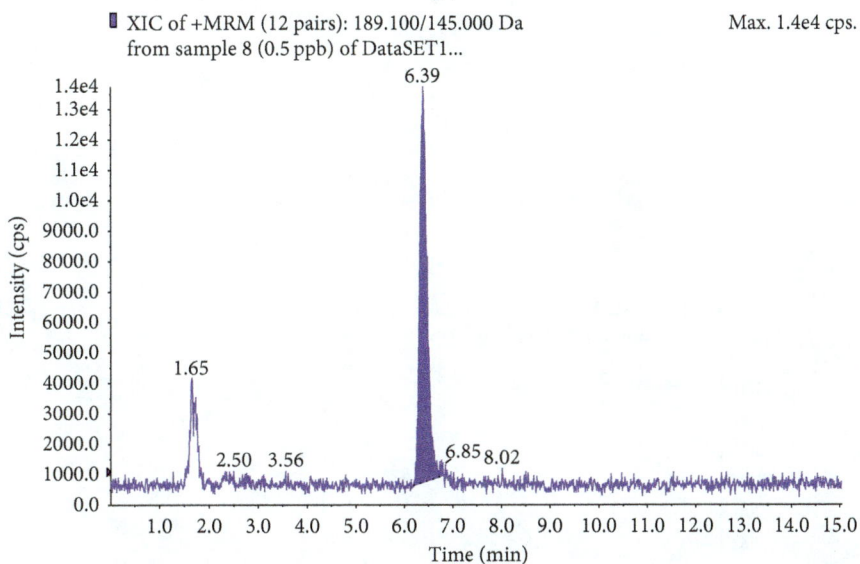

(b)

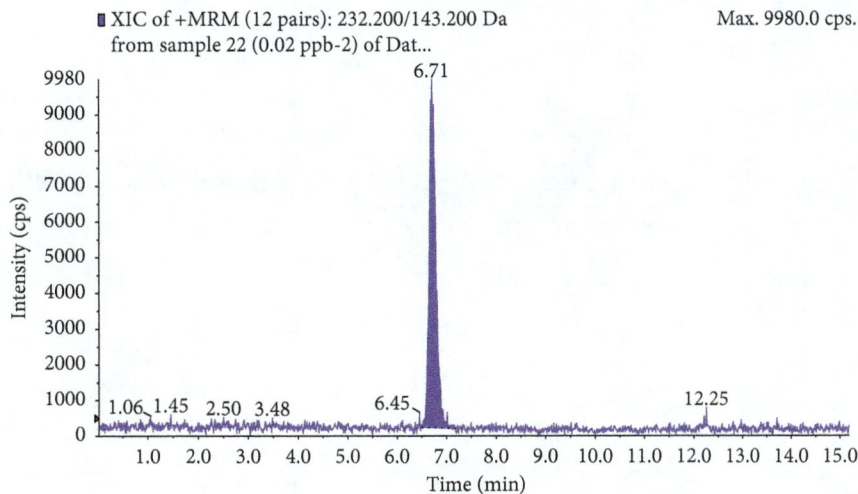

(c)

FIGURE 2: Continued.

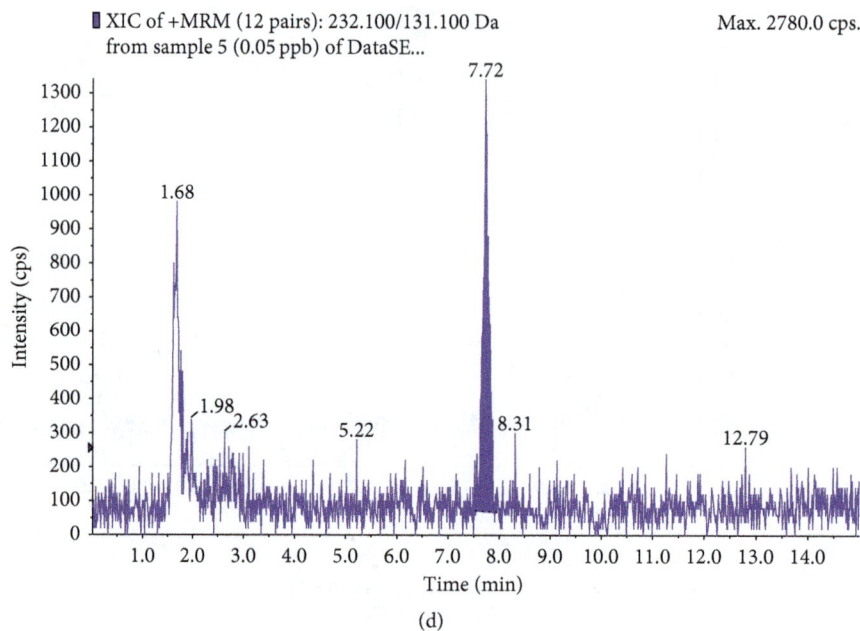

(d)

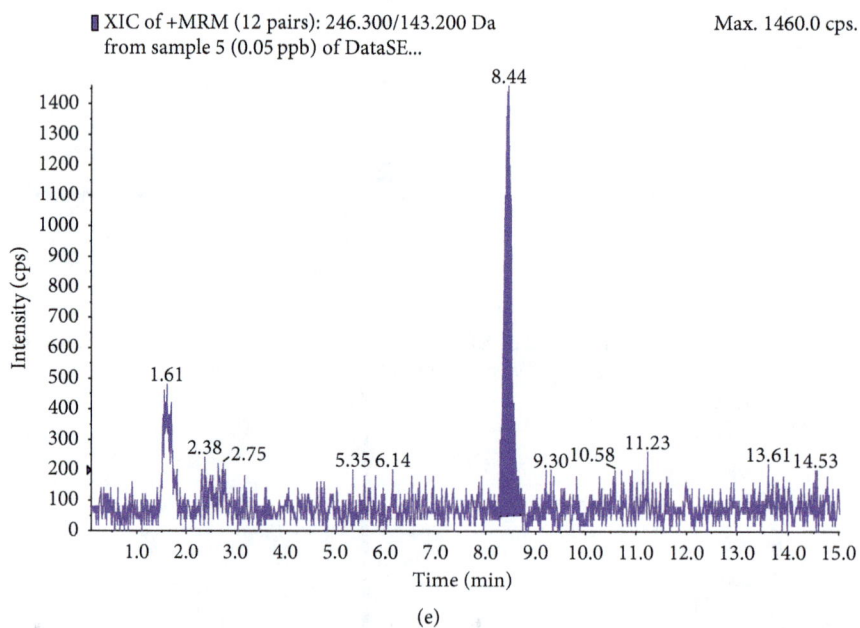

(e)

FIGURE 2: Continued.

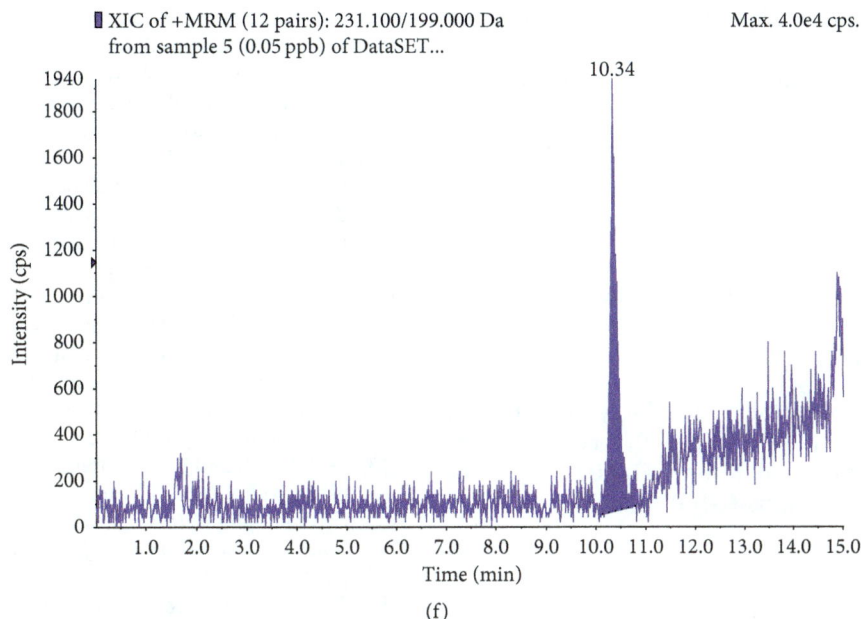

FIGURE 2: LC-ESI-MS/MS chromatograms in a positive-ion mode of (a) QCA (175.3 > 129.2), (b) MQCA (189.1 > 145.0), (c) DOLQ (232.2 > 143.2), (d) QCA-glycine (232.1 > 131.1), (e) MQCA-glycine (246.3 > 143.2), and (f) DCBX (231.1 > 199.0) fortified in the liver of swine.

TABLE 2: LC-MRM parameters of MQCA, QCA, DOLQ, DCBX, MQCA-glycine, and QCA-glycine.

Analytes	Average of ion ratios of matrix-matched standard solution (RSD%)	Maximum permitted tolerances	Average of ion ratios of spiked porcine tissues (RSD%)
MQCA	0.60	0.50–0.72 (1 ± 20%)	54
QCA	0.65	0.54–0.78 (1 ± 20%)	62
DOLQ	0.45	0.36–0.54 (1 ± 25%)	38
DCBX	0.52	0.43–0.62 (1 ± 20%)	48
MQCA-glycine	0.46	0.37–0.58 (1 ± 25%)	42
QCA-glycine	0.40	0.32–0.50 (1 ± 25%)	32

TABLE 3: Calibration curves of the analytes fortified in the swine liver.

Compound	Calibration curve	Coefficient of correlation r	Linear range (μg/kg)
QCA	$y = 52357x + 12354$	0.9998	0.5–50
MQCA	$y = 99929x + 19199$	0.9997	0.5–50
DCBX	$y = 270179x + 73776$	0.9998	0.05–50
QCA-glycine	$y = 138502x + 58623$	0.9991	0.05–50
DOLQ	$y = 182083x + 61571$	1	0.02–50
MQCA-glycine	$y = 222520x + 55169$	0.9997	0.05–50

TABLE 4: Validation parameters related to the method of QCA, MQCA, DCBX, DOLQ, QCA-glycine, and MQCA-glycine in the swine liver.

Parameter	QCA	MQCA	DCBX	DOLQ	QCA-glycine	MQCA-glycine
Transition	175.3/129.2	189.1/145.0	231.1/199.0	232.2/143.2	232.1/131.1	246.3/143.2
LOD (μg/kg)	0.10	0.25	0.02	0.01	0.02	0.02
LOQ (μg/kg)	0.50	0.50	0.05	0.02	0.05	0.05
CCα (μg/kg)	0.10	0.30	0.02	0.02	0.02	0.02
CCβ (μg/kg)	1.10	1.33	0.15	0.08	0.12	0.15
Precision (RSD%)						
Intraday ($n = 5$)	5.5	6.6	6.0	6.8	9.2	7.6
Interday ($n = 5$)	5.0	5.5	6.4	7.3	7.1	7.4
Recovery (%)	85.6 ± 3.4	79.1 ± 6.7	85.5 ± 4.6	91.1 ± 6.3	87.3 ± 7.4	89.1 ± 8.1

the carcinogenic and mutagenic metabolites (i.e., DCBX and DOLQ) should also be monitored for regulatory control. Therefore, this quantitative and confirmatory method can be used for routine analysis for regulating the use/misuse of CBX and OLQ in food animal production and for residue the depletion study of CBX and OLQ.

Conflicts of Interest

The authors declare that they have no conflicts of interest.

Acknowledgments

This work was supported by the National Key Research and Development Program (2017YFD0501401) and the Risk Assessment of Unknown and Known Hazard Factors of Livestock and Poultry Products (GJFP2017008).

References

[1] Y. Zhao, G. Cheng, H. Hao et al., "In vitro antimicrobial activities of animal-used quinoxaline 1,4-di-*N*-oxides against mycobacteria, mycoplasma and fungi," *BMC Veterinary Research*, vol. 12, p. 186, 2016.

[2] D. Li, C. Dai, X. Yang, B. Li, X. Xiao, and S. Tang, "GADD45a regulates olaquindox-induced DNA damage and S-phase arrest in human hepatoma G2 cells via JNK/p38 pathways," *Molecules*, vol. 22, no. 1, p. 124, 2017.

[3] M. Kimura, S. Mizukami, Y. Watanabe, N. Onda, T. Yoshida, and M. Shibutani, "Aberrant cell cycle regulation in rat liver cells induced by post-initiation treatment with hepatocarcinogens/ hepatocarcinogenic tumor promoters," *Experimental and Toxicologic Pathology*, vol. 68, no. 7, pp. 399–408, 2016.

[4] A. Ihsan, X. Wang, W. Zhang et al., "Genotoxicity of quinocetone, cyadox and olaquindox in vitro and in vivo," *Food and Chemical Toxicology*, vol. 59, pp. 207–214, 2013.

[5] Commission Regulation (EC) No. 2788/98, "Amending Council Directive 70/524/EEC concerning additives in feedingstuffs as regards the withdrawal of authorisation for certain growth promoters," *Official Journal of the European Union*, vol. L347, p. 3, 1998.

[6] FAO/WHO, *Joint Expert Committee on Food Additives: Evaluation of Certain Veterinary Drug Residues in Food*, Technical Series, vol. 851, World Health Organization and Food and Agriculture Organization of the United Nations, Geneva, Switzarland, 1995.

[7] FDA, *Codification of a Revised Tolerance for Residues of Carbadox in Edible Tissues, NADA 041–061 MECADOX® 10–Supplemental Approval*, Food and Drug Administration, Silver Spring, MD, USA, http://www.fda.gov/AnimalVeterinary/Products/Approved-AnimalDrugProducts/FOIADrugSummaries/ucm064222.htm, 1998.

[8] J. O. Boison, S. C. Lee, and R. G. Gedir, "A determinative and confirmatory method for residues of the metabolites of carbadox and olaquindox in porcine tissues," *Analytica Chimica Acta*, vol. 637, no. 1-2, pp. 128–134, 2009.

[9] FAO/WHO, *Joint Expert Committee on Food Additives: Evaluation of Certain Veterinary Drug Residues in Food*, Technical Series, vol. 799, World Health Organization and Food and Agriculture Organization of the United Nations, Geneva, Switzarland, 1990.

[10] FAO/WHO, *Joint Expert Committee on Food Additives: Evaluation of Certain Veterinary Drug Residues in Food*, Technical Series, vol. 918, World Health Organization and Food and Agriculture Organization of the United Nations, Geneva, Switzarland, 2003.

[11] G. M. Binnendijk, M. M. L. Aerts, H. J. Keukens, and U. A. Brinkman, "Optimization and ruggedness testing of the determination of residues of carbadox and metabolites in products of animal origin. Stability studies in animal tissues," *Journal of Chromatography A*, vol. 541, pp. 401–410, 1991.

[12] D. W. M. Sin, L. P. K. Chung, M. M. C. Lai, S. M. P. Siu, and H. P. O. Tang, "Determination of quinoxaline-2-carboxylic acid, the major metabolite of carbadox, in porcine liver by isotope dilution gas chromatography-electron capture negative ionization mass spectrometry," *Analytica Chimica Acta*, vol. 508, no. 2, pp. 147–158, 2004.

[13] M. J. Lynch and S. R. Bartolucci, "Confirmatory identification of carbadox-related residues in swine liver by gas-liquid chromatography/mass spectrometry with selected ion monitoring," *Association of Official Analytical Chemists*, vol. 65, pp. 66–70, 1982.

[14] T. Sniegocki, M. Gbylik-Sikorska, A. Posyniak, and J. Zmudzki, "Determination of carbadox and olaquindox metabolites in swine muscle by liquid chromatography/mass spectrometry," *Journal of Chromatography B*, vol. 944, pp. 25–29, 2014.

[15] M. J. Hutchinson, P. B. Young, and D. G. Kennedy, "Confirmation of carbadox and olaquindox metabolites in porcine liver using liquid chromatography-electrospray, tandem mass spectrometry," *Journal of Chromatography B*, vol. 816, no. 1-2, pp. 15–20, 2005.

[16] M. J. Hutchinson, P. Y. Young, S. A. Hewitt, D. Faulkner, and D. G. Kennedy, "Development and validation of an improved method for confirmation of the carbadox metabolite, quinoxaline-2-carboxylic acid, in porcine liver using LC-electrospray MS-MS according to revised EU criteria for veterinary drug residue analysis," *Analyst*, vol. 127, no. 3, pp. 342–346, 2002.

[17] W. L. Souza Dibai, F. J. de Alkimin Filho, F. A. da Silva Oliveira et al., "HPLC-MS/MS method validation for the detection of carbadox and olaquindox in poultry and swine feedingstuffs," *Talanta*, vol. 144, pp. 740–744, 2015.

[18] A. Merou, G. Kaklamanos, and G. Theodoridis, "Determination of carbadox and metabolites of carbadox and olaquindox in muscle tissue using high performance liquid chromatography-tandem mass spectrometry," *Journal of Chromatography B*, vol. 881-882, pp. 90–95, 2012.

[19] Commission Decision 2002/657/EC, "Implementing Council Directive 96/23/EC concerning the performance of analytical methods and the interpretation of results," *Official Journal of the European Union*, vol. L221, p. 8, 2002.

[20] Commission of the European Communities, *CRLs View on State of the Art Analytical Methods for National Residue Control Plans, CRL Guidance Paper*, Vol. 7, European Commission, Brussels, Belgium, December 2007.

11

Determination of Phosphorus in Soil by ICP-OES using an Improved Standard Addition Method

Jing Yang,[1] **Jingwen Bai** ⓘ**,**[1] **Meiyu Liu,**[1] **Yang Chen,**[1] **Shoutong Wang,**[1] **and Qiyong Yang**[2]

[1]*Institute of Mountain Hazards and Environments, Chinese Academy of Sciences, Chengdu, Sichuan, China*
[2]*Institute of Karst Geology Chinese Academy of Geological Sciences, Guilin, Guangxi, China*

Correspondence should be addressed to Jingwen Bai; jwbai@imde.ac.cn

Academic Editor: Pablo Richter

In this study, an improved standard addition method (ISAM) was developed for the determination of phosphorus in soil by ICP-OES based on the conventional standard addition method (CSAM) and calibration curve method (CCM). Certified standard soils were analyzed by the proposed ISAM method. The values obtained by ISAM method agreed with the certified values. Additionally, the results obtained by ISAM method were compared with those determined by the other two methods (CSAM and CCM). All the values obtained by the ISAM agreed with those from the other two methods. The detection limit, quantification limit, and recovery rate of each method were calculated, and the recovery rates of soil samples and the blank were all within the range of 90%–110%. Finally, the proposed method was applied to determine phosphorous in soil samples from Guangnan County, Yunnan Province, China, and the meadow soil from Qinghai-Tibet Plateau, China. The relative errors between the results from ISAM and CCM were all within 10%, and t-test showed that the results between ISAM and CCM had no significant difference ($P > 0.05$). Therefore, the proposed method overcame the matrix effect in some extent and was an acceptable method for the rapid and accurate batch analysis of P content in soil sample, especially batch samples with obvious matrix effect.

1. Introduction

Determination of total phosphorus (P) in soils is very important in agriculture as the amount of phosphorus is related to crop production. Recently, research on the biogeochemical cycle has become a significant focus in the fields of soil science, ecology, and geology. However, some of the research has focused on the phosphorus cycle [1–5]. Therefore, a reliable and rapid method is essential for the determination of phosphorus in soil.

Molybdenum blue method is the traditional technique for determination of phosphorus in soil [6–8]. This method has some disadvantages including complicated steps and procedures, being influenced by human operation, and low analytical efficiency. It is not suitable for the rapid and batch determination of soil samples. Inductively coupled plasma optical emission spectroscopy (ICP-OES) is an essential, modern analytical tool that has a high sensitivity and a broad linear range. In recent years, ICP-OES has been used in determination of phosphorous in geological samples [9], high-

purity nickel [10], and biodiesel B100 from feedstock [11]. There are several quantitative methods used in ICP-OES mainly included calibration curve method (CCM) [12–15] and conventional standard addition method (CSAM) [16, 17]. The CCM is simple, but it is the susceptible to errors by fluctuations in operating conditions and the matrix effect. However, internal standard has been used to minimize errors caused by instrumental drift and to reduce chemical matrix effect [18–20]. CSAM is useful when the sample has obvious matrix effect and the matrix-matching method cannot be used. However, it is time-consuming and required a complex procedure.

Therefore, an improved standard addition method (ISAM), based on the CSAM and CCM, was proposed in this study for the determination of total phosphorus in soil. In this method, soil mixture solution was determined by the CSAM. The value of slope was calculated from the obtained calibration curve. All the samples were measured directly by ICP-OES for the determination of P, and the response signal values were obtained. Then, the total P value of each test sample could be calculated. Be similar to CSAM, the improved method overcomes the matrix effect. And be similar to CCM, ISAM was simple and rapid. So it is suitable for the rapid and accurate analysis of batch samples, especially the large number samples with obvious matrix effect. The difference between each analysis strategy is summarized in Table 1.

To verify the reliability of the novel method, 4 types of certified standard soils were analyzed, and the results were compared with those determined by the CCM and the CSAM. The values obtained were in close agreement with the certified values and the results from CCM and the CSAM. The detection limit and quantification limit were measured. Meanwhile, the proposed method was applied to determine phosphorous in soil samples from Guangnan County of Yunnan Province, China, and the meadow soil from Qinghai-Tibet Plateau, China.

2. Materials and Methods

2.1. Materials. Concentrated nitric acid (GR, guarantee reagent) and hydrofluoric acid (GR, guarantee reagent) were purchased from Sinopharm Chemical Reagent Co., Ltd., China. A stock standard solution of phosphorus (P) (1000 μg/mL) was purchased from the National Center of Analysis and Testing for Nonferrous Metals and Electronic Materials, and a stock standard solution of rhenium (Re) (1000 μg/mL) was purchased from Shanghai Macklin Biochemical Co., Ltd., China. Working standard solutions were freshly prepared from those stock solutions as required.

Four certified standard soils—GBW (E) 070041 (Institute of Soil Science, Chinese Academy of Science), GSS-5 and GSS-8 (Institute of Geophysical and Geochemical Exploration, Ministry of Geology and Mineral Resources), and GSS-14 (Institute of Geophysical and Geochemical Exploration, IGGE)—were obtained for the determination of total phosphorus. Soil samples were from Guangnan County, Yunnan Province, China, and Qinghai-Tibet Plateau, China. The soils were air-dried and passed through a 2 mm sieve, and subsamples were further ground to pass through a 0.25 mm sieve.

2.2. Apparatus. An Optima 8300 ICP-OES (PerkinElmer, USA) was applied for the analysis, and the instrumental specifications are given in Table 2. And an 18.2 MΩ deionized pure water treatment system (Millipore, USA) was used for preparing deionized water. A closed-vessel microwave digestion system (Mars 6, CEM, US) was applied to perform microwave-assisted digestion procedures for the soil samples. It is equipped with a 40-position rotor. A heating apparatus (BHW-09C, Shanghai Botong Chemical Technology Co. LTD, China) was employed to evaporate the residual acid.

2.3. Microwave Digestion. Approximately 0.15 g of a soil sample was weighed (with an accuracy of 0.0001 g) and placed in 50 mL Teflon vessel. Concentrated nitric acid (8 mL) and hydrofluoric acid (4 mL) were added. The closed vessels were introduced in a microwave oven-assisted sample digestion system. For complete digestion of the samples, a heating program comprising three steps was used. The program of the microwave dissolution is presented in Table 3. After cooling to

TABLE 1: Difference between each analysis strategy.

Method	Advantage	Disadvantage
CCM	Rapid and simple	Easily affected by fluctuations in operation conditions and the matrix effect
CSAM	Overcome the matrix effect in some extent	Time-consuming, required a complex procedure.
ISAM	Rapid and simple, overcome the matrix effect in some extent	The detection limit and the quantitation limit were higher than the other two methods

TABLE 2: Operating conditions of ICP-OES.

Observation direction	Radial
Wavelengths (nm)	P 214.914, Re 197.248
RF power (KW)	1400
Auxiliary gas flow rate (L/min)	0.5
Atomizer flow rate (L/min)	0.5
Plasma flow rate (L/min)	18
Observation distance (mm)	15
Gas	Argon

TABLE 3: Instrument parameter of microwave digestion.

Step	Ramp time (min)	Hold time (min)	Temperature (°C)	Power (W)
1	10	10	150	1650
2	5	20	175	1650
3	5	30	200	1650

room temperature, the digestion solution in Teflon vessel was evaporated to about 1 mL in a heating apparatus at 175°C. Evaporation was a necessary step since acid concentrations would have been too high for the ICP-OES. Careful operation was needed to avoid drying of the evaporation residues. Then, all solutions were transferred to a 100 mL volumetric flask and accurately diluted to the mark with deionized water. Run blanks with all the chemicals and process except the soil sample.

2.4. Preparation of Calibration Working Standards

2.4.1. Calibration Working Standards for the ISAM. Soil mixture solution was obtained from the mixture of digested soil solutions.

A stock calibration standard of P (200 mg/L) was prepared by diluting 10 mL 1000 mg/L P calibration standard solution to 50 mL with deionized water. A stock calibration standard solution of P (500 mg/L) was prepared by diluting 12.5 mL 1000 mg/L P standard calibration solution to 25 mL with deionized water. A series of calibration working standards of P for the ISAM were diluted by soil mixture solution. The concentrations of the calibration working standards are listed in Table 4.

As element Re is rare in soil, element Re was selected as an internal standard to minimize errors caused by instrumental drift and was added in situ. A 10 mg/L Re internal standard solution was prepared by diluting 5 mL of a 1000 mg/L Re standard solution to 500 mL with deionized water.

TABLE 4: Concentrations of the calibration working standards used in ISAM.

Element	Volume of soil mixture solution (mL)	Concentration of stock calibration standards (mg/L)	Volume of stock calibration standards (μL)	Concentration of P in a series of calibration standard solutions (mg/L)
P	10	200	10	0.2
		200	25	0.5
		200	50	1.0
		500	30	1.5
		500	40	2.0
		500	60	3.0

All calibration standards were then stored in the freezer at a temperature of 5°C prior to analysis and were shaken before measurement to ensure proper mixing of P in solutions.

2.4.2. Calibration Working Standards for CCM and CSAM. Calibration working standards of P in the 0–3 mg/L range were prepared by diluting the stock calibration standards with deionized water. Re (10 mg/L) was selected as an internal standard to minimize errors caused by instrumental drift and was added in situ.

All calibration standards were then stored in the freezer at a temperature of 5°C prior to analysis and were shaken before measurement to ensure proper mixing of P in solutions.

2.5. Calculation of Detection Limit (L_D) and Quantification Limit (L_Q). Once the method was set up, the detection limit (L_D) and quantification limit (L_Q) were calculated.

The L_D was calculated from the measurement of the blank sample. The blank sample was measured ten times under reproducibility conditions. The detection limit was obtained from the following expression [18]:

$$L_D = 3.29 * S, \qquad (1)$$

where S = value of standard deviation of the measurements.

The L_Q was calculated according to the IUPAC guidelines as ten times the standard deviation of the measurement, for a number of the measurements equal to ten [21]:

$$L_Q = 10 * S. \qquad (2)$$

2.6. Recovery Test. Duplicate samples of 4 certified standard soils (GSS-5, GSS-8, GSS-14, and GBW (E) 070041) and blank were spiked with 50 μL of a 500 mg/L P standard solution. The spiked samples were digested and measured in the same manner as described in Section 2.3. Run blanks with all the chemicals and process except the soil sample.

Recovery rate was calculated using the following equation:

$$\text{recovery rate} = \frac{((\text{P content in spiked sample}) - (\text{P content in unspiked sample}))}{\text{known P content added to spiked sample}}. \qquad (3)$$

3. Results and Discussion

3.1. Theory. Generally, the CSAM was used for the determination of total P in soil, and multiple standard curves were obtained as $y_1 = m_1 x_1 + b_1$, $y_2 = m_2 x_2 + b_2$, ..., $y_n = m_n x_n + b_n$, corresponding to multiple test samples. The content of each test sample can be obtained from the intercept of each standard curve with the X-axis, which was b_n/m_n. When the matrix of the batch sample was the same or similar, the values of m were close to each other and could be approximated as $m_1 = m_2 = \cdots = m_n = m$. Therefore, for a series of samples with the same or similar matrix, the slopes were assumed to be the same.

In this study, an ISAM based on the CSAM and CCM was proposed. One sample was determined by CSAM. The value of slope, m, was calculated from the obtained calibration curve. Then, all the samples were measured directly by ICP-OES, and the response signal values, b_n, were obtained. Hence, the concentration of total P (C_n) in the each test sample was calculated according to the following equation:

$$C_n = \frac{b_n}{m}. \qquad (4)$$

The procedure of analysis by ISAM is shown in Figure 1. Actually, the matrix of the soil samples was not completely the same. To further reduce the error, the soil mixture solution prepared according to Section 2.4.1 was treated as sample and used to obtain the value of m using the CSAM.

3.2. The Use of Soil Mixture Solution. Total P of the 4 certified standard soils (GSS-5, GSS-8, GSS-14, and GBW (E) 070041) were determined by the proposed ISAM with standard solutions diluted by soil digestion solution of GSS-14 and with standard solutions diluted by soil mixture solution. The corresponding linear correlation relationship for the standard solutions diluted by soil digestion solution of GSS-14 was obtained as $y = 132.32 \, (\pm 1.58)x + 193.31 \, (\pm 1.94)$, $r = 0.9998$, and that for the standard solutions diluted by soil mixture solution was obtained as $y = 95.75 \, (\pm 1.96)x + 108.30 \, (\pm 1.18)$, $r = 0.9997$. The results are presented in Table 5, and the ICP-OES spectrograms are shown in Figures 2 and 3.

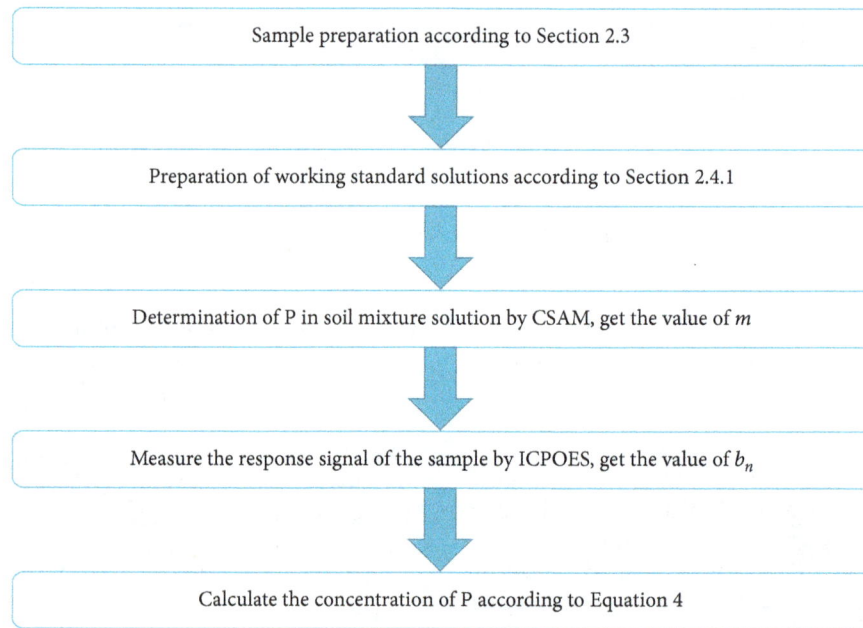

Figure 1: Procedure of analysis by ISAM.

Table 5: Comparison of the determined results of total P by ISAM with standard solutions diluted by soil digestion solution of GSS-14 and by the soil mixture solution.

Sample	Standard solutions diluted by soil digestion solution of GSS-14		Standard solutions diluted by the soil mixture solution		Certified value
	Total P value obtained by ISAM (mg/kg)	$\triangle_1{}^a$	Total P value obtained by ISAM (mg/kg)	$\triangle_2{}^b$	(mg/kg)
GSS-5	370	20	396	6	390 ± 34
GSS-8	776	1	765	10	775 ± 25
GSS-14	734	4	717	13	730 ± 28
GBW (E) 070041	544	20	530	6	524 ± 26.2

$^a\triangle_1$ = the value obtained by ISAM with the standard solutions diluted by soil digestion solution of GSS-14—certified value; $^b\triangle_2$ = the value obtained by ISAM with the standard solutions diluted by the soil mixture solution—certified value.

As shown in Table 5, the differences between the obtained value and certified value ($\triangle$) were all with the error of the certified value. The average of $\triangle_1$ was a little smaller than the average of $\triangle_2$. This indicated that the results determined by ISAM with standard solutions diluted by soil mixture solution were more close to the certified values. Thus, to some extent, the use of mixture of soil digestion solution could correct the analytical accuracy.

Operate conditions were as shown in Table 2. Analyte $1 \longrightarrow 6$: P standard solution diluted by soil mixture solution with the concentration of 0, 0.2, 0.5, 1.0, 1.5, and 2 mg/L; analyte 7, 8: GSS-5; analyte 9, 10: GSS-8; analyte 11, 12: GSS-14; and analyte 13, 14: GBW (E) 070041.

Operate conditions were as shown in Table 2. Analyte $1 \longrightarrow 6$: P standard solution diluted by soil digestion solution of GSS-14 with the concentration of 0, 0.2, 0.5, 1.0, 1.5, and 2 mg/L; analyte 7, 8: GSS-5; analyte 9, 10: GSS-8; analyte 11, 12: GSS-14; and analyte 13, 14: GBW (E) 070041.

Furthermore, as compared to Figure 3, the baselines of the spectrogram in Figure 2 were closer to each other than that in Figure 3. It indicated that the matrix of standard solutions diluted by soil mixture solution was more consistent with the matrix of soil digestion solution. From this

perspective, the matrix effect between the standard solution and the digestion solution could be reduced by using the soil mixture solution. Thus, in this study, the soil mixture solution was applied to prepare the calibration working standard solutions.

3.3. Determination of Total P Value by ISAM and Comparison with CCM and CSAM.

According to the principles discussed in Section 3.1, the concentration of P in soil mixture solution was determined by CSAM. The corresponding linear correlation relationship was obtained as follows:

$$y = 95.75(\pm 1.95)x + 108.3(\pm 1.18); \quad r = 0.9997. \quad (5)$$

According to Equation (5), the value of m, the slope of Equation (5), was 95.75. Taking GSS-5 soil for instance, the digested solution of GSS-5 was measured by ICP-OES directly, and the response signal value ($b_{\text{GSS-5}}$) was 75.10. According to Equation (4), the concentration of total P in the GSS-5 digested solution was calculated as $C_{\text{GSS-5}} = b_{\text{GSS-5}}/m = 75.10/95.75 = 0.784$ mg/L; thus, the content of P was obtained to be 391 mg/kg. The GSS-5 soil sample was measured 3 times in parallel, and the average result of GSS-5

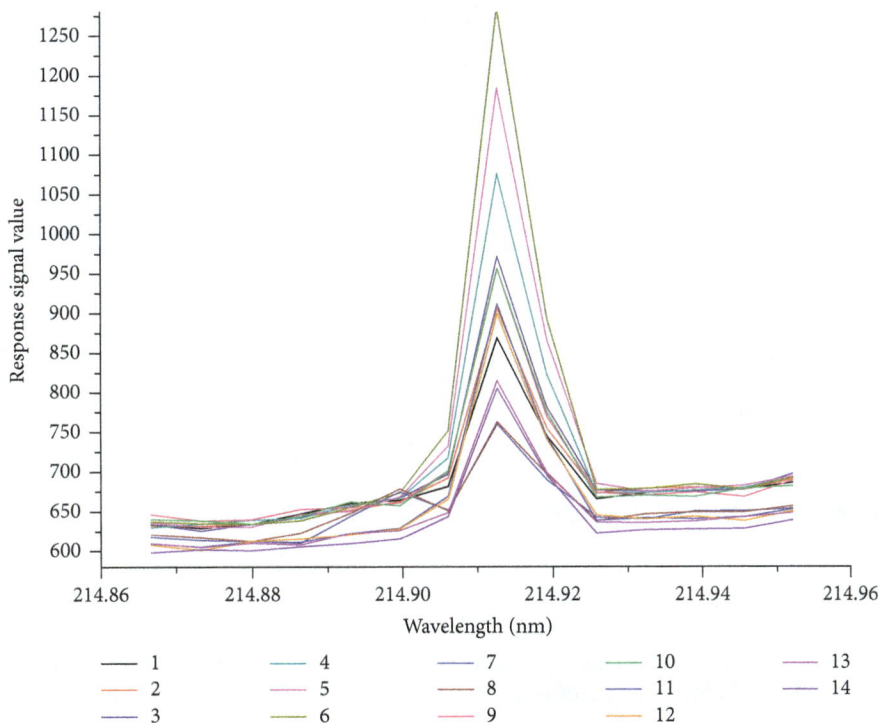

FIGURE 2: ICP-OES spectrogram of phosphorus with standard solutions diluted by soil mixture solution.

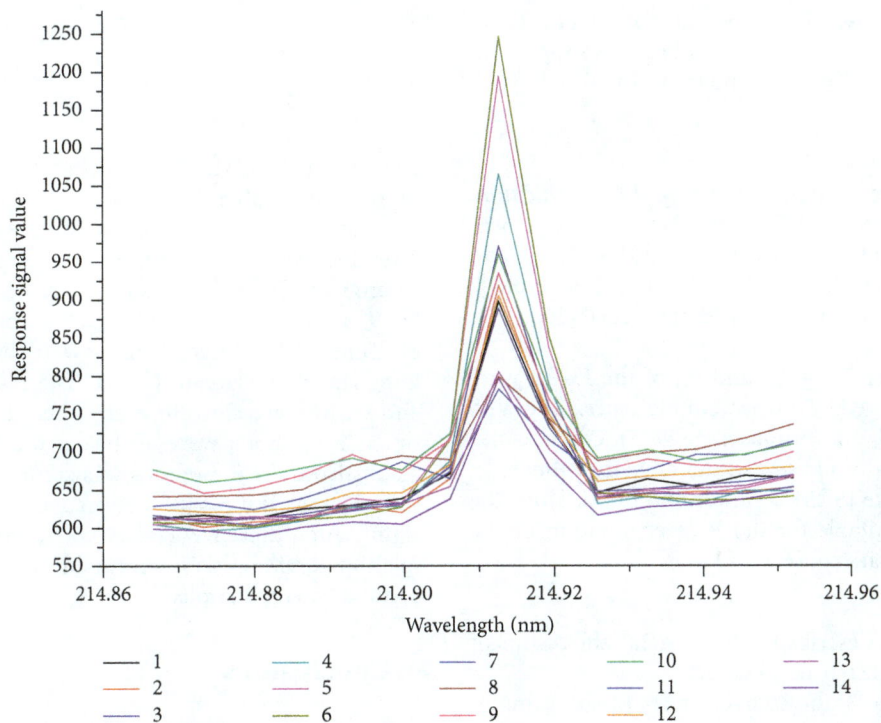

FIGURE 3: The ICP-OES spectrogram of phosphorus with standard solutions diluted by soil digestion solution of GSS-14.

was 396 mg/kg. The contents of P in other certified standard soils (GSS-8, GSS-14, and GBW (E) 070041) were obtained by the same procedure, and the results are listed in Table 6.

As shown in Table 6, the differences between the obtained and certified values ($\triangle_2$) were within the error of the certified value, indicating that the proposed method was of acceptable accuracy. Relative standard deviation (RSD) values of 1.23%–3.94% revealed relatively good precision. The ISAM combines the advantages of the CSAM and CCM, and it overcomes the matrix effect to some extent.

TABLE 6: Comparison of the determined results of ISAM, CCM, and CSAM.

| Sample | ISAM | | | CCM | | | CSAM | | | Certified value |
	Value (mg/kg)	$\triangle_2$	RSD (%)	Value (mg/kg)	$\triangle_3$[a]	RSD (%)	Value (mg/kg)	$\triangle_4$[b]	RSD (%)	(mg/kg)
GSS-5	396	6	3.94	381	9	3.44	403	13	1.34	390 ± 34
GSS-8	765	10	3.12	768	7	3.22	767	8	1.19	775 ± 25
GSS-14	717	13	1.23	739	9	1.78	745	15	1.25	730 ± 28
GBW(E)070041	530	6	3.85	509	15	3.40	509	15	1.72	524 ± 26.2

[a]$\triangle_3$ = the value obtained by CCM—certified value; [b]$\triangle_4$ = the value obtained by CSAM—certified value.

From this perspective, the ISAM is a useful and potential method for the batch analysis of P content in soil with a complex matrix.

Meanwhile, to further verify the suitability of the improved method for the determination of total P in soil, the P contents of the above certified standard soils were also determined by the CCM and CSAM. All samples were measured 3 times in parallel, and the average results are presented in Table 6. Table 6 shows that all the values obtained by ISAM, CCM, and CSAM agreed with each other. Additionally, $\triangle_2$, $\triangle_3$, and $\triangle_4$ were all within the error of the certified values. And the average of $\triangle_2$ is smaller than the average of $\triangle_3$ and $\triangle_4$. The RSDs of obtained values were all less than 5%.

Furthermore, the determination of the P in 4 certified standard soils using CSAM spent more than 108 min and that using the ISAM only spent nearly 30 min. Compared with CSAM, the proposed ISAM was an simple and time-saving method. Therefore, the proposed ISAM is a rapid and accurate method for the determination of P in batch samples.

3.4. Calibration Curve, Detection Limit, and Quantification Limit. The concentration of P in the blank solution was measured 10 times continuously using ICP-OES. The L_D and L_Q were calculated as described in Section 2.5. Table 7 presents the calibration curve, correlation coefficient, L_D, and L_Q.

As shown in Table 7, the L_D and L_Q of the ISAM were higher than that achieved by other methodologies. However, be similar to CSAM, the proposed ISAM method could overcome the matrix effect in some extent. On the other side, be similar to CCM, the ISAM is simple and rapid. Thus, the proposed ISAM is suitable for the P determination of the batch sample with matrix effect.

3.5. Recovery Test. As described in Section 2.6, the results of the recovery test are listed in Table 8.

As shown in Table 8, the recovery rates of soil samples and the blank were all within the range of 90%–110%. This indicated that the ISAM was acceptable for the determination of the total P content in soil.

3.6. Test of Soil Samples. The P content of soil samples from Guangnan County of Yunnan Province, China, and the meadow soil from Qinghai-Tibet Plateau, China, were determined by the proposed ISAM in this report and

TABLE 7: Detection and quantitation limits for three methods.

Method	Calibration curve	Correlation coefficient	Detection limit	Quantitation limit
ISAM	$y = 95.75\ (\pm1.95)$ $x + 108.3\ (\pm1.18)$	0.9996	0.092	0.279
CCM	$y = 87.92\ (\pm0.65)$ $x - 1.4\ (\pm1.08)$	0.9999	0.059	0.180
CSAM	$y = 218.5\ (\pm2.32)$ $x - 3.2\ (\pm0.2.61)$	0.9998	0.032	0.097

TABLE 8: Comparison of the recovery rate for the three methods.

Sample	ISAM Recovery rate (%)	CCM Recovery rate (%)	SAM Recovery rate (%)
GSS-5	97.7	92.4	98.4
GSS-8	105.9	105.7	96.1
GSS-14	100.0	103.2	96.4
GBW (E) 070041	98.5	92.5	102.7
Blank	98.6	100.8	96.0

conventional CCM. The results and relative errors (RE) are presented in Table 9.

The soil samples 1–20 in Table 9 were from Qinghai-Tibet Plateau, China, and the soil samples 21–40 were from Guangnan County of Yunnan Province. As shown in Table 9, the P content of soil samples from Guangnan County of Yunnan Province, China, were higher the that from Qinghai-Tibet Plateau, China. The results from the ISAM and CCM were close to each other, and the RE values between 2 methods were all below 6.23%. The difference of results from the 2 methods was compared with *t*-test. *T*-test showed that the results between ISAM and CCM had no significant difference ($P > 0.05$). Therefore, the improved method ISAM was an acceptable method to determine P content in soil sample.

4. Conclusions

A novel method ISAM, based on the CCM and CSAM, was developed to measure the content of total P in soil. The results obtained by the proposed method were compared with the certified values and those determined by conventional methods. It was justified that the ISAM was an acceptable method for evaluation of P content in soil. The developed method, combining the advantages of the CCM and CSAM, overcomes the matrix effect and is suitable for the rapid and accurate analysis of batch samples. The proposed ISAM

TABLE 9: Results of soil samples determined by ISAM and CCM.

Sample	Content of P obtained by ISAM (mg/kg)	Content of P obtained by CCM (mg/kg)	RE (%)
1	174	181	−4.14
2	152	146	4.11
3	155	159	−2.83
4	175	173	1.45
5	196	193	1.56
6	171	166	3.02
7	170	169	0.59
8	143	153	−6.23
9	173	176	−1.99
10	202	209	−3.35
11	187	183	2.19
12	182	180	1.11
13	179	182	−1.92
14	168	159	5.99
15	180	188	−4.27
16	136	141	−3.56
17	163	172	−5.23
18	189	184	3.00
19	145	154	−5.54
20	159	169	−5.64
21	402	396	1.44
22	794	780	1.81
23	422	408	3.39
24	474	451	5.12
25	1138	1164	−2.22
26	343	331	3.72
27	385	384	0.17
28	397	391	1.35
29	501	479	4.64
30	448	447	0.19
31	534	507	5.32
32	437	462	−5.49
33	1071	1079	−0.81
34	597	581	2.76
35	632	608	3.94
36	389	391	−0.46
37	374	361	3.65
38	1742	1745	−0.14
39	867	857	1.16
40	489	502	−2.59

method is a potential method for the determination of P content in soil and offers a new choice for the determination.

Conflicts of Interest

The authors declare that there are no conflicts of interest regarding the publication of this paper.

Acknowledgments

This research was supported by the China Geological Survey (DD20160324). The funder had no role in study design, data collection and analysis, decision to publish, or preparation of the manuscript.

References

[1] G. M. Filippelli, "The global phosphorus cycle: past, present, and future," *Elements*, vol. 4, no. 2, pp. 89–95, 2008.

[2] D. Defforey and A. Paytan, "Phosphorus cycling in marine sediments: advances and challenges," *Chemical Geology*, vol. 477, no. 20, pp. 1–11, 2018.

[3] K. C. Ruttenberg, "10.13–the global phosphorus cycle," *Reference Module in Earth Systems and Environmental Sciences Treatise on Geochemistry*, vol. 10, pp. 499–558, Elsevier, USA, 2nd edition, 2014.

[4] A. Thibault de Chanvalon, A. Mouret, J. Knoery, E. Geslin, O. Péron, and E. Metzger, "Manganese, iron and phosphorus cycling in an estuarine mudflat, Loire, France," *Journal of Sea Research*, vol. 118, pp. 92–102, 2016.

[5] D. B. Bate, J. E. Barrett, M. A. Poage, and R. A. Virginia, "Soil phosphorus cycling in an Antarctic polar desert," *Geoderma*, vol. 144, no. 1-2, pp. 21–31, 2008.

[6] The National Agro-Tech Extension and Service Center, *Technical Specifications for Soil Analysis*, China Agriculture Press, Beijing, China, 2009.

[7] G.-G. Liu, N.-G. Jiang, L.-D. Zhang, and Z.-L Liu, *Soil Physical and Chemical Analysis & Description of Soil Profiles*, Standards Press of China, Beijing, China, 1996.

[8] R.-K. Lu, *Analytical Methods of Soil and Agricultural Chemistry*, China Agriculture Science Press, Beijing, China, 1999.

[9] K. Asoh and M. Ebihara, "Accurate determination of trace amounts of phosphorus in geological samples by inductively coupled plasma atomic emission spectrometry with ion-exchange separation," *Analytical Chimica Acta*, vol. 779, pp. 8–13, 2013.

[10] S. Thangavel, K. Dash, S. M. Dhavile, and A. C. Sahayamn, "Determination of traces of As, B, Bi, Ga, Ge, P, Pb, Sb, Se, Si and Te in high-purity nickel using inductively coupled plasma-optical emission spectrometry(ICP-OES)," *Talanta*, vol. 131, pp. 505–509, 2015.

[11] F. D. A. Orozco, N. Kovachev, M. Á. A. Pastor, C. E. Domini, B. S. F. Band, and A. C. Hermandez, "Analysis of metals and phosphorus in biodiesel B100 from different feedstock using a flow blurring multinebulizer in inductively coupled plasma-optical emission spectrometry," *Analytica Chimica Acta*, vol. 827, pp. 15–21, 2014.

[12] K. A. Anderson and M. L. Tschirgi, "Determination of trace boron in microsamples of biological tissues," *Biological Trace Element Research*, vol. 60, no. 1-2, pp. 27–37, 1997.

[13] A. Lante, G. Lomolino, M. Cagnin, and P. Spettoli, "Content and characterization of minerals in milk and in Crescenza and Squacquerone Italian fresh cheeses by ICP-OES," *Food Control*, vol. 17, no. 3, pp. 229–233, 2006.

[14] Y. Zhang, Z.-T. Jiang, and H.-B Liu, "Determination of phosphorus in molybdenum concentrate by ICP-OES with multiple spectral fitting (MSF)," *Chinese Journal of Inorganic Analytical Chemistry*, vol. 4, no. 4, pp. 32–36, 2014.

[15] S. O. Souza, S. S. L. Costa, B. C. T. Brum, S. H. Santos, C. A. B. Garcia, and R. G. O. Araujo, "Determination of nutrients in sugarcane juice using slurry sampling and detection by ICP OES," *Food Chemistry*, 2018, In press.

[16] A. Armid, S. Ohde, R. Shinjo, and T. Toko, "Determination of uranium in pore water from coastal sediment by standard

addition ICP-MS analysis," *Journal of Radioanalytical and Nuclear Chemistry*, vol. 275, no. 1, pp. 233–237, 2008.

[17] L. Marjanovic, R. L. McCrindle, B. M. Botha, and H. J. Potgieter, "Use of a simplified generalized standard additions method for the analysis of cement, gypsum and basic slag by slurry nebulization ICP-OES," *Analytical and Bioanalytical Chemistry*, vol. 379, no. 1, pp. 104–107, 2004.

[18] M. F. Gazull, M. Rodrigo, M. Orduña, M. J. Ventura, and C. Andreu, "High precision measurement of silicon in naphthas by ICP-OES using isooctane as diluent," *Talanta*, vol. 164, pp. 563–569, 2017.

[19] M. H. Ramsey and M. Thompson, "Correlated variance in simultaneous inductively coupled plasma atomic-emission spectrometry: its causes and correction by a parameter-related internal standard method," *Analyst*, vol. 110, no. 5, pp. 519–530, 1985.

[20] S. L. dos Anjos, J. C. Alves, S. A. R. Soares et al., "Multivariate optimization of a procedure employing microwave-assisted digestion for the determination of nickel and vanadium in crude oil by ICP OES," *Talanta*, vol. 178, pp. 842–846, 2018.

[21] L. A. Currie, "Nomenclature in evaluation of analytical methods, including detect ion and quantification capabilities' (IUPAC recommendations 1995)," *Pure and Applied Chemistry*, vol. 67, no. 10, pp. 1699–1723, 1995.

12

One-Step Solid Extraction for Simultaneous Determination of Eleven Commonly used Anticancer Drugs and One Active Metabolite in Human Plasma by HPLC-MS/MS

Shouhong Gao,[1] Zhengbo Tao,[2] Jingya Zhou,[1] Zhipeng Wang,[1] Yunlei Yun,[1] Mingming Li,[1] Feng Zhang,[1] Wansheng Chen ⓘD,[1] and Yejun Miao ⓘD[3]

[1]*Department of Pharmacy, Changzheng Hospital, Second Military Medical University, Shanghai 200003, China*
[2]*Department of Orthopaedics, First Affiliated Hospital, China Medical University, 155 Nan Jing Bei Street,*
Shenyang, Liaoning 110001, China
[3]*Department of Psychiatry, Ankang Hospital, Ningbo, Zhejiang 315000, China*

Correspondence should be addressed to Wansheng Chen; chenwansheng@smmu.edu.cn and Yejun Miao; miaoyejun1968@163.com

Academic Editor: Josep Esteve-Romero

Therapeutic drug monitoring for anticancer drugs could timely reflect *in vivo* drug exposure, and it was a powerful tool for adjusting and maintaining drug concentration into a reasonable range, so that an enhanced efficacy and declined adverse reactions could be achieved. A liquid chromatography-tandem mass spectrometry method had been developed and fully validated for simultaneous determination of paclitaxel, docetaxel, vinblastine, vinorelbine, pemetrexed, carboplatin, etoposide, cyclophosphamide, ifosfamide, gemcitabine, irinotecan, and SN-38 (an active metabolite of irinotecan) in human plasma from cancer patients after intravenous drip of chemotherapy drugs. One-step solid-phase extraction was successfully applied using an Ostro sample preparation 96-well plate for plasma samples pretreated with acetonitrile containing 0.1% formic acid. Chromatographic separation was achieved on an Atlantis T_3-C_{18} column (2.1×100 mm, 3.0μm) with gradient elution using a mobile phase consisting of acetonitrile and 10 mM ammonium acetate plus 0.1% formic acid in water, and the flow rate was 0.25 mL/min. The Agilent G6410A triple quadrupole liquid chromatography-mass spectrometry system was operated under the multiple reaction monitoring mode with an electrospray ionization in the positive mode. Linear range was 25.0–2500.0 ng for paclitaxel, 10.0–1000.0 ng for docetaxel and SN-38, 100.0–10000.0 ng for vinorelbine and pemetrexed, 10.0–10000.0 ng for vinblastine and irinotecan, 1.0–1000.0 ng for cyclophosphamide and ifosfamide, 50.0–5000.0 ng for carboplatin, etoposide, and gemcitabine. Linearity coefficients of correlation were >0.99 for all analytes. The intraday and interday accuracy and precision of the method were within ±15.0% and less than 15%. The mean recovery and matrix effect as well as stability of all the analytes ranged from 56.2% to 98.9% and 85.2% to 101.3% as well as within ±15.0%. This robust and efficient method was successfully applied to implement therapeutic drug monitoring for cancer patients in clinical application.

1. Introduction

Cancer was gradually becoming an heavy burden for both developed and developing countries, and its incidence and mortality had increased awfully in recent years along with the aging of people and the deterioration of environment as well as the changing of lifestyles, for example, smoking, alcohol abuse, and massive intake of red meat. There were about 14.1 million new cancer cases and 8.2 million deaths in the worldwide in 2012 according to the GLOBOCAN report [1]. A large amount of costs must be paid on the prevention and treatment of cancer in the developed countries, and an upward trend was emerging in the developing countries.

In clinical application, the ways for treatment of cancer often included radiotherapy, chemotherapy, and surgery. Among that, chemotherapy was a potent and adjuvant method served for the surgery and sporadic cancer cells eradication. In recent years, enormous progress had been made in chemotherapy agents, and drug combinations containing multiple cytotoxic anticancer agents or biotherapy occupied the dominant status and yielded a good paradigm for cancer treatment [2, 3]. Drug combinations generally aimed at several targets or pathways to deal with the heterogeneity or multiprocess of cancer, for instance, invasion, growth, and metastasis [4]. Enhanced efficacy and shortened therapy period usually could be obtained by prescribing drug combinations, but the adverse reactions were extended based on each agent in the drug combination. Also, the therapeutic efficacy and adverse reactions were often concentration dependent [5].

The area under the concentration-time curve, steady-state concentration, and concentration over the threshold had a close relationship with the therapeutic results, and they had been considered as potent biomarkers for the individualized treatment in recent years [6]. Concentrations which are in the therapeutic window could obviously increase the responses and sometimes decrease the adverse reactions [7–10]. To guarantee the steady state of drug concentration and its location in the therapeutic window, it was necessary to carry out therapeutic drug monitoring (TDM) for the drug combinations so that clinicians could adjust the therapeutic regimen precisely [11]. TDM could be defined as the measurement of drugs in biological samples to individualize treatment by adjusting the therapeutic regimen. Many clinical practices have proved that TDM was an effective method to improve efficacy and/or reduce the toxicity [12, 13]. Many studies on LC-MS/MS method for quantification of cyclophosphamide, ifosfamide, irinotecan, etoposide, gemcitabine, carboplatin, pemetrexed, paclitaxel, vinblastine, and vinorelbine in human blood had been reported in the latest five years [14–24]. Some of them had shown advantages of shorter run time, simpler sample pretreatment, much lower limit of quantification, or more extensive linear range. But only few methods were developed for simultaneous quantification of drug combinations.

A HPLC-MS/MS method for simultaneous determination of seven anticancer agents had been developed and reported in our laboratory, but two steps of extraction hindered its high throughput in clinical application [25]. Thus, the aim of this study was to (1) optimize the pretreatment procedure for the extraction of anticancer drugs from human plasma and (2) enlarge the scope of clinical application of this method by integrating more anticancer drugs into one method based on our previously reported method. The development and validation of this new method was carried out according to the Chinese Pharmacopeia (vision 2010), and its clinical application was also verified at the same time.

2. Materials and Methods

2.1. Chemicals and Reagents. Paclitaxel, docetaxel, vinblastine, vinorelbine, pemetrexed, carboplatin, etoposide, cyclophosphamide, ifosfamide, gemcitabine, irinotecan, SN-38, and vindoline (internal standard, IS) were purchased from Sigma-Aldrich Corporation (St. Louis, MO, USA). All were corrected for purity and salt forms when weighed or diluted for standard stocks, whose chemical structures are shown in Figure 1. HPLC-grade methanol and acetonitrile were

obtained from Merck Company (Darmstadt, Germany). HPLC-grade formic acid, dimethyl sulfoxide (DMSO), and ammonium acetate were purchased from the Tedia Company, Inc. (Fairfield, OH, USA). Ultra-purified water (0.22 μm) was self-made in the laboratory by a Milli-Q reagent water system (Merck KGaA, Darmstadt, Germany) and was used throughout. Waters Ostro 96-well plate (25 mg) was used for sample pretreatment (Waters, Milford, MA, USA). Human blank plasma was donated by Shanghai Red Cross Blood Center (Shanghai, China).

2.2. LC-MS/MS Instrumentation. All experiments were carried out on an Agilent 1200 series HPLC system including an online degasser, a quatpump, an autosampler, and a column oven and interfaced to an Agilent 6410A triple-quadrupole mass spectrometer with an electrospray ionization source (Agilent Corporation, Santa Clara, CA, USA). All data were acquired and analyzed using Agilent Masshunter data processing software (version B.01.02, Agilent Corporation, Santa Clara, CA, USA).

2.3. Liquid Chromatographic Conditions. The chromatographic separation was achieved on an Atlantis T3-C_{18} analytical column (3.0 μm, 2.1 × 100 mm, Waters, Milford, MA, USA). The column was equilibrated and eluted with a mixed mobile phase consisting of acetonitrile and water containing 0.1% formic acid plus 10 mM ammonium acetate at a flow rate of 0.25 mL/min. The mobile phase was degassed automatically using the online degasser system. Mobile phase A was water containing 0.1% formic acid plus 10 mM ammonium acetate. Mobile phase B was acetonitrile. The gradient variation started with 100% A and maintained for 1 min, then switched to 100% mobile phase B at 1.01 min and lasted until 9 min, and finally switched back to 100% mobile phase A at 9.01 and lasted until 10 min, after which the system was returned to the initial condition. Under these conditions, the analytes coeluted with the IS within 9 min. The column temperature was maintained at 35°C. The injection volume was 10 μL, and the analysis time was 10.0 min.

2.4. Mass Spectrometry Conditions. The mass detection was achieved using electrospray ionization in the positive mode with the spray voltage set at 4000 V. Nitrogen was used as nebulizer gas, and nebulizer pressure was set at 40 psi with a source temperature of 105°C. Drying gas (nitrogen) was heated to 350°C and delivered at a flow rate of 10 L/min. High-purity nitrogen was used as collision gas at a pressure of about 0.2 MPa. Quantitation detection was performed in the multiple reaction monitoring (MRM) mode. The peak widths of precursors and product ions were maintained at 0.7 amu at half-height in the MRM mode.

2.5. Preparation of Standard and Quality Control (QC) Samples. The stock solutions of paclitaxel, docetaxel, vinblastine, vinorelbine, etoposide, cyclophosphamide, ifosfamide, and vindoline (IS) were individually prepared in

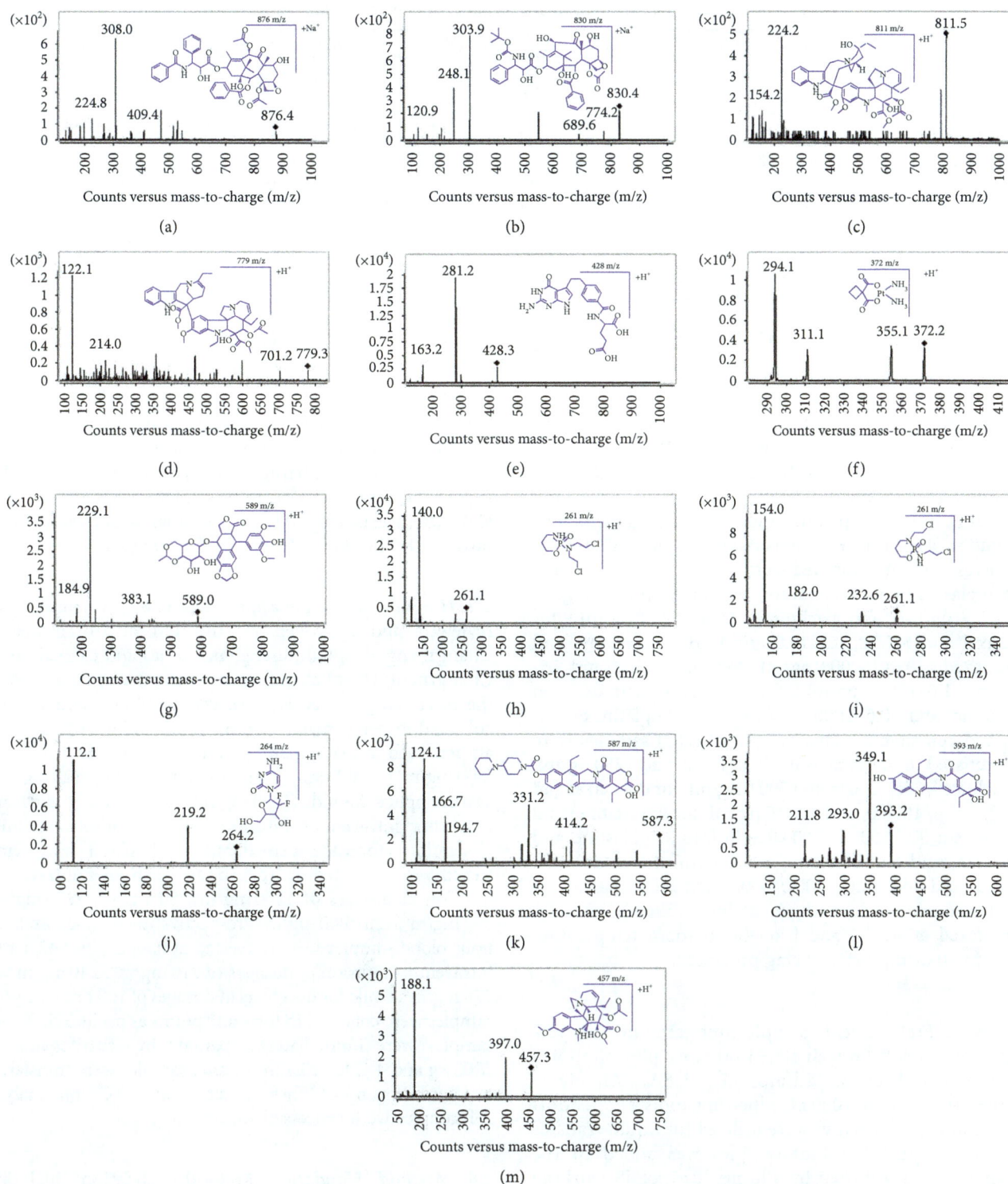

FIGURE 1: Chemical structures and multiple reaction monitoring ions of paclitaxel (a), docetaxel (b), vinblastine (c), vinorelbine (d), pemetrexed (e), carboplatin (f), etoposide (g), cyclophosphamide (h), ifosfamide (i), gemcitabine (j), irinotecan (k), SN-38 (l), and vindoline ((m), IS).

methanol, the stock solutions of pemetrexed, carboplatin, and gemcitabine were individually prepared in water, and the stock solutions of irinotecan and SN-38 were prepared in DMSO to obtain final concentrations at 1.0 mg/mL. All stock solutions were stored at −20°C. The stock solution of each analyte was further diluted with 10% methanol (V : V) to obtain a series of work solutions at concentrations of 0.25, 0.5, 1.25, 2.500, 6.25, 12.5, and 25.0 μg/mL for paclitaxel, 0.1, 0.2, 0.5, 1.0, 2.5, 5.0, and 10.0 μg/mL for docetaxel and SN-38, 1.0, 2.0, 5.0, 1.0, 25.0, 50.0, and 10.0 μg/mL for

TABLE 1: Optimized multiple reaction monitoring parameters for the analytes and IS.

Analyte	Precursor ion	Fragmentor energy (V)	Collision energy (eV)	Production ion
Paclitaxel	876.4	250	30	308.0
Docetaxel	830.4	200	20	303.9
Vinblastine	811.5	200	50	224.2
Vinorelbine	779.3	350	40	122.1
Pemetrexed	428.3	80	10	281.2
Carboplatin	372.2	80	10	294.1
Etoposide	589.0	150	10	229.1
Cyclophosphamide	261.1	80	14	140.0
Ifosfamide	261.1	80	20	154.0
Gemcitabine	264.2	80	10	112.1
Irinotecan	587.3	200	40	124.1
SN-38	393.2	150	30	349.1
Vindoline (IS)	457.3	150	20	188.1

vinorelbine and pemetrexed, 0.1, 0.5, 1.0, 5.0, 10.0, 50.0, and 100.0 μg/mL for vinblastine and irinotecan, 10.0, 50.0, 100.0, 500.0, 1000.0, 5000.0, and 10000.0 ng/mL for cyclophosphamide and ifosfamide, 0.5, 1.0, 2.5, 5.0, 10.0, 25.0, and 50.0 ng/mL for carboplatin, etoposide, and gemcitabine. Calibration standards were prepared by 10 times dilution of the corresponding combined working solutions with blank human plasma to obtain final concentrations in the range of 25.0–2500 ng/mL for paclitaxel, 10.0–1000.0 ng/mL for docetaxel and SN-38, 0.1–10.0 μg/mL for vinorelbine and pemetrexed, 10.0–10000.0 ng/mL for vinblastine and irinotecan, 1.0–1000.0 ng/mL for cyclophosphamide and ifosfamide, and 50.0–5000.0 ng/mL for carboplatin, etoposide, and gemcitabine. Quality control (QC) samples were also prepared in the same way (50.0, 250.0, and 1250 ng/mL for paclitaxel; 20.0, 100.0, and 500.0 ng/mL for docetaxel and SN-38; 200.0, 1000.0, and 5000.0 ng/mL for vinorelbine and pemetrexed; 50.0, 500.0, 5000.0 ng/mL for vinblastine and irinotecan; and 5.0, 50.0, and 500.0 ng/mL for cyclophosphamide and ifosfamide; 100.0, 500.0, and 2500.0 ng/mL for carboplatin, etoposide, and gemcitabine). The QC samples were stored at −20°C and brought to room temperature (25°C) for thawing before being processed.

2.6. Sample Pretreatment. Sample pretreatment was performed on Ostro™ 96-well plate to remove phospholipids and proteins (Waters, Milford, MA, USA). An eight-channel 100 μL and 1000 μL pipetting tools (Eppendorf AG, Hamburg, Germany) were utilized for liquid transfer steps. A 100 μL aliquot of samples was added to the designated well followed by adding 20 μL of IS working solution (100 ng/mL). A 300 μL aliquot of acetonitrile containing 1% formic acid was added to each well to precipitate the plasma proteins. All the wells were vortex mixed by aspirating 3 times with pipette. The mixtures were pulled through the cartridges to deprive the phospholipids using 96-well positive pressure processor (Waters, Milford, MA, USA) under 60 psi for 5 min. A 1 mL collection plate was used for collecting the eluents. After transferring to Eppendorf tubes, the content was evaporated under a gentle nitrogen stream at 45°C.

A 100 μL of the initial mobile phase was added to the tube, and the residual was reconstituted after vortex mixing for 3 min and then centrifuged for 10 min at 2500 ×g. The supernatant was transferred to a 1.5 mL glass autosampler vial with inserts, and 10 μL of the supernatant was injected into the HPLC-MS/MS system for analysis.

2.7. Human Sample Collection. The research protocol was reviewed and approved by the Ethical Committee of Changzheng Hospital (Shanghai, China) and carried out in Changzheng Hospital. Informed consent was signed by all the recruited patients. Human venous blood samples were collected in heparinized vacuum tubes and gently placed in an ice bath until centrifugation. Samples for irinotecan (300 mg/m^2), vinblastine (10 mg/m^2), vinorelbine (25 mg/m^2), and cyclophosphamide (1000 mg/m^2) were obtained at 0 h and 24 h after intravenous infusion. Other samples containing ifosfamide, etoposide, gemcitabine, carboplatin, and pemetrexed were collected at 0 h, 3 h, and 24 h after intravenous infusion at dosages of 1200 mg/m^2, 80 mg/m^2, 2200 mg/m^2, 330 mg/m^2, and 800 mg/m^2, respectively. For paclitaxel, venous blood samples were harvested at 0, 6, 24, and 48 h after intravenous infusion at dosages of 210 mg/m^2, 240 mg/m^2, or 270 mg/m^2, while for docetaxel at dosages of 100 or 120 mg/m^2 samples were collected in the same points as paclitaxel. Plasma samples were immediately separated by centrifugation at 2000 ×g and 4°C for 10 min. Plasma samples were transferred to 1.5 mL Eppendorf tubes and stored at −80°C until analysis. All samples were processed within 1 h.

2.8. Method Validation. Method validation including specificity, linearity, precision and accuracy, matrix effect, recovery, and stability was performed according to the Chinese pharmacopeia (version 2010).

For specificity, comparison of responses in spiked and blank samples from at least 6 lots was performed. The responses of interferences not more than 20% for analytes and 5% for IS were acceptable.

Matrix effect and recovery were assessed in three replicates at three concentration levels (low, mid, and high) for paclitaxel (50.0, 250.0, and 1250 ng/mL), docetaxel and

TABLE 2: Recovery of all the analytes after extraction with different extraction solvents.

Extraction solvent	Paclitaxel (%)	Docetaxel (%)	Vinblastine (%)	Vinorelbine (%)	Pemetrexed (%)	Carboplatin (%)	Etoposide (%)	Cyclophosphamide (%)	Ifosfamide (%)	Gemcitabine (%)	Irinotecan (%)	SN-38 (%)
Methanol	ND	ND	ND	ND	40.6	50.3	54.4	51.9	45.4	56.3	56.2	65.1
Acetonitrile	24.8	48.8	35.3	44.3	38.2	48.2	57.3	49.5	47.2	57.5	62.8	60.9
Ethyl acetate	53.4	35.3	40.6	57.5	ND	ND	88.0	74.6	71.3	85.3	64.3	68.4
Diethyl ether	93.3	46.9	47.0	82.3	ND	ND	80.4	65.3	63.9	ND	64.0	70.7
tert-Butyl methyl ether	76.3	37.0	39.4	68.9	ND	ND	32.2	67.2	67.3	ND	29.3	35.3
Ether-dichloromethane (8:2)	80.6	30.9	43.3	88.4	ND	ND	42.7	68.0	66.7	ND	56.7	60.8
Ether-dichloromethane (7:3)	52.2	74.5	65.7	80.3	ND	ND	82.4	69.5	60.3	ND	63.1	59.5
Ether-dichloromethane (5:5)	69.1	48.4	58.0	67.8	ND	ND	75.9	67.3	59.7	ND	60.7	63.3
Oasis HLB 1 cc (10 mg)	64.0	58.8	56.5	72.3	ND	ND	84.3	91.9	86.4	48.2	82.0	75.8
Ostro 96-well plate	62.4	86.3	90.4	71.2	74.4	73.9	64.7	99.1	77.7	71.4	75.4	68.5

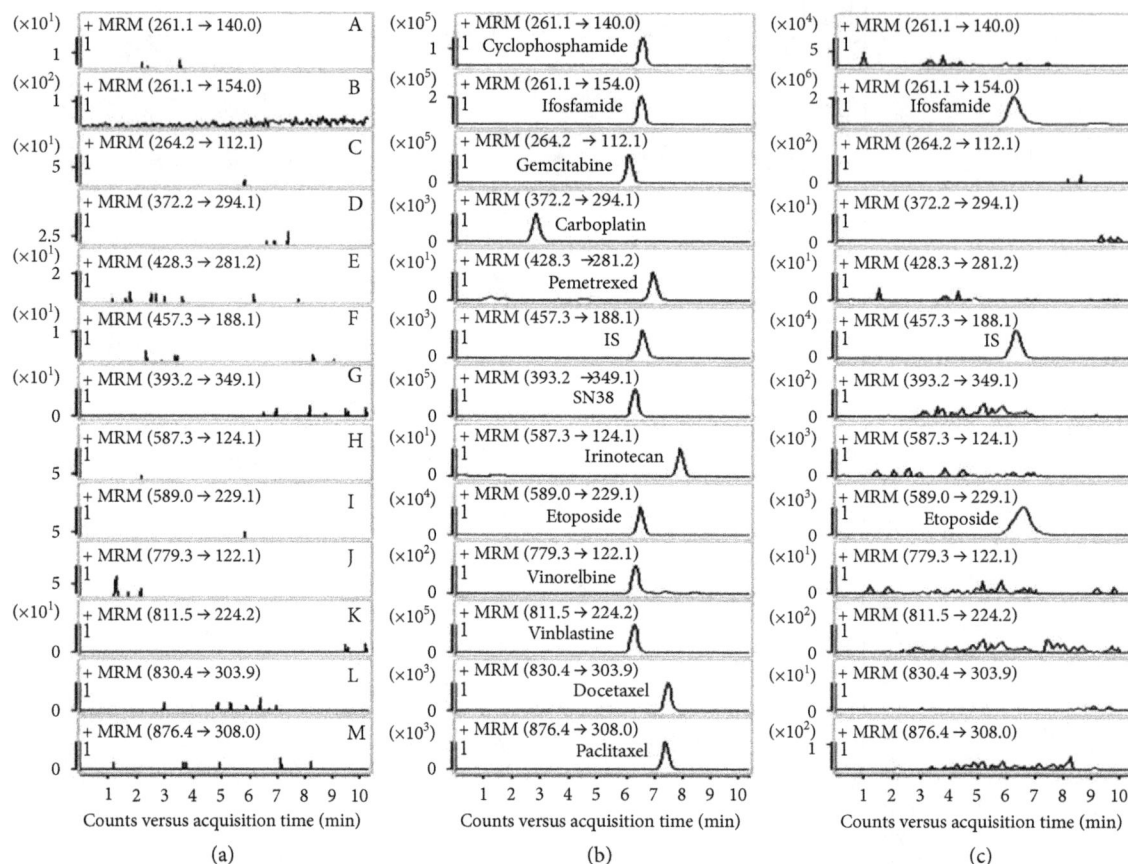

FIGURE 2: Representative MRM chromatograms of paclitaxel (A), docetaxel (B), vinblastine (C), vinorelbine (D), pemetrexed (E), carboplatin (F), etoposide (G), cyclophosphamide (H), ifosfamide (I), gemcitabine (J), irinotecan (K), SN-38 (L), and vindoline (M, IS). (a) Blank plasma sample, (b) blank plasma sample spiked with twelve analytes at LLOQ and IS, and (c) plasma sample collected from a patient at 6 h after administration of etoposide and ifosfamide at general dose.

SN-38 (20.0, 100.0, and 500.0 ng/mL), vinorelbine and pemetrexed (200.0, 1000.0, and 5000.0 ng/mL), vinblastine and irinotecan (50.0, 500.0, and 5000.0 ng/mL), cyclophosphamide and ifosfamide (5.0, 50.0, and 500.0 ng/mL), and carboplatin, etoposide, and gemcitabine (100.0, 500.0, and 2500.0 ng/mL). The matrix effect was the ratio of peak area in the spiked postextraction samples to the concentration corresponding solvent substituted samples, and the recovery was the ratio of peak area in the spiked samples concentration corresponding spiked postextraction samples. The IS was also assessed in the same way at the concentration of 100 ng/mL.

Interday and intraday precision and accuracy were also assessed in five replicates at three concentration levels (low, mid, and high). Samples were analyzed in three analytical lots in separate days (at least 2 days), and the RSD% for interday and intraday precision not more than 15% was rational. For intraday and interday accuracy, RE% (relative error) within 15% was considered acceptable.

Linearity of each analyte was evaluated in three analytical lots, and calibration curves were regressed from IS-adjusted peak area versus corresponding concentrations in at least six calibration standards using a $1/\chi^2$ weighted linear least-squares regression model. The LLOQ was set at the lowest point of the

calibration curve, which must possess a 10 times response higher than interference, and the deviation for the precision and accuracy should not be more than 20% and should be within ±20%.

Stability including long-term stability (3 months), short-term stability (24 h in an autosampler), and three frozen-thaw cycle stability was evaluated using QC samples at three levels (low, mid, and high). The calibration curve for each analyte was employed to obtain the measuring concentrations, and the deviation from nominal concentration within 15% was conformed to the criterion.

3. Results and Discussion

3.1. LC-MS/MS Optimization.
The chromatographic conditions, especially the composition of the mobile phase and types of columns, were optimized through several tests to achieve good resolution and symmetric peak shapes for analytes and the IS, as well as shorter run time. A number of C_{18} and C_8 columns, such as Zorbax SB-C_{18}, Zorbax SB-C_8, Xselect/Xbridge-C_{18}, C_8, and Atlantis T3-C_{18}, were tested. Elution of pemetrexed, carboplatin, and gemcitabine needed high percentage (at least 80%) of organic solvent in the mobile phase; however, high proportion of organic solvent

TABLE 3: Matrix effect and recovery of the analytes ($n = 3$).

Analyte	Nominal concentration (ng/mL)	Extraction recovery		Matrix effect	
		Mean (%)	RSD (%)	Mean (%)	RSD (%)
Paclitaxel	50.0	60.8	9.5	90.5	6.4
	250.0	66.1	7.2	90.3	3.5
	1250.0	62.3	6.8	92.4	3.7
Docetaxel	20.0	88.2	5.6	91.2	8.8
	100.0	87.3	6.8	89.7	6.4
	500.0	85.5	5.4	92.0	7.1
SN-38	20.0	64.1	7.9	97.3	4.3
	100.0	74.3	6.7	94.3	6.5
	500.0	72.5	5.4	101.9	5.2
Vinorelbine	200.0	69.4	7.9	96.1	1.4
	1000.0	74.6	5.4	101.3	2.1
	5000.0	70.9	7.1	94.9	3.7
Pemetrexed	200.0	70.7	6.0	93.8	6.0
	1000.0	67.7	5.4	99.6	5.8
	5000.0	73.5	5.9	93.4	5.6
Vinblastine	50.0	66.2	11.3	85.2	7.9
	500.0	77.3	9.8	85.7	8.4
	5000.0	80.1	3.8	87.9	4.2
Irinotecan	50.0	60.7	6.6	96.5	9.0
	500.0	64.2	5.9	98.5	8.1
	5000.0	75.1	6.0	97.7	7.4
Cyclophosphamide	5.0	89.9	6.2	82.2	9.9
	50.0	90.1	3.8	90.6	11.3
	500.0	98.9	3.4	90.0	4.4
Ifosfamide	5.0	72.1	5.6	99.1	11.2
	50.0	74.3	2.9	99.5	10.5
	500.0	76.6	4.8	98.0	7.6
Carboplatin	100.0	88.3	8.1	98.0	7.4
	500.0	72.9	4.7	99.7	5.1
	2500.0	72.9	4.9	95.0	7.6
Etoposide	100.0	70.1	6.3	96.9	5.5
	500.0	65.6	8.5	93.7	2.9
	2500.0	64.1	4.9	76.4	4.9
Gemcitabine	100.0	80.8	7.3	99.8	10.6
	500.0	79.3	3.2	99.9	9.8
	2500.0	70.6	8.3	96.3	7.2
IS	100.0	92	5.8	95.7	6.8

TABLE 4: Regression curves and parameters of the analytes ($n = 5$).

Analyte	Linearity range (ng/mL)	$y = a * x + b$		R^2	LLOQ (ng/mL)	LOD (ng/mL)
		a	b			
Paclitaxel	25.0–2500.0	0.0279	−0.0055	0.9907	25.0	10.0
Docetaxel	10.0–1000.0	0.0109	−0.0008	0.9963	10.0	2.0
SN-38	10.0–1000.0	1.7720	−0.1020	0.9956	10.0	1.0
Vinorelbine	100.0–10000.0	0.2455	−0.2388	0.9939	100.0	50.0
Pemetrexed	100.0–10000.0	0.3559	−0.2553	0.9911	100.0	50.0
Vinblastine	10.0–10000.0	0.6351	−0.0367	0.9944	10.0	2.0
Irinotecan	10.0–10000.0	3.1941	−0.3255	0.9942	10.0	3.0
Cyclophosphamide	1.0–1000.0	3.8840	−0.0072	0.9957	1.0	0.5
Ifosfamide	1.0–1000.0	7.2394	−0.0113	0.9951	1.0	0.5
Carboplatin	50.0–5000.0	0.0388	−0.0046	0.9974	50.0	25.0
Etoposide	50.0–5000.0	0.2673	−0.1549	0.9936	50.0	25.0
Gemcitabine	50.0–5000.0	1.1805	1.1461	0.9919	50.0	25.0

(85%) caused that the analytes could not be totally separated from the endogenous interfering materials on the SB-C$_{18}$ and C$_8$ column. Finally, the Atlantis T3-C$_{18}$ column provided sufficient retention and suitable separation medium for twelve analytes based on its stronger retention ability. Several mobile phases had been tested: 0.05% formic

TABLE 5: Intraday and interday precision and accuracy of the analytes in human plasma ($n = 5$).

Analyte	Nominal concentration (ng/mL)	Intraday ($n = 5$)			Interday ($n = 5$)		
		Measured concentration (mean ± SD, ng/mL)	Precision (%RSD)	Accuracy (%RE)	Measured concentration (mean ± SD, ng/mL)	Precision (%RSD)	Accuracy (%RE)
Paclitaxel	50.0	47.78 ± 1.26	2.6	−4.4	47.79 ± 1.69	3.5	−4.4
	250.0	250.06 ± 11.71	4.7	0.0	260.39 ± 11.21	4.3	4.2
	1250.0	1309.21 ± 46.73	3.6	4.7	1278.69 ± 78.67	6.2	2.3
Docetaxel	20.0	18.56 ± 0.40	2.2	−7.2	18.76 ± 0.44	2.3	−6.2
	100.0	96.10 ± 3.69	3.8	−3.9	101.56 ± 5.11	5.0	1.6
	500.0	538.07 ± 21.68	4.0	7.6	518.53 ± 6.09	1.2	3.7
SN-38	20.0	17.56 ± 0.92	5.2	−12.2	16.92 ± 0.42	2.5	−10.4
	100.0	103.26 ± 8.86	8.6	3.3	93.16 ± 1.86	2.0	−6.8
	500.0	546.59 ± 33.33	6.1	9.3	504.15 ± 50.82	10.1	0.8
Vinorelbine	200.0	173.52 ± 1.84	1.1	−13.2	171.59 ± 2.36	1.4	−14.2
	1000.0	1005.74 ± 5.48	0.5	0.6	992.62 ± 4.10	0.4	−0.7
	5000.0	5364.49 ± 2.62	0.1	7.3	5244.73 ± 2.17	0.1	4.9
Pemetrexed	200.0	185.20 ± 13.86	7.5	−7.4	181.32 ± 14.97	8.3	−9.3
	1000.0	951.42 ± 72.67	7.6	−4.9	955.55 ± 39.65	4.1	−4.4
	5000.0	4708.39 ± 445.31	9.5	−5.8	5230.11 ± 168.09	3.2	4.6
Vinblastine	50.0	47.51 ± 0.95	5.4	−5.0	47.08 ± 1.99	11.7	−5.8
	500.0	497.29 ± 9.82	5.0	−0.5	492.92 ± 5.17	2.7	−1.4
	5000.0	5226.19 ± 2.35	0.1	4.5	4990.19 ± 3.67	0.1	−0.2
Irinotecan	50.0	49.89 ± 1.07	5.4	−0.2	47.66 ± 0.94	5.3	−4.7
	500.0	497.83 ± 8.38	4.2	−0.4	495.69 ± 6.01	3.1	−0.9
	5000.0	5379.31 ± 178.38	3.3	7.6	5006.16 ± 196.44	3.9	0.1
Cyclophosphamide	5.0	5.00 ± 0.21	10.4	0.0	4.95 ± 0.23	11.7	−1.0
	50.0	49.27 ± 0.69	3.4	−1.5	48.39 ± 0.31	1.6	−3.2
	500.0	538.97 ± 5.80	1.2	7.8	525.12 ± 9.37	1.9	5.0
Ifosfamide	5.0	4.90 ± 0.10	5.1	−2.0	4.91 ± 0.08	4.0	−1.8
	50.0	49.04 ± 0.51	2.7	−1.9	49.16 ± 0.71	3.7	−1.7
	500.0	526.72 ± 10.80	2.1	5.3	516.36 ± 12.78	2.5	3.3
Carboplatin	100.0	98.57 ± 2.59	2.6	−1.4	103.05 ± 12.69	12.3	3.1
	500.0	479.21 ± 5.69	1.2	−4.2	477.25 ± 11.46	2.4	−4.6
	2500.0	2640.10 ± 40.11	1.5	5.6	2537.97 ± 61.41	2.4	1.5
Etoposide	100.0	99.71 ± 2.56	2.6	−0.3	98.31 ± 2.00	2.0	−1.7
	500.0	482.60 ± 16.72	3.5	−3.5	465.72 ± 14.92	3.2	−6.9
	2500.0	2711.88 ± 59.84	2.2	8.5	2633.31 ± 72.39	2.7	5.3
Gemcitabine	100.0	101.86 ± 7.55	7.4	1.9	86.53 ± 5.64	6.5	−13.5
	500.0	526.06 ± 47.53	9.0	5.2	477.32 ± 33.61	7.0	−4.5
	2500.0	2546.58 ± 100.92	4.0	1.9	2402.05 ± 106.44	4.4	−3.9

[a]RE is calculated as (mean measured concentration/spiked concentration − 1) × 100%.

acid, 0.1% formic acid, 0.05% acetic acid, 0.1% acetic acid, 5 mmol/L ammonium acetate, 10 mmol/L ammonium acetate/water solutions in combination with either methanol or acetonitrile. With methanol as organic solvent, moderate peak tailing was observed, while with acetonitrile, split peaks were shown. As the temperature rises up, the resolution improved with a narrower peak width. In conclusion, the most appropriate mobile phase was acetonitrile and 10 mmol/L ammonium acetate plus 0.1% formic acid in water at a flow rate of 0.25 mL/min, and the column temperature was kept at 30°C. The twelve analytes and the IS were at first characterized by full-scan mode and then product ions mode to ascertain their precursor ions and product ions which were utilized for constructing the MRM mode. The full-scan mode showed that the ionization of the analytes was more suitable in the positive mode. The parameters for fragmentor energy and collision energy were optimized and listed in Table 1.

3.2. Sample Pretreatment. Due to the complex composition of plasma, a sample pretreatment was often needed to remove protein and other potential interfering materials prior to LC-MS/MS analysis. Liquid-liquid extraction (LLE) with different organic solvents and protein precipitation (PPT) with acetonitrile or methanol were evaluated as sample pretreatment techniques in our earlier studies. Initially, several conventional PPT [26] and LLE procedures were investigated using different extraction solvents (methanol, acetonitrile, or mixed solvent for PPT, ethyl acetate, diethyl ether, tert-butyl methyl ether, or mixed solvent for LLE) [27–29], but no satisfactory recovery and strong matrix effect were obtained for all analytes. A previous pretreatment procedure was developed jointly using PPT by methanol for pemetrexed, gemcitabine, carboplatin and LLE by ethyl acetate for irinotecan, cyclophosphamide, ifosfamide, etoposide, and mixed ether-dichloromethane (7 : 3, V : V) for the remaining analytes to maximize recovery and

TABLE 6: Stability of analytes in human plasma ($n = 5$).

Analyte	Nominal concentration (ng/mL)	Three freeze-thaw stability		Short-term stability (24 h)		Long-term stability (3 months)	
		Precision (%RSD)	Accuracy (%RE)	Precision (%RSD)	Accuracy (%RE)	Precision (%RSD)	Accuracy (%RE)
Paclitaxel	50.0	1.9	5.9	2.8	6.7	2.4	5.8
	250.0	1.9	3.9	5.6	−1.4	3.6	3.7
	1250.0	4.3	−5.5	4.0	−4.7	3.3	−1.0
Docetaxel	20.0	1.8	6.8	3.1	4.5	2.5	5.3
	100.0	2.1	6.9	4.1	1.6	2.2	3.3
	500.0	1.9	−4.6	2.5	−3.1	2.9	−7.3
SN-38	20.0	**3.6**	−4.7	2.8	2.6	4.6	−3.1
	100.0	**2.9**	4.5	4.6	−3.0	3.5	2.9
	500.0	**4.1**	1.6	3.8	2.7	6.2	3.8
Vinorelbine	200.0	2.0	8.1	3.6	2.8	4.2	6.8
	1000.0	7.2	1.3	4.2	−0.4	6.3	0.9
	5000.0	5.4	3.0	3.3	4.9	3.2	3.5
Pemetrexed	200.0	5.0	2.6	4.7	4.8	4.9	5.2
	1000.0	2.8	2.7	6.5	−1.8	2.4	2.0
	5000.0	3.5	−6.4	4.8	−5.1	4.7	−4.7
Vinblastine	50.0	1.4	0.2	1.9	2.0	4.2	5.1
	500.0	5.0	−2.9	5.5	−3.1	4.0	3.0
	5000.0	3.5	1.5	3.9	2.9	3.2	3.3
Irinotecan	50.0	3.8	5.6	2.7	3.8	3.6	5.6
	500.0	5.5	−3.1	7.4	−1.0	4.5	2.8
	5000.0	2.8	4.3	4.4	1.3	4.0	3.9
Cyclophosphamide	5.0	7.4	1.2	1.9	4.3	5.5	−0.1
	50.0	6.1	3.7	3.0	2.8	3.6	4.8
	500.0	1.8	−4.2	2.4	−4.2	2.1	−3.7
Ifosfamide	5.0	3.5	2.4	2.6	0.2	5.3	3.5
	50.0	4.5	4.2	6.0	3.1	4.6	2.7
	500.0	3.2	−0.2	3.1	0.9	3.5	0.2
Carboplatin	100.0	7.2	4.8	6.9	−1.4	6.2	0.8
	500.0	6.7	2.6	6.6	1.0	7.4	3.1
	2500.0	1.7	−3.0	2.3	−0.7	2.1	−0.9
Etoposide	100.0	4.7	2.7	5.8	0.2	4.7	−0.3
	500.0	3.4	4.4	3.6	−2.5	4.5	3.5
	2500.0	7.3	−1.2	5.9	−4.1	6.5	−2.8
Gemcitabine	100.0	6.7	−1.4	5.6	−2.3	2.5	5.3
	500.0	7.8	−1.3	7.2	−2.0	6.7	−3.2
	2500.0	2.4	−7.2	5.4	−5.1	4.9	−4.0

reduce matrix effect of all analytes, but it was an low efficiency way for extraction of analytes. In order to increase sample throughput, the Ostro 96-well phospholipids removal plate was used, and the results displayed a shorter sample preparation time and higher efficiency. Extraction was done by the PPT using acetonitrile as the eluent solvent, and it showed lesser ion suppression compared with methanol. Table 2 shows that a one-step sample preparation with an Ostro 96-well plate proved to be simple, rapid, and high efficient for all analytes. In addition, phospholipids in matrix were deprived by the sorbent of Ostro, which also decreased the matrix interference.

3.3. Method Validation

3.3.1. Specificity. Comparison of blank and spiked human plasma chromatograms (Figure 2) indicated no significant interferences at the same retention times of the analytes and the IS.

3.3.2. Matrix Effect and Extraction Recovery. Matrix effect commonly occur due to interfering materials that coexist and coelute in the LC system in the biofluid; generally speaking, the interfering materials that were eluted together with the analytes could influence mostly the response of analytes. The PPT, which usually utilized organic solvent (methanol, acetonitrile, etc.) to strip the proteins in the biofluid by destroying the structure of them, had the lowest ability to eliminate the coexisting interfering materials. Lipids, fats, and other low molecular weight substances could not be striped by the PPT. Part selective extraction of analytes could be completed by an LLE and SPE pretreatment method, but the operation and solvent selection of the two extraction method was relatively complicated. In this study, one-step extraction by an Ostro sample preparation 96-well plate was successfully developed to obtain a high recovery and low matrix effect. The operation simply consisted of PPT and filtration and could be accomplished within 5 min, which promised a high-throughput sample

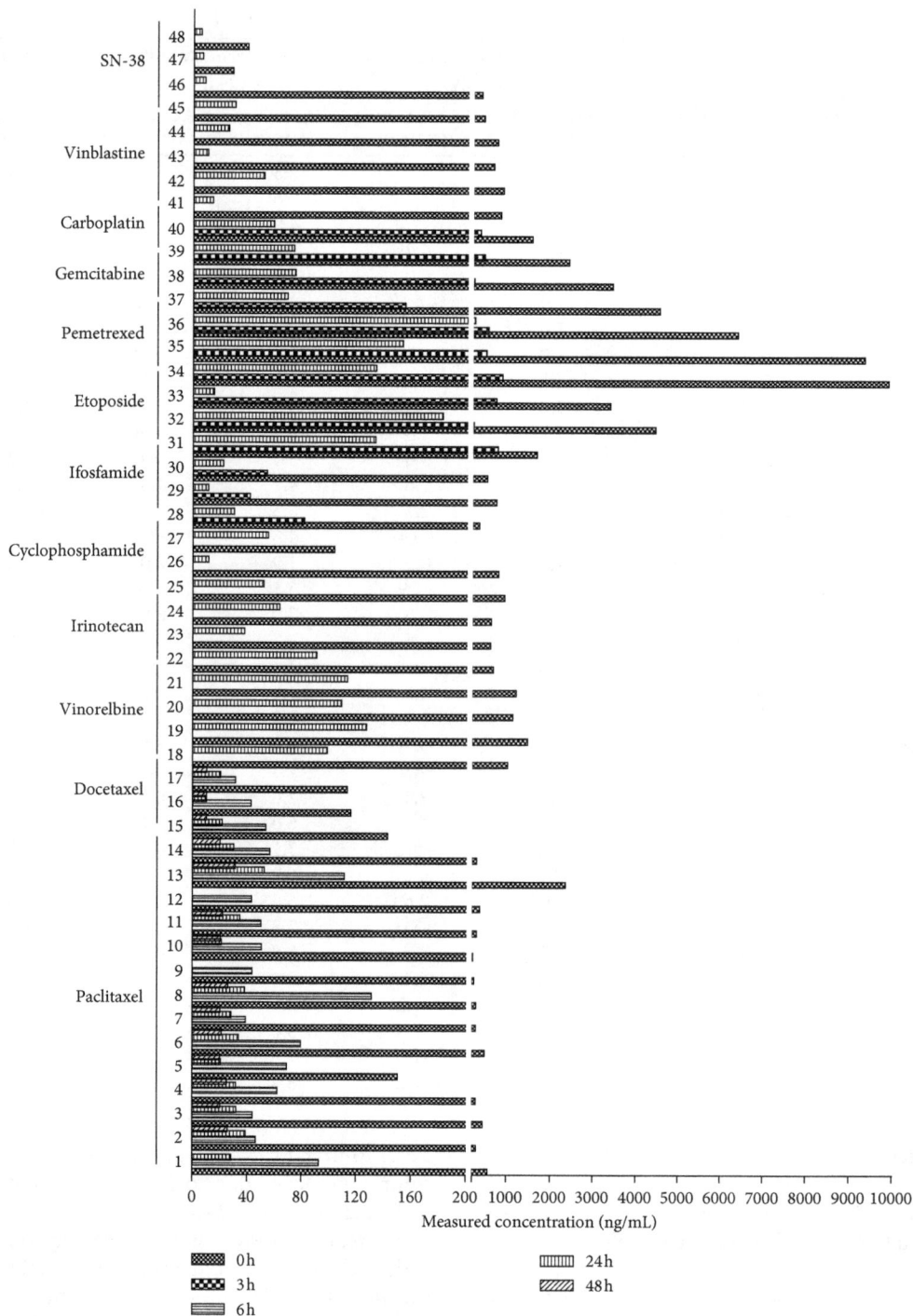

FIGURE 3: The quantification results of 132 plasma samples from 48 lung cancer patients.

pretreatment. The results showed that a matrix effect ranged from 85.2% to 101.3% and recovery ranged from 56.2% to 98.9%. The matrix effect and recovery were stable and conformed to the pharmacopeia criterion. All data are summarized in Table 3.

3.3.3. Linearity of Calibration Curves and LLOQ. Calibration curves were constructed by plotting the peak area ratios (analyte/IS) of calibration standards versus nominal

concentrations from spiked samples. The regression model was constructed based on linear regression with or without intercepts and weighing factors. The best linearity and least-squares residuals for the calibration curves were achieved with a $1/\chi^2$ weighing factor. The squares of the linear correlation coefficients were all over 0.99. Typical regression equations for the calibration curves were summarized in Table 4. The LLOQ for the determination of paclitaxel, docetaxel, vinblastine, vinorelbine, pemetrexed, carboplatin, etoposide, cyclophos-phamide, ifosfamide, gemcitabine, irinotecan, and SN-38 in human plasma, which were defined as the lowest concentration point of the calibration curves, were in accordance with the accuracy within ±20% and precision less than 20%. Five replicates of LLOQ spiked samples for each analyte were analyzed for its precision and accuracy. A signal-to-noise ratio (S/N) >10 at the LLOQ was observed for all the analytes. These LLOQs were sufficient for clinical drug monitoring of the eleven commonly used anticancer drugs and one active metabolite.

3.3.4. Precision and Accuracy. Three levels of QC samples (low, mid, and high) were chosen to analyze the interday and intraday precision and accuracy. The results showed a good precision and accuracy with intraday and interday precision less than 15% and accuracy within ±15%. Table 5 summarizes the intraday and interday precision and accuracy for the twelve analytes. All interday and intraday data were acceptable.

3.3.5. Stability. The stability of analytes in long-term cryopreservation, an autosampler, and three frozen-thaw cycles was investigated. The analytes were found to be stable in human plasma stored for 3 months at −20°C and in an autosampler at room temperature for 24 h (<5% reduction). After three freeze-thaw cycles, no obvious reductions (less than 15%) were observed for all the analytes (Table 6).

3.4. Application in Determination of Clinical Samples. Totally, 132 samples (69 from women and 63 from men) from 48 lung cancer patients were collected. The means (ranges) of age and body weight were 52.5 (18–82) years and 62.0 (51–72) kg, respectively. This entirely validated method was successfully applied for the determination of eleven commonly used anticancer drugs and one active metabolite concentration in the human plasma after administration. The results showed a big difference in patients administered with the same drugs in the same dosage, and the linearity range of this method covered all the concentration variations from different sampling points, which proved an extensive applicability of this method. All data are shown in Figure 3.

4. Conclusion

A simple, efficient, and sensitive HPLC-MS/MS method was successfully developed and was suitable for simultaneous determination of eleven commonly used anticancer drugs and one active metabolite in the human plasma from cancer patients. The analytical time was 10 min, and the LLOQ for all analytes was not more than 100 ng/mL. Owing to the application of an Ostro 96-well plate and a special column, this method achieved a simple and efficient sample preparation and separation process, which paved the way for the high throughput and extensive scope in clinical application for routine TDM of anticancer drugs to gain real-time drug exposure so that individualized therapeutic regimen may be promoted.

Conflicts of Interest

The authors declare that they have no conflicts of interest regarding the content of this article.

Authors' Contributions

The authors Shouhong Gao and Zhengbo Tao contributed equally to this work.

Acknowledgments

This work was supported by the Natural Science Foundation of Shanghai City, China (no. 17411972400), and Important Weak Subject Construction Project of Shanghai Health Science Education (2016ZB0303). The authors would like to thank Waters Company (Milford, MA, USA) for technical assistance.

References

[1] L. A. Torre, F Bray, R. L. Siegel, J. Ferlay, J. Lortet-Tieulent, and A. Jemal, "Global cancer statistics 2012," *CA: A Cancer Journal for Clinicians*, vol. 65, no. 2, pp. 87–108, 2015.

[2] D. Y. Lu, T. R. Lu, and H. Y. Wu, "Combination chemical agents with biological means in cancer therapy," *Research and Reviews in BioSciences*, vol. 7, no. 4, pp. 153–155, 2013.

[3] D. Y. Lu, E. H. Chen, J. Ding, B Xu, and T. R. Lu, "Anticancer drug combinations, a big momentum is needed," *Metabolomics*, vol. 5, no. 3, p. e139, 2015.

[4] D. Y. Lu, E. H. Chen, H. Y. Wu et al., "Anticancer drug combinations, how far we can go through?," *Anti-Cancer Agents in Medicinal Chemistry*, vol. 17, no. 1, pp. 21–28, 2017.

[5] J. Meletiadis, T. Stergiopoulou, E. M. O'Shaughnessy, J. Peter, and T. J. Walsh, "Concentration-dependent synergy and antagonism within a triple antifungal drug combination against *Aspergillus* species: analysis by a new response surface model," *Antimicrobial Agents and Chemotherapy*, vol. 51, no. 6, pp. 2053–2064, 2007.

[6] J. Meza-Junco and M. B. Sawyer, "Drug exposure: still an excellent biomarker," *Biomarkers in Medicine*, vol. 3, no. 6, pp. 723–731, 2009.

[7] D. I. Jodrell, M. J. Egorin, R. M. Canetta et al., "Relationships between carboplatin exposure and tumor response and toxicity in patients with ovarian cancer," *Journal of Clinical Oncology*, vol. 10, no. 4, pp. 520–528, 1992.

[8] A. Jakobsen, K. Bertelsen, J. E. Andersen et al., "Dose-effect study of carboplatin in ovarian cancer: a Danish Ovarian Cancer Group study," *Journal of Clinical Oncology*, vol. 15, no. 1, pp. 193–198, 1997.

[9] E. Gamelin, R. Delva, J. Jacob et al., "Individual fluorouracil dose adjustment based on pharmacokinetic follow-up compared with conventional dosage: results of a multicenter randomized trial of patients with metastatic colorectal cancer," *Journal of Clinical Oncology*, vol. 26, no. 13, pp. 2009–2105, 2008.

[10] R. Fety, F. Rolland, M. Barberi-Heyob et al., "Clinical impact of pharmacokinetically-guided dose adaptation of 5-fluorouracil: results from a multicentric randomized trial in patients with

locally advanced head and neck carcinomas," *Clinical Cancer Research*, vol. 4, no. 9, pp. 2039–2045, 1998.

[11] J. J. Lee, J. H. Beumer, and E. Chu, "Therapeutic drug monitoring of 5-fluorouracil," *Cancer Chemotherapy and Pharmacology*, vol. 78, no. 3, pp. 447–464, 2016.

[12] H. J. Klümpen, C. F. Samer, R. H. Mathijssen, J. H. Schellens, and H. Gurney, "Moving towards dose individualization of tyrosine kinase inhibitors," *Cancer Treatment Reviews*, vol. 137, no. 4, pp. 251–260, 2011.

[13] B. Gao, S. Yeap, A. Clements, B. Balakrishnar, M. Wong, and H. Gurney, "Evidence for therapeutic drug monitoring of targeted anticancer therapies," *Journal of Clinical Oncology*, vol. 30, no. 32, pp. 4017–4025, 2012.

[14] C. Shu, T. Zeng, S. Gao et al., "LC-MS/MS method for simultaneous determination of thalidomide, lenalidomide, cyclophosphamide, bortezomib, dexamethasone and adriamycin in serum of multiple myeloma patients," *Journal of Chromatography B*, vol. 1028, pp. 111–119, 2016.

[15] L. M. Torres, L. Rivera-Espinosa, J. L. Chávez-Pacheco et al., "A new method to quantify ifosfamide blood levels using dried blood spots and UPLC-MS/MS in paediatric patients with embryonic solid tumours," *PLoS One*, vol. 10, no. 11, Article ID e0143421, 2015.

[16] C. Atasilp, P. Chansriwong, E. Sirachainan et al., "Determination of irinotecan, SN-38 and SN-38 glucuronide using HPLC/MS/MS: application in a clinical pharmacokinetic and personalized medicine in colorectal cancer patients," *Journal of Clinical Laboratory Analysis*, vol. 32, no. 1, 2017.

[17] X. Gong, L. Yang, F. Zhang et al., "Validated UHPLC-MS/MS assay for quantitative determination of etoposide, gemcitabine, vinorelbine and their metabolites in lung cancer patients," *Biomedical Chromatography*, vol. 31, no. 11, 2017.

[18] E. R. Wickremsinhe, L. B. Lee, C. A. Schmalz, J. Torchia, and K. J. Ruterbories, "High sensitive assay employing column switching chromatography to enable simultaneous quantification of an amide prodrug of gemcitabine (LY2334737), gemcitabine, and its metabolite dFdU in human plasma by LC-MS/MS," *Journal of Chromatography B*, vol. 932, pp. 117–122, 2013.

[19] M. A. Fernández-Peralbo, F. Priego-Capote, M. D. Luque de Castro, A. Casado-Adam, A. Arjona-Sánchez, and F. C. Muñoz-Casares, "LC-MS/MS quantitative analysis of paclitaxel and its major metabolites in serum, plasma and tissue from women with ovarian cancer after intraperitoneal chemotherapy," *Journal of Pharmaceutical and Biomedical Analysis*, vol. 91, pp. 131–137, 2014.

[20] C. A. Crutchfield, M. A. Marzinke, and W. A. Clarke, "Quantification of docetaxel in serum using turbulent flow liquid chromatography electrospray tandem mass spectrometry (TFC-HPLC-ESI-MS/MS)," in *Methods in Molecular Biology*, vol. 1383, pp. 121–124, Humana Press, New York, NY, USA, 2016.

[21] P. Du, X. Han, N. Li et al., "Development and validation of an ultrafiltration-UPLC-MS/MS method for rapid quantification of unbound docetaxel in human plasma," *Journal of Chromatography B*, vol. 967, pp. 28–35, 2014.

[22] A. Navarrete, M. P. Martínez-Alcázar, I. Durán et al., "Simultaneous online SPE-HPLC-MS/MS analysis of docetaxel, temsirolimus and sirolimus in whole blood and human plasma," *Journal of Chromatography B*, vol. 921-922, pp. 35–42, 2013.

[23] H. Yamaguchi, A. Fujikawa, H. Ito et al., "A rapid and sensitive LC/ESI-MS/MS method for quantitative analysis of docetaxel in human plasma and its application to a pharmacokinetic study," *Journal of Chromatography B*, vol. 893-894, pp. 157–161, 2012.

[24] S. Achanta, M. Ngo, A. Veitenheimer, L. K. Maxwell, and J. R. Wagner, "Simultaneous quantification of vinblastine and desacetylvinblastine concentrations in canine plasma and urine samples using LC-APCI-MS/MS," *Journal of Chromatography B*, vol. 913-914, pp. 147–154, 2013.

[25] J. Zhou, S. Gao, F. Zhang et al., "Liquid chromatography-tandem mass spectrometry method for simultaneous determination of seven commonly used anticancer drugs in human plasma," *Journal of Chromatography B*, vol. 906, pp. 1–8, 2012.

[26] R. Cornelis, B. Heinzow, R. F. Herber et al., "Sample collection guidelines for trace elements in blood and urine. IUPAC Commission of Toxicology," *Journal of Trace Elements in Medicine and Biology*, vol. 10, no. 2, pp. 103–127, 1996.

[27] T. Storme, L. Mercier, A. Deroussent et al., "Liquid chromatography-mass spectrometry assay for quantitation of ifosfamide and its N-deschloroethylated metabolites in rat microsomal medium," *Journal of Chromatography B*, vol. 820, no. 2, pp. 251–259, 2005.

[28] T. F. Kalhorn, S. Ren, W. N. Howald, R. F. Lawrence, and J. T. Slattery, "Analysis of cyclophosphamide and five metabolites from human plasma using liquid chromatography-mass spectrometry and gas chromatography-nitrogen-phosphorus detection," *Journal of Chromatography B: Biomedical Sciences and Applications*, vol. 732, no. 2, pp. 287–298, 1999.

[29] R. Zhou, M. Frostvik-Stolt, and E. Liliemark, "Determination of etoposide in human plasma and leukemic cells by high-performance liquid chromatography with electrochemical detection," *Journal of Chromatography B: Biomedical Sciences and Applications*, vol. 757, no. 1, pp. 135–141, 2001.

13

Detection and Structural Characterization of Nucleophiles Trapped Reactive Metabolites of Limonin using Liquid Chromatography-Mass Spectrometry

Yujie Deng,[1] Yudong Fu,[2] Shumin Xu,[1] Ping Wang,[1] Nailong Yang,[1] Chengqian Li,[1] and Qing Yu (iD)[1]

[1]*Department of Endocrinology, The Affiliated Hospital of Qingdao University, 16 Jiangsu Road, Qingdao 266071, China*
[2]*Department of Ophthalmology, The Affiliated Hospital of Qingdao University, 16 Jiangsu Road, Qingdao 266071, China*

Correspondence should be addressed to Qing Yu; yuqing121985@126.com

Academic Editor: Federica Pellati

Limonin (LIM), a furan-containing limonoid, is one of the most abundant components of *Dictamnus dasycarpus* Turcz. Recent studies demonstrated that LIM has great potential for inhibiting the activity of drug-metabolizing enzymes. However, the mechanisms of LIM-induced enzyme inactivation processes remain unexplored. The main objective of this study was to identify the reactive metabolites of LIM using liquid chromatography-mass spectrometry. Three nucleophiles, glutathione (GSH), *N*-acetyl cysteine (NAC), and *N*-acetyl lysine (NAL), were used to trap the reactive metabolites of LIM in *in vitro* and *in vivo* models. Two different types of mass spectrometry, a hybrid quadrupole time-of-flight (Q-TOF) mass spectrometry and a LTQ velos Pro ion trap mass spectrometry, were employed to acquire structural information of nucleophile adducts of LIM. In total, six nucleophile adducts of LIM (M1–M6) with their isomers were identified; among them, M1 was a GSH and NAL conjugate of LIM, M2–M4 were glutathione adducts of LIM, M5 was a NAC and NAL conjugate of LIM, and M6 was a NAC adduct of LIM. Additionally, CYP3A4 was found to be the key enzyme responsible for the bioactivation of limonin. This metabolism study largely facilitates the understanding of mechanisms of limonin-induced enzyme inactivation processes.

1. Introduction

Dictamnus dasycarpus Turcz, known as Bai-Xian-Pi (BXP) in Chinese, belongs to the Rutaceae family. BXP has been widely used in Asian and European countries as an antipruritic, antidote, antibacterial, and anti-inflammatory agent. It is also used for the treatment of rubella, eczema, scabies, and jaundice. In addition, BXP displays diverse pharmacological properties, including antitumor, antiarrhythmic, antitinea, and smooth muscle-contraction activities [1]. Despite this, the safety of BXP has been questioned, and ingestion of BXP was reportedly associated with high incidence of liver injury. For instance, four cases of toxic hepatitis were reported in patients after taking a decoction made by boiling down the root of BXP [2]. In another clinical trial for the treatment of eczema,

a standard mixture containing BXP was implicated in six of thirty-three cases of severe hepatitis [3].

Quinoline alkaloids, limonoids, sesquiterpenes, coumarins, flavonoids, and steroids have been explored as the major components of BXP [1, 4–8]. Limonoids have drawn much attention, and thus far a total of 25 limonoids have been isolated and characterized from BXP [1]; among them, Limonin (LIM) is one of the most abundant limonoids found in BXP [9]. Recent studies demonstrated that LIM has great potential for inhibiting the activity of drug-metabolizing enzymes and/or transporters such as CYP enzyme isomers and P-glycoprotein [10, 11]. As far as we know, modulation of activity of drug-metabolizing enzymes and/or transporters will result in the alteration of the clearance of exogenous toxins and affects the hepatic detoxification functions.

However, the mechanisms of LIM-induced enzyme inactivation processes remain unknown.

LIM is a furan-containing component. Many xenobiotics containing a furan unit are reported to be toxic and/or carcinogenic [12], such as furosemide [13], prazosin [14], teucrin A [15–17], 8-epidiosbulbin E [18–20], and diosbulbin B [21, 22]. The toxic effects elicited by these furans are suggested to be attributed to their *cis*-enedial oxidative intermediate [12]. We hypothesized that LIM is metabolized to a *cis*-enedial intermediate (**3**, Scheme 1), an electrophilic species, which may play a critical role in enzyme inactivation activities of LIM. In this study, we present the successful characterization of a *cis*-enedial intermediate of LIM and the identification of the cytochromes P450 (CYP450) enzymes responsible for the metabolic activation of LIM.

2. Materials and Methods

2.1. Chemicals and Materials.
Mouse liver microsomes (MLMs), human liver microsomes (HLMs), recombinant human P450 enzymes, NADPH-regenerating system, glutathione (GSH), *N*-acetyl cysteine (NAC), and *N*-acetyl lysine (NAL) were purchased from BD Biosciences (Bedford, MA, USA). LIM (purity >98%) was obtained from Sigma-Aldrich (Sigma-Aldrich, St. Louis, MO, USA). Acetonitrile (ACN), methanol (MeOH), and formic acid (FA) of LC/MS grade were obtained from Fisher Scientific (Pittsburgh, PA, USA). Water was purified with a Milli-Q system (Millipore, Bedford, USA) and was passed through a 0.22 μm membrane filter before use.

2.2. Animal Studies and Sample Collection.
Experiments with mice were carried out according to the guidelines for Animal Experimentation of Qingdao University (Qingdao, China), and the protocol was approved by the Animal Ethics Committee of the institution. Female Kunming mice (20 ± 5 g) were obtained from the Experimental Animal Center of Qingdao University (Qingdao, China). Mice were housed 5 per cage and maintained in air-conditioned quarters with a room temperature of 20 ± 2°C, relative humidity of 50 ± 10%, and an alternating 12 h light/dark cycle and allowed to acclimate for at least 1 week prior to the start of the experiment. Mice were fed with standard chaw diet and water and were allowed to eat and drink *ad libitum*. LIM dissolved in dimethylsulfoxide (DMSO) was orally administered to mice at a signal dose of 5 mg/kg. Twenty-four-hour mouse urine and fecal samples were collected at room temperature by using metabolic cages. Blank urine and fecal samples were collected prior to the LIM treatment. These samples were stored under −80°C before analysis.

2.3. Sample Preparation.
One hundred and fifty microliters of ACN was added to 50 μL of urine sample, then vortexed for 3 min and centrifuged at 16,100 ×g for 10 min under 4°C. The supernatant was concentrated to dryness under a gentle stream of nitrogen gas at 45°C. The resulting residue was reconstituted with 200 μL of ACN/water (50/50, v/v) containing 2% acetic acid, followed by centrifugation at

16,100 ×g for 10 min at 4°C. Ten μL aliquot of the supernatant was injected into LC-MS/MS systems for analysis.

2.4. Microsomal Incubations.
Liver microsomes with a final concentration at 0.5 mg/mL were incubated with LIM (30 μM) for 60 min in the presence of GSH or NAC and NAL at a final concentration of 1.0 mM. The experimental incubation mixture consisted of 100 mM potassium phosphate buffer (pH 7.4), a prepared NADPH-regenerating system, and MLMs or HLMs. Stock solution of LIM was prepared in DMSO, and the final concentration of DMSO in the incubation did not exceed 1% (v/v). After preincubated at 37°C for 15 min in a water bath, the reactions were initiated by the addition of LIM and were incubated at 37°C for another 60 min. The reactions were terminated by the addition of an equal volume of ice-cold ACN containing 2% acetic acid. The mixture was vortexed and centrifuged at 16,100 ×g for 5 min. Aliquots of supernatants were stored at −20°C until analysis. Control incubations without NADPH-regenerating system, without substrate, or without liver microsomes were performed to ensure that the formation of the metabolites was microsome- and NADPH-dependent.

2.5. Chemical Synthesis.
Five milligrams of LIM was completely dissolved in 400 μl of acetone, then 50 μL of saturated sodium bicarbonate solution, and 10 mg of Oxone were added successively to the resulting solution. The mixture was stirred for 15 min at room temperature, followed by addition of 50 mg of GSH or 22 mg of NAC, both GSH and NAC were dissolved in 500 μL of saturated sodium bicarbonate solution. The mixture was stirred for 30 min and then centrifuged; the supernatants were harvested and evaporated to dryness under a stream of nitrogen gas at 45°C. The resulting residues were reconstituted with 500 μL of pH 7.4 PBS buffer, then 5 mg of NAL was added and stirred for 30 min at 70°C, the reaction was cooled to room temperature and filtered through a 0.22 μm member filter, and then analyzed by LC-MS/MS.

2.6. Recombinant Human P450 Enzyme Phenotyping.
To determine the specific P450 enzymes involved in the formation of reactive metabolites of LIM, a total of 10 human recombinant P450s, including P450s 1A2, 2A6, 2B6, 2C8, 2C9, 2C19, 2D6, 2E1, 3A4, and 4A11 were screened. Conditions were equivalent to those of the microsomal incubations except that the microsomes were replaced by individual human recombinant P450 enzymes at a concentration of 25 pmol enzyme with a total volume of 200 μL in each incubation. Experiments were performed in triplicate.

2.7. LC-MS/MS Method.
All samples were analyzed on a Thermo-Finnigan spectra system consisting of an Ultimate 3000 degasser, an Ultimate 3000 RS pump, an Ultimate 3000 RS column compartment, and an LTQ Velos Pro ion trap mass spectrometer (Thermo Scientific, San Jose, CA) coupled with an electrospray ionization (ESI) interface. The ESI interface was operated in a positive ion polarity mode. The

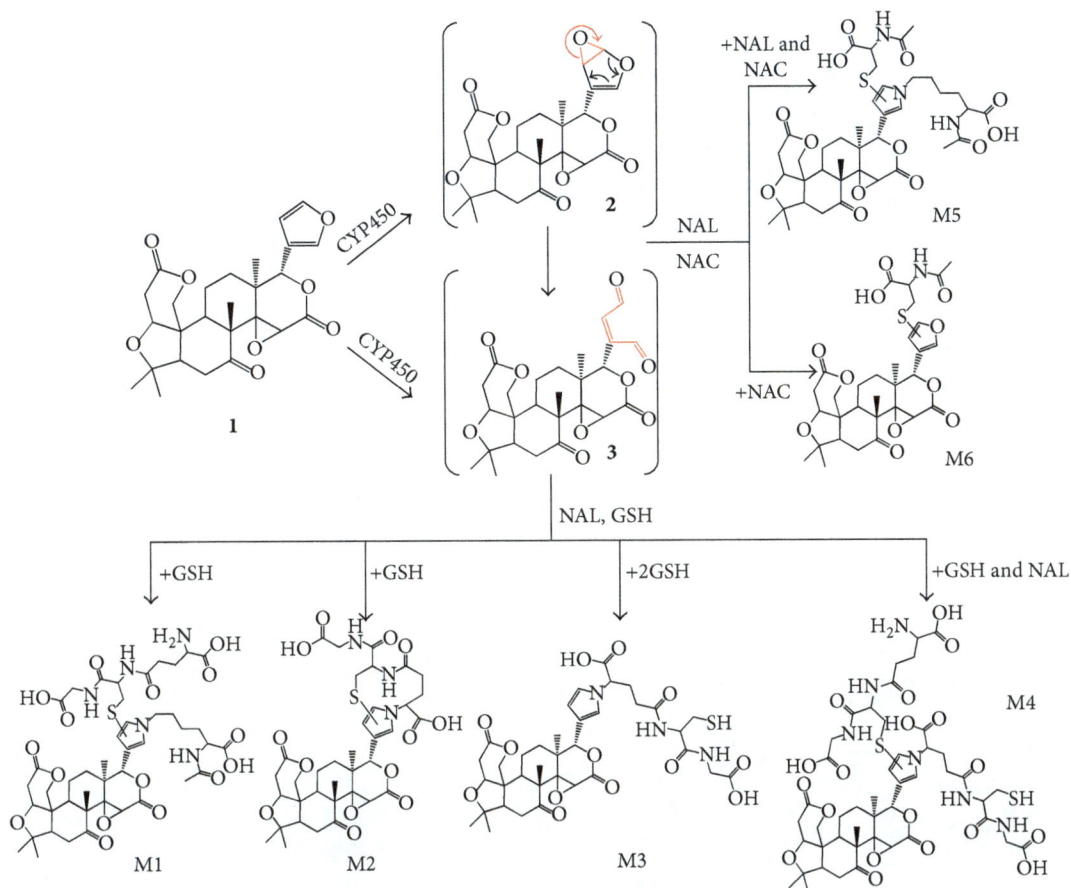

SCHEME 1: Microsomal metabolism of the furan ring of LIM and the proposed pathway for the formation of LIM-derived conjugates.

voltage on the ESI interface was maintained at approximately 4.3 kV and ESI capillary temperature was set at 300°C. Nitrogen gas was used as the sheath gas and auxiliary gas which was set at 35 and 10 units, respectively. The collision energy was set at 35 with isolation width of 2 Da for MS^2. Chromatographic separation was performed on a Phenomenex Gemini C18 column (5 μm, 3.0 mm i.d. × 150 mm; Torrance, CA, USA), and the column temperature was set at 35°C. The mobile phase was 5% aqueous MeOH with 0.1% formic acid (mobile phase A) and 95% aqueous MeOH with 0.1% formic acid (mobile phase B). The gradient was initiated at 90% A and held constant for 5 min, followed by linear increases in B to 25% from 5 to 10 min; to 60% from 10 to 30 min, to 100% from 30 to 40 min; and then held constant for 5 min. The column was then reequilibrated with 90% A for 5 min. The flow rate was set at 0.3 mL/min. The injection volume was 10 μL for each sample. Data acquisition and analysis were performed using Xcalibur 2.2 version (Thermo Electron, San Jose, CA, USA).

Samples were also analyzed on a hybrid quadrupole time-of-flight (Q-TOF) mass spectrometer (Bruker micro Q-TOF, Bremen, Germany) with an electrospray ionization interface equipped with an Agilent 1200 series rapid resolution LC system. The mass spectrum data were acquired in the positive ion mode. The parameters of ESI-MS were set as follows: capillary voltage (−4.3 kV), the nebulizer gas pressure (1.2 bar), the dry gas flow rate (8.0 L/min), and

temperature (220°C). The spectra were acquired at 2 s per spectrum in the range of m/z 100 to 1200. LC conditions were the same to those described above for the LTQ ion trap MS system. Data acquisition and analysis were performed using Bruker Daltonics data analysis 3.4 software.

3. Results

3.1. In Vitro Metabolic Activation of LIM. We proposed that the furan group of LIM played an important role in the LIM-induced inhibition of the activity of drug-metabolizing enzymes and/or transporters, and specifically this LIM is metabolized to the corresponding *cis*-enedial (Scheme 1, *cis*-enedial **3**), and the resulting electrophilic metabolites are responsible for the quench of many drug-metabolizing enzyme activities. LIM was incubated in MLMs or HLMs supplemented with GSH or NAC and NAL as trapping agents. The mixture was analyzed by a Thermo Scientific LTQ Velos ion trap mass spectrometer. Metabolites M1 and M1′ (retention times at 8.9 and 9.8 min, resp.) were detected (Figures 1(b) and 1(c)) in both HLMs and MLMs incubations by scanning of an ion pair of m/z 946 → 817, and M1 and M1′ showed identical mass fragmental patterns with indicative characteristic secondary ion signals associated the cleavage of the GSH moiety (Figure 1(e)). The product ions at m/z 928 were derived from the loss of one water molecule (−18 Da), and the product ions at m/z 871 and 817 were

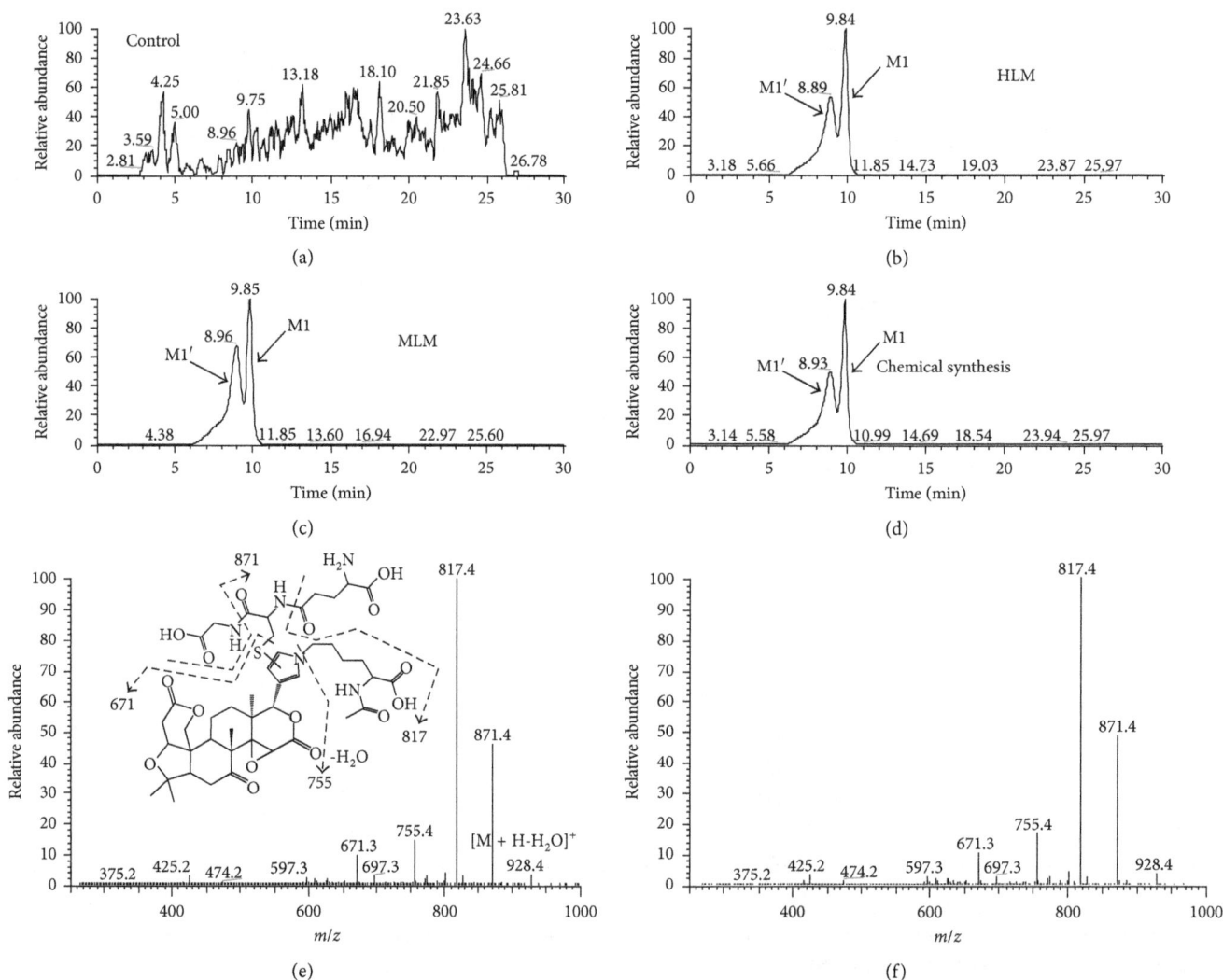

FIGURE 1: Extract ion (m/z 956 → 817) chromatograms obtained from LC-LTQ MS analysis of microsomal incubations containing LIM, GSH, NAL, and NADPH in the absence microsomes (a), or in presence of HLMs (b) or MLMs (c). (d) Extracted ion (m/z 956 → 817) chromatogram obtained from LC-LTQ MS analysis of synthetic M1 and M1′. (e) MS/MS spectrum of M1 generated in microsomal incubations (M1′ showed the same MS/MS spectrum). (f) MS/MS spectrum of synthetic M1 (synthetic M1′ showed the same MS/MS spectrum).

derived from the loss of glycine portion (−75 Da) and γ-glutamyl portion (−129 Da) from m/z 946, respectively. The mixture was also analyzed by LC/Q-TOF MS. M1 and M1′ showed their protonated molecule ion $[M + H]^+$ at m/z 946.4512 and m/z 946.4513 in positive ion mode, respectively; both of them matches the elemental composition of $C_{44}H_{60}N_5O_{16}S$. No such conjugate was detected in the microsomal incubation system with the absence of NADPH (Figure 1(a)), indicating that the formation of M1 and M1′ was mediated by the microsomal metabolism. To further characterize these two metabolites, we oxidized LIM with Oxone in acetone, followed by the addition of GSH and NAL; two major products formed in the reaction showed the same chromatographic and mass spectrometric identities (Figures 1(d) and 1(f)) as that of the products (M1 and M1′) generated in the microsomal incubations. Unfortunately, we were unable to purify enough amount of the product for nuclear magnetic resonance (NMR) characterization.

Interestingly, mono-GSH adducts of LIM were also observed by selected reaction monitoring (SRM) scanning with ion transition of m/z 758 → 683 and m/z 760 → 685 in the positive mode. Under the transition of m/z 758 → 683, metabolites M2 and M2′ (retention times at 12.9 and 12.2 min, resp.) were detected in both MLMs and HLMs incubation systems (Supplementary Figures 1(B) and 1(C)). The tandem mass spectrometry (MS/MS) spectrum of M2 and M2′ were identical, with one of the major fragmental ions at m/z 683, indicating the loss of glycine (−75 Da); the product ion at m/z 740 is postulated to arise from the elimination of H_2O (Supplementary Figure 1(E)). Further analysis by LC/Q-TOF MS demonstrated that M2 and M2′ showed their protonated molecular ions at m/z 758.2351 and 758.2353, respectively, corresponding to the formula $C_{36}H_{44}N_3O_{13}S$. On the basis of the observed mass spectrometric data, we propose that M2 and M2′ are generated by intramolecular cyclization after GSH was conjugated to

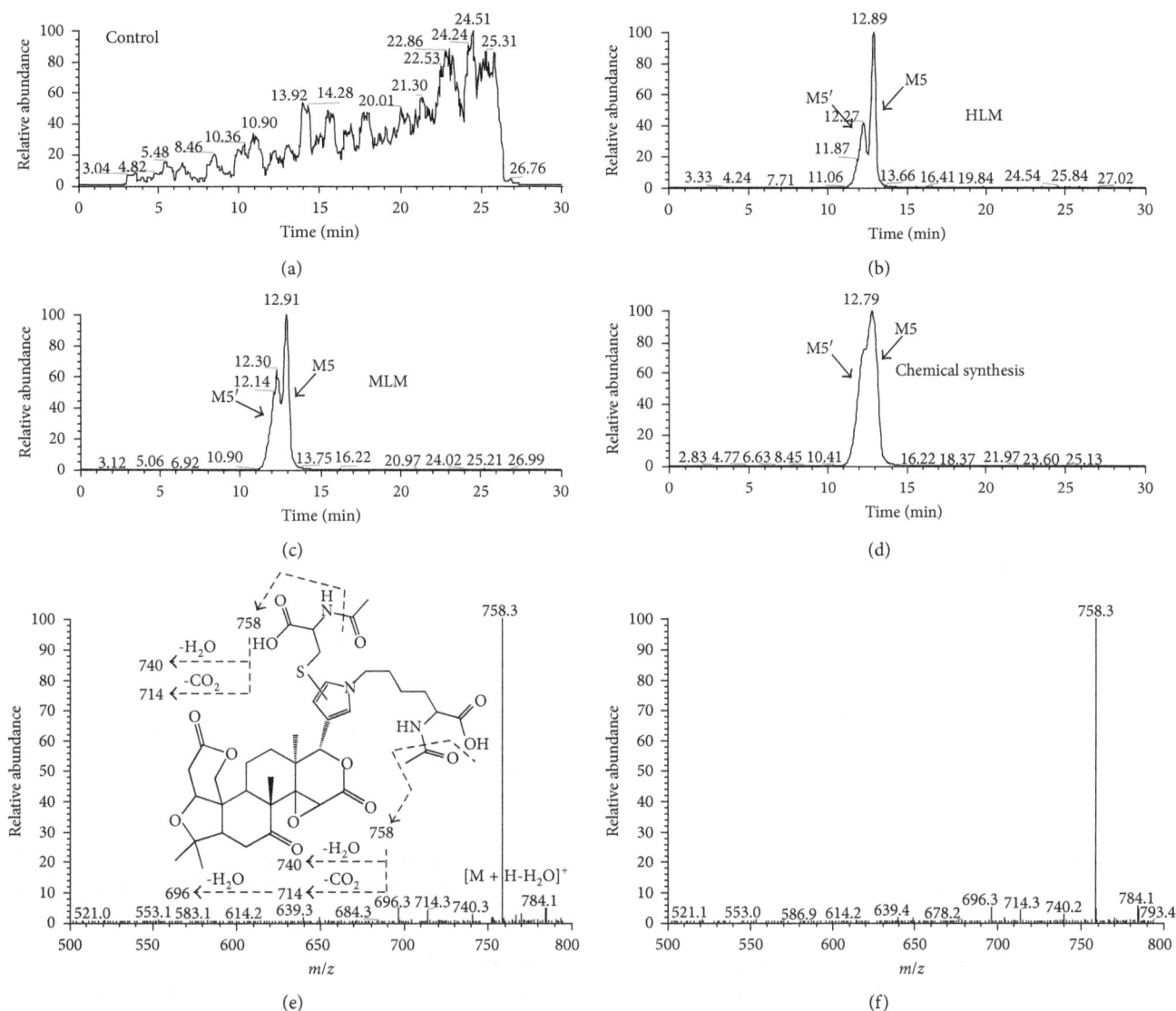

FIGURE 2: Extract ion (m/z 802 → 758) chromatograms obtained from LC-LTQ MS analysis of microsomal incubations containing LIM, GSH, NAL, and NADPH in the absence microsomes (a) or in presence of HLMs (b) or MLMs (c). (d) Extracted ion (m/z 802 → 758) chromatogram obtained from LC-LTQ MS analysis of synthetic M5 and M5′. (e) MS/MS spectrum of M5 generated in microsomal incubations (M5′ showed the same MS/MS spectrum). (f) MS/MS spectrum of synthetic M1 (synthetic M1′ showed the same MS/MS spectrum).

the *cis*-enedial intermediate which was derived from LIM. Under the transition of m/z 760 → 685, four metabolites (M3, M3′, M3″, and M3‴) with retention times at 12.9, 12.3, 11.4, and 10.6 min, respectively, were detected in HLMs incubation system by mass spectrometry (Supplementary Figure 1(B)) and two metabolites (M3 and M3′) with retention times of 12.9 and 12.3 min were observed in the MLMs incubation system; these metabolites showed identical mass fragmental patterns, provided major products ions at m/z 742, 685, 657, 614, and 552 (Supplementary Figure 2(E)). These metabolites were further analyzed by LC/Q-TOF MS, and they showed their protonated molecular ion $[M + H]^+$ at around m/z 760.2152 in the positive ion mode, all of them in agreement with the elemental composition of $C_{36}H_{46}N_3O_{13}S$. We propose that these metabolites were isomers of a pyrrole-GSH conjugate (Scheme 1). To further

characterize the metabolites detected under the transition of m/z 758 → 683 and m/z 760 → 685, we analyzed the mixture of the biomimetic oxidation of LIM described above. As expected, M2, M2′, M3, and M3′ were all detected, based on their retention times, molecular ions, and MS/MS spectra (Supplementary Figures 1(D) and 1(E); Figures 2(d) and 2(e)).

Metabolites M4 and M4′, with retention times at 9.6 and 8.5 min, were observed by scanning of an ion pair of m/z 1065 → 936 in the positive ion mode (Supplementary Figure 3(B)). The MS/MS spectrum of M4 and M4′ were identical which showed the major fragment ions associated with fragmentation of the GSH moiety (Supplementary Figure 3(E)). The product ions at m/z 990 and 936 were derived from the loss of glycinyl moiety (-75 Da) and γ-glutamyl moiety (-129 Da) from m/z 1065, respectively. This indicates the participation of GSH in the formation of

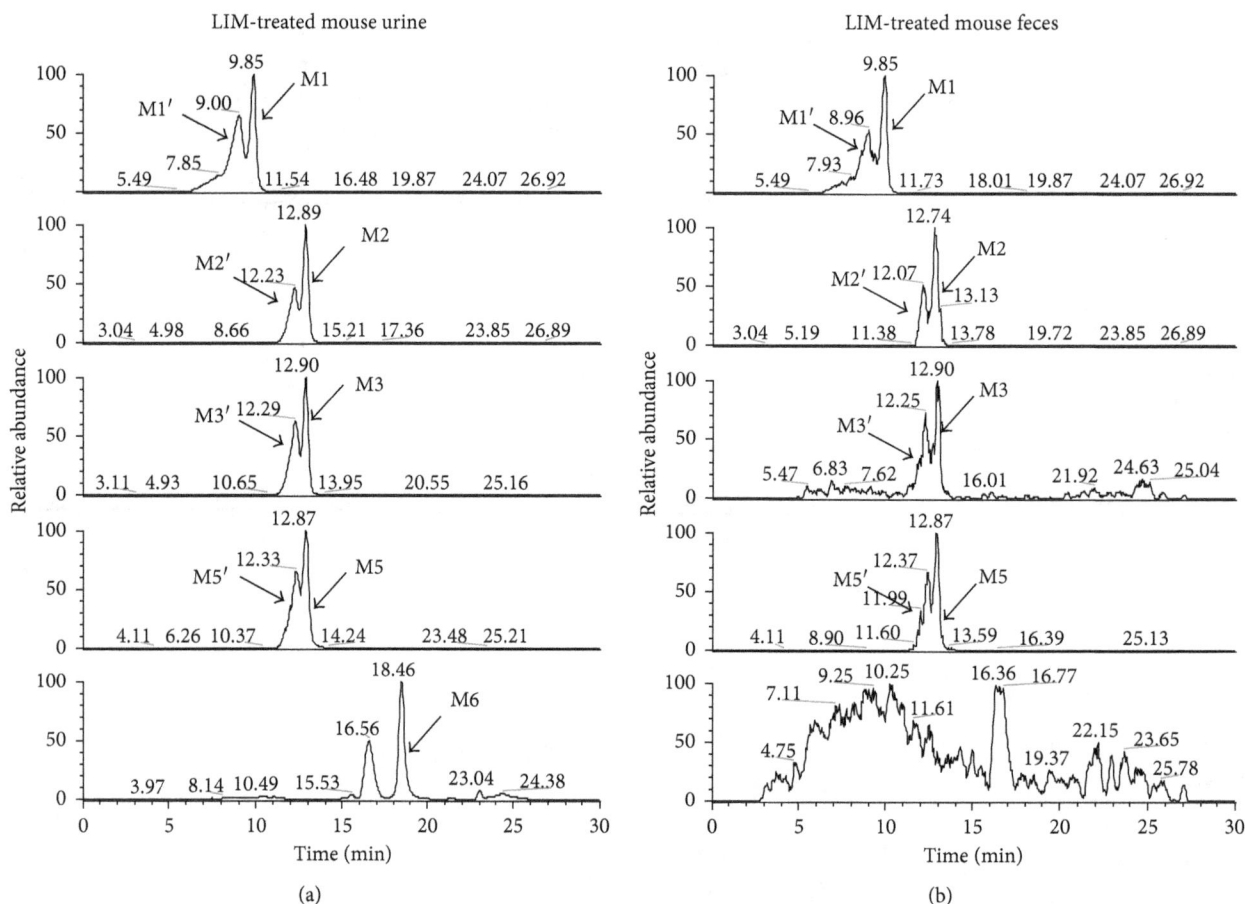

FIGURE 3: Extracted ion (m/z 946 → 817, m/z 758 → 683, m/z 760 → 685, m/z 802 → 758, and m/z 632 → 614, represent M1, M2, M3, M5, and M6, resp.) chromatograms obtained from LC-LTQ MS analysis of urine (a), and feces (b) of mice after the treatment of LIM.

M4 and M4′. Further analysis by LC-Q/TOF MS showed its molecular ion at m/z 1065.3431. The protonated molecular ions observed were consistent with the molecular mass corresponding to the elemental composition of $C_{46}H_{61}N_6O_{19}S_2$, suggesting that M4 and M4′ are derived from two molecules of GSH, which are LIM-derived di-GSH conjugates (Scheme 1).

In a parallel incubation, NAC in place of GSH was used to trap the LIM-derived *cis*-enedial intermediate. No such adducts like M1, M2, M3, and M4, which were found in GSH-fortified microsomal incubations, were observed in the NAC supplied microsomal incubation systems. Instead, we detected two metabolites (M5 and M6) most likely associated with NAC/NAL. Under the ion transition of m/z 802 → 758, M5 and M5′ were detected in both HLM and MLM incubation systems with retention times at 12.9 and 12.3 min (Figures 2(b) and 2(c)), respectively. These metabolites were further analyzed by LC/Q-TOF MS; M5 and M5′ showed their protonated molecule ion $[M + H]^+$ at m/z 802.3217 in positive ion mode, which matches the elemental composition of $C_{39}H_{52}N_3O_{13}S$. The MS/MS spectrum of M5 and M5′ identically showed the indicative characteristic neutral loss of one macular of water ($-18\,Da$) and successive losses of 44 Da and/or 18 Da. The product ions at m/z 784, 758, 740, 714, and 696 were assigned to $[M + H\text{-}H_2O]^+$, $[M + H\text{-}CO_2]^+$, $[M + H\text{-}CO_2\text{-}H_2O]^+$, $[M + H\text{-}2CO_2]^+$, and $[M + H\text{-}2CO_2\text{-}H_2O]^+$,

respectively (Figure 2(e)). The formation of M5 and M5′ were also found to be NADPH dependent (data not shown). To further characterize M5 and M5′, LIM was oxidized with Oxone in acetone, follow by reaction with NAC and NAL. Two major products formed in the reaction showed the same chromatographic and mass spectrometric identities as that of the metabolites M5 and M5′ generated in the microsomal incubation systems (Figures 2(d) and 2(f)). Besides M5 and M5′, a mono-NAC adduct of LIM (M6) was observed under ion transition of m/z 632 → 614 in positive ion mode. Fragmental ion generated from a natural loss of one macular of water ($-18\,Da$), m/z 614, was identified as the major fragmental ion; to further characterize this metabolite, we analyzed the mixture of the biomimetic oxidation of LIM described above. As expected, a product formed in the reaction showed the same chromatographic and mass spectrometric identities as that of the product (M6) generated in the microsomal incubations (Supplementary Figures 4(D) and 4(F)), and this metabolite was tentatively identified as mono-NAC adduct of LIM.

3.2. Metabolic Activation of LIM in Mice.
To investigate the bioactivation of LIM *in vivo*, urine, and fecal samples collected from LIM treated mice were monitored by a designed

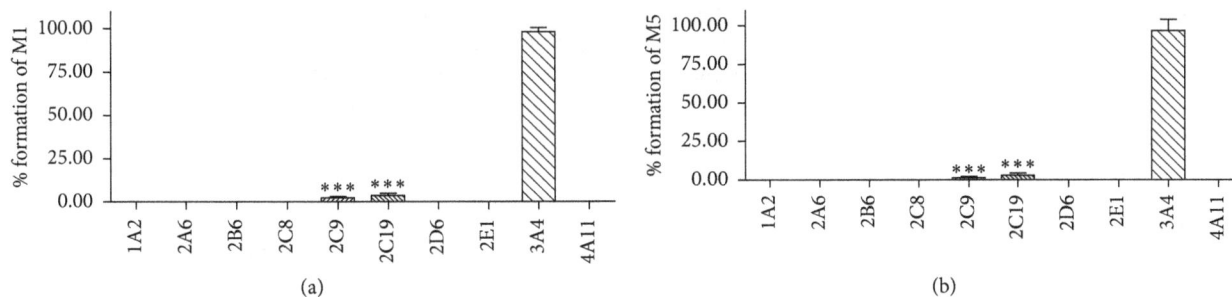

FIGURE 4: Rates of the M1 (A) and M5 (B) formation in incubations of LIM with recombinant human P450 enzymes.

selected reaction monitoring (SRM) template using LC-MS/MS. One GSH and NAL adduct (M1), two mono-GSH adducts (M2 and M3), one NAC and NAL adduct (M5), and one mono-NAC adduct (M6) were found in the urinary samples collected from the mice after given LIM (Figure 3(a)), except M6, these metabolites were also detected from the LIM-treated mouse fecal samples, while no such metabolites were observed from the blank mouse urine and fecal samples.

3.3. P450 Enzymes Responsible for the Bioactivation of LIM.

In order to identify the P450 enzymes involved in the bioactivation, LIM was incubated with individual recombinant human P450s, including P450s 1A2, 2A6, 2B6, 2C8, 2C9, 2C19, 2D6, 2E1, 3A4, and 4A11, followed by monitoring the formation of M1 and M5. The bioactivation activities of each P450s were examined, and P450 2C9, 2C19, and 3A4 displayed the metabolic activity, with 3A4 as the most potent one (Figures 4(a) and 4(b)). These results clearly confirmed that P450 2C9, 2C19, and 3A4 were the primary enzymes that were involved in the bioactivation of LIM.

4. Conclusions

The citrus bitter principle limonin was first isolated in 1841, but intensive investigation of its bioactivity did not commence until the last decade. Metabolism study of LIM is still very limited, possibly due to little attention being paid to its toxic effects. The unexpected inhibition activity of LIM towards the drug-metabolizing enzymes and/or transporters led us to investigate the metabolic activation of LIM. We hypothesized that LIM is metabolized to an electrophilic species, a cis-enedial intermediate, then reactive with the nucleophilic side residues of proteins (thiol, alcohol, phenol, carboxyl, and amine), and led to the inactivation of protein functions.

In our current study, GSH which contains two nucleophilic functional groups, including a sulfhydryl (the cysteine residue) and an amino (the glycine residue) group, NAC and NAL each contains a nucleophilic amino group in its side chain, were chosen as the trapping reagents to trap the potential LIM-derived cis-enedial intermediate in mouse and human liver microsomal incubation systems after exposure to LIM. LC-MS/MS analysis showed a total of six LIM-derived conjugates (M1–M6) with their isomers in the

microsomal incubations. On the basis of the molecular ions obtained from high-resolution mass spectrometry, six LIM-derived pyrrole derivatives are postulated to be formed in the microsomal incubation systems. Mechanistically, a sulfhydryl group and a primary amine group participate in the formation of the pyrrole derivatives. The pyrrole production possibly starts with the reaction of the cis-enedial intermediate with the sulfhydryl residue of GSH or NAC by Michael addition to form the corresponding GSH or NAC conjugate, which subsequently reacts with the amino residue of NAL to form a Schiffs base followed by intramolecular cyclization and dehydration to produce the pyrrole derivative. For the formation of M1, the sulfhydryl and amino groups came from GSH and NAL, respectively. For the generation of M2 and M3, the nucleophilic group was coming from GSH. M4 is a di-GSH conjugate, and the sulfhydryl and amino groups were acquired from the respective GSH molecules. Unlike the formation of M1-M4, for the generation M5 and M6, the sulfhydryl came from NAC, and the amino group was acquired from NAL. The in vitro findings for M1–M6 provided important evidence for the formation of cis-enedial intermediate in microsomal reactions. M1, M2, M3, M5, and M6, which were found in microsomal incubations, were also detected in the urine samples of mice given LIM (Figure 3); M1, M2, M3, and M5 were detected in LIM treated mouse fecal samples as well (Figure 3). It appears that the primary LIM-derived GSH, NAC, and/or NAL conjugates are excreted mainly through the urine with relatively trace amount they excreted through feces. However, M4, which identified as a LIM-derived di-GSH conjugate, was not detected from the mouse urine and fecal samples after LIM treatment. Nevertheless, the observation of the GSH or NAC and NAL conjugates in urine and feces infers the metabolism of LIM to the cis-enedial intermediate in vivo. In addition, bioactivation studies with individual recombinant enzymes demonstrated that P450 2C9, 2C19, and 3A4 are enzymes responsible for the bioactivation of LIM (Figure 4(a)), with P450 3A4 as the most potent one. These findings facilitate our ongoing investigation of the biochemical mechanisms of LIM-induced enzyme inactivation.

In summary, the metabolic generation of the cis-enedial intermediate from LIM was evident in both in vitro and in vivo systems. The condensation reaction of the electrophilic intermediate with GSH or NAC and NAL gave six GSH or NAC and NAL conjugates derived from the *cis*-enedial intermediate of LIM. P450 3A4 was identified as the dominant participant in the catalysis leading to the formation of the reactive *cis*-enedial intermediate. The metabolite identification work performed herein enables us to better understand the mechanisms of LIM-induced enzyme inactivation processes.

Conflicts of Interest

The author(s) declared no potential conflicts of interest with respect to the research, authorship, and/or publication of this article.

Authors' Contributions

Yujie Deng and Yudong Fu contributed equally to this study.

Acknowledgments

The authors gratefully acknowledge the financial support by National Natural Science Foundation of China (81600684) and Natural Science Foundation of Shandong Province (ZR2016HQ16) to Yujie Deng.

Supplementary Materials

Supplementary 1. Figure 1: extract ion (m/z 758 → 683) chromatograms obtained from LC-LTQ MS analysis of microsomal incubations containing LIM, GSH, NAL, and NADPH in the absence microsomes (A), or in presence of HLMs (B) or MLMs (C). (D) Extracted ion (m/z 802 → 758) chromatogram obtained from LC-LTQ MS analysis of synthetic M2 and M2′. (E) MS/MS spectrum of M2 generated in microsomal incubations (M2′ showed the same MS/MS spectrum). (F) MS/MS spectrum of synthetic M2 (synthetic M2′ showed the same MS/MS spectrum).
Supplementary 2. Figure 2: extract ion (m/z 760 → 685) chromatograms obtained from LC-LTQ MS analysis of microsomal incubations containing LIM, GSH, NAL, and NADPH in the absence microsomes (A), or in presence of HLMs (B) or MLMs (C). (D) Extracted ion (m/z 760 → 685) chromatogram obtained from LC-LTQ MS analysis of synthetic M3 and M3′. (E) MS/MS spectrum of M3 generated in microsomal incubations (M3′ showed the same MS/MS spectrum). (F) MS/MS spectrum of synthetic M3 (synthetic M3′ showed the same MS/MS spectrum).
Supplementary 3. Figure 3: extract ion (m/z 1065 → 936) chromatograms obtained from LC-LTQ MS analysis of microsomal incubations containing LIM, GSH, NAL, and NADPH in the absence microsomes (A), or in presence of HLMs (B) or MLMs (C). (D) Extracted ion (m/z 1065 → 936) chromatogram obtained from LC-LTQ MS analysis of synthetic M4 and M4′. (E) MS/MS spectrum of M3 generated in microsomal incubations (M4′ showed the same MS/MS spectrum). (F) MS/MS spectrum of synthetic M4 (synthetic M4′ showed the same MS/MS spectrum).
Supplementary 4. Figure 4: extract ion (m/z 632 → 614) chromatograms obtained from LC-LTQ MS analysis of microsomal incubations containing LIM, NAC, NAL, and NADPH in the absence microsomes (A), or in presence of HLMs (B) or MLMs (C). (D) Extracted ion (m/z 632 → 614) chromatogram obtained from LC-LTQ MS analysis of synthetic M6. (E) MS/MS spectrum of M6 generated in microsomal incubations. (F) MS/MS spectrum of synthetic M6.

References

[1] M. Lv, P. Xu, Y. Tian et al., "Medicinal uses, phytochemistry and pharmacology of the genus *Dictamnus* (Rutaceae)," *Journal of Ethnopharmacology*, vol. 171, no. 2, pp. 247–263, 2015.

[2] J. S. Jang, E. G. Seo, C. Han et al., "Four cases of toxic liver injury associated with *Dictamnus dasycarpus*," *Korean Journal Hepatolpgy*, vol. 14, no. 2, pp. 206–212, 2008.

[3] M. Sheehan and D. Atherton, "One-year follow up of children treated with Chinese medicinal herbs for atopic eczema," *British Journal of Dermatology*, vol. 130, no. 4, pp. 488–493, 1994.

[4] J. Sun, X. Wang, P. Wang, L. Li, W. Qu, and J. Liang, "Antimicrobial, antioxidant and cytotoxic properties of essential oil from *Dictamnus angustifolius*," *Journal of Ethnopharmacology*, vol. 159, no. 15, pp. 296–300, 2015.

[5] J. B. Sun, W. Qu, P. Wang, F. H. Wu, L. Y. Wang, and J. Y. Liang, "Degraded limonoids and quinoline alkaloids from *Dictamnus angustifolius* G. Don ex Sweet. and their antiplatelet aggregation activity," *Fitoterapia*, vol. 90, pp. 209–213, 2013.

[6] J. B. Sun, N. Jiang, M. Y. Lv et al., "Limonoids from the root bark of *Dictamnus angustifolius*: potent neuroprotective agents with biometal chelation and halting copper redox cycling properties," *RSC Advances*, vol. 5, no. 31, pp. 24750–24757, 2015.

[7] J. Sun, N. Jiang, M. Lv et al., "Anstifolines A and B, two dimeric furoquinoline alkaloids from the root bark of *Dictamnus angustifolius*," *RSC Advances*, vol. 6, no. 27, pp. 22550–22554, 2016.

[8] J. B. Sun, P. Wang, J. Y. Liang, and L. Chen, "Phytochemical and chemotaxonomic study on *Dictamnus angustifolius* G. Don ex Sweet (Rutaceae)," *Biochemical Systematics and Ecology*, vol. 68, pp. 74–76, 2016.

[9] Y. Jiang, S. Li, H. Chang, Y. Wang, and P. Tu, "Pressurized liquid extraction followed by high-performance liquid chromatography for determination of seven active compounds in Cortex Dictamni," *Journal of Chromatography A*, vol. 1108, no. 2, pp. 268–272, 2006.

[10] H. Iwata, Y. Tezuka, S. Kadota, A. Hiratsuka, and T. Watabe, "Mechanism-based inactivation of human liver microsomal CYP3A4 by rutaecarpine and limonin from Evodia fruit extract," *Drug Metabolism and Pharmacokinetics*, vol. 20, no. 1, pp. 34–45, 2005.

[11] Y. L. Han, H. L. Yu, D. Li et al., "Inhibitory effects of limonin on six human cytochrome P450 enzymes and P-glycoprotein in vitro," *Toxicology in Vitro*, vol. 25, no. 8, pp. 1828–1833, 2011.

[12] L. A. Peterson, "Reactive metabolites in the biotransformation of molecules containing a furan ring," *Chemical Research in Toxicology*, vol. 26, no. 1, pp. 6–25, 2012.

[13] D. P. Williams, D. J. Antoine, P. J. Butler et al., "The metabolism and toxicity of furosemide in the Wistar rat and CD-1 mouse: a chemical and biochemical definition of the toxicophore," *Journal of Pharmacology and Experimental Therapeutics*, vol. 322, no. 3, pp. 1208–1220, 2007.

[14] J. C. Erve, S. C. Vashishtha, W. DeMaio, and R. E. Talaat, "Metabolism of prazosin in rat, dog, and human liver microsomes

and cryopreserved rat and human hepatocytes and characterization of metabolites by liquid chromatography/tandem mass spectrometry," *Drug Metabolism and Disposition*, vol. 35, no. 6, pp. 908–916, 2007.

[15] J. Loeper, V. Descatoire, P. Letteron et al., "Hepatotoxicity of germander in mice," *Gastroenterology*, vol. 106, no. 2, pp. 464–472, 1994.

[16] V. De Berardinis, C. Moulis, M. Maurice et al., "Human microsomal epoxide hydrolase is the target of germander-induced autoantibodies on the surface of human hepatocytes," *Molecular Pharmacology*, vol. 58, no. 3, pp. 542–551, 2000.

[17] A. Druckova and L. J. Marnett, "Characterization of the amino acid adducts of the enedial derivative of teucrin A," *Chemical Research in Toxicology*, vol. 19, no. 10, pp. 1330–1340, 2006.

[18] D. Lin, K. Wang, X. Guo, H. Gao, Y. Peng, and J. Zheng, "Lysine-and cysteine-based protein adductions derived from toxic metabolites of 8-epidiosbulbin E acetate," *Toxicology Letters*, vol. 264, no. 15, pp. 20–28, 2016.

[19] D. Lin, X. Guo, H. Gao et al., "In vitro and in vivo studies of the metabolic activation of 8-epidiosbulbin E acetate," *Chemical Research in Toxicology*, vol. 28, no. 9, pp. 1737–1746, 2015.

[20] D. Lin, W. Li, Y. Peng et al., "Role of metabolic activation in 8-epidiosbulbin E acetate-induced liver injury: mechanism of action of the hepatotoxic furanoid," *Chemical Research in Toxicology*, vol. 29, no. 3, pp. 359–366, 2016.

[21] D. Lin, C. Li, Y. Peng, H. Gao, and J. Zheng, "Cytochrome P450-mediated metabolic activation of diosbulbin B," *Drug Metabolism and Disposition*, vol. 42, no. 10, pp. 1727–1736, 2014.

[22] W. Li, D. Lin, H. Gao et al., "Metabolic activation of furan moiety makes Diosbulbin B hepatotoxic," *Archives of Toxicology*, vol. 90, no. 4, pp. 863–872, 2016.

14

Screening and Characterizing Tyrosinase Inhibitors from *Salvia miltiorrhiza* and *Carthamus tinctorius* by Spectrum-Effect Relationship Analysis and Molecular Docking

Ya-Li Wang,[1,2] Guang Hu ⓘ,[1] Qian Zhang,[2] Yu-Xiu Yang,[2] Qiao-Qiao Li,[2] Yuan-Jia Hu ⓘ,[3] Hua Chen,[2] and Feng-Qing Yang ⓘ[2]

[1]*School of Pharmacy and Bioengineering, Chongqing University of Technology, Chongqing 400054, China*
[2]*School of Chemistry and Chemical Engineering, Chongqing University, Chongqing 401331, China*
[3]*State Key Laboratory of Quality Research in Chinese Medicine, Institute of Chinese Medical Sciences, University of Macau, Macau*

Correspondence should be addressed to Guang Hu; foxhu8201@hotmail.com and Feng-Qing Yang; fengqingyang@cqu.edu.cn

Academic Editor: Jaroon Jakmunee

Tyrosinase (TYR) is a rate-limiting enzyme in the synthesis of melanin, while direct TYR inhibitors are a class of important clinical antimelanoma drugs. This study established a spectrum-effect relationship analysis method and high-performance liquid chromatography-mass spectrometry (LC-MS) analysis method to screen and identify the active ingredients that inhibited TYR in *Salvia miltiorrhiza–Carthamus tinctorius* (Danshen–Honghua, DH) herbal pair. Seventeen potential active compounds (peaks) in the extract of DH herbal pair were predicted, and thirteen of them were tentatively identified by LC-MS analysis. Furthermore, TYR inhibitory activities of five pure compounds obtained from the DH herbal pair were validated in the test in which kojic acid served as a positive control drug. Among them, three compounds including protocatechuic aldehyde, hydroxysafflor yellow A, and tanshinone IIA were verified to have high TYR inhibitory activity (IC50 value of 455, 498, and 1214 μM, resp.) and bind to the same amino acid residues in TYR catalytic pocket according to the results of the molecular docking test. However, the other two compounds lithospermic acid and salvianolic acid A had a weak effect on TYR, as they do not combine with the active amino acid residues or act on the active center of TYR. Therefore, the developed methods (spectrum-effect relationship analysis and molecular docking) could be used to effectively screen TYR inhibitors in complex mixtures such as natural products.

1. Introduction

Tyrosinase (TYR) belongs to the type 3 copper protein family containing dinuclear copper ions and widely exists in nature from microorganisms to humans [1]. It is a critical enzyme in the synthesis process of melanin pigments, catalyzing orthohydroxylation of monophenols to o-diphenols and then to the corresponding o-quinones [2]. However, overproduction of melanin pigments becomes a problem in the cosmetic and clinical points of view, such as melasma, freckles, and melanosis [3, 4]. Due to the importance of TYR during the synthesis of melanin, blocking the activity of TYR is one of the ideal strategies to treat melanin pigment diseases currently. To date, arbutin and kojic acid are the most commonly used tyrosinase inhibitors, which often serve as positive control drugs [5, 6]. However, traditional synthetic or microbial origin tyrosine inhibitors have some drawbacks, such as long-term contact of hydroquinone can lead to skin cancer, dermatitis, and other diseases [7, 8]. In reality, it is reported that the polyphenols and flavonoids isolated from natural plants have significant tyrosinase inhibition effect and less potential side effects [9, 10]. From this point of view, it is of great importance to screen TYR inhibitors from natural products.

Salviae Miltiorrhizae Radix et Rhizoma (Danshen in Chinese, DS) is the dried root or rhizome of *Salvia miltiorrhiza*

Bunge, in which the main components are phenolic acids and diterpenes [11]. *Carthami flos* (Honghua in Chinese, HH), the dried flower of *Carthamus tinctorius* L., is generally composed of flavonoids, fatty acids, volatile oils, and polysaccharides [12]. In previous reports, the inhibitory effect of *Salvia miltiorrhiza* and *Carthamus tinctorius* on tyrosinase has been validated [13–15]; however, the active constituents with tyrosinase inhibition activity have not been clearly reported yet. Therefore, in this study, a spectrum-effect analysis method is developed to screen the active constituents that inhibit tyrosinase in Danshen–Honghua (DH) herbal pair. The spectrum-effect relationship analysis combines the chemical compositions of the fingerprint of natural products with the results of the efficacy, and is originally used to develop control standards that can truly reflect the inherent quality of products [16]. Furthermore, spectrum-effect analysis is also used to screen the active components from natural products [17]. In reality, spectrum-effect analysis shows some positive features such as reliability, time-saving capacity, and simple operation [18, 19].

In this study, the inhibition effect of DH herbal pair and single drug on tyrosinase was compared first. Then, the components in the DH herbal pair are analyzed and identified by HPLC analysis. Third, the active components in DH herbal pair were predicted by spectrum-effect analysis, and their structures were identified by LC-MS analysis. Furthermore, the TYR inhibition activities of the predicted compounds were evaluated in an *in vitro* model. Finally, molecular docking, which is a method of drug design through the characterization of the receptor and the interaction between the receptor and the drug molecules, and binding mode and affinity prediction [20], was used to confirm the binding sites of compounds with tyrosinase and to predict several possible TYR inhibitors which possess similar structure to the screened active compounds by molecular docking.

2. Materials and Methods

2.1. Chemicals and Materials. Tyrosinase (MW 128 kDa) from *Agaricus bisporus*, kojic acid (≥98%), and L-tyrosine were purchased from Sigma-Aldrich (St. Louis, MO, USA). Tyrosinase, kojic acid, and L-tyrosine were dissolved in 50 mM sodium phosphate buffer (pH 6.8) before use. The reference compounds protocatechuic aldehyde, hydroxysafflor yellow A, tanshinone IIA, lithospermic acid, and salvianolic acid A (≥98%, determined by HPLC) were obtained from PUSH Bio-Technology Co., Ltd. (Chengdu, China). HPLC-grade acetonitrile and formic acid were obtained from Beijing InnoChem Science & Technology Co., Ltd. (Beijing, China). All of the experimental water was purified by water purification system (ATSelem 1820A, Antesheng Environmental Protection Equipment Co., Ltd., Chongqing, China). All other chemicals and solvents, unless otherwise specified, were guaranteed reagent grade and purchased from Sigma-Aldrich Chemical Co. LLC. (St. Louis, MO, USA).

Crude drugs of Danshen and Honghua were both purchased from Chongqing Heping Pharmacy Co., Ltd. (Chongqing, China), in June 2017. The voucher specimens of *Salvia miltiorrhiza* Bunge (number SM2017090101) and *Carthamus tinctorius* L. (number CF2017090101) were deposited at the Pharmaceutical Engineering Laboratory in School of Chemistry and Chemical Engineering, Chongqing University, Chongqing, China.

2.2. Preparation of DH Extracts and Stock Solutions. All the dried raw DS and HH were pulverized and griddled through 50 mesh sieves (about 0.29 mm) prior to extraction. Seven different proportions of the herbs were prepared with ratios of 1 : 1, 2 : 1, 3 : 1, 5 : 1, 1 : 5, 1 : 3, and 1 : 2 (g/g) DS to HH, respectively. 20 g of DS and HH mixed powder was extracted with 200 mL water in a glass-stoppered conical flask at 75°C for 1.5 h. After extraction, the mixture was filtered through gauze, and the residue was collected and extracted with the above process for a second time. The two filtrates were combined and evaporated in a rotary evaporator (ZFQ 85A, Shanghai Medical Instrument Special Factory, Shanghai, China) at 55°C under reducing pressure to remove the solvent. The extracts were further dried by lyophilization with freezing-drying system (DZF-6050, Shanghai Jing Hong Laboratory Instrument Co., Ltd., Shanghai, China) to obtain the DH extracts at a yield of about 25% (w/w, dried extract/crude herb). All pre- and postdilution solutions were stored at 4°C. Before HPLC analysis, the sample solutions were filtered through a 0.22 μm nylon membrane filter (Shanghai Titan Scientific Co., Ltd., Shanghai, China).

The reference substances protocatechuic aldehyde, hydroxysafflor yellow A, tanshinone IIA, lithospermic acid, and salvianolic acid A and a positive control (kojic acid) were all prepared by dissolving the respective substance in methanol solution and diluted with PBS (50 mM, pH 6.8) to the required concentrations for TYR inhibitory and binding assay, respectively. All the solutions were stored at 4°C in dark before use.

2.3. HPLC and LC-MS Analysis. HPLC analysis was performed on an Agilent 1260 series liquid chromatograph system (Agilent Technologies, Palo Alto, CA, USA), which was equipped with a vacuum degasser, a binary pump, an autosampler, and a diode array detector (DAD), controlled by Agilent ChemStation software. An Agilent Zorbax SB-Aq column (250 mm × 4.6 mm, 5 μm) preceded by a Zorbax SB-C$_{18}$ guard column (12.5 × 4.6 mm, 5 μm) was adopted for the analysis. The mobile phase consisted of solvent A (0.1% formic acid aqueous solution) and solvent B (acetonitrile) using a gradient elution, which was programmed as follows: 5% B at 0–2 min, 5%–15% B at 2–10 min, 15%–22% B at 10–24 min, 22%–29% B at 24–35 min, 29%–37% B at 35–42 min, and 37%–5% B at 42–47 min. The mobile phase was set at a flow rate of 1.0 mL/min with 10 μL per sample injection. The UV detection wavelength was set at 280 nm, and the column temperature was conditioned at 37°C.

Shimadzu LC/MS-MS 8060 electrospray ionization-mass spectrometer (ESI-MS), consisting of a Triple Quadruple Detector (TSQ) as the mass detector (Shimadzu, Kyoto, Japan) and coupled with HPLC, was used for LC-MS identification. The LC conditions were the same as described previously. The ESI-MS conditions were as follows: the ESI was used in both positive and negative mode; nitrogen gas was used for desolvation at a flow rate of 3 L/min at 250°C; the temperature and flow rate of drying gas were set under 400°C and 10 L/min, respectively; the cone voltage was (+) 20 and (−) 20 V; MS data were recorded in the full-scan mode (m/z 50–1500), and MS^2 data were recorded in the range of m/z 50–1200.

2.4. TYR Inhibitory Activity Assay. The enzyme assay was performed in 96-well Corning Costar plates (Corning Incorporated, USA). 50 μL test solution and 50 μL TYR solution (800 U/mL) were mixed and incubated for 10 min at room temperature. After the incubation, 100 μL (1 mg/mL) of L-tyrosine in PBS (pH 6.8) buffer as chromogenic substrate was added to start the reaction. The absorbance was monitored at 490 nm every 30 s for 10 min with an iMark™ Microplate Absorbance Reader (Bio-Rad Laboratories, Inc., USA). PBS (50 mM, pH 6.8) buffer was prepared as blank control, and kojic acid was used as positive control. TYR inhibition activity was expressed as the inhibitory percentage of TYR:

$$\text{Inhibition percentage } (\%) = \frac{(dA/dt)_{\text{blank}} - (dA/dt)_{\text{sample}}}{(dA/dt)_{\text{blank}}} \times 100\%, \tag{1}$$

where $(dA/dt)_{\text{blank}}$ and $(dA/dt)_{\text{sample}}$ are the reaction rate of the blank and sample group, respectively. All trials were independently performed in triplicates, and the results were shown with mean value of the triplicate observations.

2.5. Spectrum-Effect Relationship Analysis. Spectrum-effect analysis was performed by transferring DH fingerprint peak area and TYR inhibition activity test results into SPSS software for canonical correlation analysis (CCA). The optimized HPLC fingerprints of seven ratios DH samples were calculated and generated by professional software named "Similarity Evaluation System for Chromatographic Fingerprint of Traditional Chinese Medicine" composed by Chinese Pharmacopoeia Committee (Version 2012). CCA was used to assess the spectrum-effect relationships between the areas of 86 peaks in fingerprint and the TYR inhibition ratios.

2.6. In Silico Molecular Docking of TYR and Identified Active Compounds. Auto Dock 4.2 program (The Scripps Research Institute, La Jolla, CA, USA) was employed for *in silico* molecular docking study to validate the binding potency of the compounds to TYR [21]. The docking operation was performed according to the following steps. First, the crystal structure file of TYR (*Agaricus bisporus* mushroom tyrosinase) complex (PDB ID = 2y9x) was downloaded [22]. The dimension grid box (90 Å × 90 Å × 102 Å) and the grid spacing of 0.619 Å were defined to enclose the active site. Second, the ligand was deleted using UCSF Chimera, and

TABLE 1: Inhibition effect of DH extracts on tyrosinase with different ratios ($n = 9$).

Sample	Inhibition rate (%)
Blank[a]	0.06 ± 0.01
Kojic acid[b]	76.83 ± 0.24
DH 1:0[b]	26.55 ± 0.17
DH 0:1[b]	34.70 ± 0.98
DH 1:1[b]	42.19 ± 0.10
DH 2:1[b]	29.93 ± 0.32
DH 3:1[b]	40.53 ± 0.28
DH 5:1[b]	46.84 ± 0.30
DH 1:5[b]	16.15 ± 1.05
DH 1:3[b]	25.72 ± 0.17
DH 1:2[b]	38.90 ± 0.09

[a]Concentration: PBS (50 mM, pH6.8); [b]concentration: 500 μg/mL.

unnecessary water molecules were removed, and hydrogen atoms were added [23]. Third, the 3D chemical structure of investigated compounds was drawn by using Microsoft office 3D and output in PDB format with minimized energy.

With the aim of docking with Autodock Vina, the grid size was set to $(x, y, z) = (50, 50, 50)$ and the grid center was set to $(x, y, z) = (10.044, 28.706, 43.443)$. In each simulation process, progress with default parameters run from Autogrid and Autodock. Lamarckian genetic algorithm (LGA) was used to find the most favorable ligand binding orientations, and the number of LGA runs is equal to 50. The interaction figures were generated, and the results of docking were recorded with binding energies and bonded residues.

2.7. Statistical Analysis. All data are presented as mean ± standard deviations (SD) of at least three different experiments. The statistical analysis was performed with SPSS (version 24, SPSS, Inc., Chicago, IL, USA).

3. Results and Discussion

3.1. Effects of DH Extracts on Tyrosinase Activity. As shown in Table 1 and Figure 1, PBS served as a blank control, while kojic acid served as a positive control. Some DH herbal pair extracts (1:5 and 1:3) showed a weaker inhibitory effect than single herbal extracts when the concentrations of tyrosinase and the sample were kept constant. However, other DH herbal pair extracts (1:1, 3:1, 5:1, and 1:2) displayed a stronger inhibitory effect than single herbal extracts, which indicated that a synergistic effect of the herbal pair may occur on the inhibition of tyrosinase activity. Therefore, DH herbal pair was used as the research object for screening their tyrosinase inhibitors.

In order to obtain the best screening performance for active compounds in the complex matrix, some important parameters of the method including incubation time, TYR concentration, and sample concentration should be optimized. According to previous research, the incubation time might be controlled at the range of 30–120 min because a short incubation time (less than 30 min) might prevent the identification of target molecules which are not firmly bound to TYR, while a long incubation time (about 120

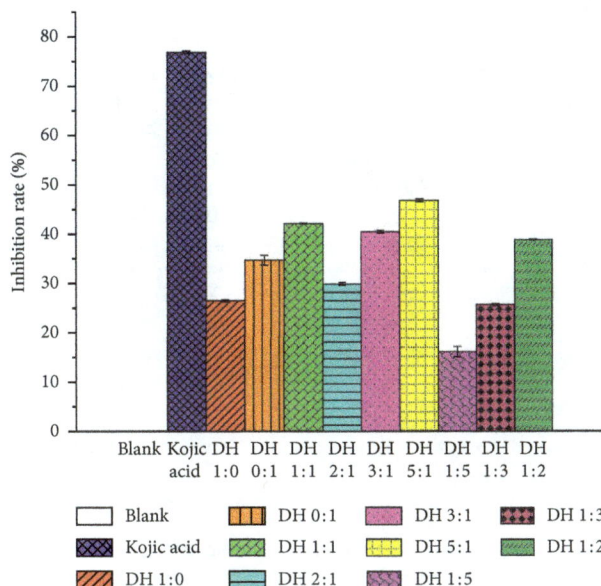

FIGURE 1: Different ratios of DH herbal pair on tyrosinase inhibition effects.

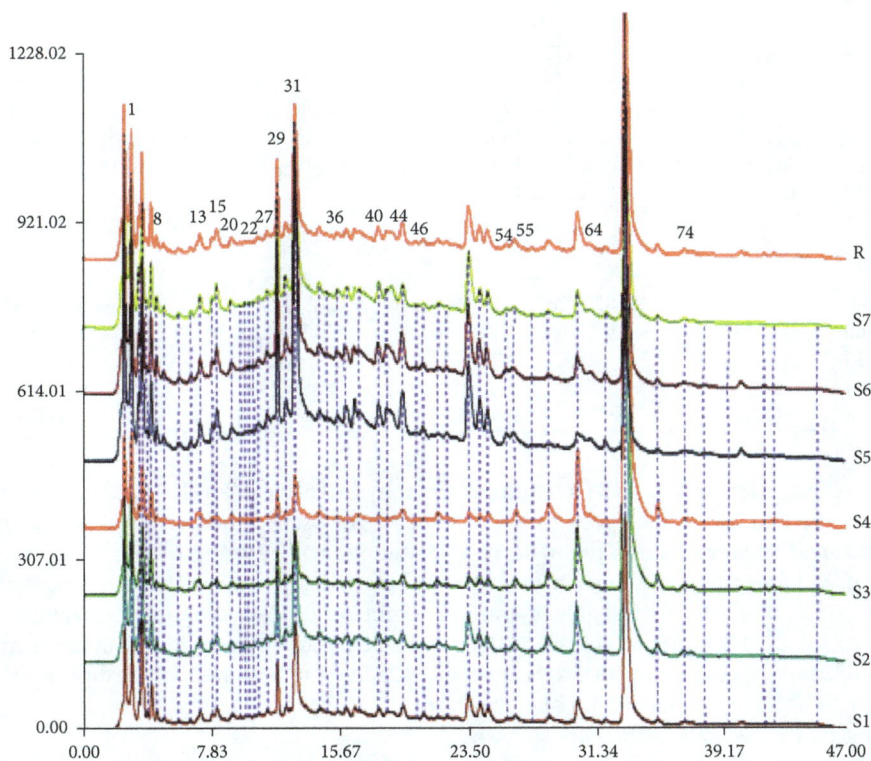

FIGURE 2: HPLC calibration fingerprints of DH herbal pair with different ratios. The chromatograms of S1–S7 are represented as follows: DH 1:1 (S1); DH 2:1 (S2); DH 3:1 (S3); DH 5:1 (S4); DH 1:5 (S5); DH 1:3 (S6); DH 1:2 (S7); control map (R).

minutes) would not make significant influence on the screening result [24, 25]. After investigation, a proper TYR concentration (800 U/mL) increased the sensitivity and number of bioactive constituents detected in the sample; meanwhile, the inhibition effect of tyrosinase is most stable by the incubation time of 40 min (data not shown). Therefore, a sufficient incubation time (40 min)

and a sufficient TYR concentration (800 U/mL) were used in this study.

3.2. Spectrum-Effect Relationship Analysis. The optimized HPLC fingerprints of DH samples with seven ratios are shown in Figure 2. A total of 86 peaks involved were detected in the

TABLE 2: Correlation coefficients between chromatogram peaks and inhibition rates.

Peak number	1	2	3	4	5	6	7	8	9	10	11	12	13
Inhibition rate	0.940*	−0.702	−0.695	−0.501	−0.753	−0.769	−0.677	0.812*	−0.721	−0.775	−0.790	−0.351	0.807*
Peak number	14	15	16	17	18	19	20	21	22	23	24	25	26
Inhibition rate	0.510	−0.878*	−0.742	−0.632	−0.713	0.714	0.919*	−0.746	0.903*	−0.129	−0.615	−0.233	−0.623
Peak number	27	28	29	30	31	32	33	34	35	36	37	38	39
Inhibition rate	0.952*	−0.544	0.908*	0.650	0.900*	−0.397	−0.015	−0.466	0.321	−0.824*	−0.795	−0.726	−0.762
Peak number	40	41	42	43	44	45	46	47	48	49	50	51	52
Inhibition rate	0.892*	−0.571	−0.790	−0.785	−0.867*	−0.745	0.891*	−0.640	−0.746	−0.708	0.135	−0.753	−0.451
Peak number	53	54	55	56	57	58	59	60	61	62	63	64	65
Inhibition rate	−0.627	−0.943*	−0.929*	−0.759	−0.778	−0.037	−0.769	0.591	−0.621	−0.741	0.786	0.824*	−0.725
Peak number	66	67	68	69	70	71	72	73	74	75	76	77	78
Inhibition rate	−0.667	−0.772	0.596	0.618	0.124	0.704	−0.712	0.654	0.887*	0.215	0.361	0.371	0.718
Peak number	79	80	81	82	83	84	85	86					
Inhibition rate	0.610	0.646	−0.733	0.565	−0.717	0.279	−0.123	−0.512					

Note. Pearson correlation, "r" represents the relevant strength; *$0.8 \leq |r| \leq 1$ means Very significant correlation.

TABLE 3: HPLC-MS/MS data of 17 predicted active compounds from DH herbal pair.

Peak No.	t_R (min)	MW	MS^1 (m/z)	MS^2 (m/z)	Formula	Structural identification
1	3.087	138	138.05	92; 78; 65	$C_7H_6O_3$	Protocatechuic aldehyde
8	3.112	198	197.05	178	$C_9H_{10}O_5$	Danshensu
13	6.491	165	166.09	120	$C_9H_{11}NO_2$	Phenylalanine
15	7.108	—	—	—	—	Unknown
20	7.952	294	295.15	277; 249	$C_{19}H_{18}O_3$	Tanshinone IIA
22	8.951	180	179.10	135	$C_9H_8O_4$	Caffeic acid
27	11.207	—	—	—	—	Unknown
29	12.386	788	787.40	625; 505; 463; 301	$C_{33}H_{40}O_{22}$	6-Hydroxykaempferol-3,6,7-O-β-D-glucoside
31	12.771	612	611.30	491; 473; 403; 353; 325; 283; 205	$C_{27}H_{32}O_{16}$	Hydroxysafflor yellow A
36	14.094	640	639.30	463; 362; 300; 255; 139	—	Unknown
40	17.091	772	773.35	695; 672; 303; 187; 112	$C_{33}H_{40}O_{21}$	6-Hydroxykaempferol 3-O-rutinoside-6-O-glucoside
44	21.352	—	—	—	—	Unknown
46	23.969	1044	1043.45	1025; 923; 863; 764; 593; 449	$C_{48}H_{52}O_{26}$	Anhydrosafflor yellow B
54	29.882	360	359.25	179; 161; 133	$C_{18}H_{16}O_8$	Rosmarinic acid
55	31.813	538	537.25	295; 253; 203	$C_{27}H_{22}O_{12}$	Lithospermic acid
64	32.528	494	493.25	295	$C_{26}H_{22}O_{10}$	Salvianolic acid A
74	36.747	718	717.35	673; 617; 519; 321	—	Salvianolic acid E

calculation of spectrum-effect relationship. CCA (canonical correlation analysis) was used to assess the spectrum-effect relationship between the areas of 86 peaks and the main parameters (inhibition rate), and the results are shown in Table 2. As suggested by the correlation coefficients, the highly relevant peaks were 1, 8, 13, 15, 20, 22, 27, 29, 31, 36, 40, 44, 46, 54, 55, 64, and 74 with Pearson relational grade more than 0.8. In other words, these 17 peaks might be the main active components of herbal pair for inhibiting tyrosinase, and further studies were performed to identify the structures of these peaks to confirm their bioactivities.

3.3. Identification of the Potential TYR-Targeted Compound by LC-MS Analysis.
HPLC-MS/MS analysis was used to identify the chemical structures of compounds in DH extracts. Based on the fragmentation behaviors, retention time and MS data (Table 3) of the peaks in the test samples, 13 compounds (protocatechuic aldehyde, danshensu, phenylalanine, tanshinone IIA, caffeic acid, 6-hydroxykaempferol-3,6,7-O-β-D-glucoside, hydroxysafflor yellow A, rosmarinic acid,

anhydrosafflor yellow B, 6-hydroxykaempferol 3-O-rutinoside-6-O-glucoside, lithospermic acid, salvianolic acid A, and salvianolic acid E) were tentatively identified, and the structures of these compounds are shown in Figure 3. Five pure reference compounds including protocatechuic aldehyde, tanshinone IIA, hydroxysafflor yellow A, lithospermic acid, and salvianolic acid A were obtained for further *in vitro* activity tests.

3.4. In Vitro Activity Tests for the Predicted Compounds.
To confirm the ability of the hit compounds with TYR inhibitory activity, *in vitro* enzymatic activity assays were performed. Five concentrations of each compound were tested, and the results are shown in Table 4. As a well-known TYR inhibitor [5], kojic acid showed strong inhibition effect with a IC50 value of 127 μM. From the results shown in Figure 4, among the five identified hit compounds, protocatechuic aldehyde, hydroxysafflor yellow A, and tanshinone IIA possessed strong TYR inhibition effects in a dose-dependent manner, with the IC50 values of 455, 498, and 1214 μM, respectively. However, lithospermic acid and salvianolic acid A did not show

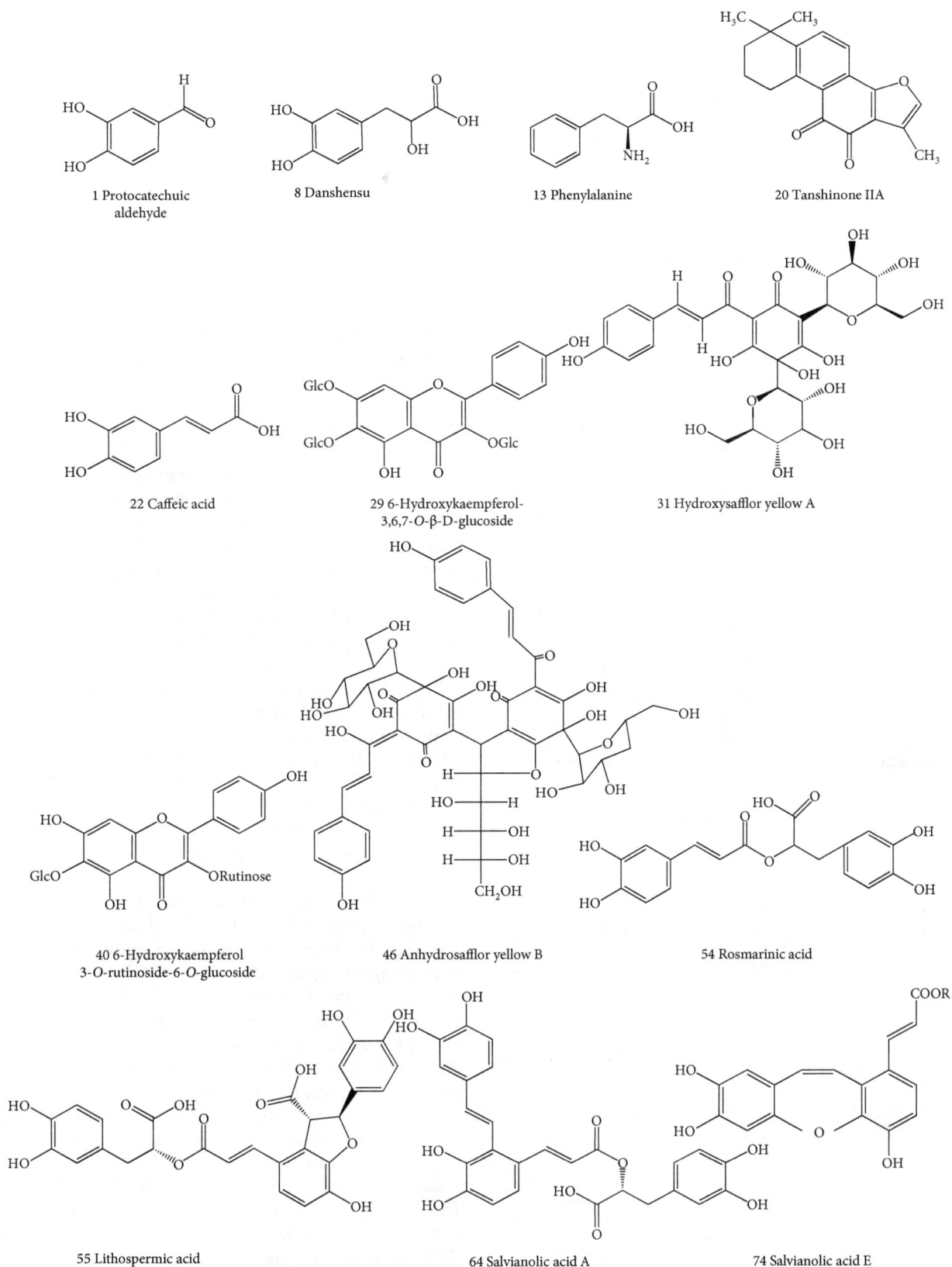

1 Protocatechuic aldehyde

8 Danshensu

13 Phenylalanine

20 Tanshinone IIA

22 Caffeic acid

29 6-Hydroxykaempferol-3,6,7-*O*-β-D-glucoside

31 Hydroxysafflor yellow A

40 6-Hydroxykaempferol 3-*O*-rutinoside-6-*O*-glucoside

46 Anhydrosafflor yellow B

54 Rosmarinic acid

55 Lithospermic acid

64 Salvianolic acid A

74 Salvianolic acid E

FIGURE 3: Chemical structures of compounds identified in DH herbal pair. The numbers of compounds are the same as the peak numbers in Figure 2.

TABLE 4: Inhibition effect of protocatechuic aldehyde, tanshinone IIA, hydroxysafflor yellow A, lithospermic acid, and salvianolic acid A on tyrosinase ($n = 9$).

Compounds	Concentration (mmol)	Inhibition rate (%)
Blank	PBS (50 mM, pH 6.8)	0.44 ± 0.02
Kojic acid	1.6	88.93 ± 0.28
	0.8	85.97 ± 0.45
	0.4	76.63 ± 1.02
	0.2	63.97 ± 2.11
	0.1	40.72 ± 0.15
Protocatechuic aldehyde	1.6	80.18 ± 0.08
	0.8	69.91 ± 2.97
	0.4	49.80 ± 1.30
	0.2	21.29 ± 1.01
	0.1	13.78 ± 0.95
Tanshinone IIA	1.6	57.88 ± 1.98
	0.8	37.26 ± 2.87
	0.4	18.23 ± 2.23
	0.2	6.03 ± 1.12
	0.1	1.98 ± 0.05
Hydroxysafflor yellow A	1.6	62.48 ± 1.34
	0.8	60.22 ± 2.28
	0.4	55.85 ± 1.80
	0.2	39.23 ± 1.12
	0.1	14.58 ± 0.55
Lithospermic acid	1.6	26.09 ± 3.28
	0.8	19.62 ± 2.22
	0.4	11.52 ± 2.64
	0.2	7.28 ± 1.01
	0.1	4.32 ± 0.21
Salvianolic acid A	1.6	17.74 ± 3.06
	0.8	14.42 ± 1.14
	0.4	8.25 ± 1.98
	0.2	5.04 ± 0.91
	0.1	1.25 ± 1.32

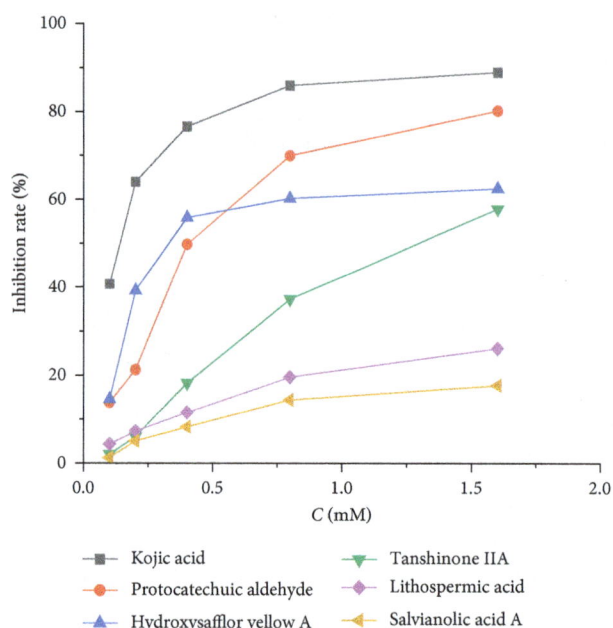

FIGURE 4: Inhibitory rate of active components to tyrosinase with different concentrations.

significant inhibitory effect on tyrosinase at a relatively high concentration (1.6 mM). The reason may be that they do not combine with the active amino acid residues or do not act on the active center of TYR, but further studies are required.

3.5. Molecular Docking of TYR and Identified Active Compounds. Molecular docking can be used to study the binding mechanism of compounds interacting with proteins. In this study, Autodock 4.2 was selected as the docking software to check out the active site of those components screened from DH extracts combined with TYR. The crystal structure of tyrosinase includes four identical parts, and one of them was used as the crystal structure of tyrosinase for computational docking analysis. Interestingly, as shown in Figures 5(a) and 5(b), protocatechuic aldehyde and hydroxysafflor yellow A could bond to the same catalytically active amino acid residues (THR324, ASN81, CYS83, GLU322, and HIS85) of TYR, which could explain the similar TYR inhibitory activity of these two compounds. Moreover, three hydrogen bonds (green line) between protocatechuic aldehyde, hydroxysafflor yellow A, and the amino acid residues were observed. As shown in Figure 5(c), tanshinone IIA was found bonding into the hydrophobic cavity of tyrosinase (blue region) and surrounded by amino acid residues VAL248, HIS244, OTR410, VAL283, SER282, PRO277, PHE264, and ARG268 of tyrosinase. The amino acid residues of the tyrosinase to which the active compound binds are shown in Table 5. The reason for the weak activity may be that lithospermic acid and salvianolic acid A do not combine with the active amino acid residues (such as GLU322 and THR324) or do not act on the active center (Cu^{2+}) of tyrosinase. As tyrosinase is a copper-containing enzyme, it is expected that potential tyrosinase inhibitors should show high binding affinity for copper ions [26].

Along with the screening result, molecular docking was also carried out on compounds (absence of pure reference substances) having similar structure to protocatechuic aldehyde, lithospermic acid, and salvianolic acid A (the structure of these compounds are shown in Figure 4), attempting to explore other active TYR inhibitors by perspective of structure-activity relationship. As shown in Figure 5(a), the important region of protocatechuic aldehyde for copper chelation was the catechol structure. Compounds that functionally chelate copper ions at tyrosinase active sites have been frequently reported as effective tyrosinase inhibitors because they are analogous to the phenolic hydroxyl substrates of tyrosinase [27, 28]. In addition, danshensu, caffeic acid, rosmarinic acid, and salvianolic acid E having a catechol structure might also be potential TYR inhibitors. The docking result is shown in Table 5. With a portion of the same active sites with screened active compounds bound to TYR, it could be found that danshensu, caffeic acid, rosmarinic acid, salvianolic acid E, anhydrosafflor yellow B, 6-hydroxykaempferol-3,6,7-O-β-D-glucoside, and 6-hydroxykaempferol 3-O-rutinoside-6-O-glucoside similar to protocatechuic aldehyde, hydroxysafflor yellow A, and tanshinone IIA might also be potential TYR inhibitors.

(a)

(b)

(c)

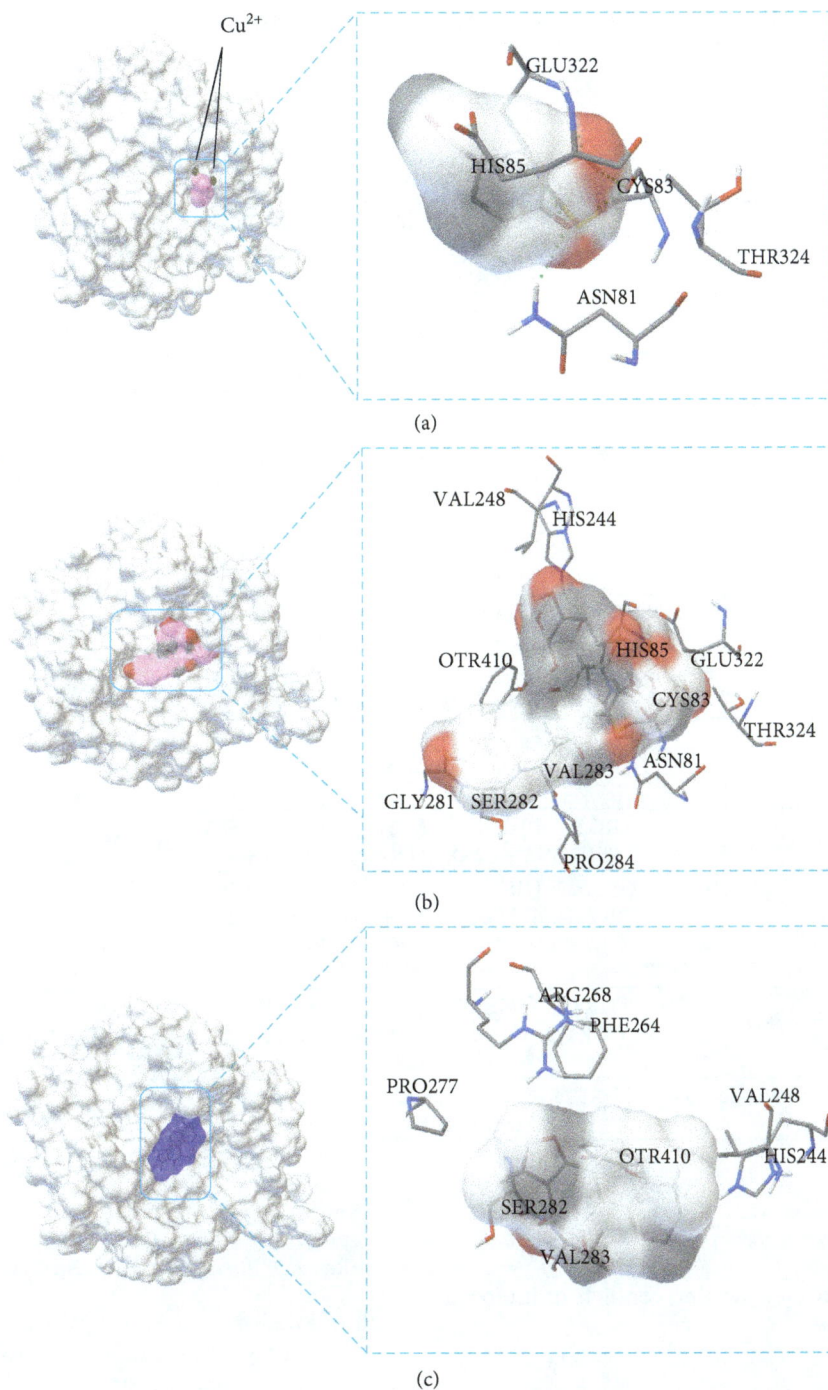

FIGURE 5: Molecular docking results of protocatechuic aldehyde (a), hydroxysafflor yellow A (b), and tanshinone IIA (c) with tyrosinase.

However, pure substances of these compounds were required for the further activity tests.

4. Conclusions

In this study, the TYR inhibitor screening methods were established, and the potential TYR inhibitory components from DH extract were screened. Combining the results of spectrum-effect analysis, LC-MS analysis, and enzymatic activity assay, three active compounds including proto-catechuic aldehyde, hydroxysafflor yellow A, and tanshinone IIA were discovered as inhibitors targeting TYR. Meanwhile, docking results showed that these compounds might bind to the same amino acid residues in TYR catalytic pocket. Additionally, other potential active TYR inhibitors such as danshensu, caffeic acid, rosmarinic acid, and salvianolic acid E, which gained similar structures with the hit compounds, might also be identified. These results proved that the proposed method could effectively screen TYR inhibitors in complex mixtures and provided a reference for the discovery of other active TYR inhibitors.

TABLE 5: Binding residues of identified compounds in DH herbal pair with tyrosinase.

Compound	Residues	Residues with hydrogen bonding
Kojic acid	ASN81, GLU322, CYS83, THR324, THR84, and HIS85	ASN81, CYS83, and HIS85
Protocatechuic aldehyde	THR324, ASN81, CYS83, GLU322, and HIS85	ASN81, CYS83, and HIS85
Hydroxysafflor yellow A	THR324, ASN81, CYS83, GLU322, HIS85, PRO284, VAL283, SER282, GLY281, OTR410, and HIS244	ASN81, CYS83, HIS85, and HIS244
Tanshinone IIA	VAL248, HIS244, OTR410, VAL283, SER282, PRO277, PHE264, and ARG268	VAL283 and SER282
Lithospermic acid	PRO270, THR261, GLY281, ARG268, PHE264, ASN260, MET257, SER282, VAL283, OTR410, VAL248, and HIS244	GLY281, ARG268, PHE264, SER282, and VAL283
Salvianolic acid A	ASN81, CYS83, HIS85, VAL283, SER282, OTR410, HIS244, VAL248, PHE264, and ARG268	CYS83, HIS85, VAL283, SER282, and HIS244
Caffeic acid	PHE264, ARG268, OTR410, GLY281, VAL283, and SER282	PHE264, ARG268, VAL283, and SER282
Danshensu	PHE264, ARG268, OTR410, GLY281, VAL283, and SER282	ARG268, VAL283, and SER282
Phenylalanine	HIS244, HIS85, THR84, CYS83, GLU322, VAL283, and ASN81	HIS85, THR84, and GLU322
Rosmarinic acid	ARG268, PHE264, VAL248, HIS244, GLU322, OTR410, HIS85, SER282, and VAL283	ARG268, GLU322, SER282, and VAL283
Salvianolic acid E	ARG268, PHE264, VAL248, HIS244, HIS85, SER282, VAL283, OTR410, and PRO284	ARG268, SER282, VAL283, and HIS244
Anhydrosafflor yellow B	GLU322, HIS244, HIS85, VAL248, VAL283, PRO284, ASN81, OTR410, ASN260, SER282, GLY281, PHE264, THR261, ARG268, and PRO277	GLY281 and ARG268
6-Hydroxykaempferol-3,6,7-O-β-D-glucoside	THR324, GLU322, ASN81, HIS85, HIS244, OTR410, VAL283, PRO284, GLY281, and SER282	GLU322, ASN81, HIS85, and HIS244
6-Hydroxykaempferol 3-O-rutinoside-6-O-glucoside	THR324, ASN81, HIS85, OTR410, VAL283, PRO284, GLY281, and SER282	ASN81, HIS85, and GLY281

Conflicts of Interest

The authors declared that they have no conflicts of interest.

Acknowledgments

This work was supported by the National Natural Science Foundation of China (81703687 and 21275169) and the Science and Technology Research Program of Chongqing Municipal Education Commission (Grant no. KJ1709219). The authors sincerely acknowledge the Science and Technology Development Fund of Macao SAR and the University of Macau for financial support through the projects FDCT013-2015-A1 and MYRG2016-00144-ICMS-QRCM for this research.

References

[1] K. U. Zaidi, A. S. Ali, S. A. Ali, and I. Naaz, "Microbial tyrosinases: promising enzymes for pharmaceutical, food bioprocessing, and environmental industry," *Biochemistry Research International*, vol. 2014, Article ID 854687, 16 pages, 2014.

[2] W. M. Chai, M. K. Wei, R. Wang, R. G. Deng, Z. R. Zou, and Y. Y. Peng, "Avocado proanthocyanidins as a source of tyrosinase inhibitors: structure characterization, inhibitory activity, and mechanism," *Journal of Agricultural and Food Chemistry*, vol. 63, no. 33, pp. 7381–7387, 2015.

[3] M. Amer and M. Metwalli, "Topical hydroquinone in the treatment of some hyperpigmentary disorders," *International Journal of Dermatology*, vol. 37, no. 6, pp. 449–450, 1998.

[4] N. T. Madan Mohan, A. Gowda, A. K. Jaiswal et al., "Assessment of efficacy, safety, and tolerability of 4-n-butylresorcinol 0.3% cream: an Indian multicentric study on melasma," *Clinical, Cosmetic and Investigational Dermatology*, vol. 9, pp. 21–27, 2016.

[5] B. K. Singh, S. H. Park, H. B. Lee et al., "Kojic acid peptide: a new compound with anti-tyrosinase potential," *Annals of Dermatology*, vol. 28, no. 5, pp. 555–561, 2016.

[6] J. S. Roh, J. Y. Han, J. H. Kim, and J. K. Hwang, "Inhibitory effects of active compounds isolated from safflower (*Carthamus tinctorius* L.) seeds for melanogenesis," *Biological and Pharmaceutical Bulletin*, vol. 27, no. 12, pp. 1976–1978, 2004.

[7] C. Y. Chen, L. C. Lin, W. F. Yang, J. Bordon, and H. M. Wang, "An updated organic classification of tyrosinase inhibitors on melanin biosynthesis," *Current Organic Chemistry*, vol. 19, no. 1, pp. 4–18, 2015.

[8] R. Sarkar, P. Arora, and K. V. Garg, "Cosmeceuticals for hyper-pigmentation: what is available?," *Journal of Cutaneous and Aesthetic Surgery*, vol. 6, no. 1, pp. 4–11, 2013.

[9] X. Hu, J. W. Wu, M. Wang et al., "2-Arylbenzofuran, flavonoid and tyrosinase inhibitory constituents of *Morus yunnanensis*," *Journal of Natural Products*, vol. 75, no. 1, pp. 82–87, 2012.

[10] J. J. Zhu, G. R. Yan, Z. J. Xu et al., "Inhibitory effects of (2′R)-2′,3′-dihydro-2′-(1-hydroxy-1-methylethyl)-2,6′-bibenzofuran-6,4′-diol on mushroom tyrosinase and melanogenesis in B16-F10 melanoma cells," *Phytotherapy Research*, vol. 29, no. 7, pp. 1040–1045, 2015.

[11] C. Qu, Z. J. Pu, G. S. Zhou et al., "Comparative analysis of main bio-active components in the herb pair Danshen-Honghua and its single herbs by ultra high performance liquid chromatography coupled to triple quadrupole tandem mass spectrometry," *Journal of Separation Science*, vol. 40, no. 17, pp. 3392–3401, 2017.

[12] L. N. Gao, Y. L. Cui, K. Yan, and C. Qiu, "Advances in studies on compatibility of Salviae Miltiorrhizae Radix et Rhizoma and Carthami flos," *Chinese Traditional and Herbal Drugs*, vol. 47, pp. 671–679, 2016.

[13] J. K. No, M. S. Kim, Y. J. Kim, S. J. Bae, J. S. Choi, and H. Y. Chung, "Inhibition of tyrosinase by protocatechuic aldehyde," *American Journal of Chinese Medicine*, vol. 32, no. 1, pp. 97–103, 2004.

[14] Y. S. Chen, S. M. Lee, C. C. Lin, C. Y. Liu, M. C. Wu, and W. L. Shi, "Kinetic study on the tyrosinase and melanin formation inhibitory activities of carthamus yellow isolated from *Carthamus tinctorius* L.," *Journal of Bioscience and Bioengineering*, vol. 115, no. 3, pp. 242–245, 2013.

[15] A. Gunia Krzyzak, J. Popiol, and H. Marona, "Melanogenesis inhibitors: strategies for searching for and evaluation of active compounds," *Current Medicinal Chemistry*, vol. 23, no. 31, pp. 3548–3574, 2016.

[16] C. S. Zhu, Z. J. Lin, M. L. Xiao, H. J. Niu, and B. Zhang, "The spectrum-effect relationship a rational approach to screening effective compounds, reflecting the internal quality of Chinese herbal medicine," *Chinese Journal of Natural Medicines*, vol. 14, no. 3, pp. 177–184, 2016.

[17] Y. Y. Hou, X. L. Cao, L. Y. Dong et al., "Bioactivity-based liquid chromatography-coupled electrospray ionization tandem ion trap/time of flight mass spectrometry for β2AR agonist identification in alkaloidal extract of *Alstonia scholaris*," *Journal of Chromatography A*, vol. 1227, pp. 203–209, 2012.

[18] Q. F. Zheng, Y. L. Zhao, J. B. Wang et al., "Spectrum-effect relationships between UPLC fingerprints and bioactivities of crude secondary roots of *Aconitum carmichaelii* Debeaux (Fuzi) and its three processed products on mitochondrial growth coupled with canonical correlation analysis," *Journal of Ethnopharmacology*, vol. 153, no. 3, pp. 615–623, 2014.

[19] L. M. Zhuo, J. J. Peng, Y. L. Zhao et al., "Screening bioactive quality control markers of QiShenYiQi dripping pills based on the relationship between the ultra-high performance liquid chromatography fingerprint and vascular protective activity," *Journal of Separation Science*, vol. 40, no. 20, pp. 4076–4084, 2017.

[20] M. A. Soares, M. A. Almeida, C. Marins Goulart, O. A. Chaves, A. Echevarria, and M. C. C. de Oliveira, "Thiosemicarbazones as inhibitors of tyrosinase enzyme," *Bioorganic and Medicinal Chemistry Letters*, vol. 27, no. 15, pp. 3546–3550, 2017.

[21] G. M. Morris, R. Huey, W. Lindstrom et al., "AutoDock4 and AutoDockTools4: automated docking with selective receptor flexibility," *Journal of Computational Chemistry*, vol. 30, no. 16, pp. 2785–2791, 2009.

[22] Y. J. Wang, G. W. Zhang, J. K. Yan, and D. M. Gong, "Inhibitory effect of morin on tyrosinase: insights from spectroscopic and molecular docking studies," *Food Chemistry*, vol. 163, pp. 226–233, 2014.

[23] D. W. Banner and P. Hadvary, "Crystallographic analysis at 3.0-a resolution of the binding to human thrombin of four active site-directed inhibitors," *Journal of Biological Chemistry*, vol. 266, pp. 20085–20093, 1991.

[24] J. Cao, J. J. Xu, X. G. Liu, S. L. Wang, and L. Q. Peng, "Screening of thrombin inhibitors from phenolic acids using enzyme-immobilized magnetic beads through direct covalent binding by ultrahigh-performance liquid chromatography coupled with quadrupole time-of-flight tandem mass spectrometry," *Journal of Chromatography A*, vol. 1468, pp. 86–94, 2016.

[25] X. X. Yang, F. Xu, D. Wang et al., "Development of a mitochondria-based centrifugal ultrafiltration/liquid chromatography/mass spectrometry method for screening mitochondria-targeted bioactive constituents from complex matrixes: herbal medicines as a case study," *Journal of Chromatography A*, vol. 1413, pp. 33–46, 2015.

[26] L. S. Chaves, M. C. C. de Barros, C. M. R. de Oliveira et al., "Biological interactions of fluorinated chalcones: stimulation of tyrosinase activity and binding to bovine serum albumin," *Journal of Fluorine Chemistry*, vol. 199, pp. 30–38, 2017.

[27] I. E. Orhan and M. T. H. Khan, "Flavonoid derivatives as potent tyrosinase inhibitors-a survey of recent findings between 2008 and 2013," *Current Topics in Medicinal Chemistry*, vol. 14, no. 12, pp. 1486–1493, 2014.

[28] Y. J. Kim and H. Uyama, "Tyrosinase inhibitors from natural and synthetic sources: structure, inhibition mechanism and perspective for the future," *Cellular and Molecular Life Sciences*, vol. 62, no. 15, pp. 1707–1723, 2005.

15

Pharmacokinetics of Eight Flavonoids in Rats Assayed by UPLC-MS/MS after Oral Administration of *Drynariae rhizoma* Extract

Zhan-Ling Xu (ID),[1] Ming-Yue Xu (ID),[2] Hai-Tao Wang (ID),[3] Qing-Xuan Xu,[4] Ming-Yang Liu (ID),[2] Chun-Peng Jia (ID),[2] Fang Geng (ID),[2] and Ning Zhang (ID)[1]

[1]Key Laboratory of Chinese Materia Medica, College of Pharmacy, College of Jiamusi,
 Heilongjiang University of Chinese Medicine, Harbin, Heilongjiang 150040, China
[2]Key Laboratory of Photochemistry Biomaterials and Energy Storage Materials of Heilongjiang Province,
 College of Chemistry & Chemical Engineering, Harbin Normal University, Harbin 150025, China
[3]Pharmacy Department, Harbin Hospital of Traditional Chinese Medicine, Harbin 150076, China
[4]Crop Academy of Heilongjiang University, Harbin 150080, China

Correspondence should be addressed to Fang Geng; gengfang1980@163.com and Ning Zhang; zhangning0454@163.com

Academic Editor: Antony C. Calokerinos

As a traditional Chinese medicine, *Drynariae rhizoma* (Kunze ex Mett.) J. Sm. has been used to treat osteoporosis and bone resorption for 2500 years. Based on the previous study and literature references, flavonoids were proved to be the most abundant and main active compounds of *Drynariae rhizoma* for osteoporosis treatment. In order to make good and rational use of *Drynariae rhizoma* in future, a rapid, sensitive, and selective ultraperformance liquid chromatography-mass spectrometry (UPLC-MS/MS) method was developed to investigate the pharmacokinetics of eight main flavonoids in rat plasma after oral administration of the *Drynariae rhizoma* extract, including neoeriocitrin, luteolin-7-*O*-β-D-glucoside, astragalin, naringin, eriodictyol, luteolin, naringenin, and kaempferol. Plasma samples' pretreatment involved a solid-phase extraction column. The separation was performed on an ACQUITY UPLC™ BEH C_{18} column with a gradient mobile-phase system of acetonitrile and 1% acetic acid in water. The detection was performed using a triple quadrupole tandem mass spectrometer equipped with an electrospray ionization interface (ESI) by multiple reaction monitoring (MRM) in the positive ion mode. All calibration curves exhibited good linearity ($r^2 > 0.9990$) over the measured ranges. The intraday and interday precisions (RSD) were within 13.87%, and the accuracy (RE) ranged from −14.57% to −0.25% at three quality control levels. Extraction recovery, matrix effect, and stability were satisfactory. The pharmacokinetic characteristics of the eight flavonoids of interest were clearly elucidated.

1. Introduction

The traditional Chinese medicine *Drynariae rhizoma* (Kunze ex Mett.) J. Sm., commonly known as Gu-Sui-Bu, is a fern plant widely distributed in southern China. *Drynariae rhizoma* is effective for the treatment of osteoporosis and bone resorption [1]. *Drynariae rhizoma* contains various types of chemical constituents, including flavonoids, triterpenes, phenolic acids, and their glycosides. But flavonoids and their glycosides are the most abundant constituents of *Drynariae rhizoma* [2, 3]. Furthermore,

flavonoids showed protective activities of osteoporosis, bone fractures, oxidative damage, and inflammation [4–10]. Total flavonoids in *Drynariae rhizoma* could activate the estrogen receptors and have the trend of replacing estrogen for clinical use [11, 12]. Flavonoids have been regarded as the principle constituents that contribute to the bioactivities of *Drynariae rhizoma*.

In the bioactive research of *Drynariae rhizoma*, the key issue is how to study the effective substance of *Drynariae rhizoma* to play a key role in osteoporosis. Wang suggested a conceptual framework for illuminating the absorbed bioactive compounds in herb medicines [13]. In general, absorbed bioactive compounds more possibly play a part in the therapeutic effect *in vivo* after oral administration. Thus, it is necessary to measure the absorbed bioactive compounds in plasma to understand the effective substances in the herb. In our pilot study, eight flavonoids were detected in plasma after oral administration of *Drynariae rhizoma* extract (DRE), including neoeriocitrin, luteolin-7-*O*-β-D-glucoside, astragalin, naringin, eriodictyol, luteolin, naringenin, and kaempferol (Figure 1), which could contribute to the therapeutic effect of the DRE.

As far as we know, no study of the pharmacokinetic characteristics of the eight flavonoids in rats after oral administration of the DRE has been reported. Up to now, only one or two flavonoids or their metabolites of the DRE were determined in rat plasma, which could not fully reflect the drug metabolism process in the body after oral administration of DRE to humans or a model animal [14]. Therefore, it is necessary to establish an appropriate analysis method to characterize the pharmacokinetics of DRE *in vivo*.

In the present study, a rapid, sensitive, and selective UPLC-MS/MS method was developed and validated for the simultaneous determination of eight flavonoids from the DRE in rat plasma. Plasma samples were pretreated with the C_{18} SPE column. The detection was performed using a triple quadrupole tandem mass spectrometer equipped with an electrospray ionization interface (ESI) by multiple reaction monitoring (MRM) in the positive ion mode. The newly described UPLC-MS/MS method was validated and successfully applied to pharmacokinetic study of eight flavonoids in rats after oral administration of DRE.

2. Materials and Methods

2.1. Materials and Reagents. *Drynariae rhizoma* naturally grown in Jiangxi Province, China, was purchased from Anguo Herb Market (Hubei, China). Neoeriocitrin, luteolin-7-*O*-β-D-glucoside, astragalin, naringin, eriodictyol, luteolin, naringenin, and kaempferol and the internal standard (IS, quercetin) were purchased from Bailingwei Technology Co., Ltd. (Beijing, China) with purities >98%. Methanol and acetonitrile of the HPLC grade were obtained from Dikma (Richmond Hill, NY, USA). Acetic acid (HPLC grade) was purchased from Scharlau Chemie S. A. (Barcelona, Spain). Water was purified by redistillation and a Milli-Q® ultrapure water system (Millipore, Bedford, MA, USA). Other chemicals and solvents used were of analytical grade.

2.2. Instruments and Analytical Conditions. Liquid chromatography analysis was performed on an ACQUITY UPLC^{TM} system (Waters Corp., Milford, MA, USA), which included a cooling autosampler, column oven, and two pumps. The chromatographic separation was performed using an ACQUITY UPLC^{TM} BEH C_{18} column (2.1 × 50 mm, 1.7 μm) at 35°C. The mobile phase consisted of 1%

acetic acid in water as the aqueous phase (A) and 100% acetonitrile as the organic phase (B). The gradient elution program was 0–3 min, 10–30% B; 3–4.5 min, 30–40% B; 4.5–6 min, 40–70% B; and 6-7 min, 70–90% B. The flow rate was 0.3 ml/min, the temperature of the autosampler was 4°C, and the sample injection volume was 5 μl.

Mass spectrometric detection was performed using a Waters® Micromass® Quattro Premier™ XE triple quadrupole tandem mass spectrometer equipped with an electrospray ionization interface in the positive ion mode. The ESI + source operation optimal parameters were capillary voltage 3.5 kV, source temperature 120°C, and desolvation temperature 300°C. Nitrogen was used as the desolvation and cone gas with flow rates of 600 and 30 L·h^{-1}, respectively. Argon was used as the collision gas at a pressure of approximately 2.61×10^{-3} mbar. Quantitative analysis was performed in multiple reaction monitoring (MRM) mode, and the parent ion, daughter ion, cone voltage, and collision energies of eight analytes and the IS were optimized. Table 1 shows product ion $[M + H]^+$ MS spectra of the eight flavonoids of interest and the IS and their fragmentation pathways. All data collected in the centroid mode were processed using MassLynx™ NT 4.1 software with the QuanLynx™ program (Waters Corp.). The structure of the eight analytes are displayed in Figure 1.

2.3. Preparation of Standard and Quality Control Samples. Stock solutions of the standards for eight flavonoids and quercetin (IS) solution were prepared in a 10 ml volumetric flask. Each standard stock solution of 100 μg/ml was prepared in methanol. The internal standard (quercetin) stock solution of 200 ng/ml was also dissolved with methanol. The plasma calibration standard solutions were prepared at concentrations in the range of 3.749–3749 ng/ml for neoeriocitrin, 1.856–4230 ng/ml for luteolin-7-*O*-β-D-glucoside, 1.317–1400 ng/ml for astragalin, 1.237–6370 ng/ml for naringin, 0.135–1040 ng/ml for eriodictyol, 2.742–3780 ng/ml for luteolin, 0.121–1210 ng/ml for naringenin, and 5.328–1209 ng/ml for kaempferol.

Three concentrations (low, middle, and high) of each analyte solution in drug-free plasma (40.4, 202, 2020 ng/ml for neoeriocitrin, 5.04, 50.4, 504 ng/ml for luteolin-7-*O*-β-D-glucoside, 5.1, 102, 1020 ng/ml for astragalin, 10.2, 408, 6120 ng/ml for naringin, 1.02, 40.8, 816 ng/ml for eriodictyol, 10.1, 40.4, 1010 ng/ml for luteolin, 5.2, 41.6, 208 ng/ml for naringenin, and 1.06, 42.4, 424 ng/ml for kaempferol) were used for quality control (QC) evaluation in UPLC-MS/MS analysis. All the solutions were stored at 4°C before use.

2.4. Preparation and Quality Assessment of DRE. The dried roots of *Drynariae rhizoma* (1 kg) were powdered and extracted three times under reflux in 15 times volume of methanol for every 2 hours. The solution was filtered and evaporated under reduced pressure in a Rotavapor R-3 rotary evaporator (Buchi Ltd., Labortechnik AG, Switzerland). Subsequently, the concentrated extract was dried in an oven, and the final weight of the DRE was 198.9 g with a yield

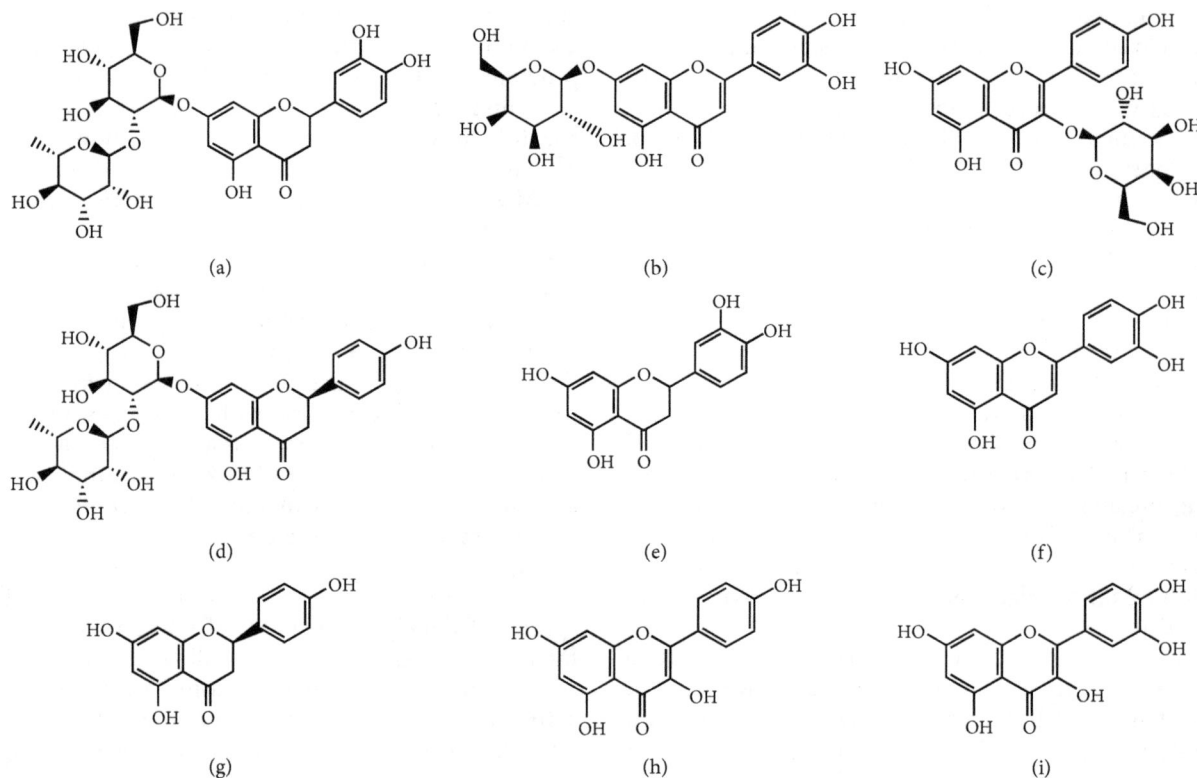

FIGURE 1: Chemical structures of eight analytes and IS. (a) Neoeriocitrin. (b) Luteolin-7-O-β-D-glucoside. (c) Astragalin. (d) Naringin. (e) Eriodictyol. (f) Luteolin. (g) Naringenin. (h) Kaempferol. (i) Quercetin.

TABLE 1: Precursor ion and product ion transition and parameters of the analytes used in this study.

Analytes	Retention time (min)	Precursor ion species	MRM transition		Cone (V)	Collision (V)	Dwell time (ms)
			Precursor ion	Product ion			
Neoeriocitrin	2.12	$[M + H]^+$	597.5	289.2	25	30	0.2
Luteolin-7-O-β-D-glucoside	2.33	$[M + H]^+$	449.1	287.1	20	20	0.2
Astragalin	2.62	$[M + H]^+$	449.1	287.1	30	30	0.2
Naringin	2.73	$[M + H]^+$	581.5	273.2	20	25	0.2
Eriodictyol	3.34	$[M + H]^+$	289.2	153.1	30	25	0.2
Luteolin	3.63	$[M + H]^+$	287.1	153.1	30	30	0.2
Naringenin	4.29	$[M + H]^+$	273.2	153.1	20	30	0.2
Kaempferol	4.41	$[M + H]^+$	287.1	153.1	30	30	0.2
Quercetin (IS)	3.42	$[M + H]^+$	303.2	153.1	20	25	0.2

of 19.9%. The contents of the main flavonoid constituents in DRE were quantitatively determined with the method described above, with the results of neoeriocitrin (5.1 mg/g), luteolin-7-O-β-D-glucoside (1.2 mg/g), astragalin (0.96 mg/g), naringin (8.5 mg/g), eriodictyol (0.66 mg/g), luteolin (0.11 mg/g), naringenin (2.71 mg/g), and kaempferol (0.14 mg/g), respectively.

2.5. Preparation of Plasma. Aliquots of plasma samples (400 μl) were placed in a 10 ml centrifuge tube; 50 μl of internal standard solution (200 ng/ml) and 50 μl acetic acid were added. The mixture was vortexed for 1.0 min and then loaded onto an activated SPE C_{18} column (the

SPE C_{18} column was activated by 3 ml of methanol before loading the sample, and the excess was washed off with 5 ml of purified water). After rinsing with 2 ml purified water, the SPE column was eluted with 4.5 ml of methanol and the eluate was evaporated to dryness under a gentle stream of nitrogen. The residue was reconstituted in 140 μl methanol and centrifuged at 16000 × g for 15 min; then, a 5 μl aliquot was injected into the UPLC-MS/MS system for analysis.

2.6. Method Validation. The method was validated in terms of specificity, linearity, lower limit of detection (LLOD), lower limit of quantification (LLOQ), accuracy and precision, extraction recovery, stability, and matrix effects based on the method validation procedure in the previous work [15].

2.6.1. Specificity.
We compared the chromatograms of blank plasma from six individual rats with those of corresponding plasma samples spiked with eight mixed standard samples and IS and plasma samples after oral administration of DRE.

2.6.2. Linearity and Quantification.
Calibration curves for the eight standards were constructed by plotting the peak area ratios of each analyte to that of the IS versus the corresponding concentration. The lower limit of detection (LLOD) was defined as the lowest concentration with a signal-to-noise ratio of $3:1$. The lower limit of quantification (LLOQ) was determined as the lowest concentration of the calibration curve with a signal-to-noise ratio of $10:1$.

2.6.3. Precision and Accuracy.
Intraday precision and interday precision and accuracy were assessed for low-, middle-, and high-concentration QC samples in six replicates on the same day and once a day for three consecutive days, respectively. Each tested sample was related to the calibration curve. The precision was calculated as the relative standard deviation (RSD%) and the accuracy as relative error (RE%). These results indicated that the precision and accuracy of the method were within acceptable limits. The intraday precision and interday precision and accuracy values for lowest acceptable reproducibility concentrations were denied as being within ±15%, and the precision and accuracy were within 80%–120%.

2.6.4. Recovery and Matrix Effect.
The recovery and matrix effects of analytes were determined for low-, middle-, and high-concentration QC samples with six replicates. Extraction recovery was determined at the three QC levels by comparing the peak areas of analytes between plasma samples spiked with analytes before and after extraction. Matrix effects were evaluated at the three QC levels by comparing the peak areas of analytes obtained from plasma samples spiked with analytes after extraction to those of pure standard solutions at the same concentrations; the acceptable range was 80–120%.

2.6.5. Stability.
The stability of determination was assessed by analysis of low-, middle-, and high-concentration QC samples with three replicates. The short-term stability was assessed by analyzing the QC samples kept in the autosampler (4°C) for 36 h. To assess long-term stability, the QC samples were stored at −20°C for 30 days. Freeze-thaw stability was evaluated by subjecting the QC plasma samples to three complete freeze/thaw cycles from −20°C to room temperature.

2.7. Pharmacokinetic Study of DRE

2.7.1. Animals.
Female Sprague Dawley rats (280–350 g) were supplied by the Animal Safety Evaluation Center of Heilongjiang University of Chinese Medicine (Harbin, Heilongjiang). All protocols for animal experiments were approved in accordance with the Regulations of Experimental Animal Administration issued by the State Commission of Science and Technology of the People's Republic of China. The rats were housed at 24 ± 2°C, and relative humidity was 60 ± 5%, with a 12 h-12 h light-dark cycle. Water and food were supplied freely.

2.7.2. Drug Administration and Sampling.
The UPLC-MS/MS method was successfully applied in a pharmacokinetic study of eight flavonoids in rats after oral administration of the DRE. The dosing solutions were freshly prepared DRE administrated via an oral gavage to the rats at a single dose of 4 g/kg. The animals had free access to water during the experiment. A series of blood samples were collected in 1.5 ml heparinized polythene tubes from the suborbital venous lexus of each rat at 0.08, 0.33, 0.5, 0.67, 1, 2, 4, 6, 8, 12, and 24 h after administration. The blood samples were centrifuged at 4500 × g for 10 min, and the supernatants were collected immediately and stored at −20°C until analysis.

2.7.3. Data Analysis.
Pharmacokinetic parameters were determined using a noncompartmental model and analyzed using pharmacokinetic software WinNonlin Standard Edition, version 1.1. Data are shown as the mean ± standard deviation (SD) for each parameter.

3. Results and Discussion

3.1. Optimization of LC-MS Conditions

3.1.1. Optimization of Mass Spectrometric Conditions.
To optimize the mass spectrometry conditions, appropriate mixed standard solutions of neoeriocitrin, luteolin-7-O-β-D-glucoside, astragalin, naringin, eriodictyol, luteolin, naringenin, kaempferol, and quercetin were monitored across a full scan in both positive and negative modes. The signal intensity and fragment stability in the positive mode were better than those in the negative mode for all the analytes. Therefore, the positive mode was used for analysis. The parent ion and daughter ion for each compound were obtained (Figure 1), and the cone voltage and collision energy for the eight analytes and the IS (quercetin) were optimized. The molecular weight, parent ion, daughter ion, cone voltage, collision energy, and retention time of the eight flavonoids and the IS are shown in Table 1.

3.1.2. Chromatographic Conditions.
To achieve symmetric peak shape and short running time for the simultaneous analysis of the eight analytes, we tested various mobile phase conditions to achieve good separation of the analytes. The

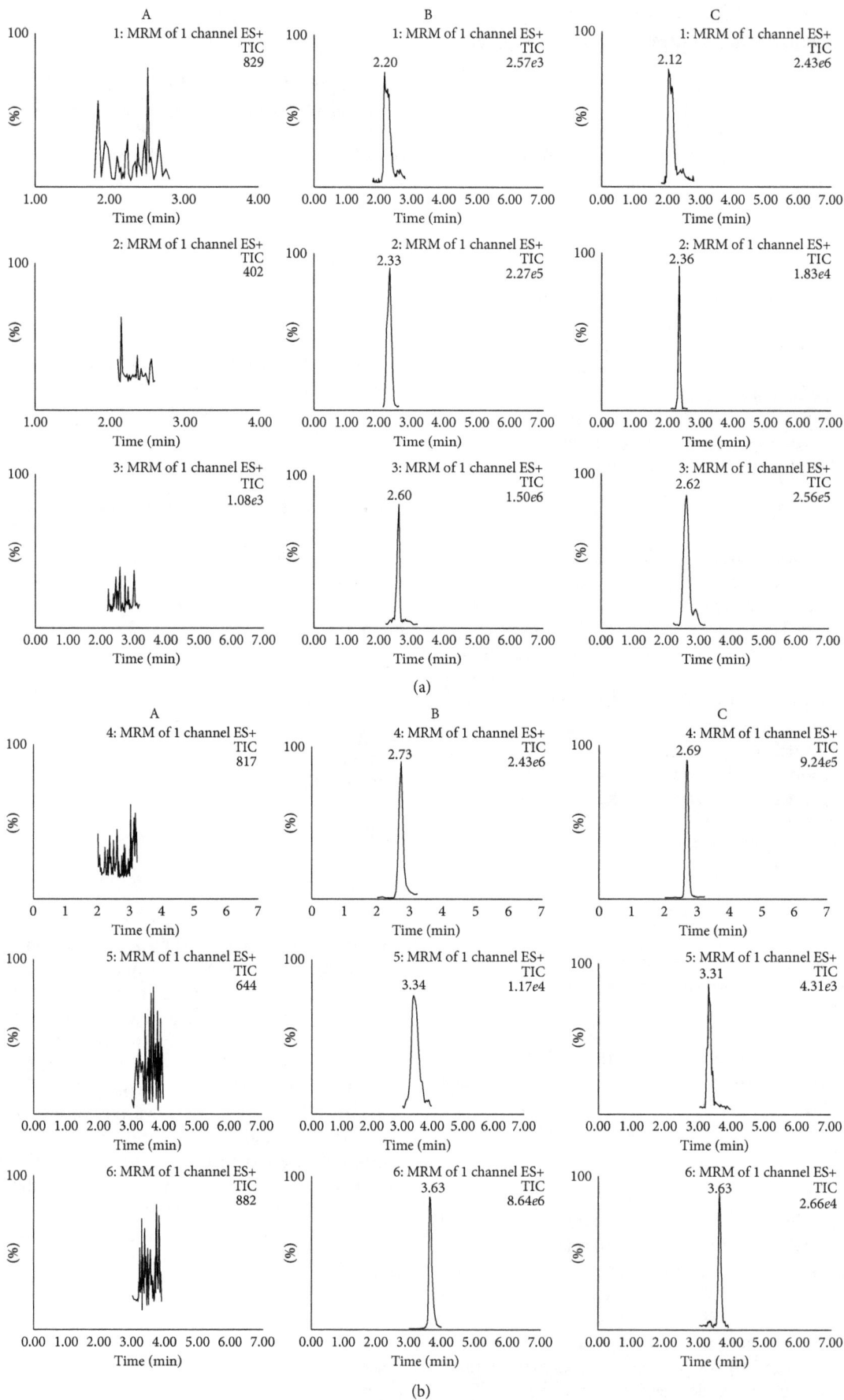

(a)

(b)

FIGURE 2: Continued.

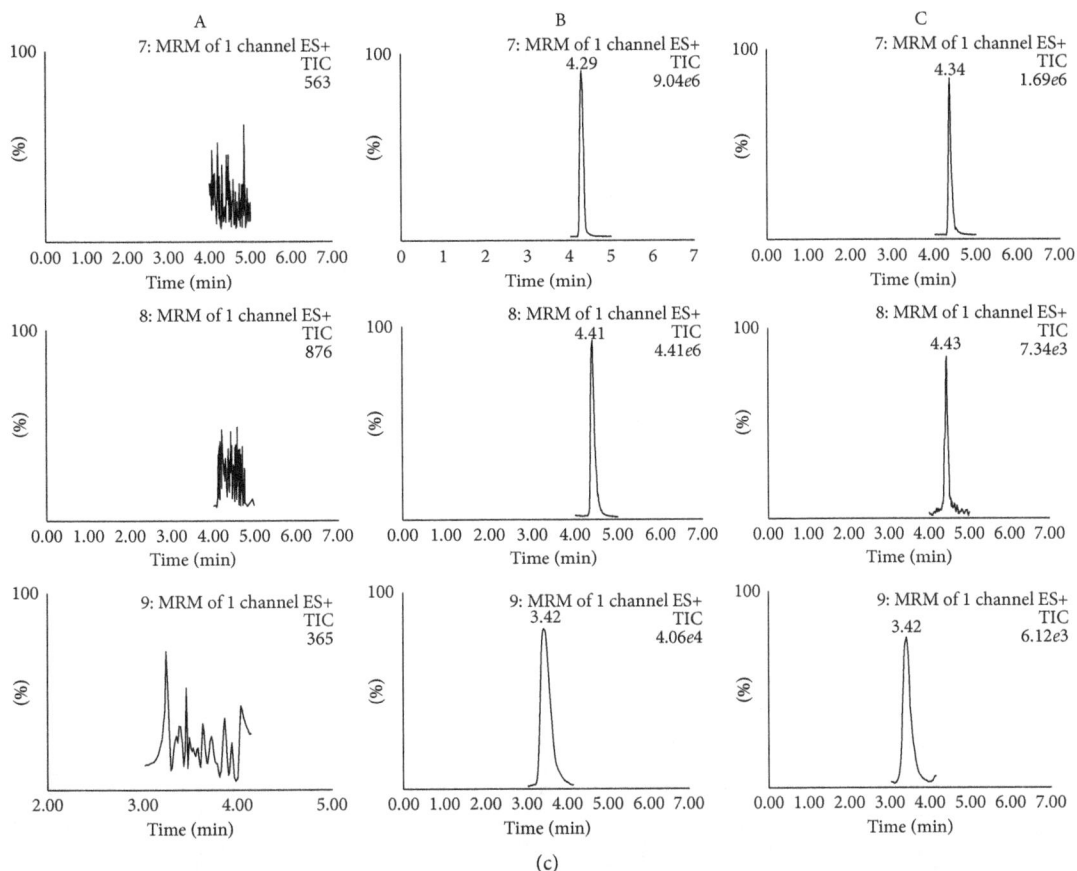

FIGURE 2: Representative MRM chromatograms of eight analytes and quercetin (IS). (a) Blank plasma. (b) Blank plasma spiked with the eight analytes and IS. (c) Plasma sample 1 h after oral administration of DRE (4 g/kg) (mean ± SD, $n = 6$).

mobile phase we finally optimized as 1% acetic acid-water as the aqueous phase (A) and 100% acetonitrile as the organic phase (B). In the optimized UPLC-MS/MS conditions for simultaneous determination of the eight compounds, all analytes were eluted rapidly with 5.0 min.

3.2. Method Validation

3.2.1. Specificity. The retention times of neoeriocitrin, luteolin-7-O-β-D-glucoside, astragalin, naringin, eriodictyol, luteolin, naringenin, kaempferol, and IS were approximately 2.20 min, 2.36 min, 2.60 min, 2.69 min, 3.31 min, 3.63 min, 4.34 min, 4.43 min, and 3.42 min, respectively. The representative chromatograms of blank plasma, blank plasma spiked with reference standards and IS, and plasma obtained after the oral administration of DRE are shown in Figure 2. Under the established optimal chromatographic conditions, no significant interfering peaks were observed at the analyte elution times, and no interference occurred between IS and the eight analytes.

3.2.2. Linearity and Calibration Curve. All calibration curves exhibited good linearity ($r^2 \geq 0.9990$) over the measured ranges. The calibration curves were linear over the concentrations in the range of 3.75–3749 ng/ml for neoeriocitrin, 1.86–4230 ng/ml for luteolin-7-O-β-D-glucoside,

1.32–1400 ng/ml for astragalin, 1.24–6370 ng/ml for naringin, 0.14–1040 ng/ml for eriodictyol, 2.74–3780 ng/ml for luteolin, 0.12–1210 ng/ml for naringenin, and 5.33–1209 ng/ml for kaempferol, respectively. The linearity regression equation, correlation coefficients, and linear ranges of the eight analytes are shown in Table 2.

3.2.3. Precision and Accuracy. The precision and accuracy of the UPLC-MS/MS method (Table 3) were within acceptable limits. The intraday precision and interday precision (R.S.D.) were within 13.87%, and the accuracy (RE) ranged from −14.79% to −0.25% at three quality control levels.

3.2.4. Recovery and Matrix Effect. The matrix effect and recovery results (Table 4) indicated that no endogenous substances from plasma significantly influenced the ionization of analytes. The matrix effects were within an acceptable range (80.23%–98.99%), and the mean extraction recoveries of the eight analytes and IS were greater than 89.45%.

3.2.5. Stability. The results of stability experiments (Table 5) showed that no significant degradation occurred. The concentrations of the eight analytes measured in the stability study were within −5.66%–3.56%. The data indicated

TABLE 2: Regression data, LLOQ, and LLOD of eight analytes.

Analytes	Calibration curves	Correlation coefficient	Linear range (ng/ml)	LLOQ (ng/ml)	LLOD (ng/ml)
Neoeriocitrin	$y = 0.132x + 11.60$	0.9996	3.75–3749	3.75	0.94
Luteolin-7-O-β-D-glucoside	$y = 12.48x + 339.3$	0.9994	1.86–4230	1.86	0.62
Astragalin	$y = 28.87x + 287.0$	0.9992	1.32–1400	1.32	0.33
Naringin	$y = 6.032x + 635.9$	0.9991	1.24–6370	1.24	0.41
Eriodictyol	$y = 35.71x + 258.9$	0.9993	0.14–1040	0.14	0.05
Luteolin	$y = 12.21x + 434.0$	0.9990	2.74–3780	2.74	0.91
Naringenin	$y = 6.269x + 128.9$	0.9991	0.12–1210	0.12	0.3
Kaempferol	$y = 7.25x - 79.66$	0.9991	5.33–1209	5.33	1.78

TABLE 3: Precision and accuracy for the determination of eight analytes ($n = 6$).

Analytes	Concentration spiked (ng/mL)	Intraday			Interday		
		Concentration measured (ng/mL)	Precision (RSD%)	Accuracy (RE%)	Concentration measured (ng/mL)	Precision (RSD%)	Accuracy (RE%)
Neoeriocitrin	40.4	37.31 ± 1.16	3.11	−7.65	36.86 ± 1.28	3.47	−8.76
	202	180.26 ± 10.01	5.55	−13.32	176.01 ± 8.92	5.06	−10.38
	2020	1859.18 ± 76.89	4.14	−7.96	1831.64 ± 65.66	3.58	−9.32
Luteolin-7-O-β-D-glucoside	5.04	4.12 ± 0.29	7.15	−13.32	4.21 ± 0.38	9.09	−11.44
	50.4	42.55 ± 5.9	13.87	−14.57	45.24 ± 2.69	6.19	−10.23
	504	470.31 ± 13.13	2.79	−6.68	473.18 ± 9.42	1.99	−6.11
Astragalin	5.1	4.12 ± 0.47	11.5	−12.24	4.19 ± 0.4	9.54	−12.94
	102	87.59 ± 8.87	10.13	−14.12	86.62 ± 8.86	10.22	−10.07
	1020	878.88 ± 12.3	1.39	−13.83	840.07 ± 34.9	4.22	−7.64
Naringin	10.2	9.96 ± 0.9	9.01	−2.35	9.53 ± 0.84	8.81	−6.59
	408	400.75 ± 19.57	4.88	−0.8	392.72 ± 12.22	3.09	−2.79
	6120	5989.39 ± 147.05	2.46	−0.37	5935.48 ± 127.9	2.15	−1.27
Eriodictyol	1.02	0.87 ± 0.08	9.46	−13.82	0.87 ± 0.07	8.42	−12.61
	40.8	40.1 ± 4.11	10.25	−0.25	39.63 ± 2.65	6.66	−1.41
	816	728.55 ± 55.67	7.64	−13.26	767.88 ± 35.79	4.74	−8.58
Luteolin	10.1	8.76 ± 0.81	9.29	−13.23	8.69 ± 0.6	6.86	−13.93
	40.4	40.75 ± 3.26	8.00	0.85	39.86 ± 4.24	10.66	−1.34
	1010	925.16 ± 53.96	5.83	−8.39	904.44 ± 65.45	7.26	−10.45
Naringenin	5.2	4.43 ± 0.48	10.72	−14.74	4.33 ± 0.46	10.55	−14.79
	41.6	37.35 ± 2.10	5.63	−10.21	37.93 ± 1.79	4.75	−8.81
	208	181.93 ± 5.97	3.28	−12.53	181.81 ± 8.24	4.54	−12.58
Kaempferol	1.06	0.9 ± 0.08	8.97	−14.25	0.92 ± 0.1	11.06	−13.67
	42.4	37.2 ± 3.06	8.23	−12.26	38.25 ± 2.3	6.06	−9.77
	424	382.89 ± 12.01	3.14	−9.69	376.99 ± 12.36	3.28	−11.08

that all analytes in the rat plasma were stable after storage at −20°C for 30 days, after three freeze/thaw cycles, and after storage in the autosampler (4°C) for 36 h.

3.3. Pharmacokinetic Studies.

The UPLC-MS/MS method was successfully applied in a pharmacokinetic study of eight flavonoids after oral administration of DRE at a dose of 4 g/kg body weight. Mean plasma concentration-time plots are shown in Figure 3. The major pharmacokinetic parameters are listed in Table 6.

After oral administration to rats, all the analytes were absorbed from the gastrointestinal tract and detected in plasma at 5 min. However, the two highest abundant flavonoids in the DRE, naringin and neoeriocitrin, showed high maximum plasma concentration (C_{max}) (4414.18 ± 360.38 ng/ml and 1490.98 ± 124.54 ng/ml) and high area under the plasma concentration-time curve from 0 h to infinity AUC (0–∞) (8760.77 ± 347.83 and 3340.34 ± 237.36). Neoeriocitrin still exhibited a shortest T_{max} to reach the maximum drug concentration (T_{max} = 0.33 h). In the plasma concentration time profile of naringin, another small peak could be seen at 8 h (Figure 3), which is in agreement with the literature where pure naringin was orally administrated to rats [16]. And this phenomenon may be due to the enterohepatic circulation of naringin in rats, which was also reported for other glycosides.

Naringenin, the aglycone of naringin, was absorbed into blood with T_{max} at 6 h and eliminated with $T_{1/2}$ at 3.2 h after oral administration of DRE, which was significantly different from the other compounds (T_{max} < 2 h). The slow elimination may be due to the fact that orally administrated naringin can be metabolized into naringenin and naringenin glucuronide [17]. Pharmacokinetic characters of naringenin were very close to a previous report [14]. The plasma concentrations of luteolin-7-O-β-D-glucoside,

TABLE 4: Extraction recovery and matrix effect of eight analytes (mean ± SD, $n = 6$).

Analytes	Concentration spiked (ng/mL)	Concentration measured (ng/mL)	Extraction recovery (%)		Matrix effect (%)	
			Mean (%)	RSD (%)	Mean (%)	RSD (%)
Neoeriocitrin	40.4	36.64 ± 1.34	90.78 ± 0.99	2.70	90.59 ± 1.69	4.61
	202	173.87 ± 8.38	84.22 ± 7.95	4.54	82.96 ± 8.80	5.10
	2020	1817.87 ± 60.05	91.00 ± 50.84	2.77	88.98 ± 69.26	3.85
Luteolin-7-O-β-D-glucoside	5.04	4.26 ± 0.43	82.90 ± 0.54	13.03	86.08 ± 0.31	7.08
	50.4	46.59 ± 1.09	91.26 ± 1.47	3.20	93.62 ± 0.71	1.50
	504	474.61 ± 7.57	93.81 ± 6.30	1.33	94.52 ± 8.83	1.85
Astragalin	5.1	4.22 ± 0.36	80.29 ± 0.29	6.99	85.13 ± 0.44	10.13
	102	86.14 ± 8.85	84.96 ± 11.27	13.01	83.94 ± 6.43	7.51
	1020	820.66 ± 46.2	80.23 ± 54.57	6.67	80.68 ± 37.82	4.60
Naringin	10.2	9.31 ± 0.81	94.85 ± 0.75	7.83	87.73 ± 0.86	9.59
	408	388.71 ± 8.55	95.80 ± 7.54	1.93	96.63 ± 9.64	2.47
	6120	5908.53 ± 118.32	98.51 ± 118.35	2.00	98.05 ± 118.30	2.01
Eriodictyol	1.02	0.86 ± 0.07	82.54 ± 0.06	6.87	81.43 ± 0.08	8.93
	40.8	39.4 ± 1.92	98.99 ± 2.27	5.70	97.02 ± 1.57	4.03
	816	787.55 ± 25.85	93.34 ± 27.70	3.53	94.17 ± 24.00	3.03
Luteolin	10.1	8.66 ± 0.49	86.12 ± 0.54	6.20	85.30 ± 0.44	5.09
	40.4	39.41 ± 4.73	97.24 ± 4.65	11.85	97.87 ± 4.80	12.15
	1010	894.08 ± 71.2	89.43 ± 63.90	7.07	87.62 ± 78.49	8.87
Naringenin	5.2	4.27 ± 0.45	81.60 ± 0.52	12.19	82.76 ± 0.38	8.75
	41.6	38.22 ± 1.63	90.02 ± 2.55	6.82	93.74 ± 0.70	1.80
	208	181.76 ± 9.38	85.73 ± 9.89	5.54	89.03 ± 8.86	4.79
Kaempferol	1.06	0.92 ± 0.11	93.71 ± 0.09	9.54	80.50 ± 0.13	14.65
	42.4	38.78 ± 1.93	91.05 ± 2.26	5.84	91.87 ± 1.60	4.10
	424	374.04 ± 12.54	88.47 ± 10.87	2.90	87.96 ± 14.20	3.81

TABLE 5: Stability evaluation results ($n = 6$).

Analytes	Spiked (ng/ml)	Autosampler (4°C, 36 h)		Long-term (−20°C, 30 days)		Freeze-thaw (−20°C-room temperature)	
		Measured	RE (%)	Measured	RE (%)	Measured	RE (%)
Neoeriocitrin	40.4	40.29 ± 1.99	−1.00	40.5 ± 2.82	−1.21	39.8 ± 1.92	−2.92
	202	198.79 ± 11.22	2.58	206.55 ± 12.4	1.72	205.16 ± 23.24	1.36
	2020	2001.91 ± 21.3	−0.90	2012.25 ± 27.81	−0.37	2032.19 ± 45.28	0.61
Luteolin-7-O-β-D-glucoside	5.04	5.02 ± 0.18	−3.12	4.99 ± 0.07	−2.34	5.01 ± 0.21	−1.26
	50.4	49.44 ± 1.55	−1.38	49.37 ± 3.07	−3.64	49.55 ± 8.55	−2.45
	504	494.39 ± 10.55	−1.90	493.07 ± 13.07	−2.16	491.55 ± 18.55	−2.47
Astragalin	5.1	5.02 ± 0.18	−1.63	4.99 ± 0.07	−2.18	5.01 ± 0.21	−1.79
	102	101.10 ± 4.29	−2.79	99.7 ± 3.76	−4.13	102.04 ± 4	−1.88
	1020	1006.41 ± 30.28	3.56	1002.46 ± 35.15	1.22	1000.65 ± 34.59	1.04
Naringin	10.2	10.5 ± 0.14	−4.02	10.3 ± 0.09	−0.27	9.9 ± 0.18	−1.11
	408	401.01 ± 8.13	−0.24	400.90 ± 9.65	−0.51	403.52 ± 11.42	−0.11
	6120	6008.15 ± 38.45	−5.66	6010.28 ± 22.59	−3.22	6009.12 ± 43.76	−4.99
Eriodictyol	1.02	0.96 ± 0.06	−1.30	1.06 ± 0.06	−4.22	0.99 ± 0.07	−2.84
	40.8	38.83 ± 2.5	−0.44	39.87 ± 1.83	2.28	39.00 ± 2.32	−2.64
	816	834.09 ± 23.38	−0.70	837.06 ± 26.59	−0.34	824.21 ± 18.77	−1.88
Luteolin	10.1	10.2 ± 0.04	3.54	9.97 ± 0.09	−0.13	9.96 ± 0.1	−1.04
	40.4	37.85 ± 1.42	0.94	38.08 ± 1.5	1.55	39.39 ± 1.9	−0.29
	1010	1000.91 ± 21.3	−1.91	1006.25 ± 27.81	−0.25	1016.19 ± 45.28	0.64
Naringenin	5.2	5.1 ± 0.06	0.92	5.1 ± 0.11	0.89	5.08 ± 0.11	0.66
	41.6	41.50 ± 3.05	−1.20	41.33 ± 3.09	−1.60	41.13 ± 3.11	−2.06
	208	207.10 ± 4.29	−4.79	208.70 ± 5.76	−8.13	204.04 ± 8	−3.76
Kaempferol	1.06	1.05 ± 0.14	−3.02	1.03 ± 0.09	−0.49	1.09 ± 0.18	−1.71
	42.4	40.59 ± 1.39	−1.07	41.15 ± 2.32	−1.34	41.8 ± 1.58	−2.67
	424	419.01 ± 7.42	−0.44	427.9 ± 3.74	−0.46	419.52 ± 8.323	−0.31

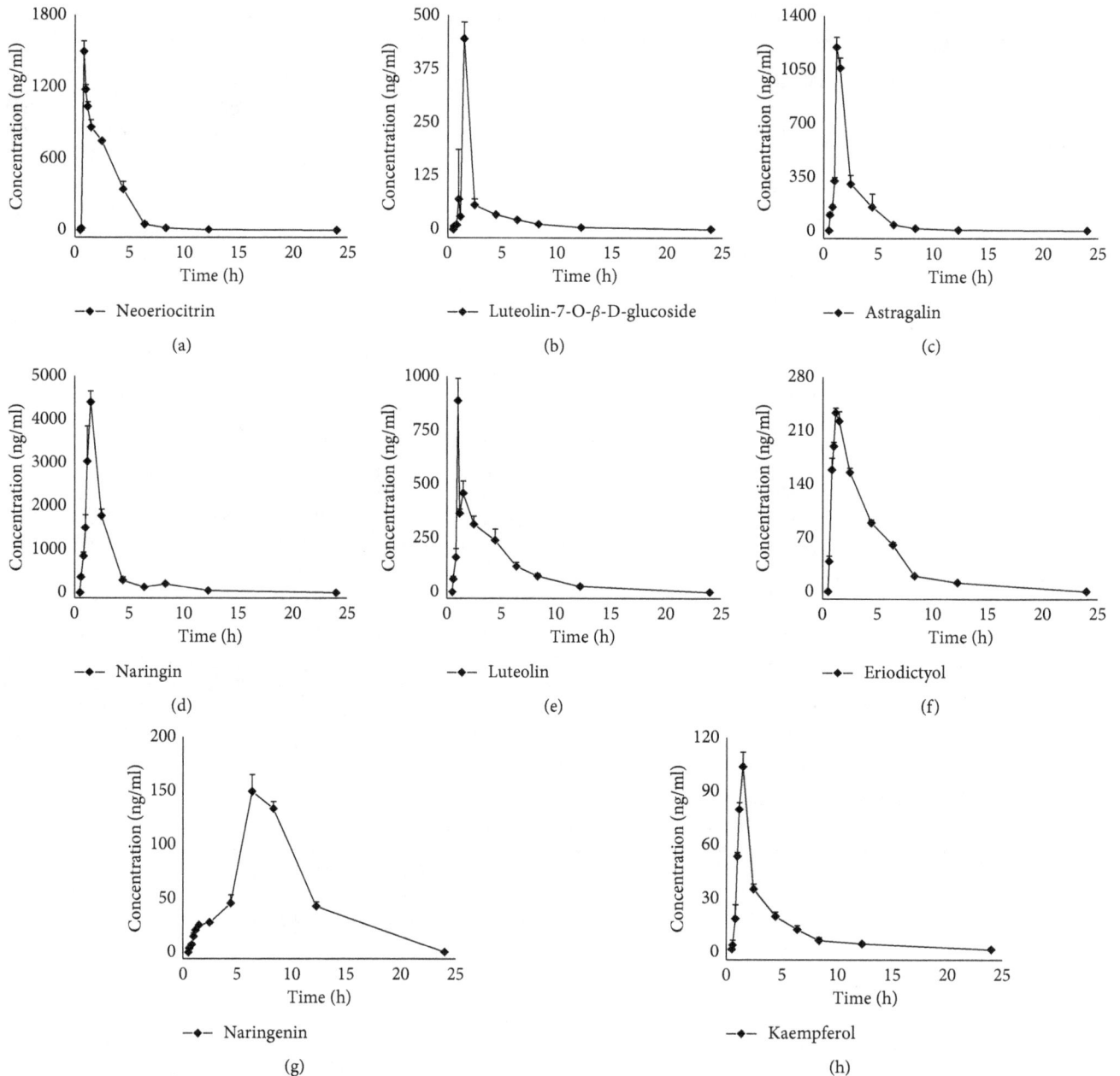

FIGURE 3: Mean plasma concentration-time curve of eight analytes in rats after oral administration of DRE.

TABLE 6: Pharmacokinetic parameters of eight analytes in rat plasma after oral administration of DRE (mean ± SD, n=6).

Analytes	AUC (0–t) (ng·h/ml)	AUC (0–∞) (ng·h/ml)	T_{max} (h)	$T_{1/2}$ (h)	C_{max} (ng/kg)	MRT (0–t) (h)
Neoeriocitrin	3327.71 ± 238.61	3340.34 ± 237.36	0.33	1.69 ± 0.28	1490.98 ± 124.54	2.16 ± 0.15
Luteolin-7-O-β-D-glucoside	550.97 ± 44.68	565.98 ± 56.06	1	2.34 ± 0.61	440.37 ± 52.16	2.55 ± 0.28
Astragalin	1999.53 ± 338.21	2012.24 ± 331.76	0.67	1.94 ± 0.6	1203.63 ± 90.89	1.97 ± 0.25
Naringin	8669.19 ± 321.61	8760.77 ± 347.83	1	2.51 ± 0.94	4414.18 ± 360.38	2.36 ± 0.08
Eriodictyol	894.09 ± 33.5	929.83 ± 26.85	0.67	2.49 ± 0.19	232.76 ± 8.52	3.24 ± 0.14
Luteolin	2040.99 ± 219.28	2144.07 ± 173.66	1	2.62 ± 0.64	895.98 ± 140.34	3.51 ± 0.11
Naringenin	946.98 ± 49.06	1164.21 ± 58.65	6	3.21 ± 0.54	153.87 ± 17.39	6.89 ± 0.04
Kaempferol	232.81 ± 29.32	244.43 ± 34.91	1	2.86 ± 0.58	102.82 ± 11.61	2.84 ± 0.32

eriodictyol, and kaempferol were lower, with small area under curve (AUC) and C_{max}. The T_{max} values for luteolin-7-O-β-D-glucoside and luteolin were 1 h and 0.5 h, respectively, which were in line with a previous report [18, 19]. Luteolin is the metabolite of luteolin-7-O-β-D-glucoside, the plasma concentration of the former is greater than that of the latter. This may due to the conversion from glycoside to aglucone after oral DRE.

Compared with synthetic drugs, ingredients of plant medicines are complex and have synergistic activity. In this work, we found that most of the analytes reached their maximum concentrations around 2 h after oral administration of DRE. This phenomenon may allow the DRE to exert its greatest pharmacological activity in clinical application.

4. Conclusions

In summary, a rapid, sensitive, and selective UPLC-MS/MS method was successfully applied in a pharmacokinetic study for analyzing the active compounds in Chinese herbal medicines. As the main active constituents of DRE, the major pharmacokinetic parameters of eight flavonoids were described for first time. The high efficiency, sensitivity, and selectivity of this multiple components quantitation were guaranteed by UPLC coupled ESI-MS. MRM quantitative analysis mode and two best MRM transitions enhanced the accuracy for each analyte. Results showed that this method could be applied for pharmacokinetic study of the DRE. The present results may provide useful information for better understanding of the absorption of the major constituents of DRE, as well as their potential clinical value.

Abbreviations

UPLC-MS/MS:	Ultraperformance liquid chromatography tandem mass spectrometry
DRE:	*Drynariae rhizoma* extract
MRM:	Multiple reaction monitoring
IS:	Internal standard
ESI+:	Positive ion mode of electrospray ionization
ACN:	Acetonitrile
MeOH:	Methanol
QC:	Quality control
LLOD:	Lower limit of detection
LLOQ:	Lower limit of quantification
S/N:	Signal-noise ratio
RSD:	Relative standard deviations
RE:	Relative error.

Conflicts of Interest

The authors declare that they have no conflicts of interest with respect to this study.

Acknowledgments

This work was financially supported by the National Natural Science Foundation of China (81673621 and 81673581); Science and Technology Improvement Program of Harbin Normal University (14XYG-03); Scientific Research Fund of Heilongjiang University of Chinese Medicine (2015 xy04); and Provincial Universities Technological Achievements Research and Cultivation Fund (TSTAU-C2018020).

References

[1] Chinese Pharmacopoeia Commission, *Pharmacopoeia of the People's Republic of China*, vol. 1Beijing, China, 2015.

[2] L. L. Liu, Q. U. Wei, and J. Y. Liang, "Progress on chemical constituents and biological activities of *Drynaria fortunei*," *Strait Pharmaceutical Journal*, vol. 24, no. 1, pp. 4–7, 2012.

[3] X. Qiao, X. Lin, and Y. Liang, *Comprehensive Chemical Analysis of the Rhizomes of Drynaria fortunei by Orthogonal, Pre-Separation and Liquid Chromatography Mass Spectrometry*, Springer-Verlag, Berlin, Germany, 2014.

[4] S. H. Song, Y. K. Zhai, C. Q. Li et al., "Effects of total flavonoids from *Drynariae rhizoma* prevent bone loss in vivo and in vitro," *Bone Reports*, vol. 5, pp. 262–273, 2016.

[5] F. Yang, Y. H. Sun, and Z. J. LV, "Effect of *Rhizoma Drynariae* flavones on alveolar bone quality in patients with osteoporosis," *China Journal of Oral and Maxillofacial Surgery*, vol. 16, no. 1, pp. 34–40, 2017.

[6] H. C. Kuo, H. C. Chang, W. C. Lan, F. H. Tsai, J. C. Liao, and C. R. Wu, "Protective effects of *Drynaria fortunei* against 6-hydroxydopamine-induced oxidative damage in B35 cells via the PI3K/AKT pathway," *Food and Function*, vol. 5, no. 8, pp. 1956–1965, 2014.

[7] Z. W. Jiang, J. Q. Zeng, and F. Huang, "Effects of flavonoids of *rhizoma Drynariae* on tibia distraction osteogenesis efficacy in rat model," *China Journal of Traditional Chinese Medicine and Pharmacy*, vol. 33, no. 2, pp. 661–663, 2018.

[8] S. Song, Z. Gao, X. Lei et al., "Total flavonoids of *Drynariae rhizoma* prevent bone loss induced by hindlimb unloading in rats," *Molecules*, vol. 22, no. 7, p. 1033, 2017.

[9] H. Guo, Q. Du, and J. F. Rebhun, "Chinese herbal combination of epimedium, *Drynariae rhizoma*, and salvia miltiorrhiza extracts prevents osteoporosis in ovariectomized rats," *International Journal of Clinical and Experimental Medicine*, vol. 10, no. 7, pp. 10607–10615, 2017.

[10] G. Zhang, F. Geng, T. Zhao et al., "Biocompatible symmetric Na-ion microbatteries with sphere-in-network heteronanomat electrodes realizing high reliability and high energy density for implantable bioelectronics," *ACS Applied Materials and Interfaces*, vol. 10, no. 49, pp. 42268–42278, 2018.

[11] N. Zhang, Z. Hu, Z. Zhang et al., "Protective role of naringenin against Aβ25-35-caused damage via ER and PI3K/Akt-Mediated pathways," *Cellular and Molecular Neurobiology*, vol. 38, no. 2, pp. 549–557, 2017.

[12] Z. Y. Yang, T. Kuboyama, K. Kazuma, K. Konno, and C. Tohda, "Active constituents from *Drynaria fortunei* rhizomes on the attenuation of Aβ25-35-induced axonal atrophy," *Journal of Natural Products*, vol. 78, no. 9, pp. 2297–2300, 2015.

[13] X. Wang, J. Liu, A. Zhang, X. Zhou, and H. Sun, "Serum pharmacochemistry of TCM screening the bioactive components from moutan cortex,"in *Serum Pharmacochemistry of Traditional Chinese Medicine*, Chapter 21, pp. 287–302, Academic Press, Cambridge, MA, USA, 2017.

[14] X. H. Li, Z. L. Xiong, S. Lu, Y. Zhang, and F. M. Li, "Pharmacokinetics of naringin and its metabolite naringenin in rats

after oral administration of *Rhizoma Drynariae* extract assayed by UPLC-MS/MS," *Chinese Journal of Natural Medicines*, vol. 8, no. 1, pp. 40–46, 2010.

[15] M. Xu, Z. Xu, Q. Xu et al., "UPLC-MS/MS method for the determination of 14 compounds in rat plasma and its application in a pharmacokinetic study of orally administered xiaoyao powder," *Molecules*, vol. 23, no. 10, p. 2514, 2018.

[16] T. Fang, Y. Wang, Y. Ma, W. Su, Y. Bai, and P. Zhao, "A rapid LC/MS/MS quantitation assay for naringin and its two metabolites in rats plasma," *Journal of Pharmaceutical and Biomedical Analysis*, vol. 40, no. 2, pp. 454–459, 2006.

[17] K. Ishii, T. Furuta, and Y. Kasuya, "Determination of naringin and naringenin in human urine by high-performance liquid chromatography utilizing solid-phase extraction," *Journal of Chromatography B: Biomedical Sciences and Applications*, vol. 704, no. 1-2, pp. 299–305, 1997.

[18] X. Wang, Q. Wang, and M. E. Morris, "Pharmacokinetic interaction between the flavonoid luteolin and γ-hydroxybutyrate in rats: potential involvement of monocarboxylate transporters," *AAPS Journal*, vol. 10, no. 1, pp. 47–55, 2008.

[19] R. Yin, X. Chen, F. Han, Z. Shen, W. Cheng, and K. Bi, "LC-MS determination and pharmacokinetic study of luteolin-7-O-β-d-glucoside in rat plasma after administration of the traditional Chinese medicinal preparation kudiezi injection," *Chromatographia*, vol. 67, no. 11-12, pp. 961–965, 2008.

16

Isolation and Determination of Fomentariol: Novel Potential Antidiabetic Drug from Fungal Material

Nevena Maljurić [ID],[1] **Jelena Golubović** [ID],[1] **Matjaž Ravnikar** [ID],[2] **Dušan Žigon,**[3] **Borut Štrukelj,**[2] **and Biljana Otašević**[1]

[1]*Department of Drug Analysis, Faculty of Pharmacy, University of Belgrade, Vojvode Stepe 450, Belgrade 11221, Serbia*
[2]*Chair for Pharmaceutical Biology, Faculty of Pharmacy, University of Ljubljana, Aškerčeva cesta 7, Ljubljana 1000, Slovenia*
[3]*Jožef Stefan Institute, Jamova 39, Ljubljana 1000, Slovenia*

Correspondence should be addressed to Jelena Golubović; golub@pharmacy.bg.ac.rs

Academic Editor: Pablo Richter

Diabetes mellitus is one of the leading world's public health problems. Therefore, it is of a huge interest to develop new antidiabetic drugs. Apart from traditional therapy of diabetes, nowadays, importance is given to natural substances with antidiabetic potential. *Fomes fomentarius* is a mushroom widely used for different purposes, due to its range of already confirmed activities. Fomentariol is a constituent of *Fomes fomentarius*, responsible for its antidiabetic potential. In that respect, it is important to develop a method for isolation and quantification of fomentariol from fungal material, which will be simple and efficient. Multistep, complex extraction applied in the previously reported studies was avoided with ethanol, providing rapid single-step extraction. The presence of fomentariol in ethanolic extract was confirmed by high-resolution mass spectrometry. Semipreparative HPLC method was developed and applied for isolation from ethanol extract and purification of the active compound fomentariol. It was a gradient reversed-phase method with a mobile phase consisting of acetonitrile and 0.1% formic acid in water and total run time of 15 minutes. The amount of 6.5 mg of high-purity fomentariol was determined by quantitative NMR with toluene as internal standard. The isolated and determined amount of substance can be further used for the quantitative estimation of activity of fomentariol.

1. Introduction

Fomes fomentarius is a mushroom of the family Polyporaceae, native to the north of the temperate zone of the northern hemisphere. Although this mushroom is firstly described in the 5th century BC by Hippocrates and has been traditionally used worldwide for different purposes, not much is published about its medicinal usage. Recent studies demonstrated that extracts of *F. fomentarius* exerted antidiabetic, antioxidant, anti-inflammatory, antinociceptive, antibacterial, and cytotoxic activities, either by unknown mechanisms [1, 2] or by virtue of the active principles other than fomentariol [3, 4].

Fomentariol (Figure 1) is a constituent of *F. fomentarius*. Although it was recognized long ago [5], its activity is insufficiently investigated. Seo et al. [6] isolated fomentariol and demonstrated its antioxidant activity.

On the other hand, our experiments with α-glucosidase and dipeptidyl peptidase-4 showed antidiabetic potential of fomentariol. Diabetes mellitus is a major public health problem that is approaching epidemic proportions globally. Development of new antidiabetic drugs is of huge interest to mankind. Natural substances with antidiabetic potential can be a very valuable alternative or supplement to the conventional therapy of diabetes.

The purpose of this study was to develop a simple and efficient method for the isolation and determination of fomentariol from the fungal material, in order to quantify its antidiabetic activity, that is, to determine its EC50. Reference standard substance is not commercially available, and the synthesis would be quite challenging [7]. Complementary analytical methods, such as mass spectrometry (MS), high-performance liquid chromatography

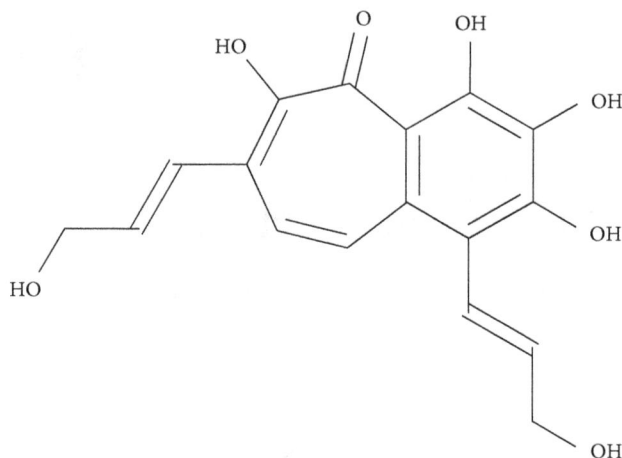

FIGURE 1: Structure of fomentariol.

(HPLC), and nuclear magnetic resonance spectroscopy (NMR), can be applied in order to confirm identity, obtain high-purity substance, and determine the quantity of the isolated substance, respectively.

Seo et al. [6] only investigated the mechanism of its antioxidant activity, without quantitative examination of this activity. Furthermore, extraction and isolation of fomentariol in the previous studies published by Seo et al. [6] consisted of multistep extraction employing several solvents, followed by several different chromatographic procedures. First isolation of fomentariol from the natural source also included multisolvent extraction [5]. Therefore, the important goal of this paper was also to propose a new procedure in order to simplify the extraction and isolation as much as possible, retaining the high efficiency.

The developed method must be robust and easily applied by analysts, allowing isolation and quantification of the compound in a reasonable time period.

2. Materials and Methods

2.1. Chemicals and Reagents. Ethanol (99.5%) used for the extraction of fomentariol was obtained from Sigma-Aldrich Chemie GmbH (Taufkirchen, Germany), while dichloromethane (99.8%) was obtained from Honeywell Riedel-de Haën (Seelze, Germany). Acetonitrile (99.8%) and formic acid (98%) were also purchased from Sigma-Aldrich Chemie GmbH, as well as methanol-d4 (99.9%). Toluene (99.5%) obtained from POCH (Gliwice, Poland) was used for quantitative NMR determination. Purified water was obtained from a Simplicity 185 purification system (Millipore, Billerica, MA, USA). Before use, the sample was filtered through 0.22 μm nylon membranes (Agilent Technologies, Santa Clara, USA). All reagents used were of analytical grade except water and acetonitrile, which were of HPLC grade.

2.2. Identification and Isolation of Fomentariol

2.2.1. Sample Preparation. Fungal material was characterized and kindly provided by Professor Franc Pohleven from the

Department of Wood Science and Technology at the Biotechnical Faculty, University of Ljubljana. The fungal material was chopped and added to 50 mL of ethanol. Ethanolic extract was incubated for 24 hours at room temperature. After 24 hours, the extract was paper filtered. In order to evaporate the solvent, the extract was left overnight at room temperature. Dry extract was reconstituted in the solvent which consisted of acetonitrile and water (50 : 50, v/v) and was filtered through a 0.22 μm nylon filter.

2.2.2. Identification by Means of High-Resolution Mass Spectrometry Analysis. Fomentariol was identified by mass measurements run on a hybrid quadrupole time-of-flight mass spectrometer Q-TOF Premier provided with an orthogonal Z-spray ESI interface (Waters Micromass, Manchester, UK). Mass spectrometer was coupled to Waters Acquity ultra-high-performance liquid chromatography (UPLC) (Waters, Milford, USA) system based on a binary pump. The separation was achieved on Luna® Omega C18 HPLC column (1.6 μm, 100 × 2.1 mm i.d., Phenomenex Inc., USA), with temperature set to 40°C. Compressed nitrogen (99.999%, Messer Slovenia) was used as both the drying and the nebulising gas. The nebulizer gas flow rate was set to approximately 20 L/h and the desolvation gas flow rate to 600 L/h. A cone voltage of 20 V and a capillary voltage of 3.0 kV were used in positive ion mode, while 2.5 kV was used in negative ionization mode. The desolvation temperature was set to 300°C and the source temperature to 100°C. Elemental composition was determined with the mass resolution of approximately 9000 full width of the peak at half its maximum (FWHM) height. MS spectra were acquired in centroid mode over an m/z range of 50–1000 in scan time 0.2 s and interscan time 0.025 s. For MS/MS experiments, argon (99.995%, Messer Slovenia) was used as collision gas at a pressure of approximately 2×10^{-5} mbar in the collision cell. Collision energies of 15 V were applied to generate product ion spectra. MS/MS spectra were acquired in centroid mode as well, over the same m/z range and scan time. The detector potential was set to 2100 V. The mobile phase consisted of acetonitrile (A) and 0.1% formic acid in water (B). The gradient started with 95% B, which was decreased to 5% in 6 min and returned to initial ratio in 0.05 min, followed by re-equilibration, giving a total run time of 7 min. The flow rate was 0.3 mL·min^{-1}. The operating software MassLynx v 4.1 (Waters Micromass, Manchester, UK) was used for data analysis.

2.2.3. Isolation of Fomentariol. The isolation of fomentariol was performed on its ethanol extract on *Dionex Ultimate 3000* HPLC system, equipped with a PDA detector. Chromatographic separation was achieved on Hypersil Gold semipreparative HPLC column (Thermo Fisher Scientific Inc., 5 μm, 150 × 10 mm). Injection volume was 100 μL. The mobile phase consisted of acetonitrile (A) and 0.1% formic acid in water (B). The HPLC method was the one applied for the identification, only transferred from UPLC to semipreparative conditions. The gradient started with 95% B, which was decreased to 5% in 13 min, and returned to

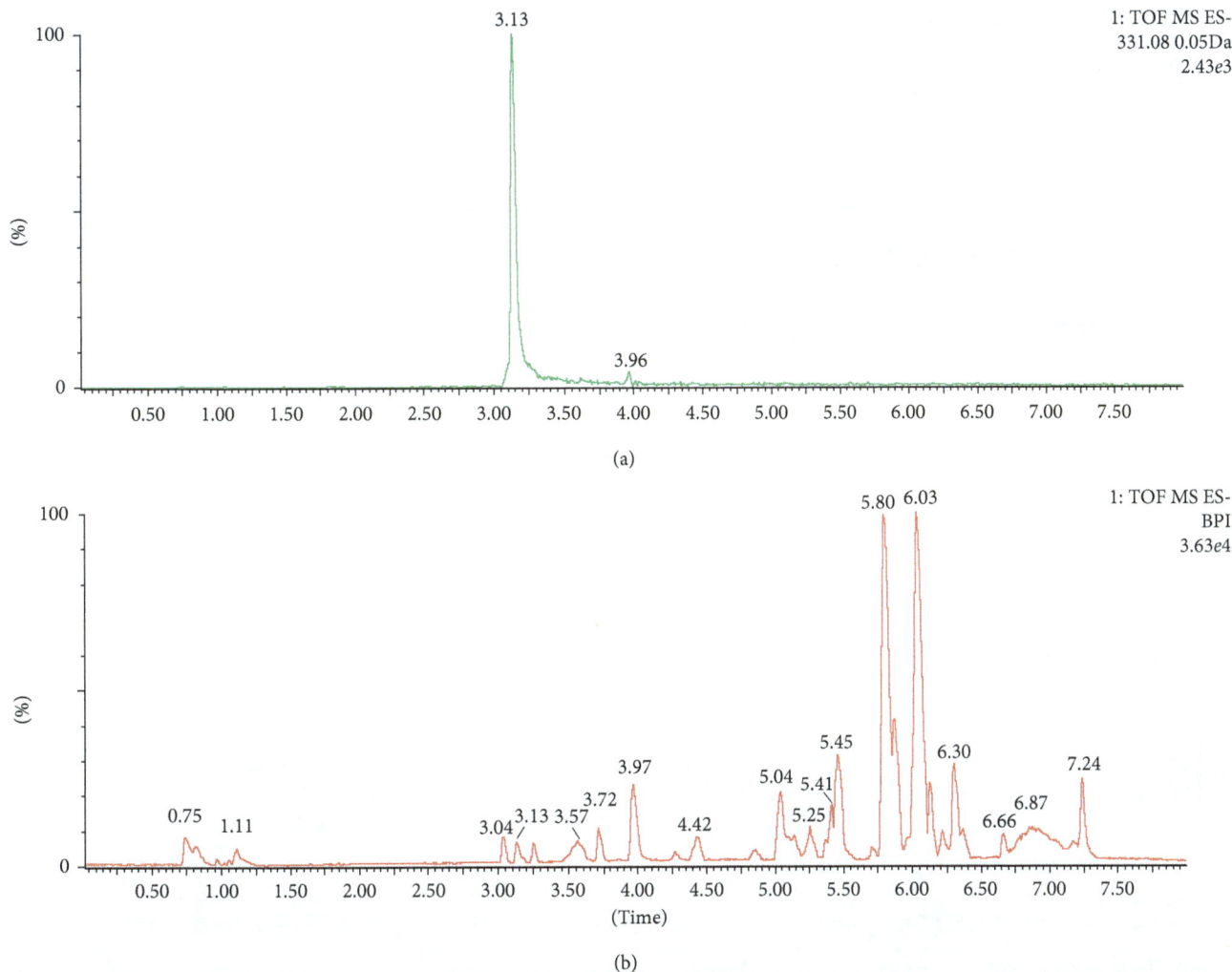

FIGURE 2: Extracted ion chromatogram of [M-H]$^-$ of fomentariol at m/z 331 (a) and total ion chromatogram (b) of the *F. fomentarius* ethanolic extract.

initial ratio in 0.05 min, followed by re-equilibration, giving a total run time of 15 minutes. The flow rate was 2 mL·min^{-1}. The detection was performed at 326 nm, and UV-Vis spectra of the major peaks were recorded. Fractions were collected according to the retention of fomentariol, at the time frame 9.3–10 min.

2.3. Quantitative NMR Determination. The solvent in the collected fractions of fomentariol was evaporated to dry using Rotavapor R-114 (Büchi, Flawil, Switzerland), and the dry sample was dissolved in methanol-d4 (99.9%). The NMR spectra were recorded on a Bruker Ascend 400 (400 MHz) spectrometer (Billerica, USA). Chemical shifts are given in parts per million (δ) downfield from tetramethylsilane as the standard used for system calibration. The quantity of the compound was calculated by the relative ratio of the integral values of the target peaks of fomentariol to the ones of toluene, internal standard of known amount.

3. Results and Discussion

3.1. Extraction Optimization. Firstly, there are certain prerequisites in order to preserve the activity of the

mushroom. The mushroom needs to be fresh, although it could be used for some period of time if it is kept frozen. When analyzing the potential activity of the mushroom components, the first step is to choose the right extraction solvent, suitable for the active compound. Our goal was to find a single solvent suitable for the extraction and avoid multistep, complex extraction applied in the previous studies [2, 5, 6]. The first try was with dichloromethane, which resulted in a poor extraction. Ethanol provided much more efficient extraction. The extraction efficiency was estimated by means of UPLC-MS method, which will be described in the next chapter. We noticed that the color of the extract can serve as a simple screening test, since the ethanol extract was orange to red colored, indicating the presence of fomentariol [3]. Furthermore, ethanol can be a convenient solvent for the future formulation of the drug product.

3.2. Identification of Fomentariol by High-Resolution Mass Spectrometry. The presence of fomentariol in ethanol extract was confirmed by high-resolution mass

Elemental Composition

File Edit View Process Help

Single Mass Analysis
Tolerance = 10.0 PPM / DBE: min = -1.5, max = 50.0
Element prediction: Off
Number of isotope peaks used for i-FIT = 3
Monoisotopic Mass, Even Electron Ions
765 formula(e) evaluated with 9 results within limits (all results (up to 1000) for each mass)
Elements Used:

Mass	Calc. Mass	mDa	PPM	DBE	Formula	i-FIT	i-FIT (Norm)	C	H	N	O
331.0820	331.0818	0.2	0.6	10.5	C17 H15 O7	90.9	4.8	17	15		7
	331.0823	-0.3	-0.9	3.5	C2 H11 N12 O8	99.6	13.5	2	11	12	8
	331.0831	-1.1	-3.3	15.5	C18 H11 N4 O3	91.1	5.1	18	11	4	3
	331.0809	1.1	3.3	-1.5	C H15 N8 O12	100.2	14.2	1	15	8	12
	331.0804	1.6	4.8	16.5	C14 H7 N10 O	86.0	0.0	14	7	10	1
	331.0836	-1.6	-4.8	8.5	C3 H7 N16 O4	99.0	12.9	3	7	16	4
	331.0791	2.9	8.8	11.5	C13 H11 N6 O5	91.2	5.2	13	11	6	5
	331.0850	-3.0	-9.1	2.5	C6 H15 N6 O10	97.3	11.3	6	15	6	10
	331.0850	-3.0	-9.1	13.5	C4 H3 N20	98.7	12.7	4	3	20	

F NEG 357 (3.133)

1: TOF MS ES-
1.31e+004

For Help, press F1

Start | MassLynx - COPY DE... | Chromatogram - [F N... | Spectrum - [F NEG] | Q-Tof Premier - c:\ma... | Fomentariol LC-MS in ... | Elemental Composi... | 3:32 PM

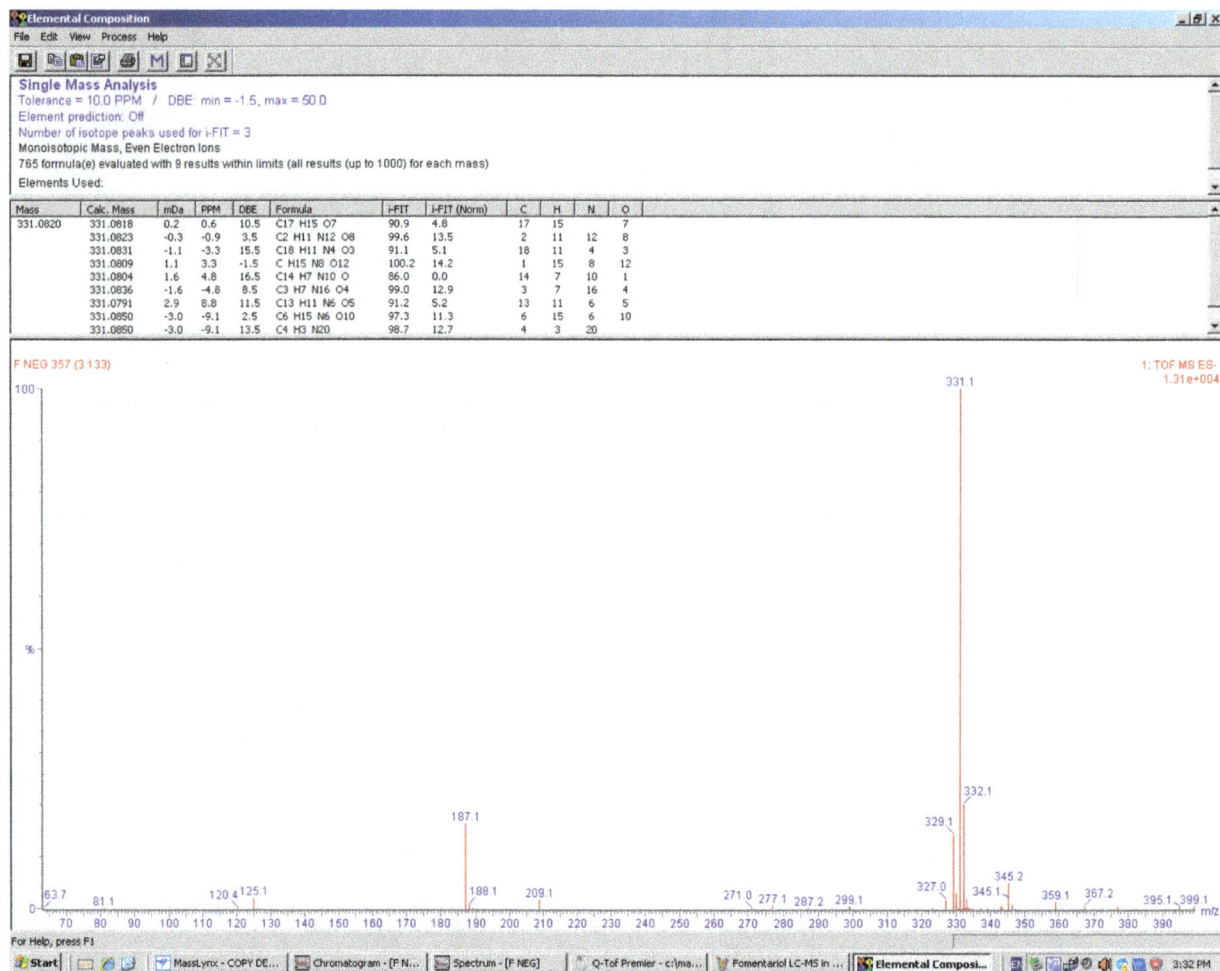

FIGURE 3: Electrospray ionization mass spectrum of fomentariol in negative ionization mode and results of mass measurement for elemental composition of [M-H]⁻ at m/z 331.

spectrometry (HRMS), comparing the obtained mass to charge ratio (m/z) with the theoretical one. When running UPLC-MS analysis, fomentariol peak was detected at 3.13 min retention time, as shown in Figure 2. Fomentariol was detected in both ionization modes, but higher signal intensity was obtained in the negative mode. Exact mass of the [M-H]⁻ ion under the peak of interest was m/z 331.0820. Elemental composition analysis proposed molecular formula of fomentariol as the first hit (Figure 3). This is also in accordance with the theoretical mass provided by ChemSpider®, which is 332.0896 Da. To our knowledge, no MS/MS spectra of fomentariol can be found in literature. We provided hereby the product ion spectrum under the collision energy of 15 V (Figure 4). All of these information combined served to confirm that the peak of interest corresponds to fomentariol.

3.3. Isolation of Fomentariol Using Semipreparative HPLC. Developing a simple and efficient method for isolation of fomentariol is crucial for its quantitative determination in order to access its activity. Fomentariol is a moderately polar substance, so a reverse-phase column was selected as the stationary phase, while gradient composition of acetonitrile and 0.1% formic acid in water was used as the mobile phase. UV-Vis spectra of the major peaks in chromatograms were recorded using PDA and used as a primary identification tool. The ethanol extract of the fungi was used to isolate fomentariol on preparative scale, using HPLC, as described. Semipreparative HPLC method was developed and applied for analyzing the ethanol extract of the fungi for the purpose of isolation and purification of the active compound fomentariol. Figure 5 represents the zoomed peak corresponding to fomentariol. The peak had a flat-top shape, indicating the desired saturation of the column. Fractions were collected according to the retention of fomentariol, at the time frame 9.3–10.0 min. The collected fractions were further evaporated and used for the quantitative determination of fomentariol. Color intensity of the collected fractions was an additional confirmation of the accurate fraction collection.

3.4. Quantitative NMR Determination. In NMR analysis, the ratio of the number of atomic nuclei in a compound corresponds to the ratio of the areas of the peaks in the

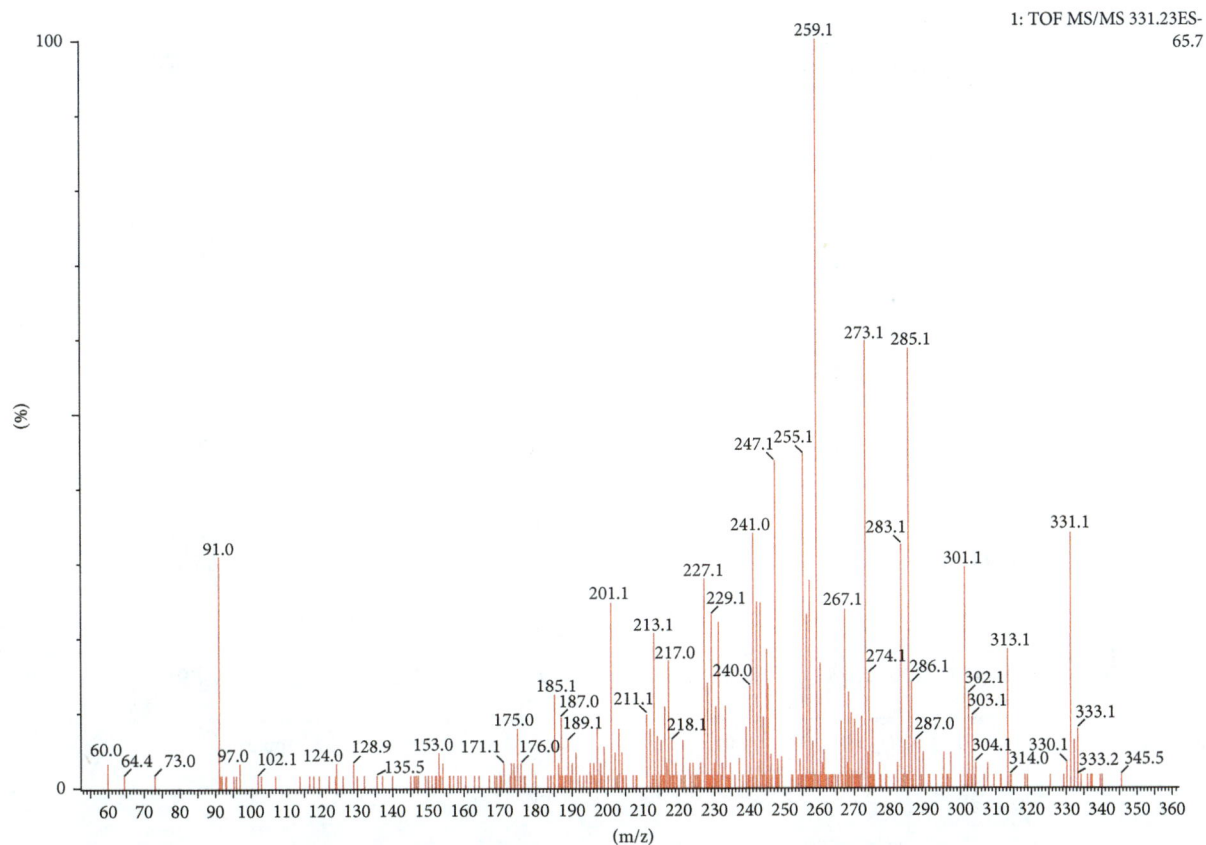

FIGURE 4: Product ion spectrum under the collision energy of 15 V.

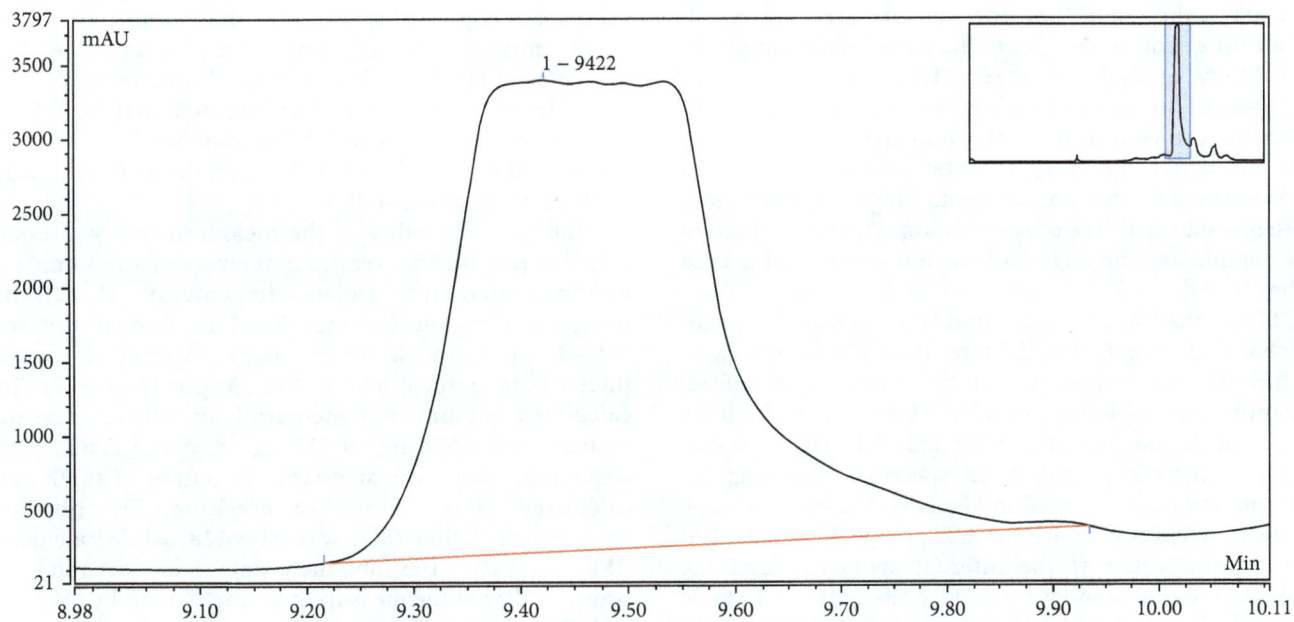

FIGURE 5: The zoomed peak of fomentariol under the semipreparative HPLC conditions.

spectrum. Therefore, the unknown amount of the compound of interest can be determined by performing a quantitative analysis. When the amount is determined by

^{1}H NMR, the sample is mixed with an internal standard having a known purity and dissolved in a deuterated solvent. The relationship between the areas of the spectral

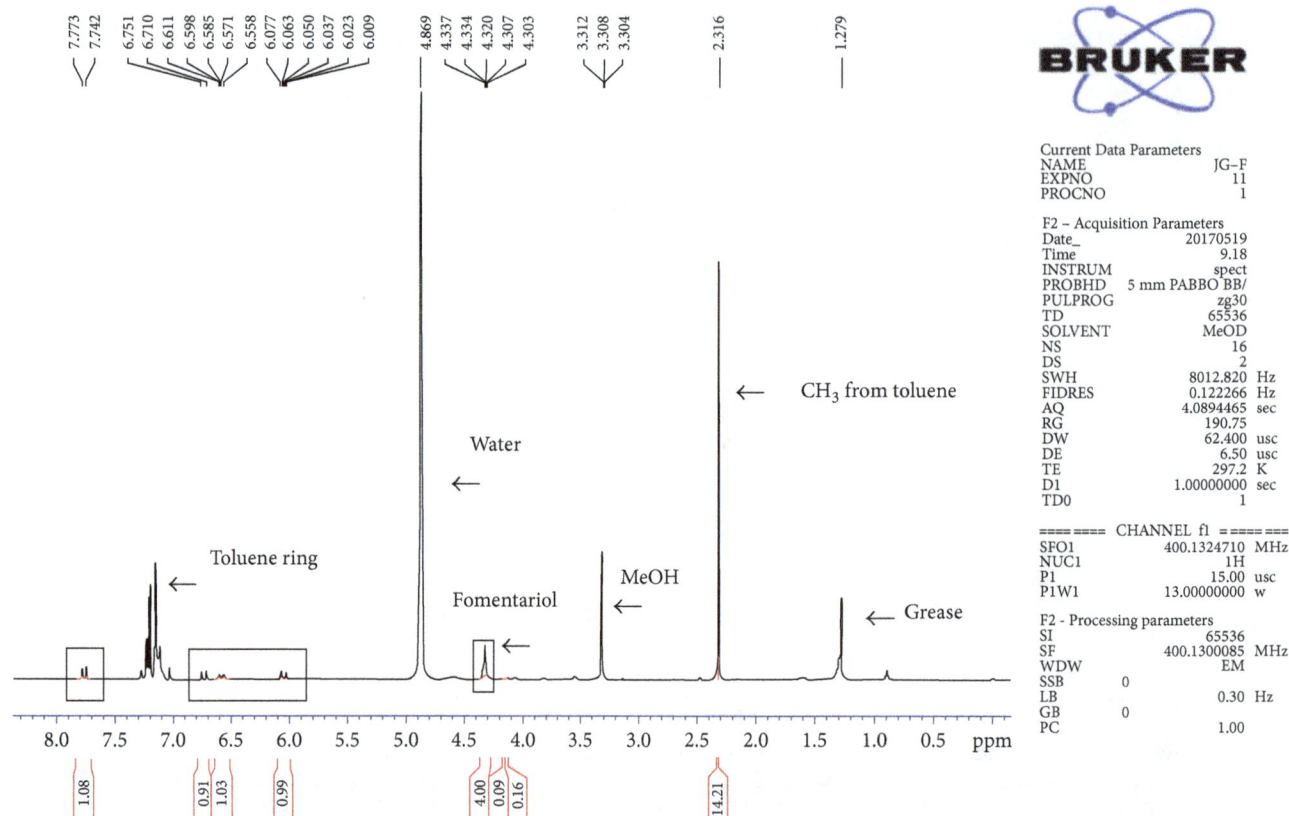

FIGURE 6: ^{1}H NMR spectrum of fomentariol and toluene as internal standard.

peaks originated from the sample and the standard, the number of protons, the weighed masses, and the molecular weights of the sample and the standard are used to calculate the quantitative value of the purity of the sample. In quantitative analysis using ^{1}H NMR (quantitative NMR), the areas of the peaks of hydrogen atoms observed in the spectrum can be quantitatively compared. Therefore, it is possible and rather easy to determine the purity and quantity of many compounds containing a hydrogen atom with one standard. The only condition is that the signals of the sample and the standard do not overlap with each other [8–14].

Since the target compound fomentariol is polar, methanol-d4 was used as the solvent for NMR analysis, to ensure that the compound will be completely dissolved. Quantification of fomentariol by ^{1}H NMR is possible by means of the integral of a well-separated specific proton signal of the compound. NMR spectrum including the interpretation is presented in Figure 6. The integral value of the signal of fomentariol at 4.35 ppm was compared to the integral value of the internal standard signal. A suitable internal standard should preferably be a stable compound with a signal in a noncrowded region of the ^{1}H NMR spectrum. For this purpose, toluene with a signal at 2.316 ppm has been chosen. In the case of ^{1}H NMR quantitative analysis, there is no need for the construction of calibration curves in order to quantify the compounds, because integration of the peaks is always proportional to the amount of the compound. The amount of fomentariol

in the sample was calculated knowing the amount of toluene. Three protons of 8.3 mg (0.0935 mmol) of toluene gave the integral value of 14.21. The amount of toluene in the mixture was 4.7 times higher than amount of fomentariol (14.21/3). The calculated amount of fomentariol in the sample was 0.0198 mmol, that is, 6.5 mg. Signals in the spectra which corresponded to the compounds other than the known ones indicated high purity of the sample, around 90%.

The reproducibility of the measurements was established through three replicate analyses of the sample of unknown amount of isolated fomentariol [11, 12]. The integration region for the signal of fomentariol was 4.28–4.36 ppm, while for the signal of toluene, used as internal standard, it was 2.29–2.35 ppm (Figure 7). The calculated amount of fomentariol over three measurements was 1.538 mg, 1.525 mg, and 1.525 mg, subsequently. Relative standard deviation (RSD) was calculated as a measure of precision. The obtained standard deviation (SD) was 0.006128, while calculated RSD was 4%. The obtained data show a good reproducibility of the measurement carried out by ^{1}H NMR technique.

4. Conclusion

In the present paper, several analytical techniques were successfully combined in order to isolate and quantify fomentariol, as a potential antidiabetic natural drug from

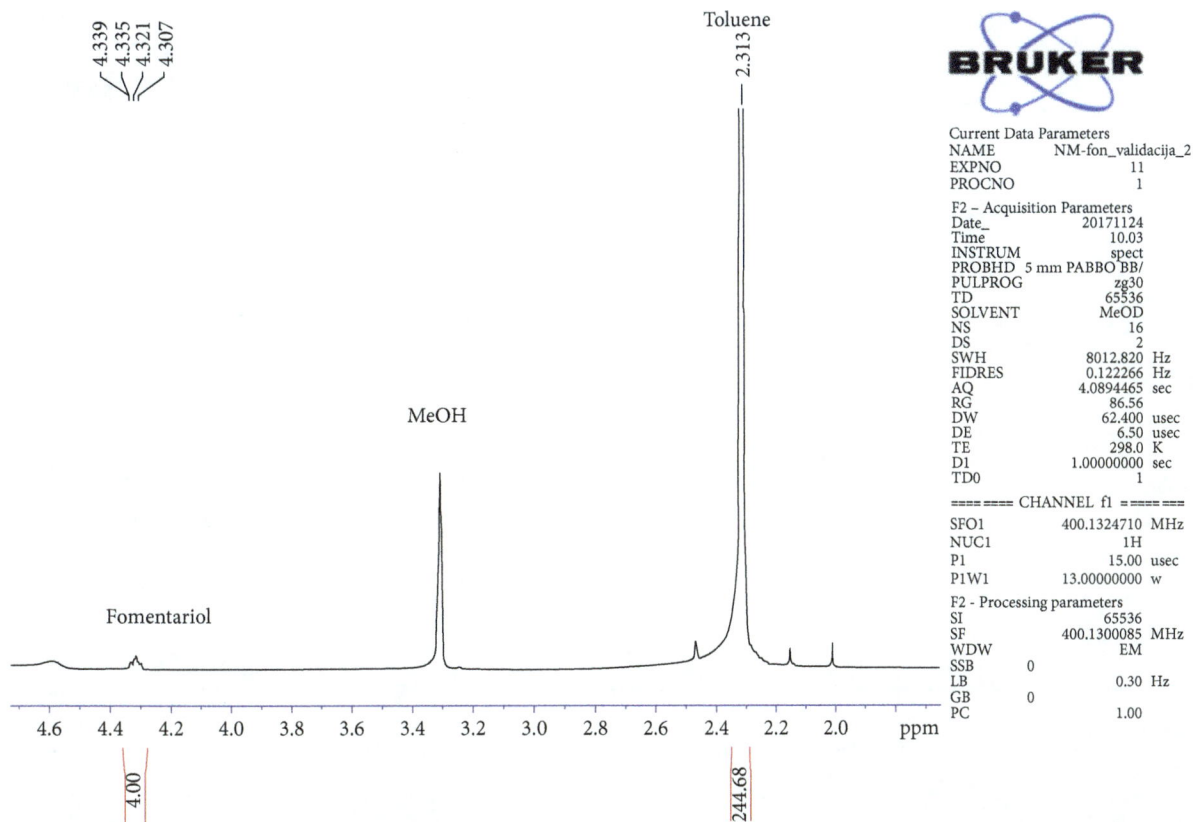

(a)

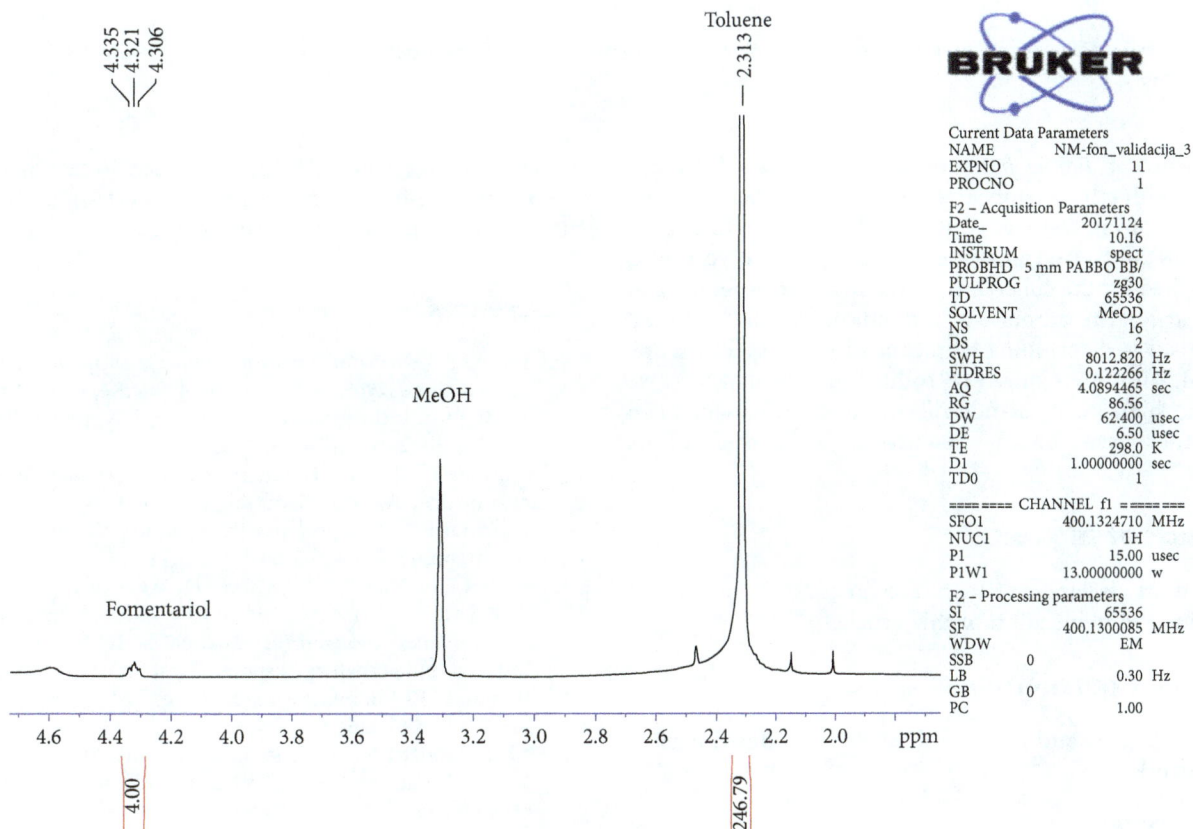

(b)

FIGURE 7: Continued.

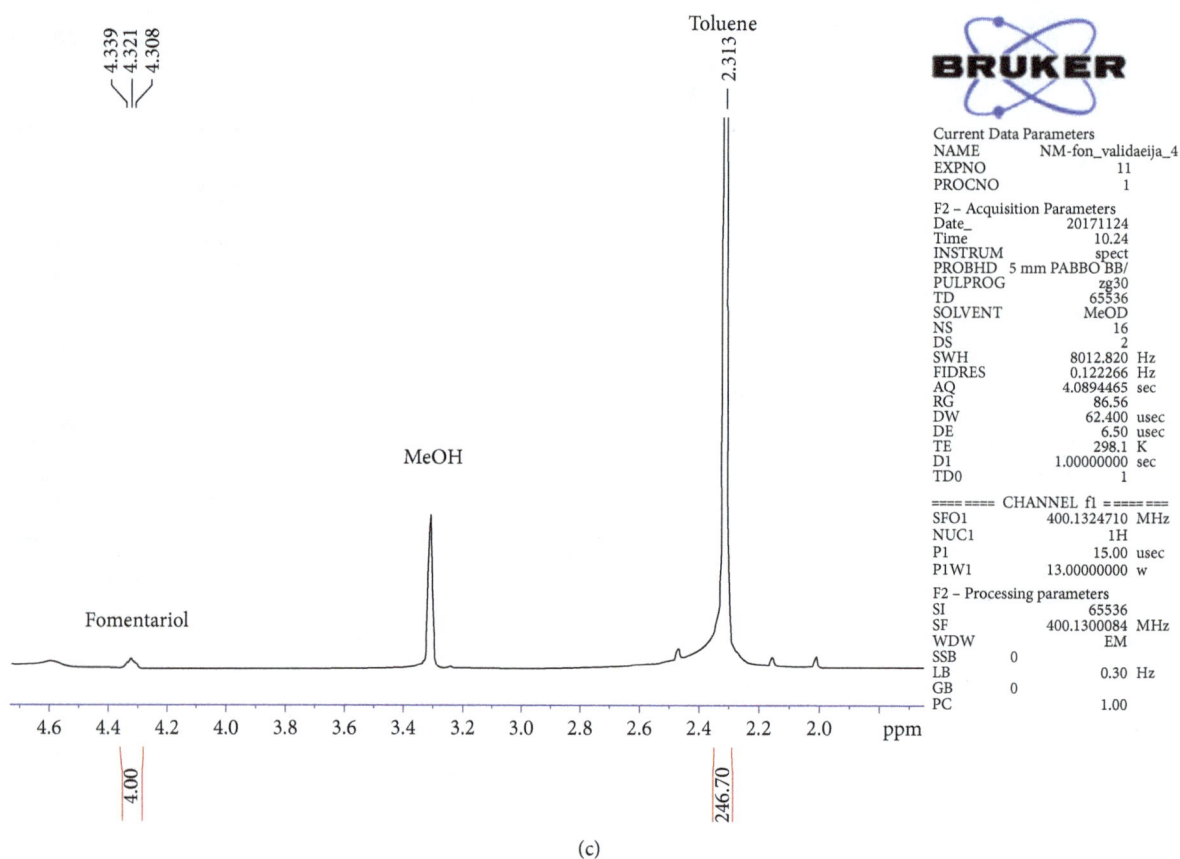

FIGURE 7: ^{1}H NMR spectra of fomentariol and toluene as internal standard recorded subsequently for the estimation of method reproducibility.

the mushroom *Fomes fomentarius*. The single-solvent extraction with ethanol appeared to have high efficiency. The substance of interest was confirmed by UPLC-HRMS. The UPLC method was transferred to semipreparative HPLC and applied for the collection of fractions corresponding to fomentariol. The amount of fomentariol in the collected fractions was determined by means of NMR spectroscopy method, with the required reproducibility. In that way, we obtained high-purity standardized fomentariol, which can be further used for the quantitative estimation of its activity.

Conflicts of Interest

The authors declare that there are no conflicts of interest regarding the publication of this article.

Authors' Contributions

Nevena Maljurić and Jelena Golubović contributed equally to this work.

Acknowledgments

This work was financially supported by Ministry of Education, Science and Technological Development of Republic of Serbia (Project no. 172033), Slovenian Research Agency (Grant no. P4-0127), and bilateral project of the two institutions (Project no. BI-RS/16-17-022).

References

[1] J. S. Lee, "Effects of *Fomes fomentarius* supplementation on antioxidant enzyme activities, blood glucose, and lipid profile in streptozotocin-induced diabetic rats," *Nutrition Research*, vol. 25, no. 2, pp. 187–195, 2005.

[2] Y. M. Park, I. T. Kim, H. J. Park et al., "Anti-inflammatory and anti-nociceptive effects of the methanol extract of *Fomes fomentarius*," *Biological and Pharmaceutical Bulletin*, vol. 27, no. 10, pp. 1588–1593, 2004.

[3] J. H. Choe, Y. J. Yi, M. S. Lee, D. W. Seo, B. S. Yun, and S. M. Lee, "Methyl 9-Oxo-(10E,12E)-octadecadienoate isolated from *Fomes fomentarius* attenuates lipopolysaccharide-induced inflammatory response by blocking phosphorylation of STAT3 in murine macrophages," *Mycobiology*, vol. 43, no. 3, pp. 319–326, 2015.

[4] M. Kolundžić, N. D. Grozdanić, M. Dodevska et al., "Antibacterial and cytotoxic activities of wild mushroom *Fomes fomentarius* (L.) Fr., *Polyporaceae*," *Industrial Crops and Products*, vol. 79, pp. 110–115, 2016.

[5] N. Arpin, J. Favre-Bonvin, and W. Steglich, "Le fomentariol: nouvelle benzotropolone isolée de *Fomes fomentarius*," *Phytochemistry*, vol. 13, no. 9, pp. 1949–1952, 1974.

[6] D. W. Seo, Y. J. Yi, M. S. Lee, B. S. Yun, and S. M. Lee, "Differential modulation of lipopolysaccharide-induced inflammatory cytokine production by and antioxidant activity of fomentariol in RAW264.7 cells," *Mycobiology*, vol. 43, no. 4, pp. 450–457, 2015.

[7] W. Steglich and L. Zechlin, "Pilzpigmente, 33: synthese des fomentariols. Eine neue methode zur darstellung von zimtalkoholen," *Chemische Berichte*, vol. 111, no. 12, pp. 3939–3948, 1978.

[8] K. Hu, W. M. Westler, and J. L. Markley, "Simultaneous quantification and identification of individual chemicals in metabolite mixtures by two-dimensional extrapolated time-zero 1H–13C HSQC (HSQC0)," *Journal of the American Chemical Society*, vol. 133, no. 6, pp. 1662–1665, 2011.

[9] E. Alvarado, *Practical Guide for Quantitative 1D NMR Integration*, University of Michigan, Ann Arbor, MI, USA, 2010.

[10] G. F. Pauli, S.-N. Chen, C. Simmler et al., "Importance of purity evaluation and the potential of quantitative 1H NMR as a purity assay: miniperspective," *Journal of medicinal chemistry*, vol. 57, no. 22, pp. 9220–9231, 2014.

[11] W. C. Brooks, N. D. Paguigan, H. A. Raja et al., "qNMR for profiling the production of fungal secondary metabolites," *Magnetic Resonance in Chemistry*, vol. 55, no. 7, pp. 670–676, 2017.

[12] M. Khatib, G. Pieraccini, M. Innocenti, F. Melani, and N. Mulinacci, "An insight on the alkaloid content of Capparis spinosa L. root by HPLC-DAD-MS, MS/MS and 1 H qNMR," *Journal of Pharmaceutical and Biomedical Analysis*, vol. 123, pp. 53–62, 2016.

[13] U. Holzgrabe, "Quantitative NMR spectroscopy in pharmaceutical applications," *Progress in Nuclear Magnetic Resonance Spectroscopy*, vol. 57, no. 2, pp. 229–240, 2010.

[14] U. Holzgrabe, R. Deubner, C. Schollmayer, and B. Waibel, "Quantitative NMR spectroscopy—applications in drug analysis," *Journal of Pharmaceutical and Biomedical Analysis*, vol. 38, no. 5, pp. 806–812, 2005.

17

Determination of Cyclopropane Fatty Acids in Food of Animal Origin by ^{1}H NMR

Veronica Lolli (iD), **Angela Marseglia** (iD), **Gerardo Palla**, **Emanuela Zanardi**, and **Augusta Caligiani** (iD)

Department of Food and Drug, University of Parma, Parco Area delle Scienze 27A, 43124 Parma, Italy

Correspondence should be addressed to Augusta Caligiani; augusta.caligiani@unipr.it

Academic Editor: Mercedes G. Lopez

Cyclopropane fatty acids (CPFAs) are unusual fatty acids of microbial origin, recently detected in milk and dairy products. CPFAs have been demonstrated to be interesting molecular markers for authentication of dairy products obtained without ensiled feeds. Moreover, they can also be recognized as a new secondary component of human diet. Information is lacking on the presence of cyclic fatty acids in other food sources. Cyclopropane fatty acids have been detected by GC-MS analysis in cheese and other animal fats in concentration ranging from 200 to 1000 mg/kg fat, but in some cases, the complex fatty acid profile and the possible presence of interfering peaks make the separation not straightforward and the quantification uneasy. Therefore, a new reliable ^{1}H NMR method was developed to detect and measure CPFA content in different foods of animal origin, based on the detection of the characteristic signals of cyclopropane ring. The ^{1}H NMR (600 MHz) method showed detection limits comparable with those of full scan GC-MS, and it allowed the identification and quantitation of the cyclopropane fatty acids in different foods.

1. Introduction

Cyclopropane fatty acids are unusual fatty acids found in microorganisms, both Gram-negative and Gram-positive, and seed oils of some tropical plants and protozoa [1, 2]. The bacterial production of cyclopropane ring is related to changes in the membrane fatty acids composition and represents one of the most important adaptive microbial responses that favours the stress tolerance of several bacteria, such as *Lactobacillus helveticus*, *L. bulgaricus*, *L. acidophilus*, and *L. sanfranciscensis* [1].

In plants, CPFAs are usually minor components, where cyclopropene fatty acids are the most abundant. They are present in Malvaceae, Sterculiaceae, and Sapindaceae, representing a significant component of *Litchi chinensis* and *Sterculia foetida* seed oils, principally sterculic acid [2].

Recently, we identified by GC-MS the presence of CPFAs (dihydrosterculic and lactobacillic acids, Figure 1) in milk and dairy products [3, 4] and more recently in meat (unpublished results).

Due to the undoubtable importance of these foodstuffs in human diet, it appears clear that a deep investigation on the dietary intake of these fatty acids and their effects on humans gains importance.

CPFAs have been also recently identified in human serum and adipose tissue [5], suggesting that they are absorbed as the other fatty acids and can exert physiological effects. Moreover, CPFAs are minor fatty acids but their presence in milk fat is in the hundred-ppm order [4], so their dietary intake may be not negligible.

CPFAs (mainly dihydrosterculic acid) also play an important role in food authentication: in fact, they were discovered in milk and dairy products from cows fed with silages, and their determination has been demonstrated to be a powerful tool for the authentication of Protected Denomination of Origin (PDO) cheeses, such as Parmigiano Reggiano, where the use of silages in cow feeding is forbidden [6]. In this context, "Consorzio del Formaggio Parmigiano Reggiano" has proposed a modification on the Production Specification Rules, including the determination of CPFAs among the official controls (UNI 11650).

Therefore, CPFAs represent an almost completely new field of research in food lipids and it is important to develop different methods of detection and quantification, in view of

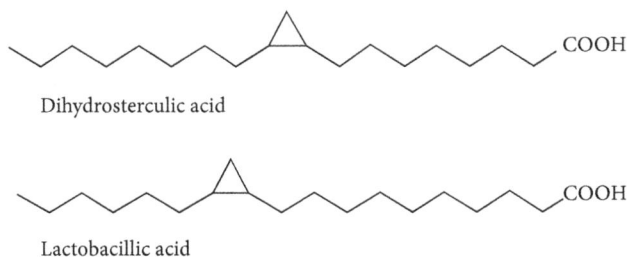

FIGURE 1: Main cyclopropane fatty acids detected in dairy products.

an expected growing body of research, both in food characterization/authentication and in food safety aspects.

Gas chromatography methods currently dominate the literature for the determination of main and secondary fatty acids in foods [7–9], and we previously applied this technique for the qualitative and quantitative determination of CPFAs in milk and dairy products [3, 4, 6]. However, gas chromatography analysis requires time-consuming sample derivatization with the risk of interfering by-products and use of large amount of solvents [10]. Moreover, in the particular case of fat from animal origin, the extreme complexity of fatty acid profile makes the separation and quantification of minor fatty acids a challenging issue. For example, more than 400 different fatty acids were detected in milk [11]. In the case of cyclopropane fatty acids, we obtained its separation in cheese fat by using apolar capillary column [6]; however, this column is not suitable for the optimal separation of fatty acids, so it is possible that changing the food matrix interferences occur. It is also possible that other cyclopropane fatty acids were present but undetectable because they were overlapped by the most abundant fatty acid signals. So, it is important to have an alternative method to confirm the cyclopropane ring presence and possibly to correctly quantify CPFAs. Moreover, the development of a rapid method that provides the necessary analytical information with minimal sample preparation would be advantageous. NMR spectroscopy is one such analytical tool that avoids sample derivatization and offers the benefit of short data acquisition times.

Nuclear magnetic resonance spectroscopy has started to represent an interesting tool to analyse biofluids and food and beverages, and in the case of lipids, it represents a reliable and fast alternative to traditional methods such as gas chromatography. This was due to the advantages of this technique as the simplicity of the sample preparation (usually it only requires the fat dissolution in deuterated chloroform) and measurement procedures, the instrumental stability, the increase of sensitivity, and modern pulse sequences, with simultaneous suppression of big signals [12].

For these reasons, the use of NMR spectroscopy has established a significant role in the analysis of lipids [13]. Several studies consider the analysis by ^{1}H NMR of triacylglycerol composition as a useful tool for both triglyceride quantitation and sample classification [14]. Minor fatty acids were also object of investigation by NMR, especially conjugated linoleic acids (CLAs) [15].

NMR could represent an ideal method to detect CPFAs due to the characteristic signals of the protons of the cyclopropane unit between −0.30 and −0.35 ppm [16], which permit their detection in a zone of ^{1}H NMR spectrum practically free from other signals. This highly shielded position of cyclopropane resonance is conventionally explained by the anisotropy of the C–C bond, just opposite to CH$_2$ group in a three-membered ring, or by an aromatic-like ring current involving the six electrons in the three C–C bonds (σ aromaticity) that shields cyclopropane protons [17].

Therefore, with the aim to investigate on the presence of CPFAs in foods, we developed a new fast and reliable quantitative ^{1}H NMR method, to be used as alternative to gas chromatographic methods and to confirm the presence of CPFAs in foods.

2. Experimental

2.1. Materials. Methanol, *n*-hexane, dichloromethane, trimethylchlorosilane, hexamethyldisilazane, 1-decanol, sodium sulphate anhydrous, sodium carbonate, deuterated chloroform, and tetracosane were from Sigma-Aldrich (Saint Louis, MO, USA), and hydrochloric acid and potassium hydroxide pellets were from Carlo Erba (Milan, Italy). Dihydrosterculic acid methyl ester was from Abcam (Cambridge).

All the solvents, standards, and reagents were of analytical grade.

Cheese, meat samples from several species animals, cured meat, and commercial fish were analysed for the content of cyclopropane fatty acids. Most of them were purchased from the market (Parma, Italy). Samples of cheese and meat produced without ensiled feeds were kindly provided from Parmigiano Reggiano Cheese Consortium and Prof. Riccardo Bozzi of the University of Florence, respectively.

2.2. Fat Extraction. Lipid extraction following the Folch method [18] was performed. 10 g of sample was homogenized with 75 mL of dichloromethane : methanol (2 : 1, v/v). The mixture was centrifuged (10 min, 3000 rpm) and filtered. This procedure was repeated three times. The three filtrates were transferred to a graduate cylinder, and a volume of about 50 mL KCl 0.88% in distilled water was added. The mixture was shaken vigorously. The final biphasic system was decanted, and the upper aqueous phase was eliminated. The lower organic phase was filtered through anhydrous sodium sulphate and collected. Lipid content was then recovered after solvent was evaporated with a rotary evaporator under vacuum.

2.3. ^{1}H NMR Analysis

2.3.1. Synthesis of Internal Standard Trimethylsilyl Decanol (TMSD). 0.2 mL of 1-decanol, 0.3 mL of trimethylchlorosilane, and 0.6 mL of hexamethyldisilyiazane were mixed in a screw cap septum vial. Mixture reacted for 1 h at 60°C, neutralized with sodium carbonate, and then dried with anhydrous sodium sulphate. Reaction mixture was diluted with 1 mL of hexane, filtered, taken to dryness in a rotary evaporator, and the residue weighed. Purity of trimethylsilyl decanol (TMSD) was confirmed by ^{1}H NMR and by GC-MS analysis in the conditions reported in Section 2.4.

2.3.2. Preparation of CPFA and TMSD Standard Solutions. Appropriate amounts of trimethylsilyl decanol (TMSD, internal standard) and CPFAs were weighed and added separately to CDCl$_3$ (10 mL) to yield two final stock solutions of about 500 mg/L each.

Adequate amounts of CPFA and TMSD stock solutions were transferred in 5 mm NMR tubes and taken to the final volume of 1 mL with CDCl$_3$ to obtain working solutions at 100, 50, 25, and 5 μg/mL of CPFAs, all containing 10 μg/mL of TMSD.

2.3.3. Preparation of Spiked Samples. 100 mg of meat fat (chicken) and cheese fat (Parmigiano Reggiano) both negative to CPFAs were spiked with the appropriate amount of CPFA and TMSD solutions and taken to the volume of 1 mL of CDCl$_3$ to obtain the same final concentrations reported above for standard solutions.

2.3.4. ^{1}H NMR Acquisition. 100 mg of fat was dissolved in 1 mL of CDCl$_3$ containing 0.01 mg of TMSD as internal standard. ^{1}H NMR spectra were recorded on a Varian INOVA-600 MHz spectrometer (Varian, Palo Alto, CA, USA), equipped with a 5 mm triple resonance inverse probe. Data were collected at 298 K, with 32 K complex points, using a 90° pulse length. 1024 scans were acquired with an acquisition time of 1.707 s and a recycle delay of 2 s. Presaturation of the fatty acids –CH$_2$– signal (1.25 ppm) was performed in order to assure a correct digitization of small signals as CPFAs. The NMR spectra were processed by MestReC software 6.0.2 (Santiago de Compostela, Spain, EU): spectra were Fourier transformed with FT size of 64k and 1 Hz line-broadening factor, manually phased and carefully baseline corrected, and referenced to the chloroform signal (7.26 ppm). Baseline correction was further manually optimized in the zone of interest (from −1 ppm to 0.7 ppm).

2.3.5. Quantitative Analysis. CPFA concentrations were obtained by integrating the peak area of the ^{1}H NMR signal at −0.35 ppm and the methyl signal of the trimethylsilyl group of the internal standard (TMSD) at 0.1 ppm.

The CPFA integral was converted in mass value (mg) according to the following formula, as previously reported [19]:

$$ACPFA \times \frac{EWCPFA}{mg\ CPFA} = ATMSD \times \frac{EWTMSD}{mg\ TMSD}, \quad (1)$$

where ACPFA = spectral area of CPFA, ATMSD = spectral area of internal standard, EWCPFA = equivalent weight of the analyte, EWTMSD = equivalent weight of internal

standard, and EW = (molecular weight/number of hydrogens in the signal).

Absolute amount of CPFAs obtained was finally expressed as mg/kg of fat.

2.3.6. Linearity and Limit of Detection and Quantification. The limit of detection (LOD) and the limit of quantification (LOQ) were calculated utilizing the S/N ratio methods, based on the determination of the peak-to-peak noise [20]. LOD and LOQ were, therefore, calculated as the concentrations of CPFAs producing a recognizable peak with a signal-to-noise ratio of, respectively, 3.3 and 10. LOD and LOQ were determined both in pure standard solution and in a sample of meat fat negative to CPFAs spiked with different concentrations of CPFAs.

2.3.7. Accuracy, Precision, and Recovery of the Method. The accuracy of the CPFA recovery was determined by assaying samples with known concentrations of CPFAs, both as pure compounds and as spiked matrix. The precision was expressed as coefficient of variation (CV%). Recovery of analytes was determined by spiking sample of fat free from CPFAs with pure dihydrosterculic acid.

2.4. Gas Chromatographic Analysis. GC-MS quantitative analysis was performed as previously reported [6]. Briefly, 200 mg of fat was dissolved in hexane (5 mL) and mixed for 1 min with 0.2 mL of KOH 10% (Carlo Erba, Milan, Italy) in methanol. After phase separation, the superior organic phase was added to internal standard (tetracosane) and injected (1 μL, split mode) on an Agilent Technologies 6890N gas chromatograph (Agilent Technologies, Palo Alto, CA, USA) coupled to an Agilent Technologies 5973 mass spectrometer (Agilent Technologies, Palo Alto, CA, USA). A low-polarity capillary column (SLB-5ms, Supelco, Bellafonte, USA) was used. The chromatogram was recorded in the scan mode (40–500 m/z) with a programmed temperature from 60°C to 280°C.

3. Results and Discussion

3.1. CPFA Signal Detection. Figure 2 depicts the characteristic upfield zone of ^{1}H NMR spectrum of dihydrosterculic acid and TMCD. ^{1}H NMR shows two individual peaks at −0.35 and 0.60 ppm for the methylene protons of the cyclopropane ring. Assignments were previously made by Knothe [16] with the aid of 2D correlations: the upfield signal is assigned to the *cis*-proton and the downfield signal to the *trans*-proton. The two methine protons of the cyclopropane ring are located at 0.68 ppm. The other CPFA proton signals are located at lower fields; for example, the four protons in alpha position with respect to the cyclopropane ring display a distinct shift at 1.17 ppm and the signal of the other two protons is observed at 1.40 ppm, within the broad methylene peak [16]. Among all these specific signals, the *cis*-methylene proton of the cyclopropane ring can be easily assigned and used for quantification

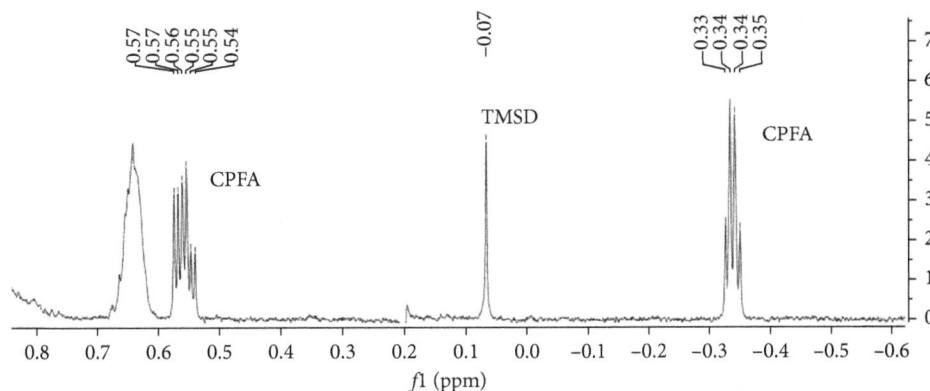

FIGURE 2: ^{1}H NMR spectrum (600 MHz, CDCl$_3$) of dihydrosterculic acid standard and trimethylsilyl decanol (TMSD) in the very upfield region of the spectrum. The CPFA signal at −0.34 ppm and the TMSD signal at 0.07 ppm were selected for quantification.

because it does not overlap with any other signal of fatty acids that could be observed in a complex food lipid ^{1}H NMR spectrum.

Because the integral of a given peak in a ^{1}H NMR spectrum is directly proportional to a corresponding number of resonant nuclei, peak areas of CPFA cis-methylene proton can be compared with the peak area of TMSD trimethylsilyl group close to 0.1 ppm and used for the determination of total CPFA content. This internal standard was specifically synthetized starting from a medium chain linear alcohol as decanol because it is not volatile, soluble in apolar solvents, and its trimethylsilyl group gives a singlet close to the CPFA selected signal. Tetramethylsilane (TMS) was commercially available but it cannot be used as quantitative internal standard due to its volatility. The synthesis of TMSD was quick and simple with a high yield (about 80%). The purity was determined by GC-MS and ^{1}H NMR (data not shown).

^{1}H NMR spectra of food fats are characterized by a dominating fatty acid methylene group peak at 1.2 ppm that is many orders of magnitude greater than those of the component of interest. This causes a number of problems: first of all, it prevents a correct digitization of small signals, hampering their observation and quantification, but the tail of this large signal can also determine a distortion of the baseline in the zone of CPFA chemical shifts. Therefore, a suppression of this signal was performed during spectra acquisition.

3.2. Quality Parameters of ^{1}H NMR Analysis. The quantitative ^{1}H NMR method was developed with the aim to determine CPFA concentration in a broad range of food fats, in particular in fats of animal origin (dairy products, meat, and fish). As a first step, the method was subjected to validation in terms of precision, accuracy, linearity, detection, and quantitation limits, following recommendations of the International Conference on Harmonization (ICH 2005: Validation of analytical procedures: text and methodology. Harmonized tripartite guideline, Q2, R1). The validation tests were performed on pure solutions of CPFA (dihydrosterculic acid) and on cheese and meat matrices naturally

free from CPFAs, spiked with dihydrosterculic acid as reported in Experimental.

To determinate accuracy and precision, solutions containing weighed amount of CPFAs were analysed by ^{1}H NMR in the experimental conditions previously reported. Measured results for the standard solutions were in agreement with the amounts weighed in the range of concentrations of 0.005–0.1 mg/mL of CPFA, a range that corresponds with the final intube concentration of the analytes in real samples of fat. The linearity was demonstrated in the same range (Figure 3). The limit of detection and the limit of quantification were calculated utilizing the S/N ratio method described above. LOD and LOQ were calculated in pure standard solutions. The instrumental quantification limit for CPFA standard (LOQ, signal to noise ratio higher than 10) in the experimental conditions reported was about 0.01 mg/mL while the limit of detection (LOD) was obtained at 0.0025 mg/mL (S/N ratio 4). The ^{1}H NMR (600 MHz) method developed showed detection limits in pure standard solutions comparable with those generally achieved by full scan GC-MS. Coefficients of variation (CV%) for three replicate measurements of each standard concentration were lower than 3%, indicating a good precision of the method.

Linearity, LOD, and LOQ were also calculated in two different matrices, cheese and chicken fat. The samples chosen for spiking were previously analysed by GC-MS and ^{1}H NMR and were found negative to CPFAs. Each matrix was spiked with four different amounts of dihydrosterculic acid as reported in Experimental. Regression curves obtained are shown in Figure 4.

The linearity is maintained, as in the case of pure standard; however, in both cases, the intercept of the regression curve indicates a matrix effect, most pronounced for chicken fat. The limit of quantification for cheese was 120 mg/kg fat (S/N ratio of 10), and the limit of quantification was found to be 50 mg/kg fat (S/N ratio of 10). In the case of chicken meat, LOQ and LOD were 180 mg/kg fat and 70 mg/kg fat, respectively. Comparing these values with those obtained for pure standard solutions, results demonstrate not negligible matrix effect both in cheese and in

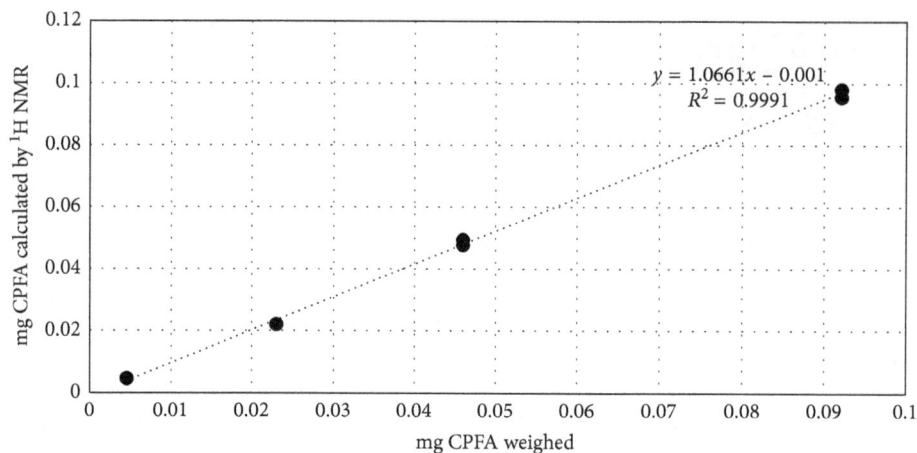

FIGURE 3: Regression curve for CPFA standard solution measured with the ^{1}H NMR method.

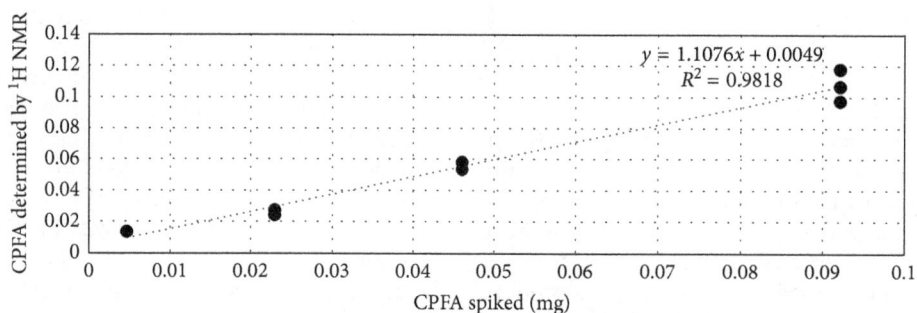

(a)

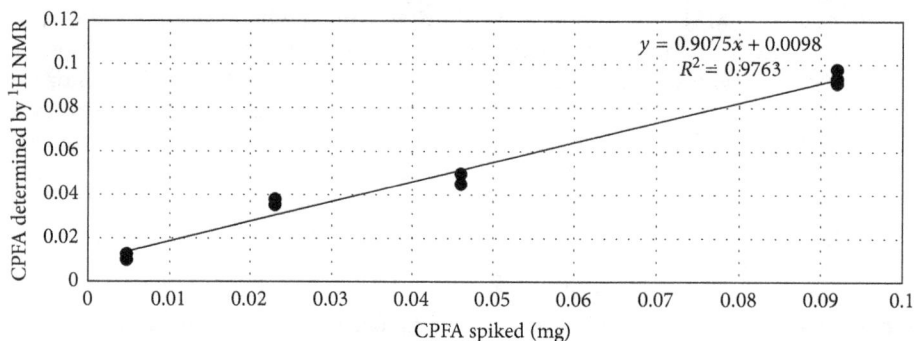

(b)

FIGURE 4: Calibration curves of CPFA spiked matrices: (a) cheese and (b) chicken meat.

meat fat, suggesting that an external calibration in a fat matrix is needed for an accurate quantification. On the contrary, the method can be easily applied for a rapid semiquantitative and qualitative analyses in both matrices.

3.3. Application to Real Samples

3.3.1. Comparison of ^{1}H NMR-Based Method and GC-MS Method. Different samples of meat, fish, and cheese were analysed both by ^{1}H NMR and GC-MS. GC-MS quantitative analysis was performed based on the method previously applied for cheese, as described in

Caligiani et al. [6]. Figure 5 shows the enlargement of the diagnostic region for CPFAs in the ^{1}H NMR spectra of lipids extracted from cheese, meat, and fish (containing the TMSD internal standard), confirming that the signal does not overlap with any other resonances representative of fatty acids.

Table 1 reports the list of the samples analysed for each food category, the number of samples negative or positive to CPFAs, and the comparison between GC-MS and ^{1}H NMR results. A reference Grana Padano cheese was specifically analysed both by GC-MS and ^{1}H NMR method to demonstrate the accuracy of the new ^{1}H NMR method. Then, five samples of Parmigiano Reggiano and five samples of Grana Padano were tested because in the case of cheese, we had

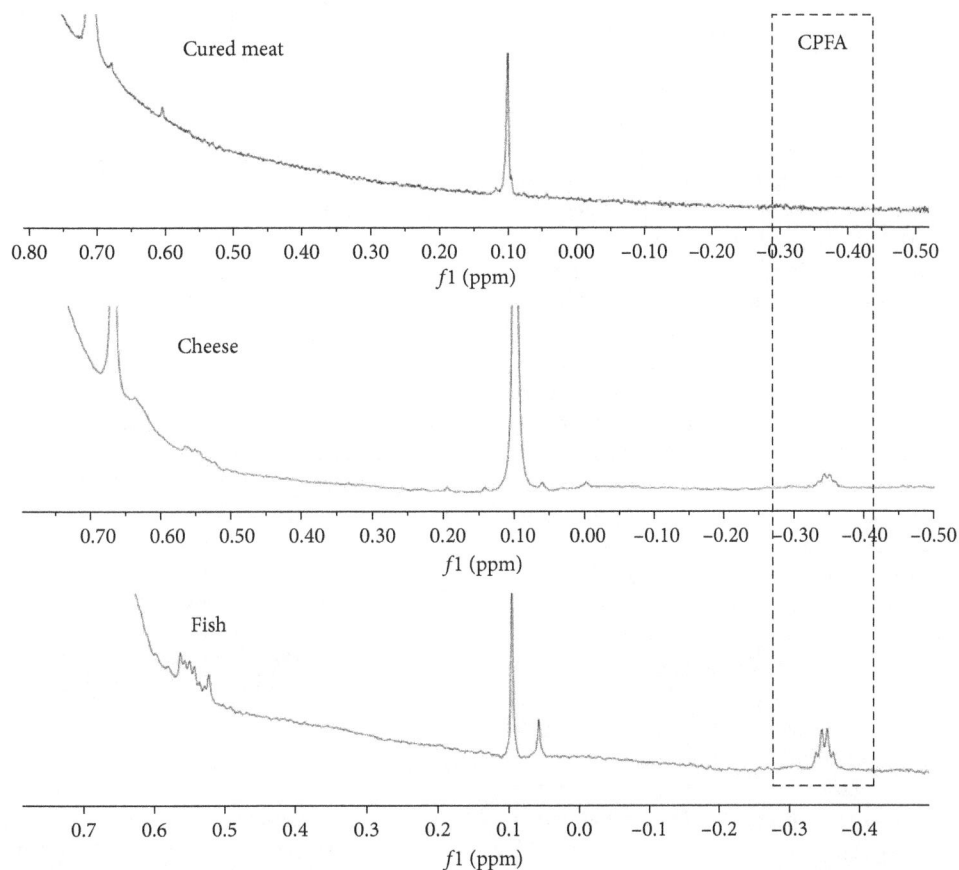

FIGURE 5: Enlargement of ^{1}H NMR 600 MHz spectra in the zone from −0.5 to 0.8 ppm, showing the signal of the cyclopropane ring (at −0.35 ppm) used for CPFA quantification in cured ham (negative to CPFAs) and cheese and fish fat (positive to CPFAs).

TABLE 1: Comparison of ^{1}H NMR and GC-MS results on the presence of CPFAs in some representative samples of fat of animal origin.

Samples	Number of analysed samples	CPFA (GC-MS) (mg/kg fat)	CPFA (^{1}H NMR) (mg/kg fat)
Cheese			
Reference cheese		600 ± 50	690 ± 60
Parmigiano Reggiano	5	<LOD	<LOD
Grana Padano	5	400–700	400–800
Meat			
Commercial bovine meat	5	200–400	300–400
Bovine meat of certified origin (not fed with silages)	2	<LOD	<LOD
Other meats (Pork and chicken)	4	<LOD	<LOD
Cured meat			
Salami	1	60	<LOD
Parma ham	1	120	<LOD
Bresaola	1	340	400
Fish			
Eel	2	180–350	400–590
Mullet	1	120	800

collected previously many data confirming the association between the use of ensiled feeds and the presence of CPFA [4, 6]. ^{1}H NMR analysis confirmed the positivity at CPFA for all samples of Grana Padano (ensiled feeds allowed) and the negativity of all Parmigiano Reggiano samples (ensiled feeds forbidden), indicating that ^{1}H NMR could be an attractive alternative technique to GC-MS to assure the authenticity of Parmigiano Reggiano and other cheeses forbidding the use of ensiled feeds in their disciplinary of productions. In the case of cheese, quantitative results suggested very good agreement between data obtained by the quantitative ^{1}H NMR analysis and previous GC analysis method.

Concerning meat, CPFAs were detected in the GC-MS profiles of most of the commercial bovine meat samples in concentrations varying from 100 to 400 mg/kg of the total fat. CPFAs were detected by ^{1}H NMR analysis in all commercial bovine meat samples previously resulted positive by the GC-MS analysis, with good agreement of the quantitative results. CPFAs were absent in two samples of certified meat from cows not fed with fermented forages, and this was evidenced by both techniques. The GC-MS analysis of other meat samples (pork and chicken) was negative to CPFAs both in GC-MS and ^{1}H NMR method. In the case of pork cured meat (salami and ham), the GC-MS analysis showed the presence of a signal at the retention time of CPFAs with concentrations of 60–100 mg/kg of the total fat. However, the corresponding analysis by ^{1}H NMR did not show the presence of cyclopropane ring, indicating the presence of an interfering peak in the GC-MS conditions adopted. This interfering peak was also resistant to oxidation as a saturated fatty acid, but it has not been identified yet and it was not easy to obtain a better separation varying chromatographic conditions. Therefore, in the case of pork, cured meat seems to be important to have the NMR confirmation of CPFA presence.

And in cured meat, the GC-MS analysis of fish samples generally showed the presence of interfering signals at the same retention time of cyclopropane fatty acids with the same corresponding mass spectrum (278 m/z), probably due to the presence of different isomers of nonadecenoic acid. Moreover, lactobacillic acid coeluted with another interfering substance with the corresponding mass spectrum of 165 m/z. This interfering peak was not resistant to oxidation and it has been suggested to be a furan fatty acid as discussed elsewhere [21, 22].

Therefore, GC-MS analysis alone was not able to confirm the presence or absence of CPFAs in fish samples, but it always required ^{1}H NMR analysis. Moreover, observing the preliminary results showed in Table 1 on three fish samples, it seems that GC-MS underestimates the content of cyclopropane fatty acids, suggesting that besides cyclopropane fatty acids with 19-carbon atom skeleton, such as dihydrosterculic and lactobacillic acids, it is possible that other CPFAs with different chain lengths occur in fish.

4. Conclusion

A new quantitative ^{1}H NMR method was developed for the determination of CPFA content in different food matrices, including dairy products, meat, and fish.

The new method reported here provides absolute quantities of CPFA (mg/kg of total fat) and shows a limit of detection comparable with those of full scan GC-MS. A complete and reliable sample analysis can be performed quickly and requires little sample preparation, reagents, and solvents. This was possible because the CPFA signal was very well defined and did not overlap with others. The role of NMR seems to be most important in meat and fish characterization because the GC-MS analysis was not able to confirm the presence of CPFAs in all the analysed samples due to the presence of interfering peaks.

Results suggested that the NMR analysis approach has potential application as a screening for quantifying cyclopropane fatty acids in meat and fish, as markers of quality and the preliminary data on few meat and fish samples presented here suggest some possible developments. For example, in the context of food authentication, cyclopropane fatty acids might be proposed, as in the case of cheese, as markers of silage feedings are able to authenticate high-quality costly meat whose producers declare the absence of silages in the feeding. This will require the construction of a robust database of meat certificated for the feeding system. This approach could also be extended to fish, to eventually distinguish farmed from wild fish.

Moreover, in the case of fish, NMR method is able to detect a higher amount of CPFAs with respect to GC-MS, indicating an important role of NMR when dietary intake of cyclopropane fatty acid has to be assessed.

Conflicts of Interest

The authors declare that there are no conflicts of interest regarding the publication of this paper.

References

[1] C. Montanari, S. L. SadoKamdem, D. I. Serrazanetti, F. X. Etoa, and M. E. Guerzoni, "Synthesis of cyclopropane fatty acids in *Lactobacillus helveticus* and *Lactobacillus sanfranciscensis* and their cellular fatty acids changes following short term acid and cold stresses," *Food Microbiology*, vol. 27, no. 4, pp. 493–502, 2010.

[2] X. Bao, S. Katz, M. Pollard, and O. John, "Carbocyclic fatty acids in plants: biochemical and molecular genetic characterization of cyclopropane fatty acid synthesis of *Sterculia foetida*," *Plant Biology*, vol. 99, no. 10, pp. 7172–7177, 2002.

[3] A. Marseglia, A. Caligiani, L. Comino, F. Righi, A. Quarantelli, and G. Palla, "Cyclopropyl and ω-cyclohexyl fatty acids as quality markers of cow milk and cheese," *Food Chemistry*, vol. 140, pp. 711–716, 2013.

[4] A. Caligiani, A. Marseglia, and G. Palla, "An overview on the presence of cyclopropane fatty acids in milk and dairy products," *Journal of Agricultural and Food Chemistry*, vol. 62, pp. 7828–7832, 2014.

[5] T. Sledzinski, A. Mika, P. Stepnowski et al., "Identification of cyclopropaneoctanoic acid 2-hexyl in human adipose tissue and serum," *Lipids*, vol. 48, no. 8, pp. 839–848, 2013.

[6] A. Caligiani, M. Nocetti, V. Lolli, A. Marseglia, and G. Palla, "Development of a quantitative GC-MS method for the detection of cyclopropane fatty acids in cheese as new molecular markers for Parmigiano Reggiano authentication," *Journal of Agricultural and Food Chemistry*, vol. 64, pp. 4158–4164, 2016.

[7] W. W. Christie, *Gas Chromatography and Lipids: A Practical Guide*, The Oily Press, Bridgwater, UK, 1989.

[8] P. Delmonte, A. R. Fardin-Kia, J. K. Kramer et al., "Evaluation of highly polar ionic liquid gas chromatographic column for the determination of the fatty acids in milk fat," *Journal of Chromatography A*, vol. 1233, pp. 137–146, 2012.

[9] J. Ecker, M. Scherer, G. Schmitz, and G. Liebisch, "A rapid GC-MS method for quantification of positional and geometric isomers of fatty acid methyl esters," *Journal of Chromatography B*, vol. 897, pp. 98–104, 2012.

[10] D. Prema, T. D. Turner, J. Jensen et al., "Rapid determination of total conjugated linoleic acid concentrations in beef by ^{1}H NMR spectroscopy," *Journal of Food Composition and Analysis*, vol. 41, pp. 54–57, 2015.

[11] M. Schröder and W. Vetter, "Detection of 430 fatty acid methyl esters from a transesterified butter sample," *Journal of the American Oil Chemical Society*, vol. 90, p. 771, 2013.

[12] G. Le Gall and I. J. Colquhoun, "NMR spectroscopy in food authentication," in *Food Authenticity and Traceability*, M. Lees, Ed., pp. 131–155, Woodhead Publishing Ltd, Cambridge, UK, 2003.

[13] A. Barison, C. W. Pereira da Silva, F. R. Campos, F. Simonelli, C. A. Lenz, and A. G. Ferreira, "A simple methodology for the determination of fatty acid composition in edible oils through ^{1}H NMR spectroscopy," *Magnetic Resonance in Chemistry*, vol. 48, pp. 642–650, 2010.

[14] W. Jakes, A. Gerdova, M. Defernez et al., "Authentication of beef versus horse meat using 60 MHz ^{1}H NMR spectroscopy," *Food Chemistry*, vol. 175, pp. 1–9, 2015.

[15] R. M. Maria, L. A. Colnago, L. A. Forato, and D. Bouchard, "Fast and simple nuclear magnetic resonance method to measure conjugated linoleic acid in beef," *Journal of Agricultural and Food Chemistry*, vol. 58, pp. 6562–6564, 2010.

[16] G. Knothe, "NMR characterization of dihydrosterculic acid and its methyl ester," *Lipids*, vol. 41, no. 4, 2006.

[17] M. Baranac-Stojanović and M. Stojanovic, "^{1}H NMR chemical shifts of cyclopropane and cyclobutane: a theoretical study," *Journal of Organic Chemistry*, vol. 78, pp. 1504–1507, 2013.

[18] J. Folch, M. Less, and G. H. Sloane, "A simple method for the isolation and purification of total lipids from animal tissues," *Journal of Biology and Chemistry*, vol. 226, pp. 497–509, 1957.

[19] J. Müller Maatsch, A. Caligiani, T. Tedeschi, K. Elst, and S. Sforza, "Simple and validated quantitative ^{1}H NMR method for the determination of methylation, acetylation, and feruloylation degree of pectin," *Journal of Agricultural and Food Chemistry*, vol. 62, pp. 9081–9087, 2014.

[20] I. Apostol, K. J. Miller, J. Ratto, and D. N. Kelner, "Comparison of different approaches for evaluation of the detection and quantitation limits of a purity method: a case study using a capillary isoelectrofocusing method for a monoclonal antibody," *Analytical Biochemistry*, vol. 385, pp. 101–106, 2009.

[21] G. Spiteller, "Furan fatty acids: occurrence, synthesis, and reactions. Are furan fatty acids responsible for the cardioprotective effects of fish diet?," *Lipids*, vol. 40, pp. 755–771, 2005.

[22] C. Truzzi, S. Illuminati, A. Annibaldi, M. Antonucci, and G. Scarponi, "Quantification of fatty acids in the muscle of Antarctic fish *Trematomus bernacchii* by gas chromatography-mass spectrometry: optimization of the analytical methodology," *Chemosphere*, vol. 173, pp. 116–123, 2017.

18

Development of a Method for Rapid Determination of Morpholine in Juices and Drugs by Gas Chromatography-Mass Spectrometry

Mengsi Cao,[1] **Pingping Zhang,**[2] **Yanru Feng,**[1] **Huayin Zhang,**[1] **Huaijiao Zhu,**[1] **Kaoqi Lian** (ID),[1,3] **and Weijun Kang** (ID)[1]

[1]*School of Public Health, Hebei Medical University, Shijiazhuang 050017, China*
[2]*Department of Reproductive Genetic Family, Hebei General Hospital, Shijiazhuang 050017, China*
[3]*Hebei Province Key Laboratory of Environment and Human Health, Shijiazhuang 050017, China*

Correspondence should be addressed to Kaoqi Lian; liankq@hebmu.edu.cn and Weijun Kang; kangwj_hebmu@126.com

Academic Editor: Serban C. Moldoveanu

A reliable derivatization method has been developed to detect and quantify morpholine in apple juices and ibuprofen with gas chromatography-mass spectrometry. Morpholine can react with sodium nitrite under acidic condition to produce stable and volatile N-nitrosomorpholine derivative. In this experiment, various factors affecting the derivatization and extraction process were optimized, including volume and concentration of hydrochloric acid, quantity of sodium nitrite, derivatization temperature, derivatization time, extraction reagents, and extraction time. The derivative was extracted with dichloromethane and determined by gas chromatography-mass spectrometry. The linearity range of morpholine was $10-500\ \mu g \cdot L^{-1}$ with good correlation, and limits of detection (LOD) and limits of quantification (LOQ) were $7.3\ \mu g \cdot L^{-1}$ and $24.4\ \mu g \cdot L^{-1}$, respectively. Low, medium, and high concentrations of morpholine were added in apple juices and ibuprofen samples to evaluate standard recovery rate and relative standard deviation. The spiked recovery rate ranged from 94.3% to 109.0%, and the intraday repeatability and interday reproducibility were 2.0%–4.4% and 3.3%–7.0%, respectively. The developed method has good accuracy and precision. This quantitative method for morpholine is simple, sensitive, rapid, and low cost and can successfully be applied to analyze the residual morpholine in apple juices and drug samples.

1. Introduction

Morpholine (tetrahydro-2H-1,4-oxazine), a heterocyclic secondary amine, is a colorless, hygroscopic, alkaline, oily liquid at normal temperature and pressure with an ammoniacal odor and is miscible with water and organic solvents in any ratio [1]. Morpholine is used as an emulsifier for protective wax coating on apples and other fruits to keep them fresh and storable [2–4]. Nowadays, more and more people like to drink fresh juice instead of fresh fruit, and some manufacturers produce fresh juice together with the pericarp to improve dietary fiber in fruit juice and economic benefits, thereby increasing the residual content of morpholine in the juice, such as apple juice. The compound also effectively suppresses the hatching process of the eggs of golden apple snails, a known pest of the rice crops in Asia, and thereby controls the reproduction of those snails to protect the rice crops [5]. Being a cyclic amine, morpholine is commonly used in pharmaceutical industries for synthesis of different active pharmaceutical substances, such as morinidazole [6], and to increase aqueous solubility of gefitinib [7]. Morpholine has been used for preparing a series of new antimicrobial and antiviral diphenyl diselenides [8]. It is also used as a reagent to prepare the morpholine derivative, 4-(2-aminoethyl) morpholine, also called AEM [9]. AME, triethylamine, and methacryloyl chloride are used to synthesize N-ethyl morpholine methacrylamide (EMA) [10]. EMA is a pH-sensitive polymer hydrogel which is used to prevent crystallization of ibuprofen [11]. Consequently, morpholine residues may be present in the production of ibuprofen. Morpholine causes irritation of eye, skin, and digestive tract and may be absorbed in the body through skin contact, inhalation, and ingestion [1]. As a result, the use of morpholine has been prohibited as an emulsifier in protective wax coating on citrus fruits, apples, and cosmetic preparations in the European Union (EU) [1, 12]. As per Health Canada Monograph [13], the no-observed-adverse-effect level (NOAEL) of morpholine is 96 mg·kg^{-1} of body weight (bw) day^{-1} and the acceptable daily intake (ADI) is 0.48 mg·kg^{-1} of bw day^{-1} [14]. Therefore, establishing a rapid and effective method to detect and quantify morpholine in fruit juices and pharmaceuticals is of primary importance.

In recent years, numerous studies have reported various analytical methods for qualitative and quantitative estimation of morpholine. These analytical methods employed various available analytical techniques, such as gas chromatography (GC) [15–17], gas chromatography-mass spectrometry (GC-MS) [18], gas-liquid chromatography-high resolution mass spectrometry (GLC-MS) [19], liquid chromatography (LC) [20], ultra performance liquid chromatography (UPLC) [21], hydrophilic interaction liquid chromatography with electrospray ionization and tandem mass spectrometry (HILIC-ESI-MS/MS) [22], and ultrahigh performance liquid chromatography-high resolution mass spectrometry (UHPLC-HRMS) [14]. However, these published methods have different disadvantages, such as tedious operation steps [14] and high cost [22]. Applying the derivatization method with 2,4-dinitrofluorobenzene (2,4-DNFB) by GC-MS to detect morpholine has better sensitivity, but has low stability [18].

This experiment was based on some of the secondary amines that could react with sodium nitrite to produce volatile N-nitrosamines (NAms) under acidic conditions [23]. We found that morpholine as a cyclic secondary amine can generate N-nitrosomorpholine (NMOR) by using sodium nitrite as the derivatization reagent under acidic condition, and NMOR which is stable and volatile can be determined by GC-MS. Our team used to establish a method to determine ketamine in urine and plasma by this derivative method and obtained good experimental results [24]. We have extensive experience about this derivatization reaction. Therefore, various factors affecting derivatization process and extraction efficiency can be optimized to develop a reliable method for rapid determination of morpholine in apple juice and drug granules through GC-MS. Compared with other existing derivatization methods, sodium nitrite and hydrochloric acid as derivatization reagents are cheap and obtained easily in this experiment. The samples only needed centrifugate and filter without complicated sample pretreatment process and was analysed rapidly by GC-MS. The consumption of organic solvents was very small in the whole test process, thereby reducing the pollution of the environment. This study established a rapid, sensitive, simple, low-cost, and reliable method to determine morpholine in apple juice and drugs, and had highly realistic application value.

2. Experimental

2.1. Chemicals. All chemicals and reagents were of analytical grade unless otherwise stated. Standard morpholine was purchased from Aladdin Reagent Co., Ltd. (Shanghai, China). The derivatization reagents of sodium nitrite (NaNO$_2$) and hydrochloric acid (HCl) were purchased from Henan Jiaozuo Three Chemical Plant (Jiaozuo, China) and Shijiazhuang Reagent Factory (Shijiazhuang, China), respectively. Dichloromethane, ethyl acetate, chloroform, n-hexane, and carbon disulfide from Xilong Chemical Factory (Shantou, China) or Tianjin General Chemical Reagent Factory (Tianjin, China) were tested to select the most optimal extraction reagent. Pure water (18.2 MΩ/cm) was obtained from Heal Force SMART-N ultrapure water system (Hong Kong).

2.2. Quantitative Methods and Quality Control Samples. Stock standard solution of morpholine (50 mg·L^{-1}) was prepared in pure water. Working calibrators at 10, 25, 50, 100, 200, 300, 400, and 500 µg·L^{-1} were prepared by diluting in pure water, and the calibration curve was fitted by linear regression method through the measurement of the peak areas corresponding to the concentrations. The acceptance criterion for the calibration curve is a correlation coefficient of 0.99 or better. Quality control (QC) samples were prepared by freshly spiking the appropriate working solution into blank apple juice and ibuprofen samples to prepare concentrations of 50, 200, and 400 µg·L^{-1} for morpholine. The series of standard solution and QC samples were freshly prepared before use.

2.3. Pretreatment of Samples. The apple juices were obtained from a local supermarket and filtered with 0.22 µm membrane filter. Ibuprofen granules were purchased from a local pharmacy and dissolved in purified water and centrifuged (10,000 rpm for 15 min) after mixing. The supernatant liquid was filtered with 0.22 µm membrane filter. All the samples were stored at 4°C.

2.4. Derivatization and Liquid-Liquid Extraction. A certain amount of morpholine stock standard solution was added to 20 mL of apple juice or ibuprofen solution in a 50 mL disposable sample pretreatment tube. The samples were centrifuged and filtered as described in the Section 2.3. To 2.0 mL of pretreated apple juice or ibuprofen solution, 200 µL of 0.05 mol·L^{-1} HCl and 200 µL of saturated NaNO$_2$ were added and vortex-mixed. The resultant solution was placed in a 10 mL glass test tube and mixed thoroughly. The mixture was heated at 40°C for 5 min on a heating block. After cooling, 0.5 mL of dichloromethane was added, and the mixture was vortex-mixed for 1 min and allowed to stand for 10 min to extract the derivative. Then, 200 µL of organic layer was transferred with a micropipette to a tipped glass tube and placed in an ice bath to prevent the volatilization of dichloromethane and the impact on experiment results.

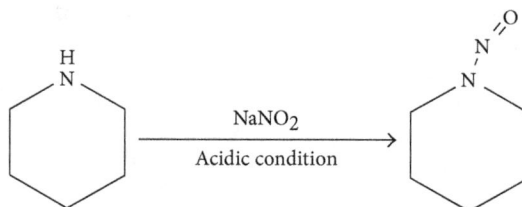

FIGURE 1: The derivatization reaction of morpholine.

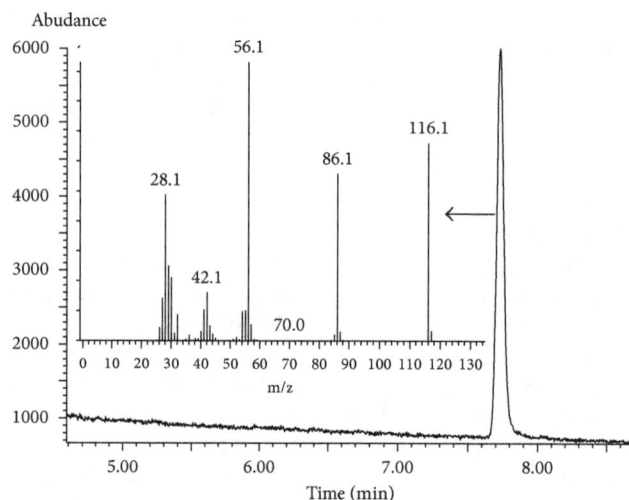

FIGURE 2: The total ion current chromatogram and mass spectra of the N-nitrosomorpholine.

Then, 1 μL of this organic layer was injected into the GC-MS with a 10 μL syringe (from Agilent).

2.5. GC-MS Analysis.

An Agilent Technologies (Little Falls, DE, USA) gas chromatograph 7890 equipped with an electronically controlled split/splitless injection port, an inert 5975C mass selective detector with electron impact (EI) ionization chamber, and a 7683B series injector/autosampler were employed for identification and quantification of N-nitrosomorpholine that was the derivative of morpholine.

The GC separation was conducted with a TM-1701 30 m × 0.32 mm I.D., 0.5 μm film thickness column (Techcomp, China). The carrier gas was helium with a constant flow rate of 2 mL·min^{-1}. The injection volume was 1 μL and was vaporized at 250°C with a 1 : 7 split ratio. The GC oven was operated with the following temperature program: initial temperature 100°C held for 4 min and programmed to 120°C at a rate of 10°C min^{-1} and held for 3 min, and then ramped at 20°C min^{-1} –250°C and held for 5 min. The total run time was 18 min.

Two different ions were selected to detect and quantify N-nitrosomorpholine (86.1, 116.1) at the selected ion-monitoring (SIM) mode. Ionization was performed by electron impact (EI) mode at 70 eV energy. The temperatures used were 280°C for the transfer line, 230°C for the ion source, and 150°C for the MS quadrupole. The solvent delay was 4.5 min.

3. Results and Discussion

3.1. Principles of Derivatization and Identification of Derivative.

Morpholine, as a secondary amine, reacts with sodium nitrite under acidic conditions to produce stable and volatile NMOR which can be determined by GC-MS. The reaction is shown in Figure 1. 2.0 mL of 400 μg·L^{-1} morpholine standard solution was used to verify the derivatization reaction. The total ion current chromatogram and mass spectra of the NMOR derivative are shown in Figure 2. Analyses of mass spectra and MS data of the derived sample proved that the derivative was NMOR.

3.2. Optimization of Derivatization and Extraction.

A rapid and low-cost derivatization technique has been developed for detection and determination of morpholine. The derivatization process of morpholine has been described in the Section 2.4. Various factors associated with derivatization and extraction process were optimized, which included concentration and dosage of hydrochloric acid (HCl), the amount of saturated sodium nitrite (NaNO$_2$), derivatization temperature, derivatization time, the extraction reagents, and extraction time.

3.2.1. Concentration and Quantity of Hydrochloric Acid.

The derivatization process was affected by the concentration and quantity of hydrochloric acid. The concentration

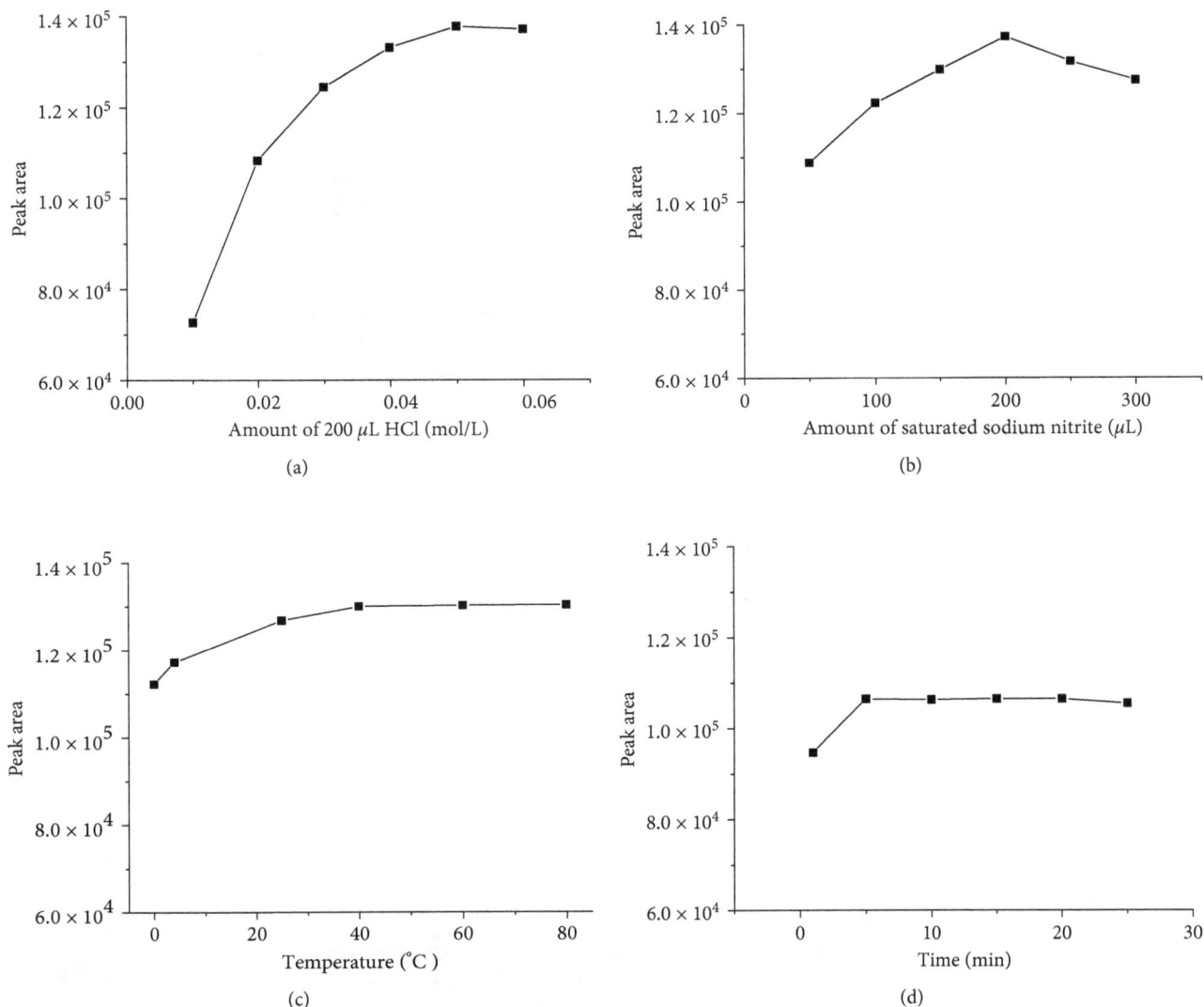

FIGURE 3: The effects of hydrochloric acid concentration and quantity (a), the amount of saturation solution of sodium nitrite (b), derivative reaction temperature (c), and time (d).

of HCl was optimized as the first step. The effects of adding 200 μL of HCl with different concentrations between 0.01 and 0.06 mol·L^{-1} are shown in Figure 3(a). The derivatization rate was found to increase with the increasing concentration of HCl in the range from 0.01 to 0.05 mol·L^{-1} and then became stable. Thus, the best result was obtained when 200 μL of 0.06 mol·L^{-1} HCl was added during the process of derivatization.

3.2.2. The Amount of Saturation Solution of Sodium Nitrite. Optimum quantity of saturation solution of sodium nitrite required for the derivatization process was determined (Figure 3(b)) by varying the addition of saturation solution of sodium nitrite in the range of 50–300 μL. The derivatization yields increased with the addition of saturation solution of sodium nitrite up to 200 μL and then became stable. Therefore, the optimum volume of saturation solution of sodium nitrite for derivatization was 200 μL.

3.2.3. Derivatization Temperature and Derivatization Time. The effects of derivatization temperature and time were tested in this experiment. The effect of temperature (0°C (ice-bath), 4°C (refrigeration), 25°C (room temperature), 40°C, 60°C, and 80°C) on derivatization was investigated. The rate of derivatization increased with reaction temperature and then became stable at 40°C (Figure 3(c)). Therefore, 40°C was selected as the optimum temperature for this experiment. Moreover, the effect of reaction time on derivatization process was investigated; the reaction time was varied between 1 and 30 min. The derivatization leveled off at 5 min (Figure 3(d)), suggesting the optimum reaction time to be 5 min.

3.2.4. Extraction Reagents and Extraction Time. Selection of suitable solvent is an important criterion for extraction of the derivative. The extraction efficiencies of n-hexane, dichloromethane, chloroform, carbon disulfide, and ethyl

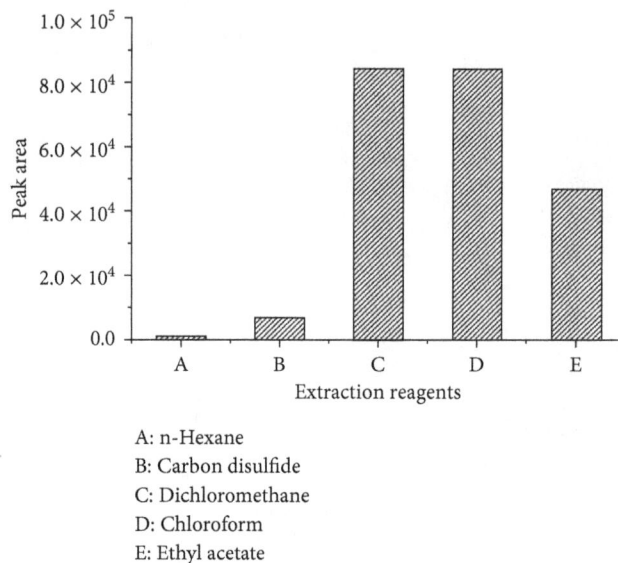

A: n-Hexane
B: Carbon disulfide
C: Dichloromethane
D: Chloroform
E: Ethyl acetate

FIGURE 4: Extraction effects of different extraction reagents.

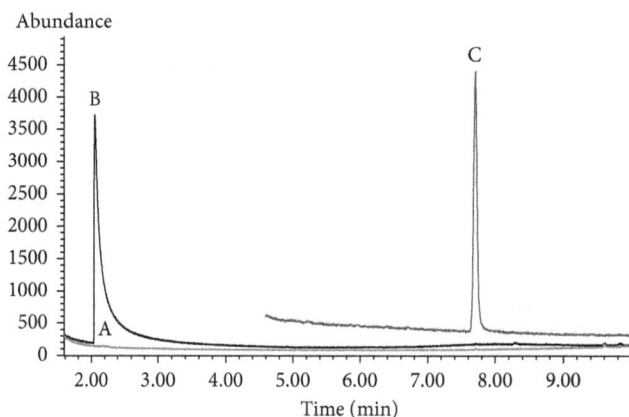

FIGURE 5: The total ion current chromatograms of morpholine resulting from different methods: direct detection of $400\,\mu g\cdot L^{-1}$ (A) and $20\,mg\cdot L^{-1}$ (B) morpholine prepared in dichloromethane and detection of $400\,\mu g\cdot L^{-1}$ morpholine prepared in pure water after the proposed derivatization (C).

acetate were evaluated as shown in Figure 4. The study revealed that dichloromethane and chloroform afforded optimum extraction of the derivative. Finally, dichloromethane was selected as the extraction reagent. 2.0 mL of dichloromethane was used to extract the derivative, and 1.5 mL organic layer was transferred to a tipped glass tube and dried with a slow stream of nitrogen at room temperature. The dried substances were dissolved in $100\,\mu L$ ethyl acetate before GC analysis. However, the experiment resulted in poor precision as indefinite derivative was blown away in the nitrogen blowing process. To improve the extraction efficiency and stabilization, 0.5 mL dichloromethane was added and vortex mixed for 1 min followed by standing for 10 min to extract the derivative.

An attempt to detect morpholine ($400\,\mu g\cdot L^{-1}$) by GC-MS without any derivatization process failed, as the method could not detect any signal of the compound (Figure 5(a)). In another attempt, morpholine produced similar signal

abundance in two different samples ($20\,mg\cdot L^{-1}$ in dichloromethane without derivatization and $400\,\mu g\cdot L^{-1}$ in pure water after derivatization) (Figures 5(b) and 5(c)). The proposed derivatization method was about 65 times more sensitive than the direct detection. Kataoka [16] had compared the effects of commonly used derivatization reagents, such as acylation, silylation, dinitrophenylation, permethylation, carbamate formation, sulfonamide formation, and phosphoamide formation for analysis of secondary amines by GC. However, sodium nitrite and hydrochloric acid are preferred as derivatization agents, since they are cheaper and easily available compared to other reagents. Sacher et al. [18] established a method based on derivatization of the amines with benzenesulfonyl chloride. However, usage of many reagents and long derivatization process (1 hour) and operation time (morpholine peak appeared at 18 min) made this method practically inconvenient. In comparison, the proposed method involves

TABLE 1: Comparison of the proposed method with previously published methods.

| Sample | The test process of sample | | | LOQ | Reference |
	Sample pretreatment	Derivatization reaction	Determination		
Apple juice and ibuprofen	Centrifugation and filtration	Sodium nitrite under acidic condition	Gas chromatography-mass spectrometry (GC-MS)	$24.4\,\mu g \cdot L^{-1}$	This work
Steam condensate	—	—	Chromatography with multimode inlet and flame ionization detection (GC-MI-FID)	$100\,\mu g \cdot L^{-1}$	[15]
Citrus and apples	15 mL 1% acetic acid in methanol	—	Hydrophilic interaction liquid chromatography with electrospray ionization and tandem mass spectrometry (HILIC-ESI-MS/MS)	$10\,\mu g \cdot kg^{-1}$	[22]
Citrus and apples	Dispersive micro-solid-phase extraction (DMSPE)	—	Ultrahigh performance liquid chromatography-high resolution mass spectrometry (UHPLC-HRMS)	$5\,\mu g \cdot kg^{-1}$	[14]

TABLE 2: Recovery and precision of three spiked levels.

Sample	Spiked concentration ($\mu g \cdot L^{-1}$)	Recovery (%)	Intraday repeatability (%)	Interday reproducibility (%)
Apple juice	50	109.0	4.4	5.2
	200	94.3	2.3	4.8
	400	98.4	3.3	5.0
Ibuprofen	50	96.0	4.4	3.3
	200	100.9	2.5	7.0
	400	107.9	2.0	5.5

less derivatization process (5 min) and operation time (morpholine peak appeared at 7.72 min).

The proposed method has many advantages compared to previously published methods [14, 15, 22] (Table 1). Use of MS detector in this study produced similar sensitivity as HILIC-ESI-MS/MS [22], and both the methods yielded better results than flame ionization detection (FID) [15]. Dawei Chen et al. [14] established a reliable method to determine morpholine residues by UHPLC-HRMS combined with dispersive micro-solid-phase extraction (DMSPE). The method is more sensitive but involves complicated sample pretreatment process and costly instrumentation.

3.3. Selection of Chromatographic Column.
In the preliminary experiment, HP-5 nonpolar chromatographic column (dimensions: 30 m × 0.32 mm × 0.25 μm; stationary phase: 5% phenyl-95% methylpolysiloxane) and TM-1701 medium polarity chromatographic column (dimensions: 30 m × 0.32 mm × 0.5 μm; stationary phase: 14% cyanopropyl phenyl-86% dimethyl polysiloxane) were employed. HP-5 column resulted in peak tailing of peak and high baseline. TM-1701 column provided better peak shape under the optimized conditions and therefore suited the experiment.

3.4. Method Validation

3.4.1. Linearity, Detection Limit, and Quantitative Limit. Calibration curve was constructed by plotting the

peak area against the concentration range from 10 to 500 $\mu g \cdot L^{-1}$ of morpholine. The linear regression equation was $A = 471.2c - 2263.8$, in which c corresponds the concentrations and A corresponds the peak areas. The results obtained a good linearity of the analytical range which was 10–500 $\mu g \cdot L^{-1}$ with the coefficient of determination (R^2) of the calibration curve for morpholine higher than 0.999. In this method, the limit of detection (LOD) and the limit of quantification (LOQ) were calculated as 3 and 10 times the S/N ratio, which indicated 7.3 $\mu g \cdot L^{-1}$ and 24.4 $\mu g \cdot L^{-1}$, respectively.

3.4.2. Accuracy and Precision. Through adding standard solution of morpholine with high concentration to the apple juice and ibuprofen blank samples, three different spiked samples with final concentration levels of 50, 200, and 400 $\mu g \cdot L^{-1}$ were obtained. The samples with each concentration level were determined on six times a day over three consecutive days. The spiked recovery rate, intraday repeatability, and interday reproducibility were 94.3%–109%, 2.3%–4.4%, and 4.8%–5.2% for apple juice spiked samples, and 96%–107.9%, 2%–4.4%, 3.3%–7% for ibuprofen spiked samples, respectively (Table 2). The results indicated that the method was suitable for determining morpholine with favourable accuracy and precision.

3.5. Application to Real Samples. Samples of apple juice and ibuprofen granules were analysed by this method under

(a)

(b)

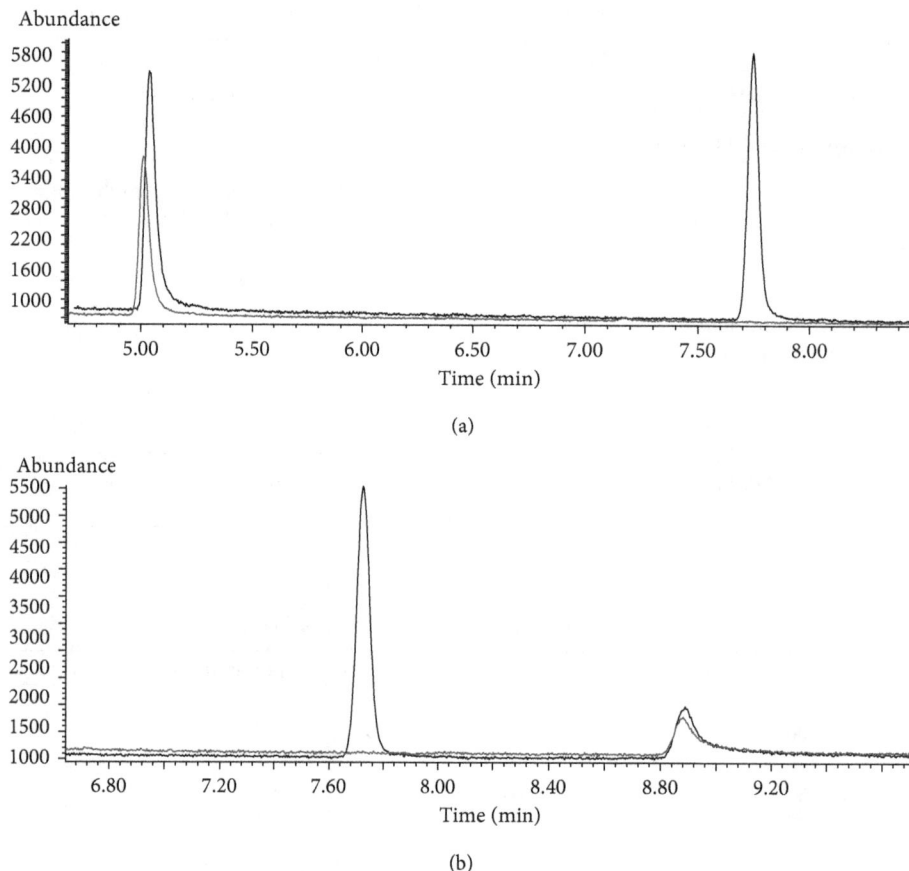

FIGURE 6: The total ion current chromatograms of apple juice samples and spiked samples (400 μg·L^{-1}) (a) and ibuprofen samples and spiked samples (400 μg·L^{-1}) (b).

optimal conditions using standard addition method. However, morpholine was not detected in any real samples. Thus, using the standard addition method, morpholine was detected in apple juice and ibuprofen granules samples. The total ion current chromatograms of the real samples of apple juice and ibuprofen granules and their spiked samples (400 μg·L^{-1}) are shown in Figure 6.

4. Conclusions

According to that morpholine could react with sodium nitrite to generate the stable and volatile N-nitrosomorpholine under acidic conditions, we established a rapid, sensitive, simple, low-cost, and reliable method to detect morpholine in apple juices and drugs. This method had been successfully analysed of spiked samples with low detection limit and favourable accuracy and precision. It can provide technical support to establish the national standards of morpholine in fruit juices and pharmaceuticals and monitor the residue of morpholine in the future.

Conflicts of Interest

The authors declare no conflicts of interest.

Authors' Contributions

Kaoqi Lian and Weijun Kang conceived and designed the experiments. Mengsi Cao, Pingping Zhang, Yanru Feng, Huayin Zhang, and Huaijiao Zhu performed the experiments. Kaoqi Lian and Mengsi Cao analysed the data and wrote the paper. All authors read and approved the final manuscript.

Acknowledgments

This work was supported by the National Natural Science Foundation of China (no. 81302471) and the Natural Science Foundation of Hebei Province (no. H2014206345).

References

[1] E. Kuchowicz and K. Rydzyński, "Risk assessment of morpholine (tetrahydro-2H-1,4-oxazine): a time for reevaluation of current occupational exposure standards?," *Applied Occupational and Environmental Hygiene*, vol. 13, no. 2, pp. 113–121, 1998.

[2] R. G. Mcguire and R. D. Hagenmaier, "Shellac coatings for grapefruits that favor biological control of *penicillium digitatumby candida oleophila*," *Biological Control*, vol. 7, no. 1, pp. 100–106, 1996.

[3] I. M. El-Gamal, T. T. Khidr, and F. M. Ghuiba, "Nitrogen-based copolymers as wax dispersants for paraffinic gas oils," *Fuel*, vol. 77, no. 5, pp. 375–385, 1998.

[4] N. S. Njombolwana, A. Erasmus, J. G. Zyl, W. Plooy, P. J. R. Cronje, and P. H. Fourie, "Effects of citrus wax coating and brush type on imazalil residue loading, green mould control and fruit quality retention of sweet oranges," *Postharvest Biology and Technology*, vol. 86, pp. 362–371, 2013.

[5] D. C. Wua, J. Z. Yua, B. H. Chena, C. Y. Lina, and W. H. Ko, "Inhibition of egg hatching with apple wax solvent as a novel method for controlling golden apple snail (*Pomacea canaliculata*)," *Crop Protection*, vol. 24, no. 5, pp. 483–486, 2005.

[6] A. C. Flick, H. X. Ding, C. A. Leverett et al., "Synthetic approaches to the 2014 new drugs," *Bioorganic & Medicinal Chemistry*, vol. 24, no. 9, pp. 1937–1980, 2016.

[7] Y. J. Wu, "Heterocycles and Medicine: a survey of the heterocyclic drugs approved by the U.S. FDA from 2000 to present," in *Progress in Heterocyclic Chemistry*, Chapter 1, vol. 24, pp. 1–53, Amsterdam, Netherlands, 2012.

[8] M. Giurg, A. Golab, J. Suchodolski et al., "Reaction of bis[(2-chlorocarbonyl)phenyl] diselenide with phenols, aminophenols, and other amines towards diphenyl diselenides with antimicrobial and antiviral properties," *Molecules*, vol. 22, no. 6, p. 974, 2017.

[9] B. Edwin, M. Amalanathan, R. Chadha, N. Maiti, S. Kapoor, and I. H. Joe, "Structure activity relationship, vibrational spectral investigation and molecular docking analysis of anti-neuronal drug 4-(2-Aminoethyl) morpholine," *Journal of Molecular Structure*, vol. 1148, pp. 459–470, 2017.

[10] D. Velasco, C. Elvira, and J. S. Román, "New stimuli-responsive polymers derived from morpholine and pyrrolidine," *Journal of Materials Science: Materials in Medicine*, vol. 19, no. 4, pp. 1453–1458, 2008.

[11] D. Velasco, C. B. Danoux, J. A. Redondo et al., "PH-sensitive polymer hydrogels derived from morpholine to prevent the crystallization of ibuprofen," *Journal of Controlled Release*, vol. 149, no. 2, pp. 140–145, 2011.

[12] M. C. Costa, T. Goumperis, W. Andersson et al., "Risk identification in food safety: strategy and outcomes of the EFSA emerging risks exchange network (EREN), 2010–2014," *Food Control*, vol. 73, pp. 255–264, 2017.

[13] Health Canada, "Archived—a summary of health hazard assessment of morpholine in wax coatings of apples," 2013.

[14] D. Chen, H. Miao, J. Zou et al., "Novel dispersive micro-solid-phase extraction combined with ultrahigh-performance liquid chromatography-high-resolution mass spectrometry to determine morpholine residues in citrus and apples," *Journal of Agricultural and Food Chemistry*, vol. 63, no. 2, pp. 485–492, 2015.

[15] J. Luong, R. A. Shellie, H. Cortes, R. Gras, and T. Hayward, "Ultra-trace level analysis of morpholine, cyclohexylamine, and diethylaminoethanol in steam condensate by gas chromatography with multi-mode inlet, and flame ionization detection," *Journal of Chromatography A*, vol. 1229, pp. 223–229, 2012.

[16] H. Kataoka, "Derivatization reactions for the determination of amines by gas chromatography and their applications in environmental analysis," *Journal of Chromatography A*, vol. 733, no. 1-2, pp. 19–34, 1996.

[17] J. H. Hotchkiss and A. J. Vecchio, "Analysis of direct contact paper and paperboard food packaging for n-nitrosomorpholine and morpholine," *Journal of Food Science*, vol. 48, no. 1, pp. 240–242, 1983.

[18] F. Sacher, S. Lenz, and H. J. Brauch, "Analysis of primary and secondary aliphatic amines in waste water and surface water by gas chromatography- mass spectrometry after derivatization with 2,4-dinitrofluorobenzene or benzenesulfonyl chloride," *Journal of Chromatography A*, vol. 764, no. 1, pp. 85–93, 1997.

[19] N. P. Sen and P. A. Baddoo, "An investigation on the possible presence of morpholine and N-nitrosomorpholine in wax-coated apples," *Journal of Food Safety*, vol. 9, no. 3, pp. 183–191, 1989.

[20] C. Lamarre, R. Gilbert, and A. Gendron, "Liquid chromatographic determination of morpholine and its thermal breakdown products in steam-water cycles at nuclear power plants," *Journal of Chromatography A*, vol. 467, pp. 249–258, 1989.

[21] R. Lindahl, A. Wästerby, and J.-O. Levin, "Determination of morpholine in air by derivatisation with 1-naphthylisothiocyanate and HPLC analysis," *The Analyst*, vol. 126, no. 2, pp. 152–154, 2001.

[22] M. J. Hengel, R. Jordan, and W. Maguire, "Development and validation of a standardized method for the determination of morpholine residues in fruit commodities by liquid chromatography–mass spectrometry," *Journal of Agricultural and Food Chemistry*, vol. 62, pp. 3697–3701, 2014.

[23] H. Kodamatani, Y. Iwaya, M. Saga et al., "Ultra-sensitive HPLC-photochemical reaction-luminol chemiluminescence method for the measurement of secondary amines after nitrosation," *Analytica Chimica Acta*, vol. 952, pp. 50–58, 2017.

[24] K. Lian, P. Zhang, L. Niu et al., "A novel derivatization approach for determination of ketamine in urine and plasma by gas chromatography–mass spectrometry," *Journal of Chromatography A*, vol. 1264, pp. 104–109, 2012.

19

Determination of Fluoroquinolones in Pharmaceutical Formulations by Extractive Spectrophotometric Methods using Ion-Pair Complex Formation with Bromothymol Blue

Trung Dung Nguyen (iD),[1] **Hoc Bau Le,**[1] **Thi Oanh Dong,**[1] **and Tien Duc Pham** (iD)[2]

[1]*Faculty of Physics and Chemical Engineering, Le Quy Don Technical University, 236 Hoang Quoc Viet, Hanoi, Vietnam*
[2]*Faculty of Chemistry, VNU-University of Science, Vietnam National University Hanoi, 19 Le Thanh Tong, Hoan Kiem, Hanoi, Vietnam*

Correspondence should be addressed to Trung Dung Nguyen; nguyentrungdung1980@gmail.com and Tien Duc Pham; tienduchphn@gmail.com

Academic Editor: Bengi Uslu

In this paper, we reported a new, simple, accurate, and precise extractive spectrophotometric method for the determination of fluoroquinolones (FQs) including ciprofloxacin (CFX), levofloxacin (LFX), and ofloxacin (OFX) in pharmaceutical formulations. The proposed method is based on the ion-pair formation complexes between FQs and an anionic dye, bromothymol blue (BTB), in acidic medium. The yellow-colored complexes which were extracted into chloroform were measured at the wavelengths of 420, 415, and 418 nm for CFX, LFX, and OFX, respectively. Some effective conditions such as pH, dye concentration, shaking time, and organic solvents were also systematically studied. Very good limit of detection (LOD) of 0.084 μg/mL, 0.101 μg/mL, and 0.105 μg/mL were found for CFX, LFX, and OFX, respectively. The stoichiometry of the complexes formed between FQs and BTB determined by Job's method of continuous variation was 1:1. No interference was observed from common excipients occurred in pharmaceutical formulations. The proposed method has been successfully applied to determine the FQs in some pharmaceutical products. A good agreement between extractive spectrophotometric method with high-performance liquid chromatography mass spectrometry (HPLC-MS) for the determination of FQs in some real samples demonstrates that the proposed method is suitable to quantify FQs in pharmaceutical formulations.

1. Introduction

Fluoroquinolones (FQs) are the important antibiotics used for the treatment of Gram-negative bacterial infections in both human and veterinary medicine. They are derivatives of 4-quinolone, which have unsubstituted or substituted piperazine ring attached at the 7-position to the central ring system of quinoline as well as fluorine atom at the 6-position. The FQs are useful to treat a variety of infections, including soft-tissue infections, respiratory infections, urinary tract infections, bone-joint infections, typhoid fever, prostatitis, sexually transmitted diseases, acute bronchitis, community-acquired pneumonia, and sinusitis [1–3].

Ciprofloxacin (CFX), which is one of the second-generated groups of synthetic FQs, can exhibit greater

intrinsic antibacterial activity and make a broader anti-bacterial spectrum. Ofloxacin (OFX) is a chiral compound that is widely used to treat above infections. Levofloxacin (LFX) is the pure (–)-(S)-enantiomer of the racemic drug substance ofloxacin. Figures 1(a)–1(c) show the chemical structures of CFX, LFX, and OFX, respectively.

Several techniques like voltammetry [4], flow injection electrogenerated chemiluminescence [5], spectrofluorometry [6, 7], spectrophotometry [8, 9], high-performance liquid chromatography [10, 11], and liquid chromatography tandem mass spectrometry [12, 13] have been used for the determination of fluoroquinolones in pharmaceutical and biological products. Among them, spectrophotometric method has several advantages such as simplicity, fast, and low cost. Spectrophotometry was successfully used for

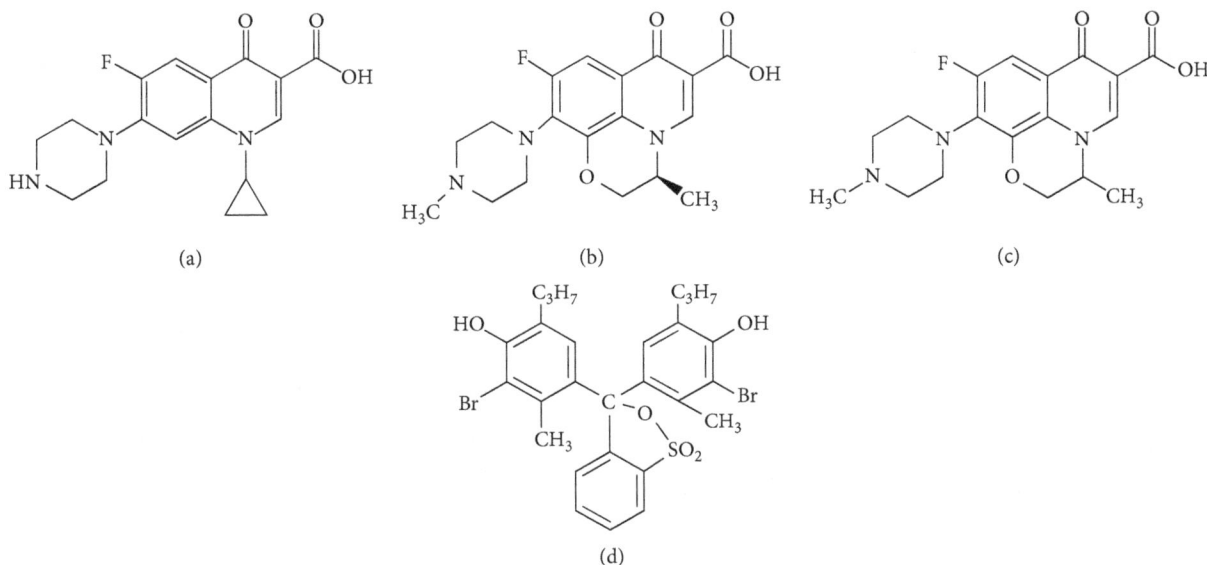

FIGURE 1: Chemical structures of ciprofloxacin (a), levofloxacin (b), ofloxacin (OFX) (c), and bromothymol blue (d).

pharmaceutical analysis, involving quality control of commercialized product and pharmacodynamic studies. Spectrophotometric methods for the determination of fluoroquinolones could be classified according to the different reactions: (i) charge-transfer complexation based on the reaction of FQs as electron donors with p-acceptors such as 2,3-dichloro-5,6-dicyano-q-benzoquinone, 7,7,8,8-tetracyanoquinodimethane, q-chloranil, q-nitrophenol, and tetracyanoethylene [7, 14–16]; (ii) oxidative coupling reaction using oxidative coupling with 3-methyl-2-benzothiazolinonehydrazone hydrochloride and cerium (IV) ammonium sulfate, Fe(III)-MBTH, tris(o-phenanthroline) iron(II), and tris (bipyridyl) iron(II) [17, 18]; (iii) ion-pair complex formation with acid-dye reagents such as Sudan III, methyl orange, supracene violet 3B, tropaeolin 000, bromophenol blue, bromothymol blue, bromocresol green, and bromocresol purple [8, 14, 19, 20]. These methods were related with some major drawbacks such as having narrow linearity range, requiring heating and close pH control, long time for the reaction to complete, and low stability of the colored product formed.

Bromothymol blue (BTB) (Figure 1(d)) is an anionic dye and that can be protonated or deprotonated to form yellow or blue, respectively. The BTB was used to make ion-pair complex, which was applied to determine many pharmaceutical compounds by extractive spectrophotometric methods [21–30]. However, the ion-pair complex between BTB and FQs has not been studied. The method based on ion-pair complexes between analytes and BTB into a suitable organic solvent is also simple, fast, and cheap.

In the previous study, we used sulphonphthalein acid including bromophenol blue, bromocresol green, and bromothymol blue to determine ciprofloxacin pharmaceutical formulations and achieved good results [31].

In this paper, for the first time, we investigated extractive spectrophotometric method based on the formation of ion-pair complexes between ciprofloxacin, levofloxacin, and

ofloxacin with BTB subsequent extraction into chloroform. Some effective conditions on the formation of complexes such as pH, shaking time, organic solvent, and the concentration of dye were systematically studied. The present method was also applied to determine FQs in some pharmaceutical formulations including tablets and infusions.

2. Experimental

2.1. Apparatus. A double beam UV-visible spectrophotometer (SP-60, Biochrom Ltd., UK) with 1.0 cm of path length quartz cells was used to measure all sample absorbances. Inolab pH-meter instrument (Germany) was used to monitor the pH of solutions. Three standard buffers were used to calibrate the electrode before measuring pH of solutions. All measurements were conducted at 25 ± 2°C controlled by air conditional laboratory.

2.2. Materials and Reagents. All chemicals used were of analytical grade and double-distilled water was used to prepare all solutions in the present study.

FQs were purchased from Sigma (Germany, with purity >99.0%), whereas bromothymol blue (BTB) was supplied by Maya-R, China, with purity >99%. The organic solvents including chloroform, dichloromethane, carbon tetrachloride, dichloroethane, benzene, toluene, and other chemicals are analytical reagents (Merck, Germany).

The following dosage forms containing FQs were purchased from local pharmacy market and employed in the study: Hasancip and Kacipro tablets equivalent to 500 mg ciprofloxacin (Hasan-Dermapharm and Dong Nam manufacturing-Trading pharmaceutical Co., Ltd, Vietnam). Ciprofloxacin infusion equivalent to 200 mg ciprofloxacin/100 ml solution for infusion (Hebei Tiancheng Pharmaceutical Co., Ltd and Shandong Hualu Pharmaceutical Co., Ltd, China). Stada and DHG tablets equivalent to 500 mg levofloxacin (Stada-VN J.V.Company and DHG pharmaceutical joint–stock company, Vietnam). Ofloxacin (200 mg/tablet) was provided by the Mekophar Chemical Pharmaceutical Company (Vietnam).

2.3. Solution Preparation.

A stock solution of FQs (1 mg/mL) in double-distilled water. The working standard solution of FQs containing 100 μg/mL was prepared by appropriate dilution. The stock solution of BTB (0.025%) was prepared in double-distilled water. All stock solutions were kept in dark bottle, stored in 4°C and could be used within one week.

2.4. Construction of Calibration Curves.

A series of 125 mL separating funnel, the volumes of working solutions of the drugs in different concentration ranges (CFX (1–35 μg/mL), LFX (0.5–25 μg/mL), and OFX (0.5–25 μg/mL) were transferred. Then, 4.0 mL of 0.025% BTB solution was added before thoroughly mixing. After that, a 10 mL of chloroform was added to each of the separating funnel. The contents were shaken for 2 min and allowed to separate the two layers. The yellow-colored chloroform layer containing the ion-pair complexes was measured at 420 nm for CFX, 415 nm for LFX, and 418 nm for OFX against the reagent blanks. At each concentration, the experiment was repeated 6 times. The colored chromogen complexes are stable for 24 h.

2.5. Sample Preparation.

Weigh and mix the contents of twenty tablets of each drug (CFX, LFX, and OFX), an accurately weighed amount of powder equivalent to 0.1 g of drugs transferred into a 100-mL beaker. A magnetic stirrer was used to completely disintegrate the powder in doubly distilled water. Then, filter through a Whatman paper (No 40) and fill up to 100 mL with doubly distilled water in a volumetric flask. The working solution of the drugs containing 100 μg/mL was prepared by dilution and determined under optimum conditions.

2.6. Validation with High-Performance Liquid Chromatography-Mass Spectrometry (HPLC-MS).

Some real samples of three FQs were determined by HPLC-MS using HPLC 20 AXL (Shimadzu, Japan) coupled with electrospray ionisation tandem mass spectrometric detection, ABI 5500 QQQ (Applied BioSystem). The chromatographic conditions are including column C18 MRC-ODS (150 mm × 2.1 mm × 3.5 μm), mobile phase containing acetonitrile (ACN) with formic acid (0.1%) in water under a flow rate of 0.5 ml/min, and gradient elution. The inject volume is 10 μL.

3. Results and Discussion

3.1. Optimum Reaction Conditions

3.1.1. Effect of Extracting Solvent.

Six organic solvents including chloroform, carbon tetrachloride, dichloromethane, dichloroethane, benzene, and toluene were used to study the effect of solvent to ion-pair formation between FQs and BTB. Figure 2 shows that chloroform is the most suitable

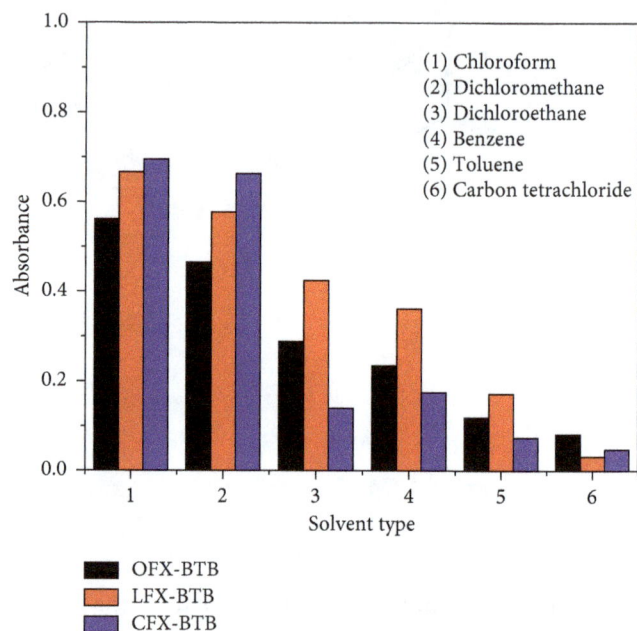

FIGURE 2: The effect of solvent on the ion-pair complex formation (10 μg/mL of fluoroquinolones (FQs) with bromothymol blue (BTB)).

solvent for the extraction of three FQs with low blank absorbance, highest absorbances, and lowest standard deviations. It implies that chloroform is the best extracting solvent to achieve a good recovery of the complexes with the shortest time to reach the equilibrium processes.

3.1.2. Effect of pH.

The pH of solution plays an important role in the complex formations. The effect of pH on the formation of ion pairs was examined by varying the pH from 2.0 to 6.0 by adjusting 1 M HCl and 1 M NaOH. The maximum absorbances were observed at pH 3.3, 3.4, and 3.5 for the complexes of BTB and OFX, CFX, and LFX, respectively (Figure 3). These pH values correspond to the initial pH of the examined drug and the dye. Therefore, it is not necessary to adjust the pH before extraction.

3.1.3. Effect of Dye Concentration.

The effect of dye concentrations was studied by adding different volumes of 0.025% BTB from 1.0 to 6.0 mL with a fixed concentration of FQs (10 μg/mL) (Figure 4). Figure 4 shows that the maximum absorbance of the complex was achieved with 4.0 mL of 0.025% of BTB in each case and excess dye did not affect the absorbance of the complex. Therefore, 4.0 mL of 0.025% of BTB is optimum dye volume and it is kept as constant for further studies.

3.1.4. Effect of Shaking Time.

The effect of shaking time on the formation and stability of the ion-pair complex was investigated by measuring the absorbance of the extracted ion associates with increasing time from 0 to 4.0 min. Figure 5 shows that the ion-pair complexes were formed

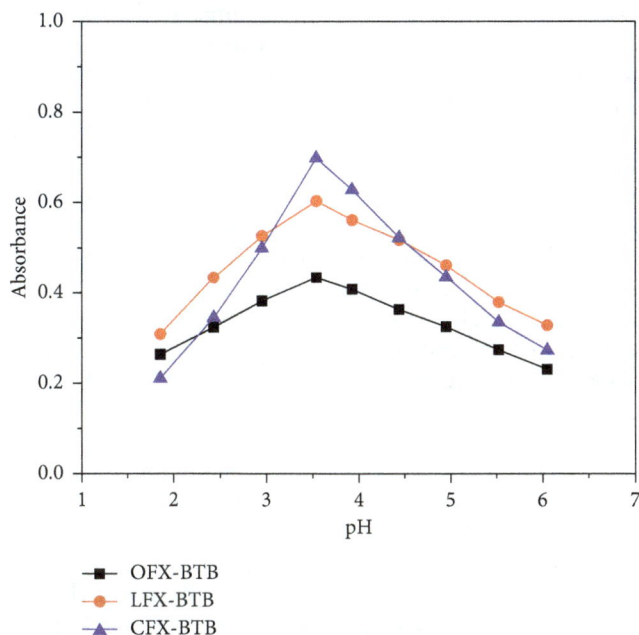

FIGURE 3: Effect of pH on the absorbances of 10 µg/mL of OFX, LFX, and CFX.

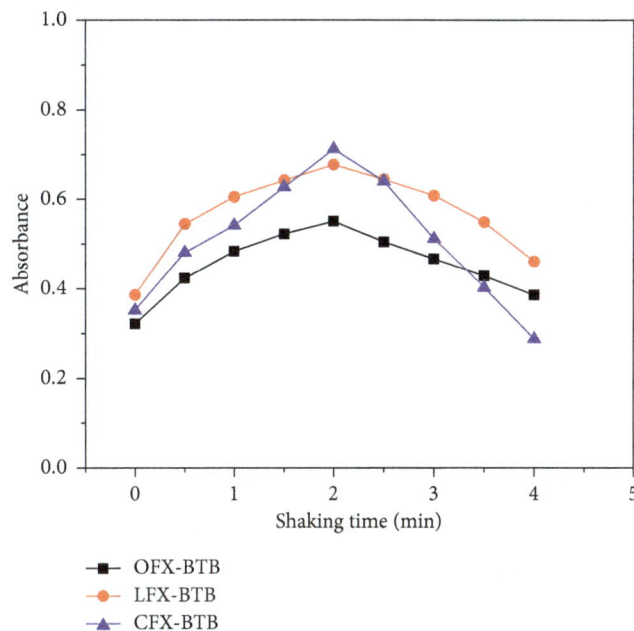

FIGURE 4: Effect of the volume of 0.025% BTB on the absorbance of 10 µg/mL of OFX, LFX, and CFX.

instantaneously with 2.0 min shaking time. Thus, 2.0 min is the optimum shaking time and it is fixed for further studies.

3.1.5. Stoichiometry of Ion-Pair Complexes.

Job's method of continuous variation of equimolar solutions was employed to evaluate stoichiometry of the complex. A 3.0×10^{-4} M standard solution of three FQs and 3.0×10^{-4} M solution of BTB were used. A series of the solutions were prepared in which the total volume of drug and reagent was kept in

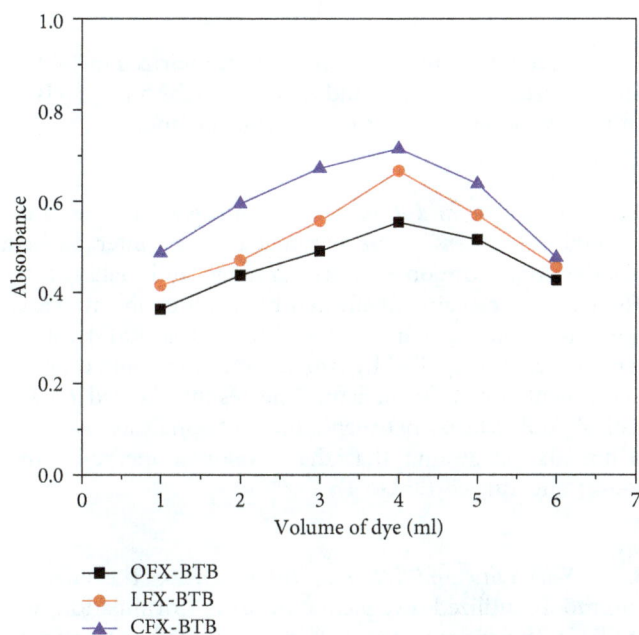

FIGURE 5: Effect of shaking time on the ion-pair complexes.

10 mL, whereas the absorbances were measured at 420, 415, and 418 nm, for CFX, LFX, and OFX, respectively. The absorbances were plotted against the mole fraction of the drugs. The stoichiometry for each drug-dye ion-pair complex was found to be 1:1 (Figure 6).

3.1.6. Mechanism of Reaction and Absorption Spectra.

Fluoroquinolones can contain a secondary amino group (CFX) and a tertiary amino group (LFX and OFX) that can be easily protonated under acidic conditions. On the one hand, the sulphonic acid group in BTB, that is, the only group undergoing dissociation in the pH range 1–5. The colour of BTB is on the basis of lactoid ring and subsequent formation of quinoid group. It is suggested that the two tautomers are plausible in equilibrium due to strong acidic nature of the sulphonic acid group. Thus, the quinoid body must predominate. Finally, the protonated fluoroquinolones form ion pairs with BTB dye that could be quantitatively extracted into chloroform. The possible reaction mechanisms are proposed and given in a scheme in Figure 7.

The absorption spectra of the ion-pair complexes, which were formed between FQs and BTB, were measured in the wavelength range 350–500 nm against the blank solution and shown in Figure 8.

Figure 8 shows that absorption maxima for CFX-BTB, LFX-BTB, and OFX-BTB in chloroform were observed at 420, 415, and 418 nm, respectively. The reagent blanks under similar conditions have insignificant absorbances. At wavelengths 420, 415, and 418 nm, absorption spectrum of BTB does not affect the absorption spectrum of ion-associate complexes of FQs. Therefore, the selectivity of the proposed method for the determination of FQs is guaranteed.

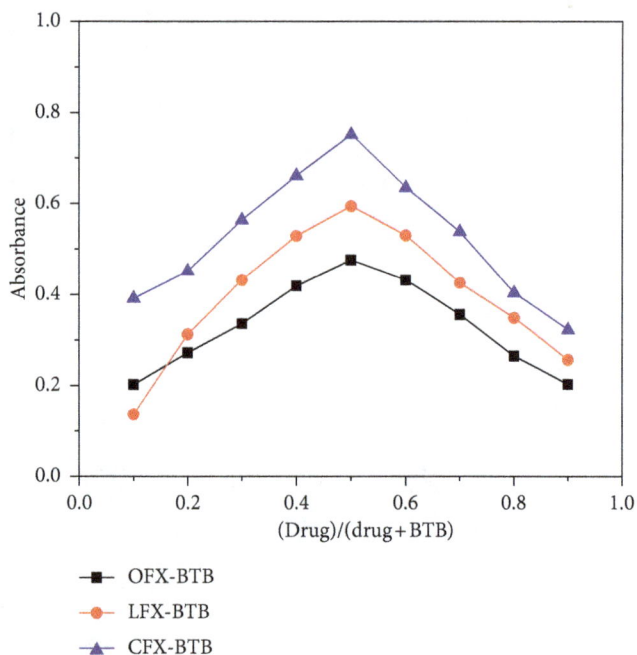

FIGURE 6: Job's method of continuous variation graph for the reaction of drug with acid dyes BTB, [drug] = [dye] = 3.0×10^{-4} M.

3.1.7. Association Constants of Ion-Pair Complexes. The equation of association constant of ion-pair complex is

$$\frac{A/A_m}{[1-(A/A_m)]^{n+2}C_{M(n)n}},\qquad(1)$$

where A and A_m are the observed absorbance and the maximum absorbance value when all the drug present is associated, respectively. C_M is the molar concentration of the drugs at the maximum absorbance and n is the stoichiometry in which BTB ion associates with drugs. The conditional stability constants (K_f) of the ion-pair complexes according to Britton [32] for the cases of FQs were calculated from the continuous variation data using the following equation:

$$K_f = \frac{A/A_m}{[1-A/A_m]^{n+2}C_M\,(n)^n}.\qquad(2)$$

The conditional stability constants (K_f) of the ion-pair complexes for FQs are indicated in Table 1.

Table 1 shows that the log K_f values of ion-pair associates for OFX-BTB, LFX-BTB, and CFX-BTB were 6.08 ± 0.46, 6.04 ± 0.58, and 5.91 ± 0.32, respectively (numbers of replicated experiments, $n = 6$). The obtained results confirmed that the ion-pair formation complexes are of high stability.

3.2. Validation of the Present Method. The proposed methods are validated according to ICH recommendations Q2(R1) [33]. The parameters that have been investigated are indicated below.

3.2.1. Linearity, Sensitivity, and Limits of Detection and Quantification. A linear relationship between the measured

absorbance and the concentration range studied for each drug as shown in Figure 9 and the correlation coefficient (R) of at least 0.997 were achieved. The limit of detection (LOD) and quantification (LOQ) of the method are determined by $3.3\,(SD/b)$ and $10\,(SD/b)$, respectively, where SD is the standard deviation of blank absorbance values and b is the slope of the calibration curve equation.

The LOD and LOQ values, slope, and intercept of linear graphs for all the drugs and analytical parameters are indicated in Table 2. The molar absorptivities and Sandell's sensitivity of each methods were calculated and these values showed that the molar absorptivity of ion-pair complexes was in the order CFX-BTB > LFX-BTB > OFX-BTB.

3.2.2. Accuracy and Precision. The accuracy and precision of the methods were determined by preparing solutions of three different concentrations of drug and analyzing them in six replicates. The precision of the proposed methods was evaluated as percentage relative standard deviation (RSD%) and accuracy as percentage relative error (RE%). The percentage relative error was calculated using the following equation:

$$RE\,(\%) = \left[\frac{(founded-added)}{added}\right] \times 100.\qquad(3)$$

The accuracy and precision were summarized in Table 3. The low values of the RSD and RE confirm the high precision and the good accuracy of the present method.

3.2.3. Robustness and Ruggedness. For the evaluation of the method robustness, some parameters were interchanged: pH, dye concentration, wavelength range, and shaking time. The capacity remains unaffected by small deliberate variations. Method ruggedness was expressed as RSD% of the same procedure applied by two analysts and using different instruments on different days. The results showed no statistical differences between different analysts and instruments, suggesting that the developed methods were robust and rugged (Table 4).

3.2.4. Selectivity and Effect of Interferences. The effect of commonly utilized excipients in drug formulation was studied. The investigated FQs were studied with various excipients such as magnesium stearate, glucose, lactose, starch, and sodium chloride which were prepared in the proportion corresponding to their amounts in the real drugs with a final dosage of $10\,\mu g/mL$ FQ. The effect of excipients on the determination of FQs was evaluated by recovery when determining FQs analyzed with the proposed method in the presence of excipient (Table 5).

The results in Table 5 show that the recoveries are in the range of 98.53–102.04, demonstrating that there is no interference of excipients when FQs in drugs are quantified by extractive spectrophotometric using ion-pair formation with BTB. In other words, the present method has a high selectivity for determining FQs in its dosage forms.

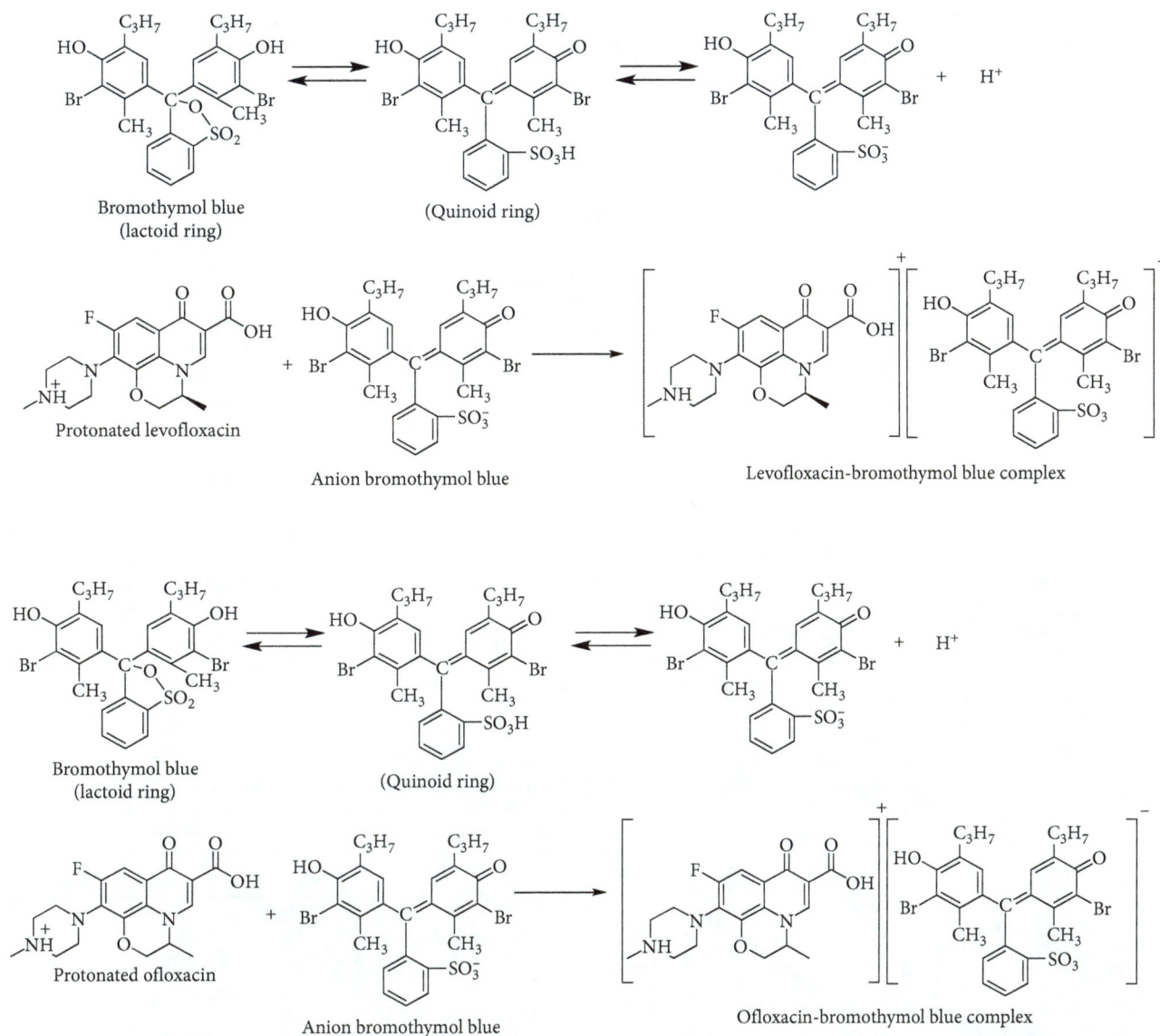

FIGURE 7: Proposal mechanism for the reaction between levofloxacin, ofloxacin, and bromothymol blue.

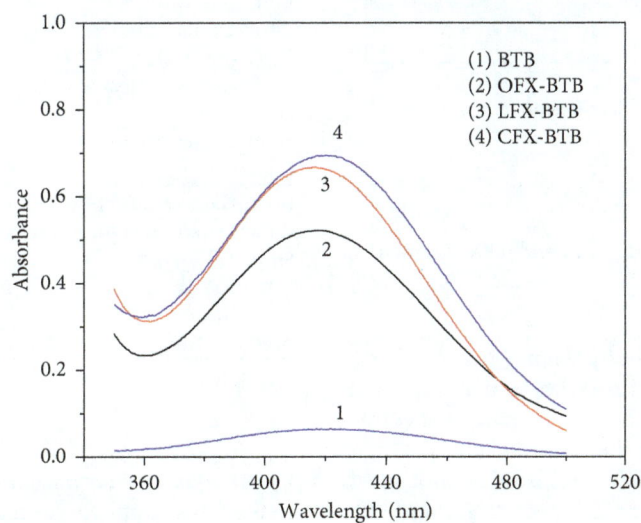

FIGURE 8: Absorption spectrum of ion-associate complexes of fluoroquinolones ($10\,\mu g/mL$) with BTB against reagent blank.

TABLE 1: The conditional stability constants (K_f) of the ion-pair complexes for FQs.

Sample	V_{drug} (mL)	V_{BTB} (mL)	A	n	$n^\wedge n$	$[1-(A/A_m)]^{n+2}$	K_f	$\log K_f$	Mean
				Ofloxacin					
1	0.25	2.25	0.202	0.1111	0.7834	0.3116	50204.7802	4.7007	**6.08**
2	0.5	2	0.272	0.2500	0.7071	0.1486	157048.9127	5.1960	
3	0.75	1.75	0.336	0.4286	0.6955	0.0512	572507.7789	5.7578	
4	1	1.5	0.419	0.6667	0.7631	0.0035	9562329.8320	6.9806	
5	1.25	1.25	0.476	1.0000	1.0000	0.0000	—	—	
6	1.5	1	0.432	1.5000	1.8371	0.0002	59414177.6247	7.7739	
7	1.75	0.75	0.356	2.3333	7.2213	0.0026	1172246.1806	6.0690	
				Levofloxacin					
1	0.25	2.25	0.137	0.1111	0.7834	0.5749	14789.8587	4.1700	**6.04**
2	0.5	2	0.313	0.2500	0.7071	0.1856	115961.3839	5.0643	
3	0.75	1.75	0.432	0.4286	0.6955	0.0426	708574.1920	5.8504	
4	1	1.5	0.559	0.6667	0.7631	0.0005	67745245.6475	7.8309	
5	1.25	1.25	0.594	1.0000	1.0000	0.0000	—	—	
6	1.5	1	0.53	1.5000	1.8371	0.0004	34165312.8945	7.5336	
7	1.75	0.75	0.426	2.3333	7.2213	0.0042	682897.0665	5.8344	
				Ciprofloxacin					
1	0.25	2.25	0.392	0.1111	0.7834	0.2112	83530.0520	4.9218	**5.91**
2	0.5	2	0.452	0.2500	0.7071	0.1265	178145.0100	5.2508	
3	0.75	1.75	0.564	0.4286	0.6955	0.0345	828480.0305	5.9183	
4	1	1.5	0.661	0.6667	0.7631	0.0036	8522213.4055	6.9306	
5	1.25	1.25	0.752	1.0000	1.0000	0.0000	—	—	
6	1.5	1	0.615	1.5000	1.8371	0.0026	4572267.1248	6.6601	
7	1.75	0.75	0.538	2.3333	7.2213	0.0043	608798.6008	5.7845	

—, not determined.

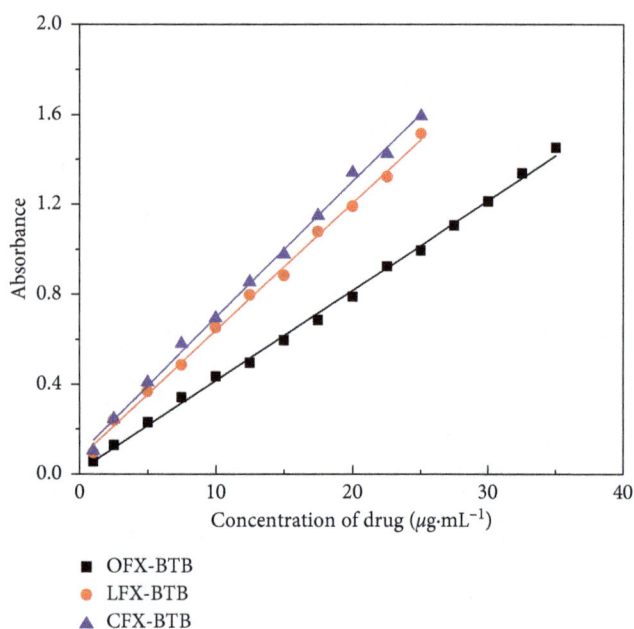

FIGURE 9: Calibration curves for OFX, LFX, and CFX at 418, 415, and 420 nm, respectively.

- ■ OFX-BTB
- ● LFX-BTB
- ▲ CFX-BTB

TABLE 2: Analytical characteristics of the proposed methods ($n = 6$).

Parameters	Proposed methods		
	Ofloxacin	Levofloxacin	Ciprofloxacin
Colour	Yellow	Yellow	Yellow
Wavelengths λ_{max} (nm)	418	415	420
pH	3.3	3.5	3.4
Stability (h)	24	24	24
Shaking time (min)	2	2	2
Stoichiometric ratio	1:1	1:1	1:1
Beer's law range (μg/mL)	1–35	0.5–25	0.5–25
Limit of detection, LOD (μg/mL)	0.105	0.101	0.084
Limit of quantitation, LOQ (μg/mL)	0.315	0.303	0.252
Molar absorptivity (L/mol.cm)	1.44×10^4	2.07×10^4	2.09×10^4
Sandell's sensitivity (μg/cm^2)	0.068	0.048	0.046
Regression equation ($Y = bx + a$), where Y is the absorbance, a is the intercept, b is the slope, and x is the concentration in μg/mL			
Slope (b)	0.040	0.057	0.061
Intercept (a)	0.0165	0.072	0.089
Correlation coefficient (R)	0.998	0.997	0.998

3.3. Comparison with Other Spectrophotometric Methods.
The proposed method compares with other reported methods. It has been observed that the extractive spectrophotometric method with BTB in the present study is of high sensitivity than other ones (Table 6). It also does not need heating, the product is stable for a longer time, and the interferences are minimum.

3.4. Analysis of Pharmaceutical Formulations.
The proposed method was applied successfully for the determination of studied drugs in the pharmaceutical formulations (tablets

TABLE 3: Evaluation of accuracy and precision of the proposed methods ($n = 6$).

Method	Additive concentration (μg/mL)	Found concentration (μg/mL)	Recovery (%)	RSD (%)	RE (%)
Ofloxacin	5.00	5.11	102.19	2.31	2.2
	10.00	10.26	102.64	1.34	2.6
	15.00	14.89	99.25	0.88	−0.73
Levofloxacin	5.00	5.16	102.70	2.03	2.8
	10.00	10.16	101.56	1.10	1.6
	15.00	14.82	98.80	0.50	−1.2
Ciprofloxacin	5.00	5.13	102.71	1.92	2.6
	10.00	9.74	97.41	0.52	−2.6
	15.00	14.60	97.33	0.57	−2.7

TABLE 4: The results of analysis of pharmaceutical preparation and standard of fluoroquinolones by two different analysts and instruments ($n = 6$).

Method	Different instruments			Different analysts		
	X	±SD	RSD (%)	X	±SD	RSD (%)
Ofloxacin-BTB pure ofloxacin ($10\,\mu$g·mL^{-1})	10.21	0.19	1.86	9.93	0.24	2.42
Mekopharm (200 mg ofloxacin per tablet)	199	0.57	0.29	196	0.76	0.39
Levofloxacin-BTB pure levofloxacin ($10\,\mu$g·mL^{-1})	10.16	0.15	1.48	10.19	0.21	2.06
Stada (500 mg levofloxacin per tablet)	498	0.61	0.12	501	0.85	0.17
Ciprofloxacin-BTB pure ciprofloxacin ($10\,\mu$g·mL^{-1})	9.85	0.18	1.83	10.12	0.25	2.47
Hasancip (500 mg ciprofloxacin per tablet)	502	0.64	0.13	497	0.92	0.19

TABLE 5: The effect of excipients on the determination of fluoroquinolones ($10\,\mu$g/mL).

Excipients	Amount of excipient added (μg/mL)	Recovery (%) ± SD		
		Ofloxacin	Levofloxacin	Ciprofloxacin
Magnesium stearate	500	102.04 ± 0.12	101.23 ± 0.089	98.53 ± 0.91
Glucose	250	100.17 ± 0.16	99.04 ± 0.14	99.08 ± 0.062
Lactose	500	99.92 ± 0.21	100.20 ± 0.12	99.73 ± 0.21
Starch	200	100.96 ± 0.24	98.89 ± 0.13	101.31 ± 0.17
Sodium chloride	500	100.13 ± 0.24	100.15 ± 0.11	99.75 ± 0.16

and infusion) and the results are presented in Table 7. Six replicated determinations were measured. Table 7 shows that satisfactory recovery data were obtained and the recovery efficiency varies from 97.41% to 101.20%, indicating high accuracy of the present method in determining real pharmaceutical samples.

3.5. *Comparison with HPLC-MS Method.* In order to validate the experimental data in determining some real drug samples, HPLC-MS was used with the conditions described on Section 2.6 according to the previously published paper [13]. The comparison between the results determined by the present method with HPLC-MS method was indicated in Table 8.

Table 8 shows a good agreement between the proposed method and HPLC-MS where the relative differences of two methods were less than 11%. Furthermore, the standard deviation of the proposed method is almost lower than that of HPLC-MS. Our results indicate that the extractive spectrophotometric determination of FQs using BTB dye in

chloroform is a very good method to quantify the FQ in pharmaceutical formulations.

4. Conclusions

We have reported a new method when using BTB as an anionic dyes for the extractive spectrophotometric determination of ciprofloxacin (CFX), levofloxacin (LFX), and ofloxacin (OFX) in different pharmaceutical drugs (tablets and infusions). The methods have the advantages of simplicity without heating, pH-adjustment, and high sensitivity. The limit of detection (LOD) values are $0.084\,\mu$g/mL for CFX, $0.101\,\mu$g/mL for LFX, and $0.105\,\mu$g/mL for OFX. No interference from common excipients was confirmed. The stoichiometry complexes of FQs and BTB determined by Job's method of continuous variation were found to be 1 : 1. The developed and validated methods are indicated as the acceptable precision and accuracy, and recovery of the drugs and suitable for routine analysis of drugs in pharmaceutical formulations. The results of some real samples by the present method that were compared with HPLC-MS method with

TABLE 6: The comparison of present study with other spectrophotometric methods.

Drug	Reagent	λ_{max} (nm)	Range of determination (μg/mL)	Molar absorptivity (L/mol·cm)	Remarks	Reference
Ciprofloxacin	Co (II) tetrathiocyanate	623	20–240	8.38×10^2	Less sensitive	[34]
	Supracene violet 3	575	2.5–30	8.62×10^3	Less sensitive	[35]
	Eosin Y	547	2–8	3.56×10^4	Less stable colour	[36]
	Merbromin	545	2–15	1.23×10^4	Addition of CN^- to inhibit Hg^{+2} ions	
	Ce(IV)- MBTH	630	10–50	—	Involves shaking time	[17]
	Tris(o-phenanthroline) iron(II)	510	0.04–7.2	3.4×10^4	Involves shaking time and heating	[18]
	Tris (bipyridyl) iron(II)	522	0.05–9	2.95×10^4	Involves shaking time and heating	
	CL	520	16–96	—	Involves shaking time and heating	[16]
	TCNE	335	0.25–15	—	Involves shaking time and heating	
	Sudan II	550	0.8–7.1	5.3×10^4		
	Congo red	517	0.5–6.0	2.83×10^4	Narrow linear range	[8]
	Gentian violet	585	0.5–10	2.21×10^4		
	Brilliant blue G	610	0.5–6.0	2.86×10^4	Narrow linear range and required pH adjustment	[37]
	Bromocresol green	412	1–20	2.28×10^4	Required pH adjustment	[14]
	BTB	420	0.5–25	2.09×10^4	Highly sensitive with wide linear dynamic ranges, no heating, and no pH adjustment	This study
Levofloxacin	Chloranilic acid	521	15–250	1.2×10^3	Less sensitive	[14]
	Bromocresol green	411	1–20	2.16×10^4	Required pH adjustment	
	Eosin Y	547	2–8	4.83×10^4	Less stable colour	[36]
	Merbromin	545	2–15	1.58×10^4	Addition of CN^- to inhibit Hg^{+2} ions	
	Cobalt (II) tetrathiocyanate	623	20–240	—	Less sensitive	[34]
	Bromophenol blue	424	1.85–31.5	1.98×10^4	Required pH adjustment	[19]
	Bromocresol green	428	1.85–25	1.82×10^4		
	BTB	415	0.5–25	2.07×10^4	Highly sensitive with wide linear dynamic ranges, no heating, and no pH-adjustment	This study
Ofloxacin	Supracene violet 3	575	2.5–25	1.09×10^4	Less sensitive	[35]
	Tropaeolin 000	485	2.5–30	8.23×10^2	Less sensitive	
	Sudan II	560	0.8–8.4	2.97×10^4		
	Congo red	530	0.5–5.5	3.29×10^4	Narrow linear range	[8]
	Gentian violet	575	0.8–11	2.51×10^4		
	Bromocresol purple	400	1.0–16.0	2.4×10^4	Required pH adjustment	[38]
	Bromocresol green	410	1.0–16.0	1.96×10^4	Required pH adjustment	
	Bromophenol blue	410	5–25	1.03×10^4	Required close pH control and involved extraction steps	[20]
	Bromothymol blue	415	2–15	2.01×10^4		
	Bromocresol purple	410	2–20	1.64×10^4		
	Bromothymol blue	415	1–35	1.44×10^4	Highly sensitive with wide linear dynamic ranges, no heating, and no pH-adjustment	This study

TABLE 7: Determination of the studied drugs in their pharmaceutical preparations using the proposed method ($n = 6$).

Pharmaceutical preparation	Hasancip tablet	Kacipro tablet	Shandong infusion	Hebei infusion	Levofloxacin Stada	Levofloxacin DHG	Ofloxacin mekopharm
Labeled amount (mg/form)	500/tablet	500/tablet	200/100 mL	200/100 mL	500/tablet	500/tablet	200/tablet
Recovery (%) ± SD	98.89 ± 0.23	101.20 ± 0.20	97.41 ± 0.42	97.69 ± 0.36	99.53 ± 0.17	101.01 ± 0.35	99.58 ± 0.46

TABLE 8: Amount of some fluoroquinolone antibiotics determined by the proposed method and HPLC-MS.

Sample	Amount (mg/tablet)		Difference (%)
	Proposed method	HPLC -MS	
Ciprofloxacin-Hasancip table	494.45 ± 11.63	446.93 ± 15.84	10.63
Ofloxacin mekopharm	199.16 ± 0.85	202.00 ± 2.72	−1.41
Levofloxacin DHG	505.05 ± 17.33	480.55 ± 54.16	5.10
Levofloxacin Stada	497.65 ± 9.24	486.04 ± 9.24	2.39

the relative differences are less than 11%, indicating that the present method is good for determination of FQs in pharmaceutical formulations.

Conflicts of Interest

The authors declare that they have no conflicts of interest.

Acknowledgments

This work was supported financially by the project of Le Quy Don Technical University, under 11/HLKT/2017 project. The authors would like to thank the National Institute for Food Control (Vietnam) for providing HPLC-MS system to validate the present method.

References

[1] V. Kapetanovic, L. Milovanovic, and M. Erceg, "Spectrophotometric and polarographic investigation of the Ofloxacin-Cu(II) complexes," *Talanta*, vol. 43, no. 12, pp. 2123–2130, 1996.

[2] Y. Khaliq and G. G. Zhanel, "Fluoroquinolone-associated tendinopathy: a critical review of the literature," *Clinical Infectious Diseases*, vol. 36, no. 11, pp. 1404–1410, 2003.

[3] G. G. Zhanel, K. Ennis, L. Vercaigne et al., "A critical review of the fluoroquinolones," *Drugs*, vol. 62, no. 1, pp. 13–59, 2002.

[4] Y. Ni, Y. Wang, and S. Kokot, "Simultaneous determination of three fluoroquinolones by linear sweep stripping voltammetry with the aid of chemometrics," *Talanta*, vol. 69, no. 1, pp. 216–225, 2006.

[5] H. Ma, X. Zheng, and Z. Zhang, "Flow-injection electrogenerated chemiluminescence determination of fluoroquinolones based on its sensitizing effect," *Luminescence*, vol. 20, no. 4-5, pp. 303–306, 2005.

[6] S. T. Ulu, "Spectrofluorimetric determination of fluoroquinolones in pharmaceutical preparations," *Spectrochimica Acta Part A: Molecular and Biomolecular Spectroscopy*, vol. 72, no. 1, pp. 138–143, 2009.

[7] L. M. Du, A. P. Lin, and Y. Q. Yang, "Spectrofluorimetric determination of certain fluoroquinolone through charge transfer complex formation," *Analytical Letters*, vol. 37, no. 10, pp. 2175–2188, 2004.

[8] A. S. Amin, M. E. Moustafa, and R. M. S. El-Dosoky, "Spectrophotometric determination of some fluoroquinolone derivatives in dosage forms and biological fluids using ion-pair complex formation," *Analytical Letters*, vol. 41, no. 5, pp. 837–852, 2008.

[9] A. M. A.-E. El-Didamony and O. Mona, "Kinetic spectrophotometric method for the determination of some fourth generation fluoroquinolones in bulk and in pharmaceutical formulations," *Journal of Saudi Chemical Society*, vol. 21, pp. S58–S66, 2017.

[10] M. I. R. M. Santoro, N. M. Kassab, A. K. Singh, and E. R. M. Kedor-Hackmam, "Quantitative determination of gatifloxacin, levofloxacin, lomefloxacin and pefloxacin fluoroquinolonic antibiotics in pharmaceutical preparations by high-performance liquid chromatography," *Journal of Pharmaceutical and Biomedical Analysis*, vol. 40, no. 1, pp. 179–184, 2006.

[11] G. Carlucci, "Analysis of fluoroquinolones in biological fluids by high-performance liquid chromatography," *Journal of Chromatography A*, vol. 812, no. 1-2, pp. 343–367, 1998.

[12] H. Ziarrusta, N. Val, H. Dominguez et al., "Determination of fluoroquinolones in fish tissues, biological fluids, and environmental waters by liquid chromatography tandem mass spectrometry," *Analytical and Bioanalytical Chemistry*, vol. 409, no. 27, pp. 6359–6370, 2017.

[13] L. Johnston, L. Mackay, and M. Croft, "Determination of quinolones and fluoroquinolones in fish tissue and seafood by high-performance liquid chromatography with electrospray ionisation tandem mass spectrometric detection," *Journal of Chromatography A*, vol. 982, no. 1, pp. 97–109, 2002.

[14] A. M. El-Brashy, M. E.-S. Metwally, and F. A. El-Sepai, "Spectrophotometric Determination of Some Fluoroquinolone Antibacterials through Charge-transfer and Ion-pair Complexation Reactions," *Bulletin of the Korean Chemical Society*, vol. 25, no. 3, pp. 365–372, 2004.

[15] L. M. Du, H. Y. Yao, and M. Fu, "Spectrofluorimetric study of the charge-transfer complexation of certain fluoroquinolones with 7,7,8,8-tetracyanoquinodimethane," *Spectrochimica Acta Part A: Molecular and Biomolecular Spectroscopy*, vol. 61, no. 1-2, pp. 281–286, 2005.

[16] S. Mostafa, M. El-Sadek, and E. A. Alla, "Spectrophotometric determination of ciprofloxacin, enrofloxacin and pefloxacin through charge transfer complex formation," *Journal of Pharmaceutical and Biomedical Analysis*, vol. 27, no. 1-2, pp. 133–142, 2002.

[17] M. Rizk, F. B., F. Ibrahim, S. M. Ahmed, and N. M. El-Enany, "A simple kinetic spectrophotometric method for the determination of certain 4-quinolones in drug formulations," *Scientia Pharmaceutica*, vol. 68, no. 2, pp. 173–188, 2000.

[18] B. S. Nagaralli, J. Seetharamappa, and M. B. Melwanki, "Sensitive spectrophotometric methods for the determination of amoxycillin, ciprofloxacin and piroxicam in pure and pharmaceutical formulations," *Journal of Pharmaceutical and Biomedical Analysis*, vol. 29, no. 5, pp. 859–864, 2002.

[19] S. Ashour and R. Al-Khalil, "Simple extractive colorimetric determination of levofloxacin by acid-dye complexation methods in pharmaceutical preparations," *Farmaco*, vol. 60, no. 9, pp. 771–775, 2005.

[20] Y. M. Issa, F. M. Abdel-Gawad, M. A. Abou Table, and H. M. Hussein, "Spectrophotometric determination of ofloxacin and lomefloxacin hydrochloride with some sulphonphthalein dyes," *Analytical Letters*, vol. 30, no. 11, pp. 2071–2084, 1997.

[21] H. A. Omara and A. S. Amin, "Extractive-spectrophotometric methods for determination of anti-Parkinsonian drug in pharmaceutical formulations and in biological samples using sulphonphthalein acid dyes," *Journal of Saudi Chemical Society*, vol. 16, no. 1, pp. 75–81, 2012.

[22] A. A. Gouda, A. S. Amin, R. El-Sheikh, and A. G. Yousef, "Spectrophotometric determination of gemifloxacin mesylate, moxifloxacin hydrochloride, and enrofloxacin in pharmaceutical formulations using acid dyes," *Journal of Analytical Methods in Chemistry*, vol. 2014, Article ID 286379, 16 pages, 2014.

[23] S. G. Nair, J. V. Shah, P. A. Shah, M. Sanyal, and P. S. Shrivastav, "Extractive spectrophotometric determination of five selected drugs by ion-pair complex formation with bromothymol blue in pure form and pharmaceutical preparations," *Cogent Chemistry*, vol. 1, no. 1, article 1075852, 2015.

[24] N. Rahman and S. N. Hejaz-Azmi, "Extractive spectrophotometric methods for determination of diltiazem HCl in pharmaceutical formulations using bromothymol blue, bromophenol blue and bromocresol green," *Journal of Pharmaceutical and Biomedical Analysis*, vol. 24, no. 1, pp. 33–41, 2000.

[25] H. E. Abdellatef, "Extractive-spectrophotometric determination of disopyramide and irbesartan in their pharmaceutical formulation," *Spectrochimica Acta Part A: Molecular and Biomolecular Spectroscopy*, vol. 66, no. 4, pp. 1248–1254, 2007.

[26] N. Rahman, S. K. Manirul Haque, S. N. H. Azmi, and H. Rahman, "Optimized and validated spectrophotometric methods for the determination of amiodarone hydrochloride in commercial dosage forms using N-bromosuccinimide and bromothymol blue," *Journal of Saudi Chemical Society*, vol. 21, no. 1, pp. 25–34, 2017.

[27] D. Taşkın, G. Erensoy, and S. Sungur, "Optimized and validated spectrophotometric determination of butamirate citrate in bulk and dosage forms using ion-pair formation with methyl orange and bromothymol blue," *Farmacia*, vol. 65, pp. 761–765, 2017.

[28] P. Govardhan Reddy, V. Kiran Kumar, V. Appala Raju, J. Raghu Ram, and N. Appala Rraju, "Novel spectrophotometric method development for the estimation of boceprevir in bulk and in pharmaceutical formulations," *Research Journal of Pharmacy and Technology*, vol. 10, p. 4313, 2017.

[29] A. Sakur and S. Affas, "Direct spectrophotometric determination of sildenafil citrate in pharmaceutical preparations via complex formation with two sulphonphthalein acid dyes," *Research Journal of Pharmacy and Technology*, vol. 10, p. 1191, 2017.

[30] K. N. Prashanth, K. Basavaiah, and K. B. Vinay, "Sensitive and selective spectrophotometric assay of rizatriptan benzoate in pharmaceuticals using three sulphonphthalein dyes," *Arabian Journal of Chemistry*, vol. 9, pp. S971–S980, 2016.

[31] T. D. Nguyen, L. Bau, L. Q. Thao, and N. Dang Dat, "Extractive spectrophotometric methods for determination of ciprofloxacin in pharmaceutical formulations using sulfonephthalein acid dyes," *Vietnam journal of chemistry*, vol. 55, no. 6, pp. 767–774, 2017.

[32] H. T. S. Britton, *Hydrogen Ions*, Chapman & Hall, 4th edition, 1952.

[33] I. T. Q. (R1), *Validation of Analytical Procedures: Text and Methodology*, (CPMP/ICH/281/95), ICH Secretariat, Geneva, Switzerland, 2010.

[34] A. M. El-Brashy, M. E.-S. Metwally, and F. A. El-Sepai, "Spectrophotometric determination of some fluoroquinolone antibacterials by ion-pair complex formation with cobalt (II) tetrathiocyanate," *Journal of the Chinese Chemical Society*, vol. 52, no. 1, pp. 77–84, 2005.

[35] C. S. P. Sastry, K. R. Rao, and D. S. Prasad, "Extractive spectrophotometric determination of some fluoroquinolone derivatives in pure and dosage forms," *Talanta*, vol. 42, no. 3, pp. 311–316, 1995.

[36] A. M. El-Brashy, M. El-Sayed Metwally, and F. A. El-Sepai, "Spectrophotometric determination of some fluoroquinolone antibacterials by binary complex formation with xanthene dyes," *Farmaco*, vol. 59, no. 10, pp. 809–817, 2004.

[37] B. G. Gowda and J. Seetharamappa, "Extractive spectrophotometric determination of fluoroquinolones and anti-allergic drugs in pure and pharmaceutical formulations," *Analytical Sciences*, vol. 19, no. 3, pp. 461–464, 2003.

[38] K. N. Prashanth, K. Basavaiah, and M. S. Raghu, "Simple and selective spectrophotometric determination of ofloxacin in pharmaceutical formulations using two sulphonphthalein acid dyes," *ISRN Spectroscopy*, Article ID 357598, 9 pages, 2013.

20

Collaborative Penalized Least Squares for Background Correction of Multiple Raman Spectra

Long Chen [iD],[1] **Yingwen Wu,**[1] **Tianjun Li,**[1] **and Zhuo Chen**[2]

[1]*Faculty of Science and Technology, University of Macau, E11 Avenida da Universidade, Taipa, Macau*
[2]*Chemistry and Chemical Engineering, College of Biology, Hunan University, Changsha 410082, China*

Correspondence should be addressed to Long Chen; longchen@umac.mo

Academic Editor: Małgorzata Jakubowska

Although Raman spectroscopy has been widely used as a noninvasive analytical tool in various applications, backgrounds in Raman spectra impair its performance in quantitative analysis. Many algorithms have been proposed to separately correct the background spectrum by spectrum. However, in real applications, there are commonly multiple spectra collected from the close locations of a sample or from the same analyte with different concentrations. These spectra are strongly correlated and provide valuable information for more robust background correction. Herein, we propose two new strategies to remove background for a set of related spectra collaboratively. Based on weighted penalized least squares, the new approaches will use the fused weights from multiple spectra or the weights from the average spectrum to estimate the background of each spectrum in the set. Background correction results from both simulated and real experimental data demonstrate that the proposed collaborative approaches outperform traditional algorithms which process spectra individually.

1. Introduction

Raman spectroscopy, which provides valuable chemical and physical information of studied samples, is widely used as an analytical tool for many applications like material identification, chemical detection, and biomedical analysis [1–3]. Peaks in Raman spectra are the fingerprints of the analyte, and the corresponding peak heights or peak areas have strong correlations with the concentration of the analyte. However, spectral interferences have a strong negative effect on the measurement of peaks, and this in the long run hinders the performance of Raman spectroscopy-based quantitative analysis [4]. Representative interferences for Raman spectra include backgrounds mainly caused by instrument fluctuations and fluorescent substances. The noises of the instrument and the occasional spikes caused by cosmic rays also deteriorate the quality of Raman spectra. As a result, some preprocessing steps should be conducted to handle the interferences in the Raman spectra. In this paper, we mainly focus on the background correction problem.

Numeric background correction algorithms have been proposed in the past decades for Raman and other spectra. For example, the wavelet transform is used as a powerful tool for background removal by decomposing the Raman signals in the frequency domain [5–7]. Because the performance of wavelet-based approaches is greatly affected by the selection of base wavelets and scales, the adaptive wavelet transform was used in [8] to obtain a multiresolution decomposition of a Raman spectrum. The low-frequency background and high-frequency noise were removed thereafter.

The iterative smoothing algorithms also play an important role in background estimation because the background is usually characterized by its smooth variation. The general procedure of this approach is continually smoothing the spectrum until the background is obtained. Many well-known smoothing filters and their enhancements have been widely used in previous studies for the purpose of iteratively removing peaks and deriving backgrounds in the spectra [9–12]. The drawback of such smoothing algorithms is the difficulty in automating their iterations, although some endeavors have been made on this issue [10, 11].

Curve fitting is to fit the background with appropriate points in the spectrum by some fidelity or loss functions like the least squares [13–15]. Such a kind of selection-then-approximation approach is very similar to the manual background estimation procedure. The simple implementation and short running time of curve fitting have made it one popular background correction method used in real applications. Different curves such as Bezier curves [16], splines [13], and polynomial functions [14, 17] have been used to fit the background. On the contrary, without specifying the curve shape of the background, the penalized least squares- (PLS-) based algorithms attempt to automatically estimate the background by a direct approximation of the spectrum with a penalization on the roughness of the approximated curve [15, 18]. In the PLS-based algorithms, one critical issue is the weight setting for different points in the spectrum. These weights are used to indicate the contribution of corresponding points to the final curve construction. Many methods have been proposed to this end [19, 20], and some automatic setting techniques are also suggested [21, 22].

Some comparisons on different baseline correction approaches have been conducted [23], and the optimal choice of background removal for the statistical analysis of spectra has been explored [24]. However, by far, there is no single automatic method that can well handle all the spectra universally and be regarded as the best. Recently, more new baseline correction algorithms like the ones based on sparse representation [25] and neural networks [26] have been proposed.

In practical applications of Raman spectroscopy, multiple measurements of a given analyte are normal practices. A set of strongly correlated spectra is then obtained although they may be generated under different environmental conditions and sampling protocols. In the quantitative Raman analysis like the mixture analysis and multivariate calibration, except multiple measurements, multiple spectra derived from the same mixture with different analyte concentrations are also correlated. Due to the varying backgrounds and random noises in the set of related spectra, the clear information like the peak locations in one spectrum may not be significant in another spectrum. How do we collaboratively use the valuable information in a set of related spectra for the purpose of spectrum preprocessing? Foist et al. first noticed this problem and proposed a method to denoise multidimensional spectral data collaboratively [27]. For background correction, few approaches have been proposed by utilizing the common characteristics shared in a set of related spectra [6, 28, 29]. For instance, the multiple spectra baseline correction (MSBC) algorithm designed in [28] assumed that the pairwise differences between the background removed spectra are small and inserted a regularization for this prior to the asymmetric least squares.

In this paper, based on PLS, we propose a new approach focusing on collaborative background correction for a set of related spectra. Specifically, our main contribution is to design two ensemble strategies to embed the weight information of PLS from multiple spectra to boost each spectrum's background correction. For PLS, the weight of each point in a spectrum denotes the contribution of the point to the final background estimation. In the first scheme, we directly use the weights derived from the average spectrum (by averaging all the related spectra) to calculate the background of each spectrum. The second scheme applies the average of weights obtained from each spectrum using traditional PLS-based approaches. By using the two schemes, our new approach utilizes the strong correlations among multiple spectra and suppresses the effect of noise and signal variation in different spectra.

To illustrate the advantage of collaboratively calculating the weights for PLS algorithms when handling several related spectra, we combine the adaptive iteratively reweighted penalized least squares algorithm (airPLS) [19] and the morphological weighted penalized least squares algorithm (MPLS) [20] with the proposed weight ensemble strategies. In the experiments on the synthetic and real Raman spectra, these enhanced PLS approaches show accurate and robust background removal capability.

2. Theory

2.1. PLS for Background Correction. The signal smoothing problem was first proposed by Whittaker in 1922 [30]. The pioneering works on applying PLS for baseline correction were conducted by Eilers et al. more than 10 years ago [31, 32]. The rationale behind PLS is to approximate the observed data by balancing the conflicts between the fidelity to original data and the roughness of fitting data.

Assume that y is a vector of the Raman spectrum and z is the fitting vector; both of them are with the length of N elements. The fitted z should keep the fidelity to y as well as the roughness of the fitted vector. F denotes the fidelity to the Raman spectrum y, which can be expressed as the sum of squares of differences between y and z:

$$F = \sum_{i=1}^{N} (y_i - z_i)^2. \tag{1}$$

R denotes the roughness of the fitting vector z, which can be expressed as the sum of squares of differences between each element of z and its neighbors:

$$R = \sum_{i=2}^{N} (z_i - z_{i-1})^2, \tag{2}$$

where the square of first differences penalty is adopted in (2) to simplify the presentation. In other cases, it is also a natural way to quantify the roughness by the square of higher-order differences.

The following equation is adopted to measure the balanced combination of fidelity and roughness:

$$Q = F + \lambda R, \tag{3}$$

where λ is a user adjustable parameter that balances the fidelity and roughness. Larger λ favours a smoother fitted vector.

In order to apply the PLS to estimate background, a weight vector w was introduced for fidelity; its element w_i can be regarded as a weight that depicts the reliability of point i as a part of background. Then, F is changed to

$$F = \sum_{i=1}^{N} w_i (y_i - z_i)^2. \qquad (4)$$

To solve the minimization problem of (3), we get a linear system by equating the partial derivatives of Q to zero ($\partial Q/\partial z = 0$), and the matrix form of the obtained linear system is as follows:

$$\left(\text{diag}(w) + \lambda D^T D \right) z = \text{diag}(w) y, \qquad (5)$$

where $\text{diag}(w)$ is a diagonal matrix with w on its diagonal and D is the derivative of an identity matrix. Finally, we solve the fitting vector as follows:

$$z = \left(\text{diag}(w) + \lambda D^T D \right)^{-1} \text{diag}(w) y. \qquad (6)$$

There have been some proposed methods for the weight calculation in PLS. To control the smoothness of the fitted vector iteratively, the airPLS method [19] calculates the weight vector w in an adaptive way. w_i in each iteration t is obtained as follows:

$$w_i = \begin{cases} 0, & y_i \geq z_i^{t-1}, \\ e^{t(y_i - z_i^{t-1})/|d^t|}, & y_i < z_i^{t-1}. \end{cases} \qquad (7)$$

The vector d^t consists of negative elements obtained from the subtraction between y and z^{t-1} in the tth iteration step. The fitted vector z^{t-1} in the previous $t-1$ iteration step is a candidate of the baseline. If the value of the signal y_i is greater than the candidate, it can be seen as a part of the peak, of which the weight is set to zero. If not, the weight is calculated as (7). When the iteration count reaches the maximum or when the following termination criterion is satisfied, the iteration will stop and the final weight vector is used for PLS to generate the background:

$$|d^t| < 0.001 \times |y|. \qquad (8)$$

Unlike airPLS which adjusts the weights adaptively, the MPLS method [20] directly calculates the weight vector by applying the mathematical morphology operations on the spectrum to remove the peaks and generate a rough background firstly. The morphology operation involves an object spectrum y and a plane structuring element E. The transformation is an opening operation which consists of dilation and erosion. To refine the background, the local minimum points between peak areas are selected as meaningful background points with weight 1, and the remaining points are set with a weight of 0. The weighted PLS is then applied to get the final background.

In these cases, we can collaboratively estimate each spectrum's background by comprehensively considering the valuable information shared in the whole set of spectra. Specifically, for the weighted PLS-based approaches, we design two schemes to utilize the global information in the set of spectra for the weight calculation.

The simplest information fusion approach for a set of highly related spectra is to average them. More formally, given a set of m spectra y^1, y^2, $\ldots$, y^m, we calculate the average spectrum as follows:

$$y^{\text{avg}} = \sum_{i=1}^{m} y^i. \qquad (9)$$

By averaging, the effect of noise on spectra is suppressed, and the average spectrum can be regarded an informative representation of the set of related spectra. With the average spectrum, we can apply some traditional weighed PLS-based approaches to calculate the weights of different points that denote the reliabilities of their reliabilities as some parts of background. Then, the obtained weight vector w^{avg} is used for each single spectrum's background removal. For a more accurate fusion, we may set weights for different spectra in the summation of (9). For example, the high-quality spectrum with a higher signal-noise ratio may take a higher weight to contribute more to the final average spectrum. But in our experiments, we find the simple summation produces good results as well.

The second scheme to fuse the information from multiple spectra is to ensemble the weight vectors of all the spectra. For the spectrum y^i, we first use some traditional weighted PLS-based approaches to calculate the weight vector w^i for it. Then, the weight vectors for all the related spectra are combined into one as follows:

$$w^{\text{com}} = \sum_{i=1}^{m} w^i, \qquad (10)$$

where the combined weight vector w^{com} will be used as the final weight vector in weighted PLS-based background estimation of each spectrum.

To illustrate the application of the two schemes proposed above, we improved the airPLS and MPLS methods by modifying their final weight vectors. In these two PLS-based background estimation methods, no matter the weight vector for points in the spectrum is adaptively adjusted in the iteration (the airPLS case) or the weight vector is directly obtained from morphology operations, the final step to estimate the background is a linear regression step that solves (5) by using (6) with the determined weights.

Given a set of m spectra y^1, y^2, $\ldots$, y^m, we have the following 4 enhanced PLS approaches to estimate each spectrum's background.

2.2. Collaborative Weighted Penalized Least Squares for Multiple Spectra.

As discussed in Introduction, in practical applications of Raman spectroscopy, we may collect strongly related spectra that are from either the same kind of material or the solution with different proportional concentrations.

2.2.1. Average Spectrum-Based airPLS (AS airPLS).

In this method, we first derive the average spectrum y^{avg} from (9). Then, the airPLS is applied over the y^{avg}. But here, we only record the weight vector derived at the last iteration of airPLS using (7). This weight vector is denoted as w^{avg}. Now, the background of y^i is calculated by using (6), in which the weight vector w is w^{avg} and the spectrum y is y^i.

2.2.2. Combined Weight-Based airPLS (CW airPLS). This method first applies airPLS for each spectrum. We only record the final weight vector for each spectrum obtained in the last iteration of airPLS. Then, the combined weight vector is calculated by using (10), and we use this weight vector to derive each spectrum's background by using (6).

2.2.3. Average Spectrum-Based MPLS (AS MPLS). This method is very similar to AS airPLS. The only difference is that w^{avg} is calculated by applying MPLS on the average spectrum y^{avg}.

2.2.4. Combined Weight-Based MPLS (CW MPLS). This method is similar to CW airPLS. The difference is that the initial weight vector for each spectrum is obtained by MPLS instead of airPLS.

Because the weight vector is also extensively used in other weighted PLS-based background removal approaches [33–35], including some recently proposed enhancements of airPLS [18], the average spectrum and combined weight-based schemes proposed in this paper can be easily adopted by these approaches to collaboratively process a set of related spectra.

3. Experimental

To verify the performance of collaborative approaches proposed in this paper on background removal of a set of related Raman spectra, we compare them with traditional approaches on simulated spectra and real Raman spectra.

3.1. Simulated Data. For the simulated data, each spectrum s^i in the set S consists of the pure signal p^i, background b^i, and noise n^i:

$$s^i = p^i + b^i + n^i. \qquad (11)$$

As our methods are to process multiple correlated spectra, the simulated pure signal p^i is a mixture of three pure spectra illustrated in Figure 1(a) that represent three chemical components. The concentrations of different components are randomly drawn from 0% to 100% in the mixture. By doing this, the simulated pure signals show variations of the peaks. Figure 1(b) depicts some simulated pure signals with different concentrations of components. In addition, the background b^i in (11) is generated by exponential, polynomial, sigmoid, or sine curves with a random amplitude. Finally, the Gaussian white noise n^i is added to the summation of the background and pure signal. Altogether, we randomly generate 30 simulated spectra for each type of backgrounds, as shown in Figures 1(c)–1(f). Their corresponding backgrounds are plotted.

We compare the performance of our improved methods (AS airPLS, AS MPLS, CW airPLS, and CW MPLS) with one of the original airPLS and MPLS methods. Unlike the collaborative AS- and CW-based methods, the airPLS and MPLS process the spectra in the simulated data separately. In addition, the multiple spectra baseline correction (MSBC) algorithm proposed in [28] is also included in comparison.

In these simulated data, as the ground truth backgrounds are known, we can evaluate the performance of each method by calculating the mean squared error (MSE) or the relative error (RE) between the real background b_i and the estimated background $\widetilde{b}_i$:

$$\text{MSE} = \frac{1}{N} \sum_{i=1}^{N} \left(b_i - \widetilde{b}_i \right)^2, \qquad (12)$$

$$\text{Relative error} = \frac{1}{N} \sum_{i=1}^{N} \frac{\left| b_i - \widetilde{b}_i \right|}{b_i} \%. \qquad (13)$$

According to (12) or (13), for each spectrum in the 30 simulated spectra, we can get one MSE or RE value. For the multispectra, we compare the average MSE or average RE of the 30 simulated spectra. The smaller value of MSE or RE denotes the estimated background is more similar to the ground truth. Therefore, the smaller MSE or RE indicates the better background correction method.

As shown in Table 1, our AS- and CW-based methods can obtain much better RE than the traditional approaches. The bar plot in Figure 2 visualizes the results in Table 1. From these results, we can see the multiple spectra baseline correction (MSBC) algorithm performs similar to or a little better than the approaches relying on separate spectrum processing. But its performance is worse than the collaborative approaches proposed in this paper.

To have a better comparison, we also plot several representative pairs of ground truth backgrounds (the simulated sine curves) and the estimated ones obtained by different methods in Figure 3. The zoomed-in inserts from $600 \, \text{cm}^{-1}$ to $800 \, \text{cm}^{-1}$ are provided in Figure 3 for better visualization. It is pretty obvious that the estimated backgrounds obtained from our enhanced approaches (the red and green lines) are much closer to the real background (the black line).

Because the simulated data are spectra of some mixtures of three chemical components, we can conduct a regression analysis on the preprocessed spectra to predict the concentration values of different components. Here, the principal component regression (PCR) is applied on 29 simulated spectra, and the concentrations of three components in the remainder spectrum are predicted thereafter. We do this 30 times by choosing each simulated spectrum as the testing one once. The root mean squared error (RMSE) is used as the prediction error to evaluate different methods' performance. As shown in Table 2, the spectra preprocessed by the enhanced PLS show higher prediction accuracy when compared with other approaches.

For all the six PLS methods, the smoothness parameter λ should be adjusted to get a good estimation of

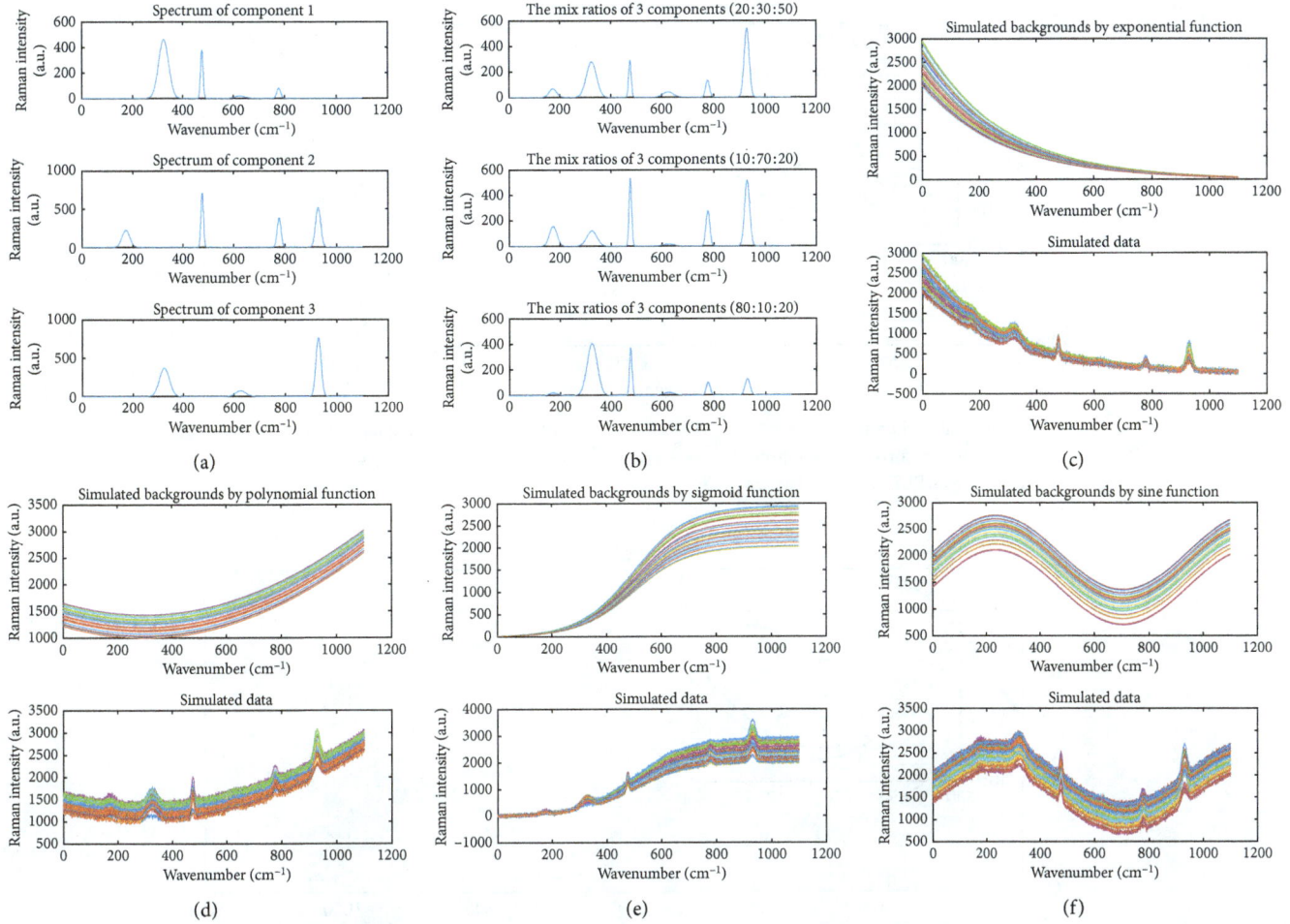

FIGURE 1: Simulated data. (a) Spectra of three chemical components. (b) Three sampling mixture signals with different concentrations of components. (c–f) The simulated spectra and their backgrounds.

TABLE 1: Relative error for different backgrounds.

	Exponential (%)	Polynomial (%)	Sigmoid (%)	Sine (%)
airPLS	21.3	1.9	24.1	1.5
MPLS	19.8	2.2	26.3	1.3
MSBC	19.6	2.1	26.3	2.1
AS airPLS	3.7	0.3	5.3	0.3
CW airPLS	1.5	0.2	2.1	0.3
AS MPLS	3.6	0.3	4.6	0.6
CW MPLS	3.5	0.5	7.0	0.8

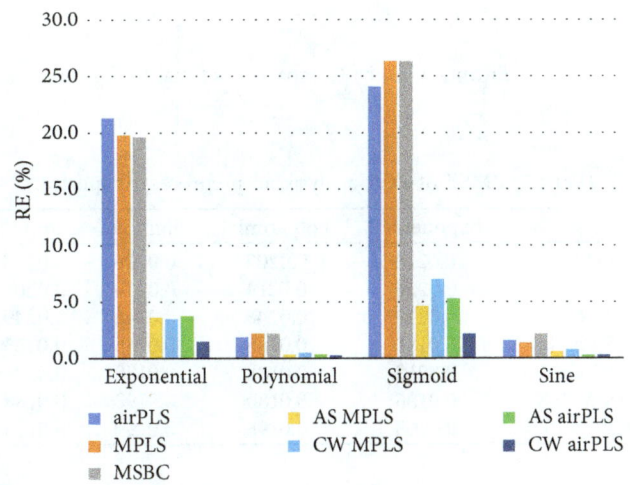

FIGURE 2: Relative error of different methods on different backgrounds.

background. If λ is too small, the estimated background would fit the peaks too much. However, if λ is too large, the estimated background would be too smooth to catch the fluctuation of backgrounds. So, it is necessary to delve further into the value of λ. As we have mentioned before, we compare the performance of different methods by calculating the MSE between real and estimated backgrounds. Here, we design the experiments with various λ and compare the MSE so as to find the optimum λ. λ is varied in the log scale as is recommended in Eilers' paper [31]. Therefore, we test with λ values from 10^2 to 10^8 and

discard the λ value with bad estimated background at first glance. Through comprehensive consideration, we let λ to change from 10^4 to 10^7 and record the MSE value for each

(a)

(b)

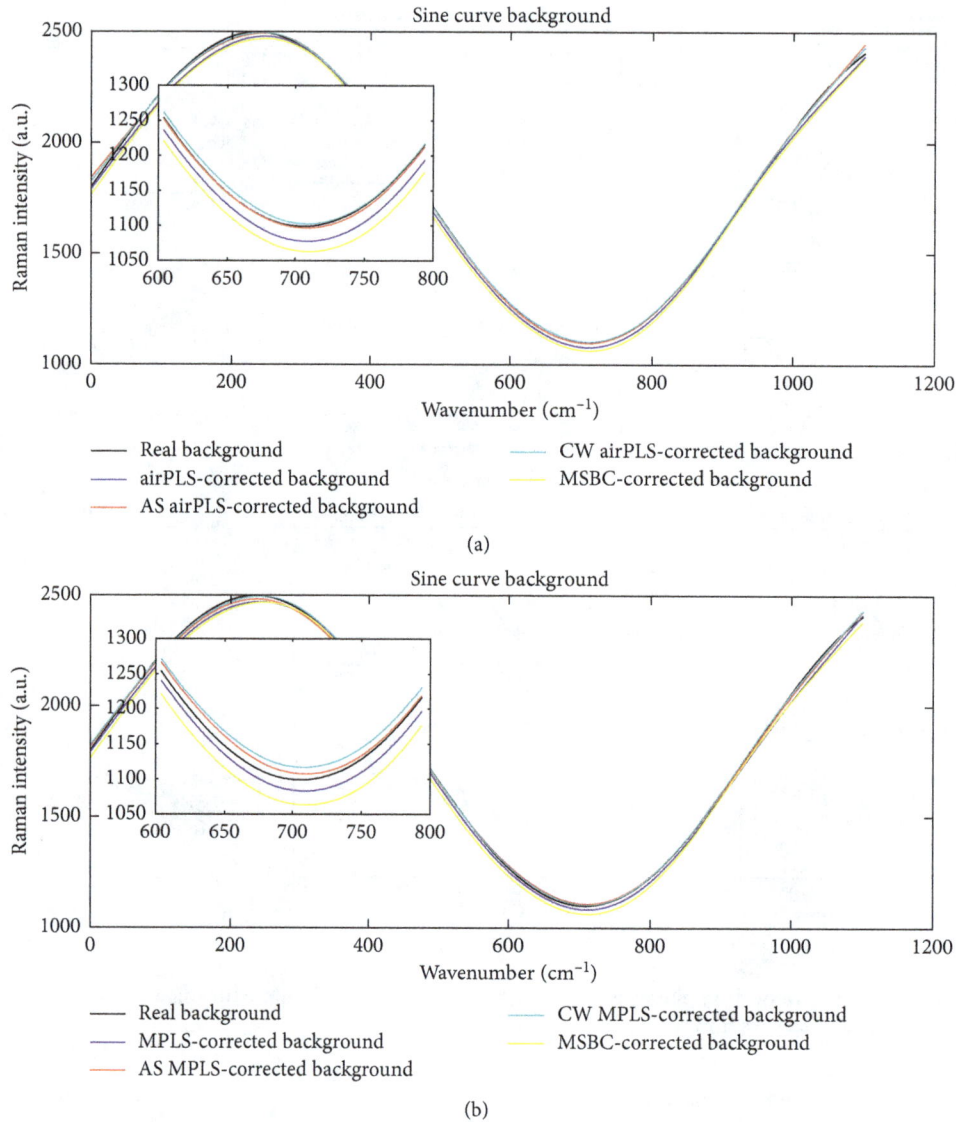

FIGURE 3: (a, b) Comparison of real backgrounds and estimated backgrounds obtained by different methods.

TABLE 2: RMSE of PCR analysis on preprocessed spectra.

	Exponential	Polynomial	Sigmoid	Sine
airPLS	0.0216	0.0202	0.0206	0.0208
MPLS	0.0226	0.0214	0.0224	0.0208
MSBC	0.0249	0.0248	0.0248	0.0249
AS airPLS	0.0181	0.0185	0.0185	0.0185
CW airPLS	0.0219	0.0188	0.0189	0.0189
AS MPLS	0.0186	0.0180	0.0190	0.0188
CW MPLS	0.0203	0.0196	0.0199	0.0193

method. The MSE results of each method with various λ are displayed in Figure 4. Overall, our collaborative methods can always obtain lower MSE compared with original methods, which means that these methods can get a better estimation of background. The optimal results appear around $\lambda = 10^{5.2}$, and similar performance can be obtained between $10^{4.8}$ and $10^{5.8}$. It seems that our methods

are not too sensitive to λ. What it means is that our methods are relatively robust to the choice of λ.

3.2. Real Raman Spectra. Graphene-isolated-Au-nanocrystal (GIAN) is a unique and stable nanostructure and has been utilized for different biomedical applications, such as sensitive Raman imaging, drug loading for chemotherapy, and photothermal therapy [36]. Herein, we utilized the GIAN to verify the efficiency of proposed background correction approaches. The Raman spectra of GIANs were collected via a static scan in the region of 200–2200 cm^{-1}. The spectrum was saved every 0.01 second, and 45 numbers of spectra signals were collected as shown in Figure 5. Because the set of spectra is collection from the same sample in a short time, although the varying backgrounds make the spectra show different intensities, the background-removed spectra should show the same intensity ideally. As shown in Figure 6,

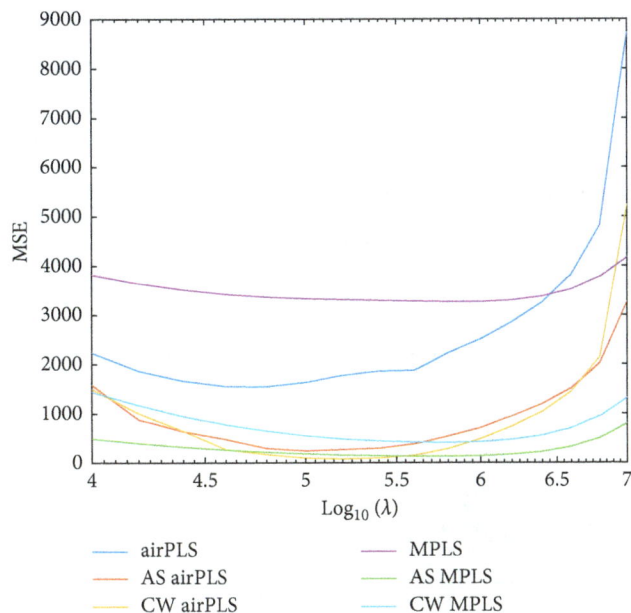

FIGURE 4: MSE of each PLS method for different λ.

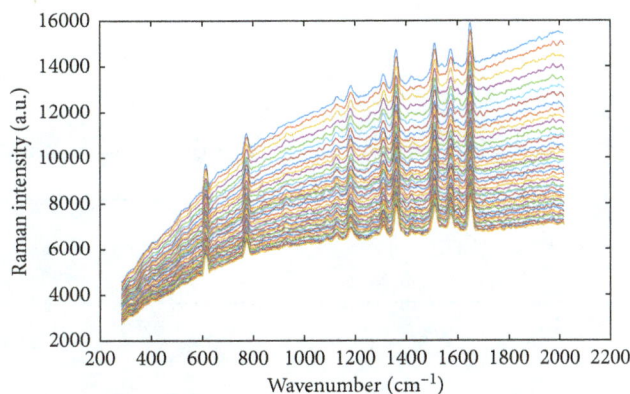

FIGURE 5: A set of Raman spectra for GIANs collected in a short time.

the processed spectra by AS- and CW-enhanced PLS methods show more homogeneous features of spectra. This is especially observable in the zoomed region from $1500 \, cm^{-1}$ to $1600 \, cm^{-1}$.

To quantitatively measure the consistency of background-removed spectra, the average intensity variance of all the processed spectra is calculated. A smaller variance means a closer relation of the background-removed spectra, which indicates a relative better background correction result. From Table 3, we see the variances of the AS and CW methods are smaller than those of the original methods, which implies the better performance of collaborative methods.

As a commonly used approach to quantify the background correction [19, 20, 35], principal component analysis (PCA) is also conducted on the matrix of spectra before and after background removal. In PCA, the score of a data point along with a principal component (PC) is the distance from the origin to the data point's projection on this PC. Figure 7 shows the score plot of the original spectra according to the

first two principal components. The PCA score plots of background-corrected spectra are shown in Figure 8. For the real Raman data used in this experiment, the discrepancy in spectra is mainly caused by the baseline interference. The effective removal of baselines should increase similarity in processed spectra, which is reflected by a tighter PCA cluster [16]. In Figure 8, the convex hulls of data are also highlighted in score plots to illustrate the compactness of PCA clusters. Clearly, the enhanced PLS methods show tighter convex hulls, and this confirms their advantages over classical approaches.

4. Conclusion

Considering the valuable information in the whole set of related spectra, we propose to remove their backgrounds collaboratively. Based on penalized least squares, we improve the background correction methods for a set of correlated Raman spectra by designing two collaborative weighting

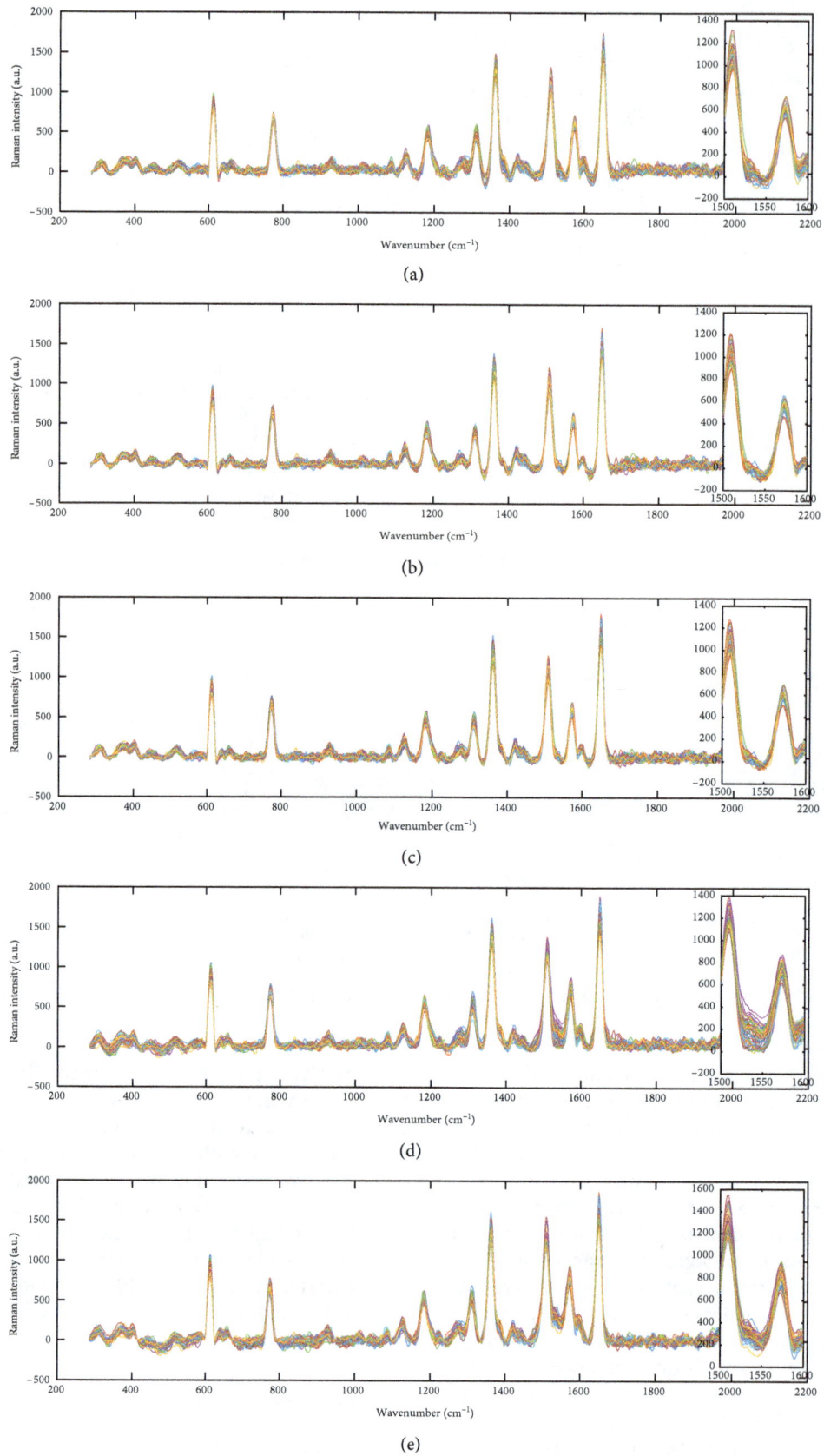

(a)

(b)

(c)

(d)

(e)

Figure 6: Continued.

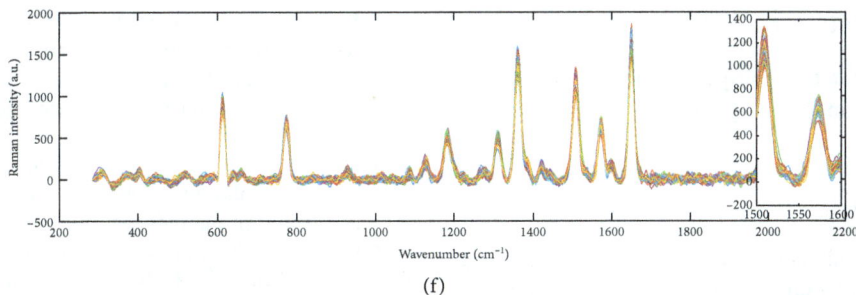

(f)

FIGURE 6: Spectra of GIAN with background removed by airPLS (a), AS airPLS (b), CW airPLS (c), MPLS (d), AS MPLS (e), and CW MPLS (f).

TABLE 3: Average intensity variance of background-corrected GIAN data.

Method	Variance
airPLS	1101.02
AS airPLS	**880.67**
CW airPLS	914.54
MPLS	1635.78
AS MPLS	1530.35
CW MPLS	**988.65**

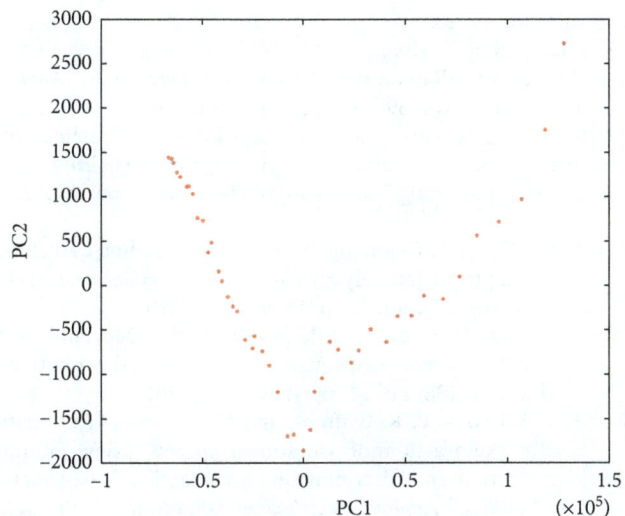

FIGURE 7: Score plot for original spectra.

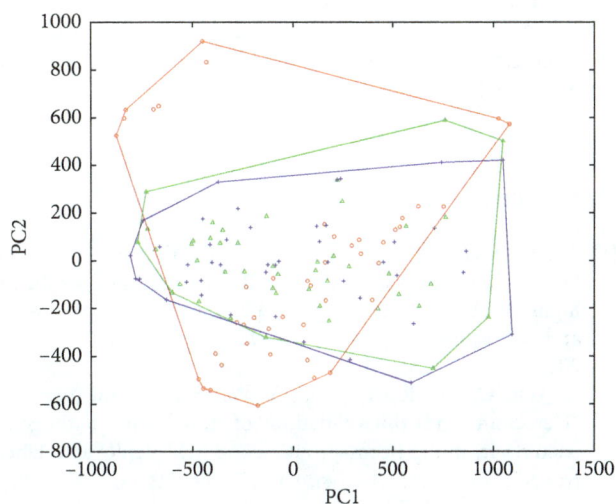

○ airPLS-corrected spectra
△ AS airPLS-corrected spectra
+ CW airPLS-corrected spectra

(a)

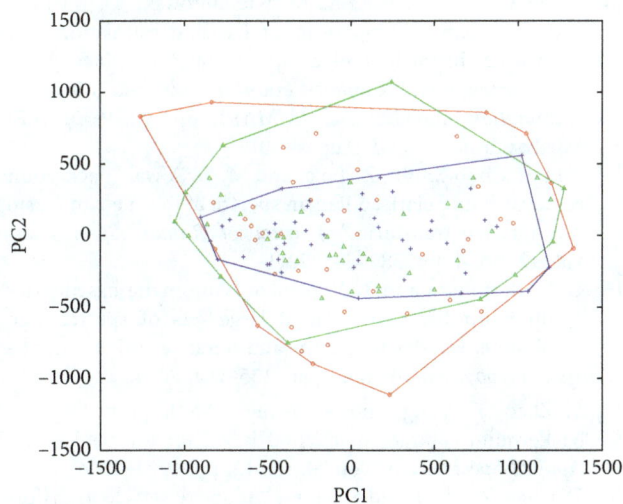

○ MPLS-corrected spectra
△ AS MPLS-corrected spectra
+ CW MPLS-corrected spectra

(b)

FIGURE 8: (a, b) Score plots for corrected spectra.

schemes for background estimations. The average spectrum (AS) method fuses all the considered spectra into an average one to calculate the weights or contributions of different points to the background. The combined weight (CW) method averages the weights derived from all spectra for background estimation. To illustrate the performance of such collaborative approaches on background correction problems, we apply the AS and CW versions of airPLS and MPLS to process simulated and real Raman spectra. The results demonstrate the collaborative approaches are much better than traditional approaches that process spectra individually.

Conflicts of Interest

The authors declare that there are no conflicts of interest regarding the publication of this paper.

Acknowledgments

This work was financially supported by the Science and Technology Development Fund of Macao S.A.R (067/2014/A and 097/2015/A3), the National Key Basic Research Program of China (no. 2013CB932702), the Research Fund for the Program on National Key Scientific Instruments and Equipment Development of China (no. 2011YQ0301241402), the National Natural Science Foundation of China (nos. 21522501, 21521063, and 61673405), and the Hunan Innovation and Entrepreneurship Program.

References

[1] S. S. R. Dasary, P. C. Ray, A. K. Singh, T. Arbneshi, H. Yu, and D. Senapati, "A surface enhanced Raman scattering probe for highly selective and ultra sensitive detection of iodide in water and salt samples," *Analyst*, vol. 138, no. 4, pp. 1195–1203, 2013.

[2] X. Wu, C. Xu, R. A. Tripp, Y.-W. Huang, and Y. Zhao, "Detection and differentiation of foodborne pathogenic bacteria in mung bean sprouts using field deployable label-free SERS devices," *Analyst*, vol. 138, no. 10, pp. 3005–3012, 2013.

[3] Y. Zou, L. Chen, Z. Song et al., "Stable and unique graphitic Raman internal standard nanocapsules for surface-enhanced Raman spectroscopy quantitative analysis," *Nano Research*, vol. 9, no. 5, pp. 1418–1425, 2016.

[4] W. Ilewicz, M. Kowalczuk, M. Niezabitowski, D. Buchczik, and G. Adam, "Comparison of baseline estimation algorithms for chromatographic signals," in *Proceedings of 2015 20th International Conference onMethods and Models in Automation and Robotics (MMAR)*, pp. 925–930, IEEE, Międzyzdroje, Poland, August 2015.

[5] J. Li, L. Choo-Smith, Z. Tang, and M. G. Sowa, "Background removal from polarized Raman spectra of tooth enamel using the wavelet transform," *Journal of Raman Spectroscopy*, vol. 42, no. 4, pp. 580–585, 2011.

[6] C. G. Bertinetto and T. Vuorinen, "Automatic baseline recognition for the correction of large sets of spectra using continuous wavelet transform and iterative fitting," *Applied Spectroscopy*, vol. 68, no. 2, pp. 155–164, 2014.

[7] F. Zhao, J. Wang, and A. Wang, "An improved spectral background subtraction method based on wavelet energy," *Applied Spectroscopy*, vol. 70, no. 12, pp. 1994–2004, 2016.

[8] D. Chen, Z. Chen, and E. Grant, "Adaptive wavelet transform suppresses background and noise for quantitative analysis by Raman spectrometry," *Analytical and Bioanalytical Chemistry*, vol. 400, no. 2, pp. 625–634, 2011.

[9] H. Ren, Z. Chen, X. Zhang et al., "Rapid and quantitative determination of S-adenosyl-L-methionine in the fermentation process by surface-enhanced Raman scattering," *Journal of Analytical Methods in Chemistry*, vol. 2016, Article ID 4910630, 6 pages, 2016.

[10] K. Chen, H. Wei, H. Zhang, T. Wu, and Y. Li, "A Raman peak recognition method based automated fluorescence subtraction algorithm for retrieval of Raman spectra of highly fluorescent samples," *Analytical Methods*, vol. 7, no. 6, pp. 2770–2778, 2015.

[11] K. Chen, H. Zhang, H. Wei, and Y. Li, "Improved Savitzky–Golay-method-based fluorescence subtraction algorithm for rapid recovery of Raman spectra," *Applied Optics*, vol. 53, no. 24, pp. 5559–5569, 2014.

[12] H. G. Schulze, R. B. Foist, K. Okuda, A. Ivanov, and R. F. B. Turner, "A model-free, fully automated baseline-removal method for Raman spectra," *Applied Spectroscopy*, vol. 65, no. 1, pp. 75–84, 2011.

[13] S. He, S. Fang, X. Liu et al., "Investigation of a genetic algorithm based cubic spline smoothing for baseline correction of Raman spectra," *Chemometrics and Intelligent Laboratory Systems*, vol. 152, pp. 1–9, 2016.

[14] C. Gallo, V. Capozzi, M. Lasalvia, and G. Perna, "An algorithm for estimation of background signal of Raman spectra from biological cell samples using polynomial functions of different degrees," *Vibrational Spectroscopy*, vol. 83, pp. 132–137, 2016.

[15] L. Martin, A. Barcaru, M. J. Sjerps, and G. Vivó-Truyols, "Leveraging probabilistic peak detection to estimate baseline drift in complex chromatographic samples," *Journal of Chromatography A*, vol. 1431, pp. 122–130, 2016.

[16] Y. Liu, X. Zhou, and Y. Yu, "A concise iterative method using the Bezier technique for baseline construction," *Analyst*, vol. 140, no. 23, pp. 7984–7996, 2015.

[17] J. Liu, J. Sun, X. Huang, G. Li, and B. Liu, "Goldindec: a novel algorithm for Raman spectrum baseline correction," *Applied Spectroscopy*, vol. 69, no. 7, pp. 834–842, 2015.

[18] S.-J. Baek, A. Park, Y.-J. Ahn, and J. Choo, "Baseline correction using asymmetrically reweighted penalized least squares smoothing," *Analyst*, vol. 140, no. 1, pp. 250–257, 2015.

[19] Z.-M. Zhang, S. Chen, and Y.-Z. Liang, "Baseline correction using adaptive iteratively reweighted penalized least squares," *Analyst*, vol. 135, no. 5, pp. 1138–1146, 2010.

[20] L. Zhong, D.-J. Zhan, J.-J. Wang et al., "Morphological weighted penalized least squares for background correction," *Analyst*, vol. 138, no. 16, pp. 4483–4492, 2013.

[21] A. T. Weakley, P. R. Griffiths, and D. E. Aston, "Automatic baseline subtraction of vibrational spectra using minima identification and discrimination via adaptive, least-squares thresholding," *Applied spectroscopy*, vol. 66, no. 5, pp. 519–529, 2012.

[22] L. G. Johnsen, T. Skov, U. Houlberg, and R. Bro, "An automated method for baseline correction, peak finding and peak grouping in chromatographic data," *Analyst*, vol. 138, no. 12, pp. 3502–3511, 2013.

[23] Ł. Komsta, "Comparison of several methods of chromatographic baseline removal with a new approach based on quantile regression," *Chromatographia*, vol. 73, no. 7-8, pp. 721–731, 2011.

[24] K. H. Liland, T. Almøy, and B.-H. Mevik, "Optimal choice of baseline correction for multivariate calibration of spectra," *Applied Spectroscopy*, vol. 64, no. 9, pp. 1007–1016, 2010.

[25] X. Ning, I. W. Selesnick, and L. Duval, "Chromatogram baseline estimation and denoising using sparsity (beads)," *Chemometrics and Intelligent Laboratory Systems*, vol. 139, pp. 156–167, 2014.

[26] A. Mani-Varnosfaderani, A. Kanginejad, K. Gilany, and A. Valadkhani, "Estimating complicated baselines in analytical signals using the iterative training of Bayesian regularized artificial neural networks," *Analytica Chimica Acta*, vol. 940, pp. 56–64, 2016.

[27] R. B. Foist, H. G. Schulze, A. Jirasek, A. Ivanov, and R. F. B. Turner, "A matrix-based two-dimensional regularization algorithm for signal-to-noise ratio enhancement of multidimensional spectral data," *Applied Spectroscopy*, vol. 64, no. 11, pp. 1209–1219, 2010.

[28] J. Peng, S. Peng, A. Jiang, J. Wei, C. Li, and J. Tan, "Asymmetric least squares for multiple spectra baseline correction," *Analytica Chimica Acta*, vol. 683, no. 1, pp. 63–68, 2010.

[29] Y. Wu, Q. Gao, and Y. Zhang, "A robust baseline elimination method based on community information," *Digital Signal Processing*, vol. 40, pp. 53–62, 2015.

[30] E. T. Whittaker, "On a new method of graduation," *Proceedings of the Edinburgh Mathematical Society*, vol. 41, pp. 63–75, 1922.

[31] P. H. C. Eilers, "A perfect smoother," *Analytical Chemistry*, vol. 75, no. 14, pp. 3631–3636, 2003.

[32] P. H. C. Eilers and H. F. M. Boelens, "Baseline correction with asymmetric least squares smoothing," Leiden University Medical Centre Report, vol. 1, no. 1, Leiden University Medical Centre, Leiden, Netherlands, 2005.

[33] H. Ruan and L. K. Dai, "Automated background subtraction algorithm for Raman spectra based on iterative weighted least squares," *Asian Journal of Chemistry*, vol. 23, no. 12, p. 5229, 2011.

[34] P. J. Cadusch, M. M. Hlaing, S. A. Wade, S. L. McArthur, and P. R. Stoddart, "Improved methods for fluorescence background subtraction from Raman spectra," *Journal of Raman Spectroscopy*, vol. 44, no. 11, pp. 1587–1595, 2013.

[35] Z.-M. Zhang, S. Chen, Y.-Z. Liang et al., "An intelligent background-correction algorithm for highly fluorescent samples in Raman spectroscopy," *Journal of Raman Spectroscopy*, vol. 41, no. 6, pp. 659–669, 2010.

[36] X. Bian, Z.-L. Song, Y. Qian et al., "Fabrication of graphene-isolated-Au-nanocrystal nanostructures for multimodal cell imaging and photothermal-enhanced chemotherapy," *Scientific Reports*, vol. 4, no. 1, 2014.

21

Pharmacokinetics and Bioavailability Study of Tubeimoside I in ICR Mice by UPLC-MS/MS

Lianguo Chen,[1] Qinghua Weng,[1] Feifei Li,[1] Jinlai Liu,[1] Xueliang Zhang (iD),[1] and Yunfang Zhou (iD)[2]

[1]*Wenzhou People's Hospital, The Third Clinical Institute Affiliated to Wenzhou Medical University, Wenzhou 325000, China*
[2]*Laboratory of Clinical Pharmacy, The People's Hospital of Lishui, The Sixth Affiliated Hospital of Wenzhou Medical University, Lishui 323000, China*

Correspondence should be addressed to Xueliang Zhang; 854867181@qq.com and Yunfang Zhou; zyf2808@126.com

Lianguo Chen and Qinghua Weng equally contributed to this work.

Academic Editor: Josep Esteve-Romero

The aim of this study is to establish and validate a rapid, selective, and sensitive ultra-performance liquid chromatography-tandem mass spectrometry (UPLC-MS/MS) method to determine tubeimoside I (TBMS-I) in ICR (Institute of Cancer Research) mouse whole blood and its application in the pharmacokinetics and bioavailability study. The blood samples were precipitated by acetonitrile to extract the analytes. Chromatographic separation was performed on a UPLC BEH C18 column (2.1 mm × 50 mm, 1.7 μm). The mobile phase consisted of water with 0.1% formic acid and methanol (1 : 1, v/v) at a flow rate of 0.4 mL/min. The total eluting time was 4 min. The TBMS-I and ardisiacrispin A (internal standard (IS)) were quantitatively detected by a tandem mass spectrometry equipped with an electrospray ionization (ESI) in a positive mode by multiple reaction monitoring (MRM). A validation of this method was in accordance with the US Food and Drug Administration (FDA) guidelines. The lower limit of quantification (LLOQ) of TBMS-I was 2 ng/mL, and the calibration curve was linearly ranged from 2 to 2000 ng/mL ($r^2 \geq 0.995$). The relative standard deviation (RSD) of interday precision and intraday precision was both lower than 15%, and the accuracy was between 91.7% and 108.0%. The average recovery was >66.9%, and the matrix effects were from 104.8% to 111.0%. In this assay, a fast, highly sensitive, and reproducible quantitative method was developed and validated in mouse blood for the first time. The absolute availability of TBMS-I in the mouse was only 1%, exhibiting a poor oral absorption.

1. Introduction

Tubeimoside I (TBMS-I), a triterpenoid saponin, is derived from the traditional Chinese bulb of *Bolbostemma paniculatum* (Maxim.). It was often used for the treatment of poisonous snake bite and inflammation [1]. In the past couple of years, additional attention was drawn to TBMS-I as it was reported to be a potential anticancer agent and appeared to be effective against several types of cancer, such as gliomas, breast cancer, colon cancer, and non-small cell lung cancer [2–5].

Pharmacokinetic studies are important in drug research and development which can provide systemic concentrations and exposure times of the drug for predicting a diverse range

of efficacy- and toxicity-related events. To systematically examine the preclinical pharmacokinetic studies of TBMS-I in a reproducible and precise manner, a sensitive, fast, and validated analytical method for the determination of TBMS-I in biological fluids is imperative. However, the research on pharmacokinetics of TBMS-I lags behind compared to its pharmacological studies. Up to January of 2018, there was only one study report that published high-performance liquid chromatography coupled with tandem mass spectrometry (LC-MS/MS) for the determination of TBMS-I, which has been applied to the pharmacokinetic study in rats in 2007 [6]. But it had several drawbacks, such as long analysis time (more than 6 min) and low sensitivity (20 ng/mL), especially

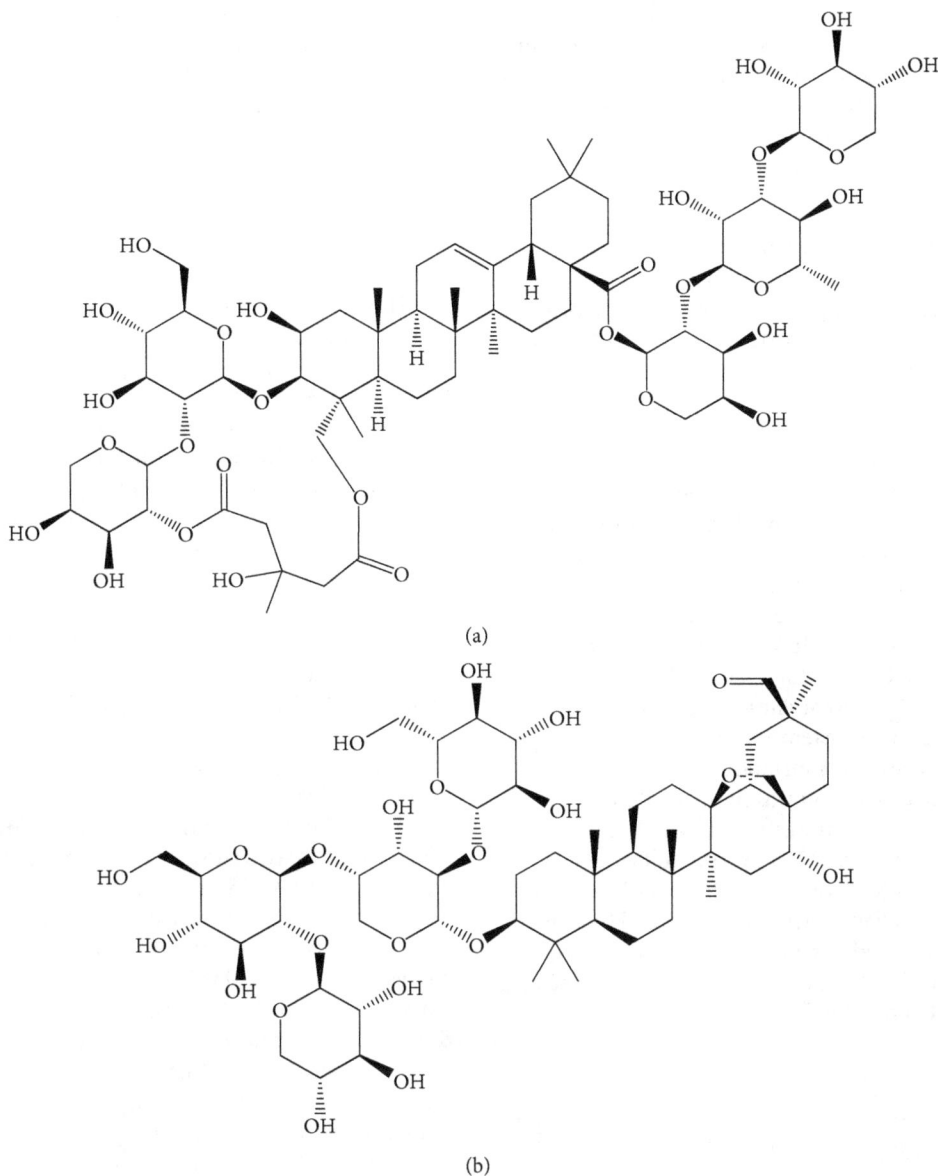

FIGURE 1: Chemical structures of tubeimoside I (a) and ardisiacrispin A (b).

requiring a large volume of plasma (100 μL), which make it unsuitable for the serial blood sampling in mice pharmacokinetic evaluation [7]. In the last couple of years, the UPLC technique has attracted more and more concern along with the development of analysis techniques [8]. Compared with LC-MS/MS, the UPLC-MS/MS method was faster, more sensitive, and with higher sample throughput [9, 10]. Meanwhile, its strong ability of isolating was more suitable for the analysis of the metabolism *in vivo* of the complex traditional Chinese medicine and complex compound [11, 12].

To the best of our knowledge, the profile of toxicity or pharmacokinetics of some drugs is alterable in different species [13–16], so it is not reasonable that these pharmacokinetic data are used directly in mice [17]. The mouse was chosen as the animal model to study the pharmacokinetics of TBMS-I in our study not only because of the reason mentioned above but also because the mouse is the most frequently used species for the preclinical efficacy [18, 19], toxicology [20], biodistribution [21], and pharmacokinetic [22] studies to evaluate a potential anticancer agent, particularly with a limiting drug supply or specialized animal models in the early new drug discovery stage [7, 23].

Thus, we established a rapid, sensitive, and selective UPLC-MS/MS method to quantitate the concentration of TBMS-I directly in the mouse utilizing low-volume whole blood after intravenous and oral administration in this study for the first time. A validation of this method was in accordance with the FDA guidelines. This method was successfully applied to the pharmacokinetics and bioavailability of TBMS-I in mice.

2. Materials and Methods

2.1. Experimental Materials. TBMS-I (purity >98%; Figure 1(a)) and ardisiacrispin A (internal standard, purity >98%; Figure 1(b)) were purchased from Chengdu Mansite Biotechnology Co., Ltd. (Chengdu, China). HPLC-grade methanol and acetonitrile were bought from Merck (Darmstadt, Germany). HPLC-grade formic acid was supplied by Tedia (Ohio, USA). A Milli-Q system (Millipore, Bedford, USA) is used for generating ultrapure water. The ICR mice (male, weight 20–22 g, $n = 12$) obtained were from the Laboratory Animal Center of Wenzhou Medical University (Wenzhou, China).

2.2. UPLC and Mass Spectrometric Conditions.

The determination of analytes was carried out using the ACQUITY UPLC I-Class system equipped with a triple-quadrupole mass spectrometer (Waters Corp., Milford, MA, USA). MassLynx 4.1 software (Waters Corp.) was used to collect data and control the system.

TBMS-I and IS were separated on a UPLC BEH C18 column (2.1 mm × 50 mm, 1.7 μm) with a stable temperature of 40°C. The mobile phases A and B were methanol and water with 0.1% formic acid, respectively. The details of gradient elution were as follows: the percentage of methanol was kept at 10% from 0 to 0.2 min and it reached 80% within 1.3 min; then, it was kept at the same percentage for 0.5 min and subsequently it turned back to 10% for another 0.5 min, and finally it was maintained at 10% for 2.5 min. The flow rate was set at 0.4 mL/min, and the total elution time was 4.0 min.

The mass spectrometer system for analysis was equipped with an electrospray source ionization (ESI) in a positive mode. The quantitative detection was performed in a multiple reaction monitoring mode at transitions m/z 1319.7 → 1187.6 for TBMS-I (collision voltage 12 V and cone voltage 30 V) and m/z 1083.5 → 407.1 for IS (collision voltage 72 V and cone voltage 100 V). The capillary voltage was 2.3 kV. High-purity nitrogen as curtain gas and drying gas was set at 50 L/h and 800 L/h, respectively. The temperature of the ion source and dissolvent was 150°C and 400°C, respectively.

2.3. Preparation of Stock Solutions, Quality Control (QC) Samples, and Calibration Standards (CS).

TBMS-I and IS were separately dissolved in methanol at a final concentration of 1.0 mg/mL as stock solutions. The working standard solutions were diluted from the stock solution using methanol. The standard working solution of IS was diluted with acetonitrile to the concentration of 50 ng/mL. All solutions were stored at 4°C prior to analysis.

CS samples were prepared by diluting blank mouse blood into corresponding standard working solutions. A series of concentrations of standard solutions were prepared with TBMS-I stock solutions and serially diluted by using methanol. The final concentrations of TBMS-I were from 2 to 2000 ng/mL, including 2, 5, 10, 20, 50, 100, 200, 500, 1000, and 2000 ng/mL.

Low-, mid-, and high-level QC samples of TBMS-I were similarly prepared at finial concentrations of 3, 190, and 1900 ng/mL, respectively. All the solutions were stored at −20°C until processed.

2.4. Sample Preparation.

A 20 μL aliquot of the mouse blood sample and 100 μL of acetonitrile containing 50 ng/mL IS were added into 1.5 mL EP tubes [24]. After vortexing for 1 min, the specimens were centrifugated (13000 rpm) for 10 min at 4°C. Then, 80 μL of the supernatant was collected; subsequently, a 2 μL of the supernatant was injected into the UPLC-MS/MS system for analysis.

2.5. Method Validation.

A validation of this method was in accordance with the FDA guidelines, including selectivity, linearity, precision, accuracy, recovery, matrix effect, and stability [25].

The chromatograms of blank mouse blood, blank blood spiked with TBMS-I and IS, and the real sample from mouse after dosing were used to estimate the selectivity of the UPLC-MS/MS method.

Calibration curves were generated by analyzing different concentrations of calibration samples on three consecutive days. The linear regressions of the peak area ratios (y) of each TBMS-I to the corresponding IS versus the nominal concentration (x) of TBMS-I were fitted over the range 2–2000 ng/mL. Linearity was evaluated covering the concentration range 2–2000 ng/mL.

The interday precision, intraday precision, and accuracy were estimated by determining three concentrations of quality control samples ($n = 6$) on the same day and on three days in a row.

The recovery was calculated by comparison of the peak areas of TBMS-I and IS in the extracted low (3 ng/mL), middle (190 ng/mL), and high (1900 ng/mL) concentrations of QC samples with those of the extracted blank blood spiked with TBMS-I and IS at corresponding concentrations.

Matrix effects were tested by comparison of the peak areas of these new working solutions with those of the corresponding standard solutions diluted with methanol : 0.1% formic acid (1 : 1, v/v) at equivalent concentrations, and this peak area ratio is defined as the matrix effect.

The stability of TBMS-I was tested under four conditions: storage in an autosampler at 4°C, storage at room temperature for 2 hours, storage at −20°C for a month, and three complete freeze-thaw cycles (from −20°C to room temperature).

The stability of TBMS-I in mouse blood was obtained by comparing the areas of the newly configured QC samples with the corresponding three concentrations (3, 190, and 1900 ng/mL) of standard samples.

2.6. Pharmacokinetic Study.

Twelve mice were randomly and equally divided into two groups (A and B). Mice in the group A were injected sublingually with 5 mg/kg TBMS-I, and mice in the group B were given TBMS-I orally at a final concentration of 20 mg/kg. The study protocol was approved by the Animal Care and Use Committee of Wenzhou Medical University. Mice were allowed to receive standard food and water ad libitum in a temperature-controlled room (25°C) with a 12-hour on and 12-hour off light cycle before the experiment.

Blood samples (20 μL) were obtained from an individual mouse by tail vein bleeding in 1.5 mL tubes at 0 (prior to dosing), 0.0833, 0.5, 1, 1.5, 2, 3, 4, 8, 12, and 24 h after dosing. Six separate mice were used for sample collection and

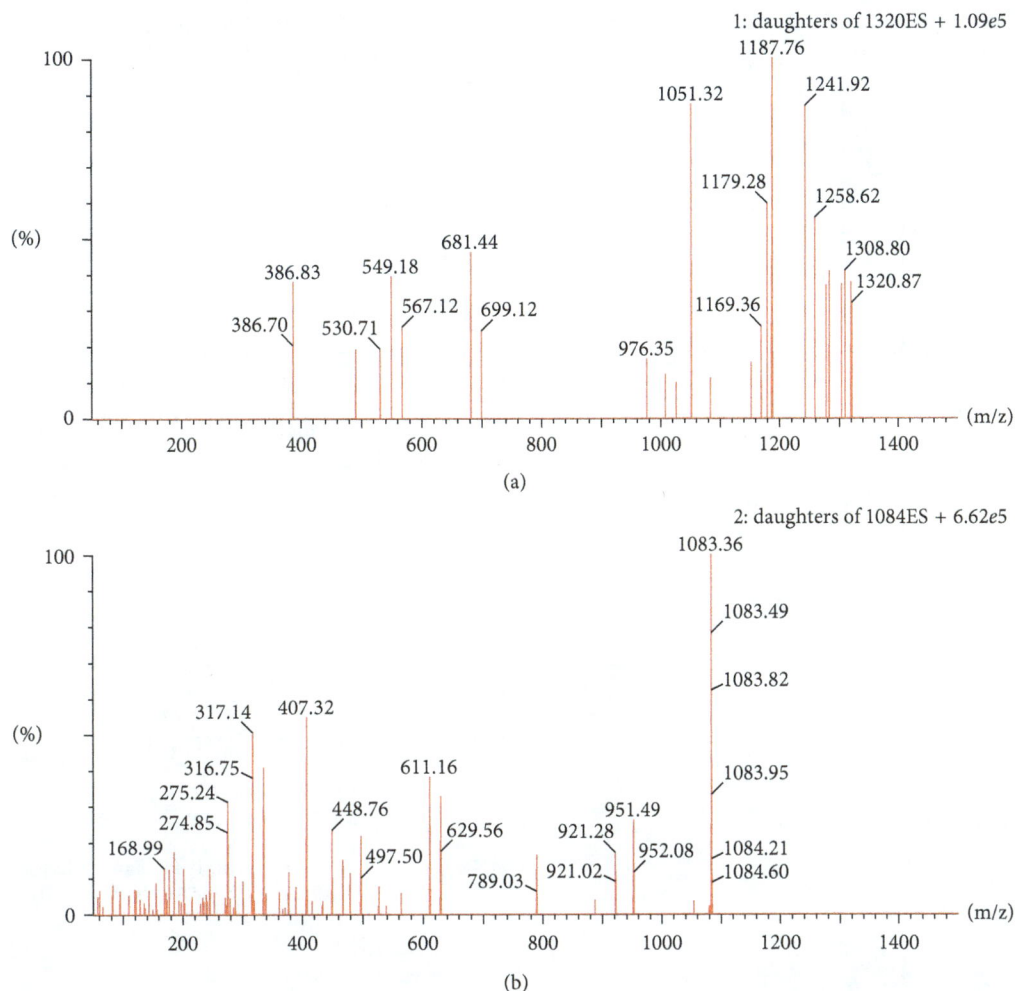

FIGURE 2: Mass spectrum of tubeimoside I (a) and ardisiacrispin A (b).

analysis at each time point. All the blood samples were directly stored at −20°C until analysis. DSA 2.0 pharmacokinetic software (China Pharmaceutical University, China) was used to calculate the main pharmacokinetic parameters, including the area under the time-concentration curve (AUC), half-life ($t_{1/2}$), the maximum of blood concentration(C_{max}), blood clearance rate (CL), apparent volume of distribution (V), and mean retention time (MRT). Bioavailability was calculated by absolute bioavailability $= 100\% \times \mathrm{AUC_{po}} \cdot D_{iv}/(\mathrm{AUC_{iv}} \cdot D_{po})$, where $\mathrm{AUC_{iv}}$ and $\mathrm{AUC_{po}}$ are the AUC of the drug from $(0-\infty)$ after intravenous and oral administration, and D_{iv} and D_{po} are the single dosage of TBMS-I for the intravenous and oral administration, respectively.

3. Results and Discussion

3.1. Method Optimization.
The mode of electronic source ionization (positive- or negative-ion mode) selection was often tested in a methodological study [26–30]. In this study, we chose the positive mode for the higher response achieved. According to the optimized results of mass spectrometric conditions, we can see that the daughter ions at m/z 1187.8 and m/z 407.3 were the strongest and the most stable among

abundant fragment ions produced, respectively, by TBMS-I and IS, which was presented in Figure 2. Thus, we selected m/z 1319.7 → 1187.8 and m/z 1083.4 → 407.3 for TBMS-I and IS, respectively.

In order to wash out the endogenous compounds as much as possible and avoid endogenous interference, the mobile phase was optimized [31, 32]. Several mobile phases were investigated on the ACQUITY BEH C18 column to obtain a perfect separation and a more symmetrical peak shape [33], including acetonitrile and water with 0.1% formic acid, acetonitrile and 10 mmol/L ammonium acetate solution (0.1% formic acid), methanol and water (0.1% formic acid), and methanol and 10 mmol/L ammonium acetate solution (0.1% formic acid). Among this, the mobile phase containing the mixture of methanol and water (including 0.1% formic acid) was chosen in this study for the best mass spectrometry peak and retention time using gradient eluting.

Proteins and other potential interference would affect the analysis of the mass spectrometry system [34, 35]. Therefore, an effective and simple sample preparation was a key point for establishing the UPLC-MS/MS method of TBMS-I. Liquid-liquid extraction (LLE) has the advantages

(a)

(b)

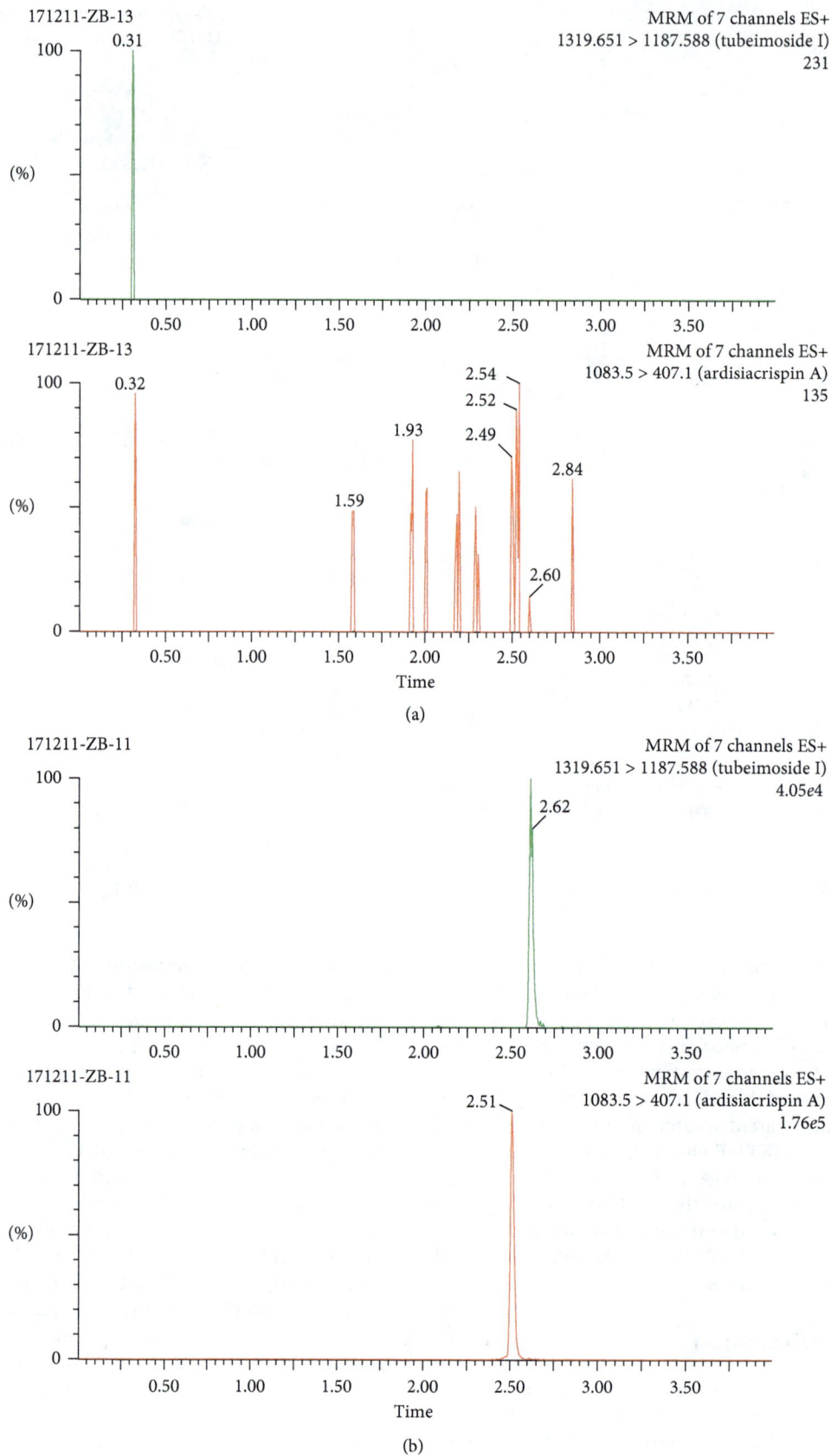

FIGURE 3: Continued.

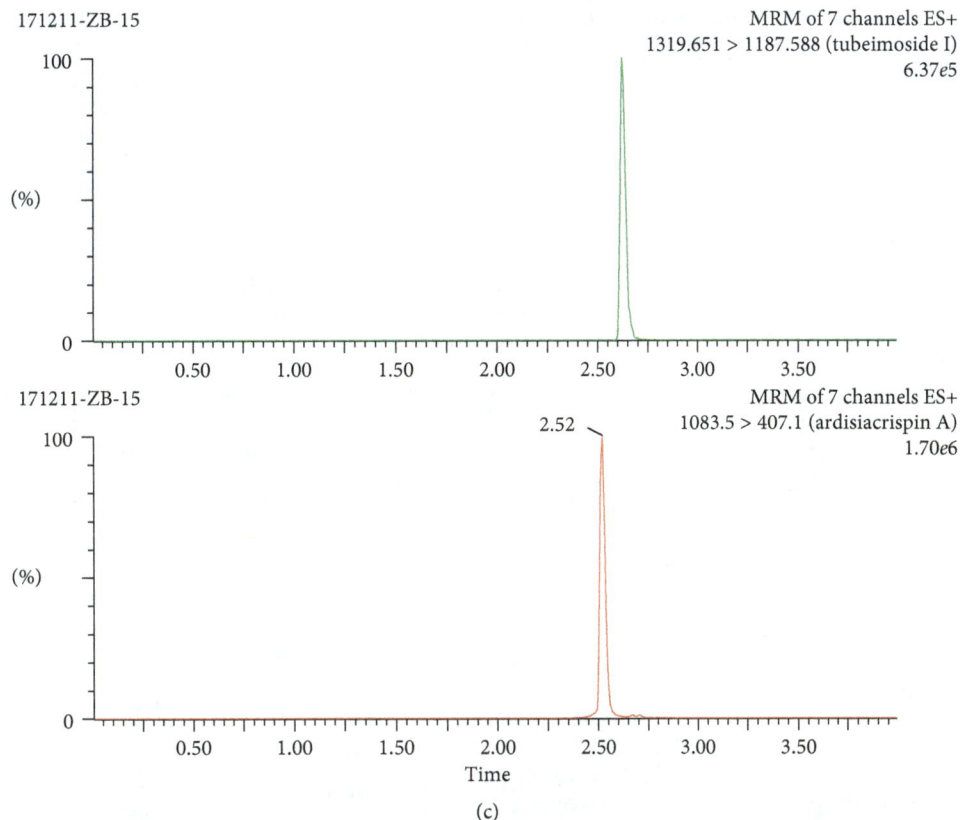

FIGURE 3: Spectrogram of the TBMS-I and IS: (a) a blank extract, (b) a blank extract with tubeimoside I and IS, and (c) the blood samples after administration spiked with IS.

of the high extraction rate and low limit of quantification [36], but this method required a long time for evaporation of the extracting solvent and large sample volumes. The number of blood samples, which can be taken from a mouse (~20 g), is limited. Thus, it is not easy to get enough plasma after centrifuging for liquid-liquid extraction at each point (10 total time points in 24 h) by tail vein transection bleeding. Taking these factors into consideration, only 20 μL of blood samples was collected at different time points, and a one-step protein precipitation procedure for whole blood was chosen in our study following the example of the previous literature [37]. The small sample volume requirement further supports a serial blood sampling and enables entire pharmacokinetics from a single mouse which significantly reduces the numbers of mice used and inaccuracy of the pharmacokinetics because of individual differences [7, 23]. Our method provides a simple, direct, and high-throughput assay for measuring TBMS-I because of the simple sample processing. The following precipitating solvents and their mixtures in different combinations and ratios were tested: methanol, acetonitrile, and acetonitrile-methanol. The results indicated that acetonitrile was a good precipitating reagent for the best recoveries for the analytes. Considering that blood samples are more complex than plasma, 20 μL of the blood sample was mixed with 5 volumes of acetonitrile, which can not only provide higher recoveries and less matrix effect but also provide a sufficient supernatant volume for multiple injections for analysis. The level of TBMS-I in the

supernatant obtained from the blood after protein precipitation and centrifugation is high enough to be detected by UPLC-MS/MS because the LLOD is 0.7 ng/mL and LLOQ is 2 ng/mL for TBMS-I, which will contribute to the assay of lower concentration of TBMS-I at the last time point for sample collection.

Internal standard was also an important task for establishing this method [38–40]. Tubeimoside I and ardisiacrispin A had a similar structure, so the retention time and the way of ionization of them are similar. In addition, ardisiacrispin A was a good choice for IS in our study because of its robustness, stability, absence of matrix effects, and reproducible extraction.

3.2. Method Validation

3.2.1. Selectivity. Figure 3 presents the ion chromatogram of a blank extract, a blank extract with TBMS-I and IS, and a blood sample from the caudal vein spiked with IS. The peaks of TBMS-I and IS appeared at 2.62 and 2.52 min, respectively. No interfering peaks were found at or close by the retention times of TBMS-I and IS. The total runtime was 4.0 minutes.

3.2.2. Linearity. The regression equation of the calibration curve of TBMS-I was $y = 0.00027776x + 0.0000866688$ (y represents the value of the peak area ratio of TBMS-I and IS and x represents the concentrations of TBMS-I in blood).

TABLE 1: Accuracy, precision, matrix effect, and recovery of the TBMS-I in mouse blood ($n = 6$).

Concentration (ng/mL)	Precision (RSD %)		Accuracy (%)		Matrix effect (%)	Recovery (%)
	Intraday	Interday	Intraday	Interday		
3	13.4	14.5	105.8	108.0	105.2	78.4
190	11.2	10.6	94.4	91.7	104.8	68.6
1900	5.3	8.4	101.3	104.1	111.0	66.9

TABLE 2: The stability of TBMS-I under various storage conditions ($n = 3$).

Concentration (ng/mL)	Autosampler (4°C, 12 h)		Ambient 2 h		−20°C 30 d		Freeze-thaw	
	Accuracy	RSD	Accuracy	RSD	Accuracy	RSD	Accuracy	RSD
3	96.0	3.8	107.5	5.1	92.6	14.2	113.5	8.8
190	106.2	6.7	108.0	4.8	109.9	9.4	111.0	11.2
1900	104.1	5.5	95.8	4.4	91.7	6.0	90.3	7.5

The correlation coefficient r^2 was 0.9976, which showed a good linearity. The LLOQ was 2 ng/mL with the signal-to-noise ratio (S/N ratio) of 10 for the determination of TBMS-I in mouse blood, and the lower limit of detection was 0.7 ng/mL with the S/N ratio of 3.

3.2.3. Precision, Accuracy, Recovery, and Matrix Effect. Table 1 shows the results of the precision, accuracy, recovery, and matrix effect. The RSD of interday precision and intraday precision was no more than 14% and 15%, respectively. The accuracy was in the range of 91.7% to 108.0% at each QC level. All of the recoveries were above 66.9%, and matrix effects were between 104.8% and 111.0%. These data suggest that this method was satisfied with the pharmacokinetic study of TBMS-I.

3.2.4. Stability. The blood samples under the different storage conditions mentioned above ($n = 3$) were carried out the stability experiment (results shown in Table 2). In this study, the variations of each condition were within 14% and RSD was under 15%, which indicated a reliable stability behavior of TBMS-I under the different storage conditions.

3.3. Pharmacokinetic Studies. Time-concentration curve of TBMS-I after oral and intravenous administration is shown in Figure 4. The pharmacokinetic parameters were calculated according to the noncompartment model (results are presented in Table 3). The $t_{1/2z}$ was 2.3 ± 0.5 h for oral administration and 6.8 ± 5.6 h for intravenous administration, respectively. The T_{max} was 1.8 ± 1.3 h after oral administration. The absolute availability was only 1.0%. These results in mice were similar to that of rats described by Liang et al. [6].

4. Conclusions

A novel UPLC-MS/MS method for the quantitative measurement of TBMS-I in mouse blood has been developed and validated. The application of this method for the

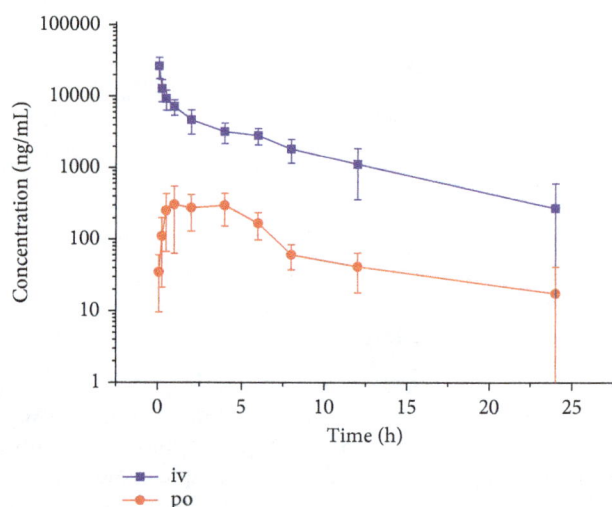

FIGURE 4: Mean blood concentration of TBMS-I after sublingual intravenous administration at the dose of 3 mg/kg and gavage of 15 mg/kg.

TABLE 3: The pharmacokinetic parameters of TBMS-I after oral and intravenous administration ($n = 6$).

Parameters	Unit	iv (5 mg/kg)	po (20 mg/kg)
AUC $(0-t)$	ng/mL·h	51205.8 ± 13134.0	2023.9 ± 1145.5
AUC $(0-\infty)$	ng/mL·h	59370.3 ± 21468.0	2051.8 ± 1106.5
MRT $(0-t)$	h	5.1 ± 1.8	4.6 ± 2.4
MRT $(0-\infty)$	h	8.9 ± 7.0	4.9 ± 2.0
$t_{1/2z}$	h	6.8 ± 5.6	2.3 ± 0.5
$CL_{z/F}$	L/h/kg	0.1	16.3 ± 17.4
T_{max}	H	—	1.8 ± 1.3
$V_{z/F}$	L/kg	0.4	53.9 ± 56.9
C_{max}	ng/mL	26192.5 ± 8491.9	429.6 ± 164.9

determination of TBMS-I extracted from only 20 μL of whole blood using a simple one-step protein precipitation procedure within 4 min was more sensitive, more convenient, and faster than traditional and commonly used analytical techniques. It is clear, in addition, that there are potentially large savings in the amount of compound and number of animals required for early pharmacokinetic studies. In the present study, the UPLC/MS/MS method has been successfully applied to the pharmacokinetic investigations of TBMS-I in mice after sublingual intravenous and intragastric administration. The oral bioavailability of tubeimoside I in mice is 1%, which indicates that tubeimoside I is not easily absorbed into blood circulatory system through the gastrointestinal tract.

Conflicts of Interest

The authors declare that they have no conflicts of interest.

Acknowledgments

This research was supported by the Science and Technology Foundation of Wenzhou (Y20160540 and Y20150183) and a grant from the National Natural Science Foundation of China (81702388). The authors thank Wenzhou Medical University Analytical and Testing Center for excellent technical support.

References

[1] T. X. Yu, R. D. Ma, and L. J. Yu, "Structure-activity relationship of tubeimosides in anti-inflammatory, antitumor, and antitumor-promoting effects," *Acta Pharmacologica Sinica*, vol. 22, no. 5, pp. 463–468, 2001.

[2] Y. Gu, C. Korbel, C. Scheuer, A. Nenicu, M. D. Menger, and M. W. Laschke, "Tubeimoside-1 suppresses tumor angiogenesis by stimulation of proteasomal VEGFR2 and Tie2 degradation in a non-small cell lung cancer xenograft model," *Oncotarget*, vol. 7, no. 5, pp. 5258–5272, 2016.

[3] Q. Bian, P. Liu, J. Gu, and B. Song, "Tubeimoside-1 inhibits the growth and invasion of colorectal cancer cells through the Wnt/beta-catenin signaling pathway," *International Journal of Clinical and Experimental Pathology*, vol. 8, no. 10, pp. 12517–12524, 2015.

[4] G. Jia, Q. Wang, R. Wang et al., "Tubeimoside-1 induces glioma apoptosis through regulation of Bax/Bcl-2 and the ROS/Cytochrome C/Caspase-3 pathway," *OncoTargets and Therapy*, vol. 8, pp. 303–311, 2015.

[5] Y. Peng, Y. Zhong, and G. Li, "Tubeimoside-1 suppresses breast cancer metastasis through downregulation of CXCR4 chemokine receptor expression," *BMB Reports*, vol. 49, no. 9, pp. 502–507, 2016.

[6] M. J. Liang, W. D. Zhang, C. Zhang et al., "Quantitative determination of the anticancer agent tubeimoside I in rat plasma by liquid chromatography coupled with mass spectrometry," *Journal of Chromatography B*, vol. 845, no. 1, pp. 84–89, 2007.

[7] J. Kim, J. S. Min, D. Kim et al., "A simple and sensitive liquid chromatography-tandem mass spectrometry method for trans-epsilon-viniferin quantification in mouse plasma and its application to a pharmacokinetic study in mice," *Journal of Pharmaceutical and Biomedical Analysis*, vol. 134, pp. 116–121, 2017.

[8] J. H. Xiong, H. Ye, Y. X. Lin et al., "Determining concentrations of icotinib in plasma of rat by UPLC method with ultraviolet detection: applications for pharmacokinetic studies," *Current Pharmaceutical Analysis*, vol. 13, no. 4, pp. 340–344, 2017.

[9] L. G. Chen, S. P. Yang, Z. Z. Liu et al., "Pharmacokinetic study of macitentan in rat plasma by ultra performance liquid chromatography-tandem mass spectrometry," *Latin American Journal of Pharmacy*, vol. 34, no. 7, pp. 1411–1416, 2015.

[10] L. G. Chen, Z. Wang, S. Wang, T. Li, Y. Pan, and X. Lai, "Determination of apremilast in rat plasma by UPLC-MS-MS and its application to a pharmacokinetic study," *Journal of Chromatographic Science*, vol. 54, no. 8, pp. 1336–1340, 2016.

[11] T. F. Yuan, S. T. Wang, and Y. Li, "Quantification of menadione from plasma and urine by a novel cysteamine-derivatization based UPLC-MS/MS method," *Journal of Chromatography B*, vol. 1063, pp. 107–111, 2017.

[12] M. Sarkar, R. G. Grossman, E. G. Toups, and D. S. Chow, "UPLC-MS/MS assay of riluzole in human plasma and cerebrospinal fluid (CSF): application in samples from spinal cord injured patients," *Journal of Pharmaceutical and Biomedical Analysis*, vol. 146, pp. 334–340, 2017.

[13] M. Mueller, C. Maldonado-Adrian, J. Yuan, U. D. McCann, and G. A. Ricaurte, "Studies of (+/-)-3,4-methylenedioxymethamphetamine (MDMA) metabolism and disposition in rats and mice: relationship to neuroprotection and neurotoxicity profile," *Journal of Pharmacology and Experimental Therapeutics*, vol. 344, no. 2, pp. 479–488, 2013.

[14] K. Tatum-Gibbs, J. F. Wambaugh, K. P. Das et al., "Comparative pharmacokinetics of perfluorononanoic acid in rat and mouse," *Toxicology*, vol. 281, no. 1–3, pp. 48–55, 2011.

[15] P. L. Toutain, A. Ferran, and A. Bousquet-Melou, "Species differences in pharmacokinetics and pharmacodynamics," in *Handbook of Experimental Pharmacology*, pp. 19–48, Springer, Berlin, Germany, 2010.

[16] Y. Hu and D. E. Smith, "Species differences in the pharmacokinetics of cefadroxil as determined in wildtype and humanized PepT1 mice," *Biochemical Pharmacology*, vol. 107, pp. 81–90, 2016.

[17] Z. Li, C. Chen, D. Ai et al., "Pharmacokinetics and tissue residues of hydrochloric acid albendazole sulfoxide and its metabolites in crucian carp (*Carassius auratus*) after oral administration," *Environmental Toxicology and Pharmacology*, vol. 33, no. 2, pp. 197–204, 2012.

[18] S. M. Butler, M. A. Wallig, C. W. Nho et al., "A polyacetylene-rich extract from Gymnaster koraiensis strongly inhibits colitis-associated colon cancer in mice," *Food and Chemical Toxicology*, vol. 53, pp. 235–239, 2013.

[19] K. C. Park, S. W. Kim, J. H. Park et al., "Potential anti-cancer activity of N-hydroxy-7-(2-naphthylthio) heptanomide (HNHA), a histone deacetylase inhibitor, against breast cancer both in vitro and in vivo," *Cancer Science*, vol. 102, no. 2, pp. 343–350, 2011.

[20] H. Yamazaki, H. Suemizu, M. Mitsui, M. Shimizu, and F. P. Guengerich, "Combining chimeric mice with humanized liver, mass spectrometry, and physiologically-based pharmacokinetic modeling in toxicology," *Chemical Research in Toxicology*, vol. 29, no. 12, pp. 1903–1911, 2016.

[21] V. Gota, J. S. Goda, K. Doshi et al., "Biodistribution and pharmacokinetic study of 3,3′ diseleno dipropionic acid (DSePA), a synthetic radioprotector, in mice," *European Journal of Drug Metabolism and Pharmacokinetics*, vol. 41, no. 6, pp. 839–844, 2016.

[22] R. Kumar, P. S. Suresh, G. Rudresh et al., "Determination of ulixertinib in mice plasma by LC-MS/MS and its application to a pharmacokinetic study in mice," *Journal of Pharmaceutical and Biomedical Analysis*, vol. 125, pp. 140–144, 2016.

[23] A. Watanabe, R. Watari, K. Ogawa et al., "Using improved serial blood sampling method of mice to study

pharmacokinetics and drug-drug interaction," *Journal of Pharmaceutical Sciences*, vol. 104, no. 3, pp. 955–961, 2015.

[24] S. Gao, Z. Yang, T. Yin, M. You, and M. Hu, "Validated LC-MS/MS method for the determination of maackiain and its sulfate and glucuronide in blood: application to pharmacokinetic and disposition studies," *Journal of Pharmaceutical and Biomedical Analysis*, vol. 55, no. 2, pp. 288–293, 2011.

[25] US Department of Health and Human Services, *Guidance for Industry: Bioanalytical Method Validation*, F.a.D.A. US Department of Health and Human Services, Washington, DC, USA, 2013.

[26] W. Ye, R. Chen, W. Sun et al., "Determination and pharmacokinetics of engeletin in rat plasma by ultra-high performance liquid chromatography with tandem mass spectrometry," *Journal of Chromatography B*, vol. 1060, pp. 144–149, 2017.

[27] G. Y. Lin, L. F. Hu, X. Z. Yang, X. J. Pan, and X. Q. Wang, "Determination of gemcitabine in rabbit plasma by LC-ESI-MS using an allure PFP propyl column," *Latin American Journal of Pharmacy*, vol. 30, no. 3, pp. 571–575, 2011.

[28] C. Ding, W. Dong, F. F. Yang et al., "Determination of ibudilast in rabbit plasma by liquid chromatography-mass spectrometry and its application," *Latin American Journal of Pharmacy*, vol. 30, no. 10, pp. 2065–2069, 2011.

[29] S. H. Wang, Z. X. Lin, K. Su et al., "Effect of curcumin and pirfenidone on toxicokinetics of paraquat in rat by UPLC-MS/MS," *Acta Chromatographica*, vol. 30, no. 1, pp. 26–30, 2018.

[30] L. G. Chen, W. W. You, D. W. Chen et al., "Pharmacokinetic interaction study of ketamine and rhynchophylline in rat plasma by ultra-performance liquid chromatography tandem mass spectrometry," *BioMed Research International*, vol. 2018, Article ID 6562309, 8 pages, 2018.

[31] M. L. Zhang, J. Zhang, L. Y. Wan, X. L. Wu, C. C. Wen, and X. Q. Wang, "A simple and selective high performance liquid chromatography developed for determination diphenoxylate in rat plasma and tissues," *Latin American Journal of Pharmacy*, vol. 35, no. 10, pp. 2327–2330, 2016.

[32] J. Y. Guo, Q. Q. Xu, S. H. Tong et al., "The effect of transmetil on pharmacokinetics of MS-275 in rats," *Latin American Journal of Pharmacy*, vol. 33, no. 9, pp. 1567–1570, 2014.

[33] S. Wang, H. Wu, X. Huang et al., "Determination of N-methylcytisine in rat plasma by UPLC-MS/MS and its application to pharmacokinetic study," *Journal of Chromatography B*, vol. 990, pp. 118–124, 2015.

[34] B. M. Fang, S. H. Bao, S. H. Wang et al., "Pharmacokinetic study of ardisiacrispin A in rat plasma after intravenous administration by UPLC-MS/MS," *Biomedical Chromatography*, vol. 31, no. 3, article e3826, 2017.

[35] S. Wang, H. Wu, P. Geng et al., "Pharmacokinetic study of dendrobine in rat plasma by ultra-performance liquid chromatography tandem mass spectrometry," *Biomedical Chromatography*, vol. 30, no. 7, pp. 1145–1149, 2016.

[36] E. Eliassen and L. Kristoffersen, "Quantitative determination of zopiclone and zolpidem in whole blood by liquid-liquid extraction and UHPLC-MS/MS," *Journal of Chromatography B*, vol. 971, pp. 72–80, 2014.

[37] M. K. Nielsen and S. S. Johansen, "Determination of olanzapine in whole blood using simple protein precipitation and liquid chromatography-tandem mass spectrometry," *Journal of Analytical Toxicology*, vol. 33, no. 4, pp. 212–217, 2009.

[38] C. C. Wen, S. H. Wang, X. L. Huang et al., "Determination and validation of hupehenine in rat plasma by UPLC-MS/MS and its application to pharmacokinetic study," *Biomedical Chromatography*, vol. 29, no. 12, pp. 1805–1810, 2015.

[39] C. Wen, Q. Zhang, Y. He, M. Deng, X. Wang, and J. Ma, "Gradient elution LC-MS determination of dasatinib in rat plasma and its pharmacokinetic study," *Acta Chromatographica*, vol. 27, no. 1, pp. 81–91, 2015.

[40] W. Q. Tian, J. Z. Cai, Y. Y. Xu et al., "Determination of xanthotoxin using a liquid chromatography-mass spectrometry and its application to pharmacokinetics and tissue distribution model in rat," *International Journal of Clinical and Experimental Medicine*, vol. 8, no. 9, pp. 15164–15172, 2015.

Permissions

List of Contributors

Federica Bianchi and Maria Careri
Department of Chemistry, Life Sciences, and Environmental Sustainability, University of Parma, Parco Area delle Scienze 17/A, 43124 Parma, Italy

Nicolò Riboni
Department of Chemistry, Life Sciences, and Environmental Sustainability, University of Parma, Parco Area delle Scienze 17/A, 43124 Parma, Italy
Department of Environmental Science and Analytical Chemistry, Stockholm University, 10691 Stockholm, Sweden

Leopold Ilag
Department of Environmental Science and Analytical Chemistry, Stockholm University, 10691 Stockholm, Sweden

Veronica Termopoli and Achille Cappiello
Department of Pure and Applied Sciences, LC-MS Laboratory, Piazza Rinascimento 6, 61029 Urbino, Italy

Lucia Mendez and Isabel Medina
Instituto de Investigaciones Marinas, Spanish National Research Council (IIM-CSIC), Eduardo Cabello 6, 36208 Vigo, Spain

Yinping Li
College of Chemistry, Beijing Normal University, Beijing 100875, China
College of Chemistry and Chemical Engineering, Xinjiang Normal University, Urumqi, Xinjiang 830000, China

Qing He
School of Chemical Engineering and Technology, Tianjin University, Tianjin 300350, China

Shushan Du and Shanshan Guo
Beijing Key Laboratory of Traditional Chinese Medicine Protection and Utilization, Faculty of Geographical Science, Beijing Normal University, Beijing, China

Zhufeng Geng
Beijing Key Laboratory of Traditional Chinese Medicine Protection and Utilization, Faculty of Geographical Science, Beijing Normal University, Beijing, China
Analytic and Testing Center, Beijing Normal University, Beijing 100875, China

Zhiwei Deng
Analytic and Testing Center, Beijing Normal University, Beijing 100875, China

Nguyen Phuong Thao and Pham Thanh Binh
Advanced Center for Bio-Organic Chemistry, Institute of Marine Biochemistry (IMBC), Vietnam Academy of Science and Technology (VAST), 18 Hoang Quoc Viet, Caugiay, Hanoi, Vietnam

Nguyen Thi Luyen and Nguyen Hai Dang
Advanced Center for Bio-Organic Chemistry, Institute of Marine Biochemistry (IMBC), Vietnam Academy of Science and Technology (VAST), 18 Hoang Quoc Viet, Caugiay, Hanoi, Vietnam
Graduate University of Science and Technology, VAST, 18 Hoang Quoc Viet, Caugiay, Hanoi, Vietnam

Nguyen Tien Dat
Graduate University of Science and Technology, VAST, 18 Hoang Quoc Viet, Caugiay, Hanoi, Vietnam
Center for Research and Technology Transfer, VAST, 18 Hoang Quoc Viet, Caugiay, Hanoi, Vietnam

Ta Manh Hung
National Institute of Drug Quality Control (NIDQC), 48 Hai Ba Trung, Hoankiem, Hanoi, Vietnam

Cristian Daniel Quiroz-Moreno, Jose Rafael de Almeida, Amanda Sofía Cevallos and Roldán Torres-Guiérrez
Ikiam-Universidad Regional Amazónica, Km 7 Via Muyuna, Tena, Napo, Ecuador

Noroska Gabriela Salazar Mogollón
Ikiam-Universidad Regional Amazónica, Km 7 Via Muyuna, Tena, Napo, Ecuador
Institute of Chemistry, State University of Campinas, Cidade Universitária Zeferino Vaz, 13083 970 Campinas, SP, Brazil

Paloma Santana Prata and Fabio Augusto
Institute of Chemistry, State University of Campinas, Cidade Universitária Zeferino Vaz, 13083 970 Campinas, SP, Brazil

Kátia S. D. Nunes and Felix G. R. Reyes
Department of Food Science, School of Food Engineering, University of Campinas, Rua Monteiro Lobato 80, 13083-862 Campinas, SP, Brazil

Márcia R. Assalin, José H. Vallim, Claudio M. Jonsson and Sonia C. N. Queiroz
Embrapa Meio Ambiente, 13820-000 Jaguariúna, SP, Brazil

Maurilio Gustavo Nespeca, Rafael Rodrigues Hatanaka and José Eduardo de Oliveira
Centro de Monitoramento e Pesquisa da Qualidade de Combustíveis, Biocombustíveis, Petróleo e Derivados (Cempeqc), São Paulo State University (UNESP), R. Prof. Francisco Degni 55 Quitandinha, 14800-900 Araraquara, SP, Brazil

Danilo Luiz Flumignan
Instituto Federal de Educação, Ciência e Tecnologia de São Paulo (IFSP), Campus Matão, Rua Est'efano D'avassi, 625 Nova Cidade, 15991-502 Matão, SP, Brazil

Alessia Daveri and Manuela Vagnini
Laboratorio di Diagnostica per i Beni Culturali di Spoleto, Rocca Albornoziana, Piazza B. Campello 2, 06049 Spoleto, Italy

Marco Malagodi
Laboratorio Arvedi di Diagnostica Non Invasiva, Universit`a di Pavia, via Bell'Aspa 3, 26100 Cremona, Italy

Maydla dos Santos Vasconcelos, Wilson Espíndola Passos, Magno Aparecido Gonçalves Trindade and Rozanna Marques Muzzi
Faculty of Exact Sciences and Technology, Federal University of Grande Dourados, Dourados, MS, Brazil

Caroline Honaiser Lescanos
Faculty of Medical Sciences, State University of Campinas, Campinas, SP, Brazil

Ivan Pires de Oliveira
Institute of Chemistry, State University of Campinas, Campinas, SP, Brazil

Anderson Rodrigues Lima Caires
Institute of Physics, Federal University of Mato Grosso do Sul, Campo Grande, MS, Brazil

Jin Li and Yuhong Li
Tianjin State Key Laboratory of Modern Chinese Medicine, Tianjin University of Traditional Chinese Medicine, Tianjin 300193, China

Kun-ze Du, Xinrong Guo and Yan-xu Chang
Tianjin State Key Laboratory of Modern Chinese Medicine, Tianjin University of Traditional Chinese Medicine, Tianjin 300193, China
Tianjin Key Laboratory of Phytochemistry and Pharmaceutical Analysis, Tianjin University of Traditional Chinese Medicine, Tianjin 300193, China

Heying Zhang, Wei Qu and Shuyu Xie
MOA Laboratory for Risk Assessment of Quality and Safety of Livestock and Poultry Products, Huazhong Agricultural University, Wuhan, Hubei 430070, China

Lingli Huang, Zhenli Liu, Yuanhu Pan and Zonghui Yuan
MOA Laboratory for Risk Assessment of Quality and Safety of Livestock and Poultry Products, Huazhong Agricultural University, Wuhan, Hubei 430070, China
National Reference Laboratory of Veterinary Drug Residues and MAO Key Laboratory for Detection of Veterinary Drug Residues, Huazhong Agricultural University, Wuhan, Hubei 430070, China

Yanfei Tao and Dongmei Chen
National Reference Laboratory of Veterinary Drug Residues and MAO Key Laboratory for Detection of Veterinary Drug Residues, Huazhong Agricultural University, Wuhan, Hubei 430070, China

Jing Yang, Jingwen Bai, Meiyu Liu, Yang Chen and Shoutong Wang
Institute of Mountain Hazards and Environments, Chinese Academy of Sciences, Chengdu, Sichuan, China

Qiyong Yang
Institute of Karst Geology Chinese Academy of Geological Sciences, Guilin, Guangxi, China

Shouhong Gao, Jingya Zhou, Zhipeng Wang, Yunlei Yun, Mingming Li, Feng Zhang and Wansheng Chen
Department of Pharmacy, Changzheng Hospital, Second Military Medical University, Shanghai 200003, China

Zhengbo Tao
Department of Orthopaedics, First Affiliated Hospital, China Medical University, 155 Nan Jing Bei Street, Shenyang, Liaoning 110001, China

Yejun Miao
Department of Psychiatry, Ankang Hospital, Ningbo, Zhejiang 315000, China

Yujie Deng, Shumin Xu, Ping Wang, Nailong Yang, Chengqian Li and Qing Yu
Department of Endocrinology, e Affiliated Hospital of Qingdao University, 16 Jiangsu Road, Qingdao 266071, China

Yudong Fu
Department of Ophthalmology, e Affiliated Hospital of Qingdao University, 16 Jiangsu Road, Qingdao 266071, China

Guang Hu
School of Pharmacy and Bioengineering, Chongqing University of Technology, Chongqing 400054, China

Ya-Li Wang
School of Pharmacy and Bioengineering, Chongqing University of Technology, Chongqing 400054, China
School of Chemistry and Chemical Engineering, Chongqing University, Chongqing 401331, China

Qian Zhang, Yu-Xiu Yang, Qiao-Qiao Li, Hua Chen and Feng-Qing Yang
School of Chemistry and Chemical Engineering, Chongqing University, Chongqing 401331, China

Yuan-Jia Hu
State Key Laboratory of Quality Research in Chinese Medicine, Institute of Chinese Medical Sciences, University of Macau, Macau

Zhan-Ling Xu and Ning Zhang
Key Laboratory of Chinese Materia Medica, College of Pharmacy, College of Jiamusi, Heilongjiang University of Chinese Medicine, Harbin, Heilongjiang 150040, China

Ming-Yue Xu, Ming-Yang Liu, Chun-Peng Jia and Fang Geng
Key Laboratory of Photochemistry Biomaterials and Energy Storage Materials of Heilongjiang Province, College of Chemistry & Chemical Engineering, Harbin Normal University, Harbin 150025, China

Hai-Tao Wang
Pharmacy Department, Harbin Hospital of Traditional Chinese Medicine, Harbin 150076, China

Qing-Xuan Xu
Crop Academy of Heilongjiang University, Harbin 150080, China

Nevena Maljurić, Jelena Golubović and Biljana Otašević
Department of Drug Analysis, Faculty of Pharmacy, University of Belgrade, Vojvode Stepe 450, Belgrade 11221, Serbia

Matjaž Ravnikar and Borut Štrukelj
Chair for Pharmaceutical Biology, Faculty of Pharmacy, University of Ljubljana, Aškerˇceva cesta 7, Ljubljana 1000, Slovenia

Dušan Žigon
Jožef Stefan Institute, Jamova 39, Ljubljana 1000, Slovenia

Veronica Lolli, Angela Marseglia, Gerardo Palla, Emanuela Zanardi and Augusta Caligiani
Department of Food and Drug, University of Parma, Parco Area delle Scienze 27A, 43124 Parma, Italy

Mengsi Cao, Yanru Feng, Huayin Zhang, Huaijiao Zhu and Weijun Kang
School of Public Health, Hebei Medical University, Shijiazhuang 050017, China

Kaoqi Lian
School of Public Health, Hebei Medical University, Shijiazhuang 050017, China

Hebei Province Key Laboratory of Environment and Human Health, Shijiazhuang 050017, China

Pingping Zhang
Department of Reproductive Genetic Family, Hebei General Hospital, Shijiazhuang 050017, China

Trung Dung Nguyen, Hoc Bau Le and Thi Oanh Dong
Faculty of Physics and Chemical Engineering, Le Quy Don Technical University, 236 Hoang Quoc Viet, Hanoi, Vietnam

Tien Duc Pham
Faculty of Chemistry, VNU-University of Science, Vietnam National University Hanoi, 19 Le)anh Tong, Hoan Kiem, Hanoi, Vietnam

Long Chen, Yingwen Wu and Tianjun Li
Faculty of Science and Technology, University of Macau, E11 Avenida da Universidade, Taipa, Macau

Zhuo Chen
Chemistry and Chemical Engineering, College of Biology, Hunan University, Changsha 410082, China

Lianguo Chen, Qinghua Weng, Feifei Li, Jinlai Liu and Xueliang Zhang
Wenzhou People's Hospital, e ird Clinical Institute Affiliated to Wenzhou Medical University, Wenzhou 325000, China

Yunfang Zhou
Laboratory of Clinical Pharmacy, e People's Hospital of Lishui, e Sixth Affiliated Hospital of Wenzhou Medical University, Lishui 323000, China

Index